ENVIRONMENTAL ENGINEERING – I

WATER SUPPLY ENGINEERING

[CONTAINING 17 CHAPTERS]

ENVIRONMENTAL ENGINEERING–I

WATER SUPPLY ENGINEERING

By

Dr. B.C. PUNMIA
Formerly,
Professor and Head, Deptt. of Civil Engineering, &
Dean, Faculty of Engineering
M.B.M. Engineering College,
Jodhpur

Er. ASHOK KUMAR JAIN
Director,
Arihant Consultants,
Jodhpur

Dr. ARUN KUMAR JAIN
Assistant Professor
M.B.M. Engineering College,
Jodhpur

IN S.I. UNITS

LAXMI PUBLICATIONS (P) LTD
(An ISO 9001:2015 Company)

BENGALURU • CHENNAI • GUWAHATI • HYDERABAD • JALANDHAR
KOCHI • KOLKATA • LUCKNOW • MUMBAI • RANCHI
NEW DELHI

ENVIRONMENTAL ENGINEERING–I WATER SUPPLY ENGINEERING

Printed and bound in India

Typeset at : Arihant Consultants, Jodhpur

Second Edition : July 1998, Sept. 1998, Aug. 2001, Aug. 2002, Aug. 2003, Sept. 2004 June 2005, June 2006, *Reprint* : Dec. 2006, Aug. 2007, Feb. 2008, Jan. 2009, May 2010

Reprint Edition : Feb. 2011, May 2011, Nov. 2011, April 2012, April 2013, Dec. 2013 May 2014, July 2016, May 2018, 2019, 2021, *Edition* : 2022

ISBN : 978-81-318-0703-3

PUBLISHED IN INDIA BY

Laxmi Publications (P) Ltd.

(An ISO 9001:2015 Company)

113, GOLDEN HOUSE, GURUDWARA ROAD,

DARYAGANJ, NEW DELHI - 110002, INDIA

Telephone : 91-11-4353 2500, 4353 2501

www.laxmipublications.com

info@laxmipublications.com

Branches

Bengaluru	080-26 75 69 30
Chennai	044-24 34 47 26
Guwahati	0361-254 36 69
Hyderabad	040-27 55 53 83
Jalandhar	0181-222 12 72
Kochi	0484-405 13 03
Kolkata	033-40 04 77 79
Lucknow	0522-430 36 13
Ranchi	0651-224 24 64

C—00238/022/06

Printed at: Ajit Printing Press, Delhi.

DEDICATED

TO

LATE SHRI NIHAL CHAND JAIN

WHO TAUGHT US

THE VALUES OF LIFE

DEDICATED

TO

LATE SHRI NIHAL CHAND JAIN

WHO TAUGHT US

THE VALUES OF LIFE

Preface to the Second Edition

Man's search for pure water began in prehistoric times. The quest for pure water can benefit the life and health of every one. Earlier, water treatment methods were developed almost entirely as an art rather than as a science. However, through the modern and new concepts in water purification processes, it has been possible to bring the wide spread water borne diseases under control. The resulting improvements in public health are great.

Water purification is now confronted with a myriad of difficult problems caused due to supply sources receiving greatly pollutional loads of domestic and industrial wastes. This book is an attempt to present those essential principles and present-day practice necessary to solution of the problems of water collection, water purification and water distribution. In the second Edition, the book has been completely rewritten incorporating the latest developments and techniques on the subject. Chapter 1 of the book describes the water systems while chapter 2, 3 and 4 are on the development of water resources. Chapter 5 is devoted to the water demand and determination of water quantity. Water quality aspects have been discussed in chapter 6. Chapter 7 introduces various unit operations used for water purification and waste water treatment. Chapter 8 is on screening and aeration of water. The principles and methods of sedimentation are given in chapter 9 while chapter 10 is devoted to filtration. Chapter 11 is on disinfection of water while chapter 12 is on softening of water. Miscellaneous treatment methods are given in chapter 13. Chapter 14 and 15 are on pumps and pumping, and conveyence of water. The distribution of purified water is given in Chapter 16. Lastly, Chapter 17 describes the water supply for buildings.

This book is not a treatise or a design manual. Instead, it is a text book written for engineering students at undergraduate and postgraduate levels. A large portion of the material presented in this book has been derived from the works of others. Their contribution is greatly acknowledged. The recommendations of various Indian Standards on the subject, along with those of Manual on Water Supply and Treatment, prepared by the Central Public Health and Environmental Engineering Organisation under the Ministry of Urban Development, have been closely followed. Each chapter begins with clear statement of pertinent definitions, design principles, and theories and the design procedures. The basic principles are supplemented with numerous design examples and illustrations. Further suggestions on the improvement of the book are welcome. The authors are thankful to Shri Mool Singh Gahlot and Shri Prem Singh Sankhla for the fine laser typesetting done by them.

JODHPUR **B.C. PUNMIA**
STHAPANA, **ASHOK K. JAIN**
25TH SEP. 1995 **ARUN K. JAIN**

CONTENTS

1

Water Systems

1.1. INTRODUCTION

The five essential requirements for human existence are : (*i*) air (*ii*) water (*iii*) food (*iv*) heat and (*v*) light. Contamination of these elements may cause serious health hazards not only to man but also to animal and plant life. *Environmental Engineering* deals with all these essential elements.

The use of water by man, plants and animals is universal. Without it, there can be no life. Every living thing requires water.

Man and animals not only consume water, but they also consume vegetation for their food. Vegetation, in turn, cannot grow without water. Growth of vegetation also depends upon bacterial action, while bacteria need water in order to thrive. The bacterial action can convert vegetable matter into productive soil. New plants, which grow in this soil, grow by sucking nutrients through their roots in the form of solution in water. Thus an ecological chain is maintained. Water maintains an ecological balance – balance in the relationship between living things and environment in which they live.

The use of water is increasing rapidly with our growing population. Already there are acute shortages of both surface and under ground waters in many parts of the country. Careless pollution and contamination of the streams, lakes, reservoirs, wells and other under ground sources has greatly impaired the quality of available water. This pollution results because of improper disposal of waster water —both domestic as well as industrial. Organised community life require twin services of water supply and sewage disposal. Good sanitation cannot be maintained without adequate water supply system. Without proper disposal, the wastes of a community can create intolerable nuisance, spread diseases and create other health hazards. The planning, designing, financing and operation of water and waste water systems are complex undertakings, and they require a high degree of skill and judgement. The work of construction and maintenance of water supply and waste water disposal systems is generally undertaken by Government agencies – mostly through *Public Health Engineering* or *Environmental Engineering Departments* consisting of Civil Engineers.

1.2. HISTORICAL DEVELOPMENT

Man's search for pure water began is prehistoric times. The story of water supply begins with the growth of ancient capital cities, or religious and trade centres. In olden days, most of community settlements throughout the World were made near springs, lakes and rivers from where the water supply for drinking and irrigation purposes was obtained. Rig Veda (4000 years B.C) makes a mention of digging of wells. Similarly, Ramayana, Mahabhartha and Puranas make mention of wells as the principal source of water supply. These wells were mostly of shallow depth, dug near river banks. Water was lifted from the wells through indegenous methods. However, no water treatment or distribution works existed. Apart from India (Bharat), other major civilisations of the World, such as Greece, Egypt, Assyria etc. used wells for their settlements which were located slightly away from springs, lakes and rivers. Joseph's well at Cairo is one of the oldest deep wells excavated in rock to a depth of about 300 feet. These wells, however, caused water supply problems during periods of drought. It became necessary, therefore, to store water. Cisterns were constructed for collecting rain water while reservoirs were constructed to store water from streams and rivers during monsoon period. The stored water was conveyed to towns through masonry conduits and aqueducts. The earlier examples are the aqueducts built by Appius Claudius in about 312 B.C. for water supply to Rome. Lyons in Paris, Metz in Germany and Segovia and Serille in Spain built similar aqueducts and syphons for water supply used for drinking, bathing and other purposes. Sextus Julius Frontinus, Water Commissioner of Rome (A.D. 97) reported the existence of nine aqueducts supplying water to Rome and varying in length from 10 to over 50 miles and in cross-section from 7 to over 50 sq. ft., with an estimated aggregate capacity of 84 mgd. The great sewer, known as the *cloaca maxima* and constructed to drain the Roman Forum, is still in service.

There was practically no improvement in water supply systems in the middle ages. The earlier water supply structures got destroyed with the fall of Rome. In the ninth century, few important water supply structures were constructed by the Moors in Spain. In the twelfth century, small aqueduct was constructed in Paris. In London, spring water was brought by means of lead pipes and masonry conduits in the thirteenth century. In Germany, water works were constructed in 1412 and pumps were introduced in 1527 in Hanover. Franciscan monk constructed aqueduct of Zempola in Mexico in the middle of 16th century. In 1582, a pump was erected on the old London bridge for the supply of water from the Thames. The water was conveyed through lead pipes. In Paris, pumps operated by water power were erected in 1608. Pumps operating from steam were introduced in the 18th century in London and Paris. In the United States, spring water was conveyed by gravity to Boston in 1652. Pumps etc. were introduced at Bethlehem in 1754.

However, purposeful quality control of water supply is quite recent in origin. The scientific discoveries and engineering inventions of the eighteenth and nineteeth centuries created centralised industries to which people flocked for employment. This caused serious water supply and waste disposal problems in the industrial towns. No great schemes of water supply were started until the Industrial Revolution had well passed its first half century. The development of the large impounding reservoir was largely due to the necessity of feeding canals constructed during the first phase of the Industrial Revolution.

The first water filter was constructed in 1804 by John Gibb at Paisley in Scotland. It was a slow sand filter and worked in conjunction with a settling basin and roughening filter. Next successful filters were constructed in 1827 by Robert Thom at Greenock. In 1829, James Simpson built sizable filters for the Chelsea Water Company to improve its supply from the Thames river. By 1870, the mechanical filter of the pressure type began to be employed, the earliest being the Halliday filters installed at Crewe (1888), Bridlington and elsewhere. In 1894 pre-filters were successfully built. In the first decade of 20th century, mechanical pressure filters were introduced, Hastings being an early pioneer with Canndy filters built in 1900. In India, Calcutta was the first city where a modern water supply system was constructed in 1870.

The technique of clarification and filtration soon grew. By 1939, mechanically-sludged sedimentation tanks were in general use. The micro-strainer, for the removal of plankton from the impounded water was developed by Boucher, and was introduced by Glenfield and Kennedy in 1945. Coagulation of water with sulphate of alumina began experimentally in 1827, but was adapted practically only in 1881 to treat Bolton's water supply. Activated silica was introduced by Bayliss in U.S.A. during 1937. The first permanent use of chlorination originated under the direction of Sir Alexander Houston at Lincoln in 1905. In 1917, Paterson Engineering Company installed the first gaseous chlorinator at the Rye Common Works. Super-chlorination and dechlorination was first applied in 1922 at the Deptford works of the Metropolitan Water Board. The art of softening water was also first developed in Great Britain. The first municipal softener was constructed by Plumstead in 1854. Development of the softener took a novel turn in 1912 by the construction, at the Hooten works of the West Cheshire Water Board, of a base exchange softener.

Since India was under British occupation, water supply schemes in India were undertaken practically about the same time as in England, though with a slower rate. In 1870, a water supply system was constructed at Calcutta. Till Independence, only few cities had protected water supply systems.

1.3. SOURCES OF WATER

The following are common sources of water

(*i*) Rain water (*ii*) Surface water

(*iii*) Ground water (*iv*) Water obtained from reclamation.

1. Rain Water

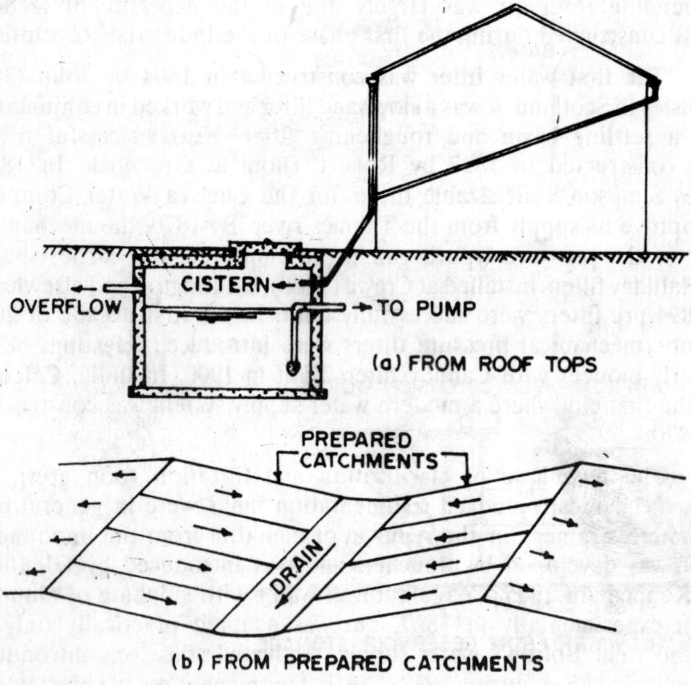

FIG. 1.1. DIRECT COLLECTION OF RAIN WATER

(*a*) ***From roofs of houses and dwellings*** : Water is stored in small underground tank or cistern, for small individual supplies (Fig. 1.1 *a*).

(*b*) ***From prepared catchments*** : The surface of catchments is made impervious by suitable lining material, and suitable slope is given so that water is stored in moderate size reservoirs. This water is used for communal supplies, mostly for drinking purposes.

2. Surface Water

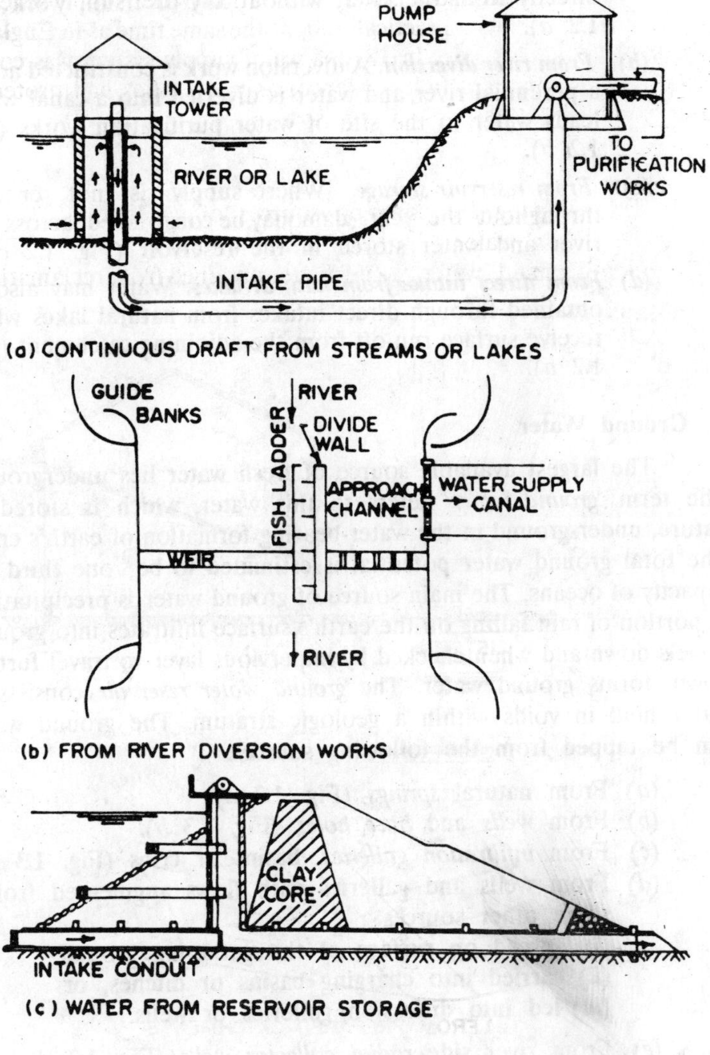

FIG. 1.2. SOURCES OF SURFACE WATER

Surface water is the one which is available as *run-off* from a catchment area, during rainfall or precipitation. This runoff flows either into streams or into undrained lakes. The runoff water flowing into streams can either be *stored* in a reservoir by constructing a dam across it, or be diverted into a water supply channel. Thus, depending upon the scheme of collection, we get surface water from the following sources.

(*a*) ***From rivers by continuous draft*** : Water may be collected directly from the river, without any diversion work (Fig 1.2 *a*).

(*b*) ***From river diversion.*** A diversion work is constructed across a perennial river and water is diverted into a canal which leads water to the site of water purification works (Fig. 1.2 *b*).

(*c*) ***From reservoir storage.*** Where supply is not ensured throughout the year, dam may be constructed across the river and water stored in the reservoir (Fig. 1.2 *c*).

(*d*) ***From direct intake from natural lakes.*** Water may also be obtained through direct intakes from natural lakes which receive surface run-off from the adjoining catchment (Fig. 1.2 *a*).

3. Ground Water

The largest available source of fresh water lies underground. The term *'ground water'* refers to this water, which is stored by nature, under-ground in the water-bearing formation of earth's crust. The total ground water potential is estimated to be one third the capacity of oceans. The main source of ground water is precipitation. A portion of rain falling on the earth's surface infitrates into ground, travels down and when checked by impervious layer to travel further down, forms ground water. The *ground water reservoir* consists of water held in voids within a geologic stratum. The ground water can be tapped from the following sources.

(*a*) From natural *springs* (Fig. 1.3 *a*).
(*b*) From *wells* and *bore holes* (Fig. 1.3 *b*).
(*c*) From *infiltration galleries*, basins or cribs (Fig. 1.3 *c*).
(*d*) From wells and galleries with flows augmented from some other sources :
 (*i*) spread on surface of the gathering ground
 (*ii*) carried into charging basins or ditches, or
 (*iii*) led into diffussion galleries or wells.

(*e*) From river side *radial collector wells* (Fig. 1.3 *d*)

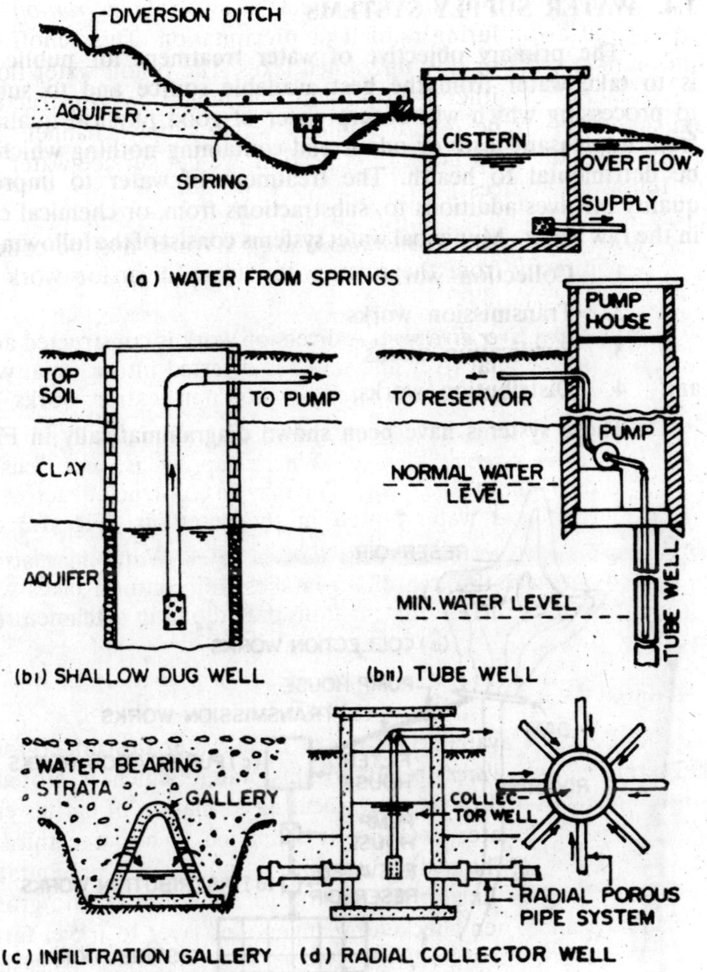

FIG. 1.3. SOURCES OF UNDERGROUND WATER.

4. Water obtained by reclamation

(*a*) ***Desalination.*** Saline or brakish water may be rendered useful for drinking purposes by installing desalination plants. The common methods used for desalination are: distillation, reverse osmosis, electrodialysis, freezing and solar evaporation.

(*b*) ***Re-use of treated waster water.*** Effluent or waste water can be treated suitably so that it may be re-used. An example of the controlled indirect re-use is the intentional artificial recharge of ground water aquifers by adequately treated waste water.

1.4. WATER SUPPLY SYSTEMS

The primary objective of water treatment for public supply is to take water from the best available source and to subject is to processing which will ensure water of good physical quality, free from unpleasant taste or odour and containing nothing which might be detrimental to health. The treatment of water to improve its quality involves additions to, substractions from, or chemical changes in the raw water. Municipal water systems consist of the following units.

1. Collection works
2. Transmission works
3. Purification works

and 4. Distribution works.

These systems have been shown diagrammatically in Fig. 1.4.

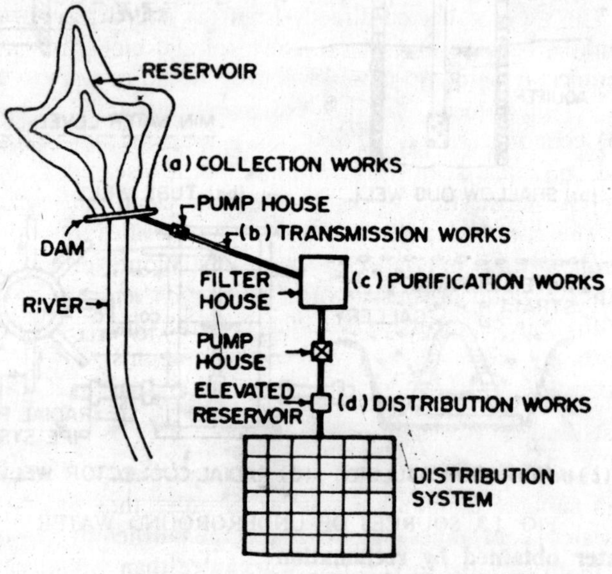

FIG. 1.4. WATER SUPPLY SYSTEMS

1. Collection Works

Collection works are meant for the development of surface water or ground water resources. For major cities, or where water requirements are large, water is collected from a surface source—mostly a river or stream. If the river is perennial, a direct intake structure can be built on the river bank. If, however, river is not perennial, a dam is built across the river so that water is stored in the reservoir. Water is then drawn from the reservoir as per needs. The collection works, therefore, consist of a storage or diversion work, and an intake structure. The planning and design of collection works have been discussed in Chapter 3.

2. Transmission Works

In many cases, the collection works may be far away from the city where water is to be supplied. In that case, water is conveyed to the city through the transmission works. These form the connecting link between the collection works and the purification works. Depending upon the topography of the area between the two sites, the transmission works may be in the form of conduits, canals or aqueducts. For simply gravity flow, canals are generally used. However, if highlands intervene, pumping may have to be resorted to.

3. Purification Works

The water collected directly from the source may not be safe for drinking because of physical, chemical and biological impurities. The municipal water works must deliver to the consumer the water that is : (1) hygienically safe, (2) esthetically attractive and palatable, and (3) economically satisfactory for its intended use. Diseases like typhoid, cholera, dysentery etc. are water borne diseases. The principal aim of the purification works is to supply clean and bacteria free water. The common components of water purification works are : (*i*) *Filtration plants* to remove objectionable colour, turbidity, bacteria and other harmful organisms, (*ii*) *Deferrization and demanganization plants* to remove excessive amounts of iron and manganese, and (*iii*) *Softening plants* to remove excessive amounts of scale forming, soap consuming ingredients like calcium and magnesium ions.

4. Distribution Works

The treated and purified water is finally sent to the consumers through suitable distribution system. In order that water may flow in the water supply pipes under pressure, the purified water is normally stored in an elevated service reservoir. More than one reservoir may be needed in large systems. There are two patterns of water distribution system : (*i*) branching pattern with dead ends, and (*ii*) grid iron pattern. The plan, topography and location of the area with respect to the service reservoir establish the type of distribution system and its character of flow. The design of distribution system has been discussed in Chapter 16.

2

Hydrology

2.1. THE WATER CYCLE

Hydrology is the science which deals with the occurrence, distribution and movement of water on the earth, including that in the atmosphere and below the surface of the earth. Water occurs in the atmosphere in the form of vapour, on the surface as water, snow or ice and below the surface as ground water occupying all the voids within a geologic stratum.

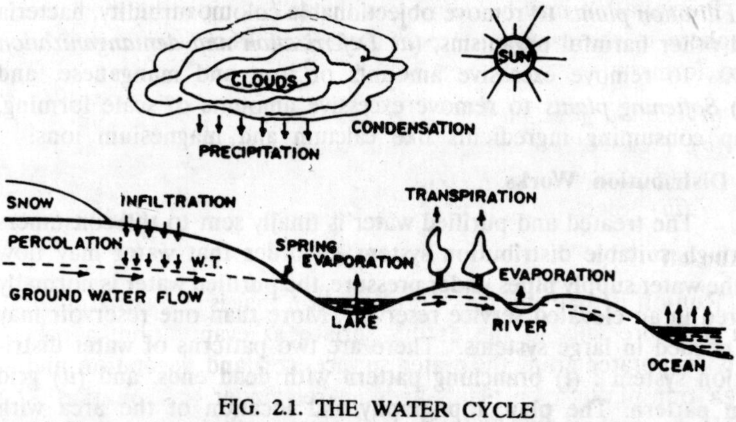

FIG. 2.1. THE WATER CYCLE

Except for the deep ground water, the total water supply of earth is in constant circulation from earth to atmosphere, and back to the earth. The earth's water circulatory system is known as the *water cycle* or the *hydrologic cycle.* Water circulates naturally through five principal realms—(*i*) oceans, (*ii*) atmosphere, (*iii*) lakes and rivers, (*iv*) ice caps and glaciers, and (*v*) underground. Hydrology concerns water and its behaviour in all these realms. *Hydrologic cycle* or the *water cycle* is the process of transfer of moisture from atmosphere to the earth in the form of precipitation, conveyance of the precipitated water by streams and rivers to ocean and lakes etc., and evaporation of water back to the atmosphere. Fig. 2.1 illustrates, diagrammatically, the complete hydrologic cycle.

The hydrologic cycle consists of the following processes:

1. Evaporation and Transpiration (*E*)

The water from the surfaces of ocean, rivers, lakes and also from the moist soil evaporates. The vapours are carried over the land by air in the form of clouds. Transpiration is the process of water being lost from the leaves of the plants from their pores. Thus, the total evaporation (*E*), inclusive of the transpiration consists of :

(*i*) Surface evaporation
(*ii*) Water surface evaporation
  (*a*) From river surface
  (*b*) From oceans
(*iii*) Evaporation from plants and leaves (transpiration)
and (*iv*) Atmospheric evaporation.

2. Precipitation (*P*)

Precipitation may be defined as the fall of moisture from the atmosphere to the earth surface in any form. Precipitation may be in two forms:

(*a*) *Liquid Precipitation* : *i.e.* rainfall.
(*b*) *Frozen Precipitation* : This consists of :
  (*i*) Snow (*ii*) Hail
  (*iii*) Sleet (*iv*) Freezing rain.

3. Run-off (*R*)

Run-off is that portion of precipitation that is not evaporated. When moisture falls to the earth's surface as evaporation, a part of it is evaporated from the water surface, soil and vegetation and through transpiration by plant, and the remainder precipitation is available as run off which ultimately runs to the ocean through surface or sub-surface streams. Thus, run off may be classified as follows :

(1) *Surface run off*

Water flows over the land and is first to reach the streams and rivers, which ultimately discharge the water to the sea.

(2) *Interflow or sub-surface run off*

A portion of precipitation infiltrates into surface soil and, depending upon the geology of the basin, runs as sub-surface run-off and reaches the streams and rivers.

(3) *Ground water flow or base flow*

It is that portion of precipitation, which after infiltration, percolates down and joins the ground water reservoir which is ultimately connected to the ocean.

Thus, the hydrologic cycle may be expressed by the following simplified equation.

$$\text{Precipitation} = \text{Evaporation} + \text{Run off}$$
$$(P) = (E) + (R)$$

provided adjustment is made for the moisture held in storage at the beginning and at the end of the period.

2.2. PRECIPITATION

To the hydrologist, precipitation is the general term for all forms of moisture emanating from the clouds and falling to the ground. The following are the essential requirements for precipitation to occur :

1. Some mechanism is required to cool the air sufficiently to cause condensation and droplet growth.

2. Condensation nuclii are also necessary for formation of droplets. They are usually present in the atmosphere in adequate quantities.

3. Large scale cooling is essential for significant amount of precipitation. This is achieved by lifting of air. Thus a meteorological phenomenon of lifting of air masses is essential to result precipitation.

Types of Precipitation

Precipitation is often classified according to the factors responsible for lifting. Broadly speaking, there are four types of precipitation.

(1) Cyclonic precipitation.
(2) Convective precipitation
(3) Orographic precipitation
(4) Precipitation due to turbulent ascent.

1. Cyclonic Precipitation

Cyclonic precipitation results from lifting of air masses converging into low pressure area or cyclone. The cyclonic precipitation may be divided into (*a*) frontal precipitation, and (*b*) non-frontal precipitation.

*(i) **Frontal Precipitation***

A border region between two adjacent air masses having different characteristics such as temperature and humidity is called a *front*. When a flow of warm and moist air mass from the south meets cold air mass of polar region, the cold air, being heavier, under-run the warm air flow in the form of flat wedge forcing the warm air aloft. The lifted warm air mass cools down at high altitudes, causing precipitation. A front may be warm front or cold front depending upon whether there is active or passive ascent of warm air mass over cold air mass.

*(ii) **Non-frontal Precipitation***

In the case of non-frontal precipitation, the moist warm air mass is stationary and the moving cold air mass meets it. Thus, due to lightness of the warm air, there is passive ascent of warm air over cold air owing to the active under-cutting. When the lifted warm air cools down at higher altitude, precipitation occurs.

2. Convective Precipitation

Convective precipitation is caused by natural rising of warmer, lighter air in colder, denser surroundings. The difference in temperature may result from unequal heating at the surface, unequal cooling at the top of the air layer, or mechanical lifting when air is forced to pass over a denser colder air masses. Convective precipitation is spotty and its intensity may vary from light shower to cloud bursts.

3. Orographic Precipitation

Orographic precipitation is due to the lifting of warm moisture laden air masses due to topographic barriers (such as mountains). All the precipitation we have in Himalayan region is because of the orographic ascent of air masses, rich in moisture content because of their long travel over oceans.

4. Precipitation due to Turbulent Ascent

Air mass is forced to rise up due to greater friction of earth surface after its travel over ocean. The air mass rises up because of increased turbulence and friction, when it ultimately condenses and precipitation occurs. Winter rainfall in Madras State is mainly due to this process.

In nature, the effects of the various types of cooling are often inter-related and the resulting precipitation cannot be identified as being of any one type.

Forms of Precipitation

The various forms of precipitation are : Drizzle, rain, glaze, sleet, snow, snow flakes, and hail.

Drizzle : When the size of water droplets is under 0.5 mm, and its intensity is <0.01 mm per hour.

Rain : When the size of the drops is more than 0.5 mm. The upper size of water drop is generally 6.25 mm, as drops greater than this tend to break up as they fall through the air.

Glaze : When the drizzle or rain freezes as it comes in contact with cold objects, it is known as glaze.

Sleet : It is frozen rain drops cooled to the ice stage while falling through air at sub-freezing temperature.

Snow : Precipitation in the form of ice crystals resulting from sublimations (i.e. water vapour changed directly to ice).

Snow flakes : Number of ice crystals fused together form snow flakes.

Hail : Hails is lumps or bulbs of ice over 5 mm diameter formed by alternate freezing or melting as they are carried up and down in highly turbulent air currents.

2.3. MEASUREMENT OF RAINFALL

Rainfall is the source of all water used for irrigation purposes and, therefore, a knowledge of its amount, character, seasons or periods and the effects produced by it is of prime importance to all whose duty is to design, carry out, improve, or maintain irrigation works. The amount of precipitation is expressed as the depth in centimeters (or inches) which falls on a level surface, and is measured by *rain-gauge.* The following are the main types of rain-gauges used for measurement of rainfall.

1. *Non-automatic rain-gauge*

This is also known as non-recording rain-gauge. Symon's rain-gauge is the instrument prescribed for use at all Government raingauge stations throughout India.

2. *Automatic rain-gauge*.

These are integrating type recording rain-gauges and are of three types :

(*i*) Weighing bucket rain-gauge. (*ii*) Tipping bucket rain-gauge. (*iii*) Float type rain-gauge.

1. Symon's Rain-gauge

Symon's rain-gauge is the most common type of non-automatic rain-gauge, and is used by Meteorological Department of Government of India. As shown in Fig. 2.2, it consists of a cylindrical vessel 127 mm (or 5″) in diameter with a base enlarged to 210 mm (or 8") diameter. The top section is a funnel provided with circular brass rim exactly 127 mm (5″) internal diameter. The funnel shank is inserted in the neck of a receiving bottle which is of 75 to 100 mm (3″ to 4″) diameter. A receiving bottle of rain-gauge has a capacity of about 75 to 100 mm of rainfall and as during a heavy rainfall

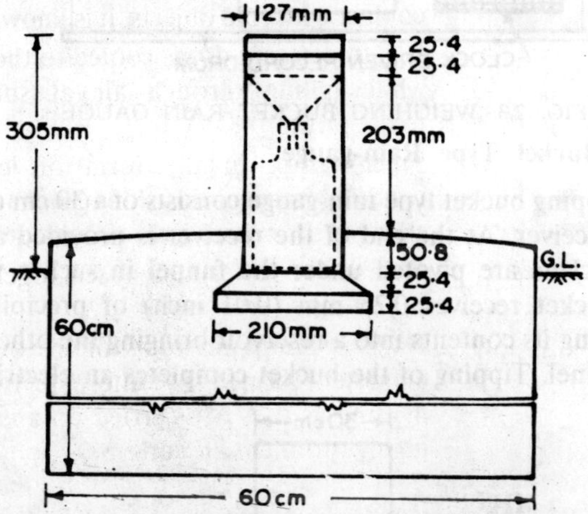

FIG. 2.2. SYMON'S RAIN-GAUGE (ADOPTED BY METEOROLOGICAL DEPARTMENT OF GOVT. OF INDIA)

this quantity is frequently exceeded, the rain should be measured 3 or 4 times in a day on days of heavy rainfall lest the receiver fill should overflow. A cylindrical graduated measuring glass is furnished with each instrument, which reads to 0.2 mm. The rainfall should be estimated to the nearest 0.1 mm.

2. Weighing Bucket Type Rain-gauge

Self-recording gauges are used to determine rates of rainfall over short periods of time. The most common type of self-recording gauge is the *weighing bucket type*, as shown in Fig. 2:3.

The weighing bucket rain-gauge essentially consists of a receiver bucket supported by a spring or lever balance or any other weighing mechanism. The movement of the bucket due to its increasing weight is transmitted to a pen which traces the record on a clock driven chart.

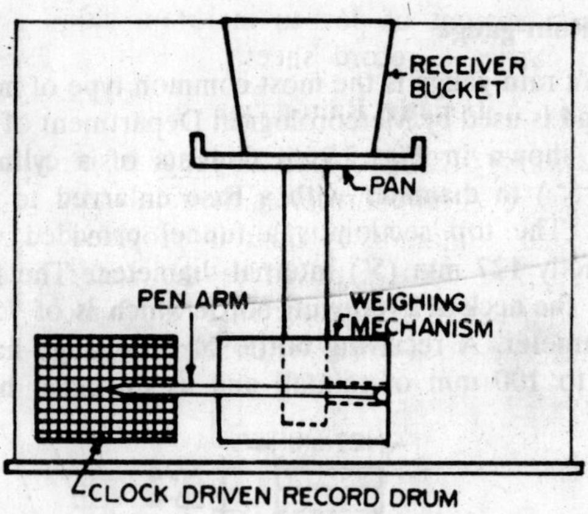

FIG. 2.3. WEIGHING BUCKET RAIN GAUGE

3. Tipping Bucket Type Rain-gauge

The tipping bucket type rain-gauge consists of a 30 cm diameter sharp edge receiver. At the end of the receiver is provided a funnel. A pair of buckets are pivoted under the funnel in such a way that when one bucket receives 0.25 mm (0.01 inch) of precipitation it tips, discharging its contents into a reservoir bringing the other bucket under the funnel. Tipping of the bucket completes an electric circuit

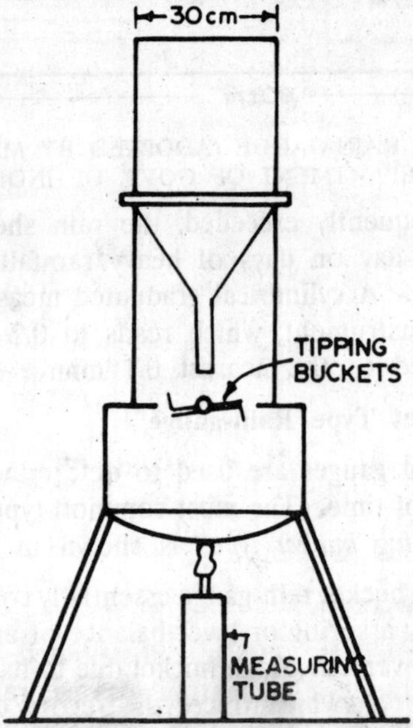

FIG. 2.4. TIPPING BUCKET TYPE RAIN-GAUGE

causing the movement of pen to mark on clock driven revolving drum which carries a record sheet.

4. Float Type Automatic Rain-gauge

The working of a float type rain-gauge is similar to the weighing bucket type gauge. A funnel receives the rain water which is collected in a rectangular container. A float is provided at the bottom of the container. The float is raised as the water level rises in the container, its movement being recorded by a pen moving on a recording drum actuated by a clock-work. When the water level in the container

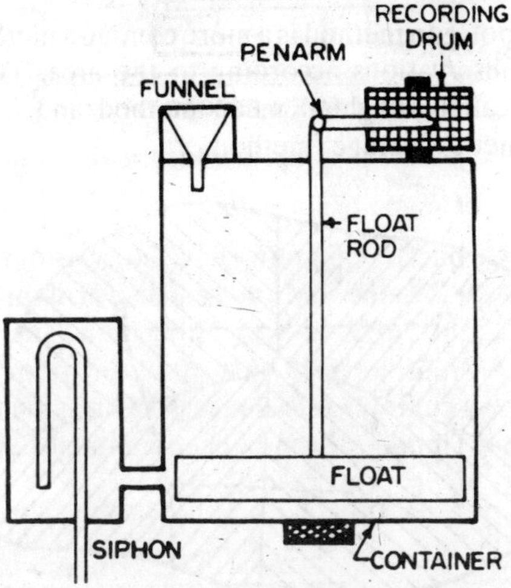

FIG. 2.5. FLOAT TYPE AUTOMATIC RAIN-GAUGE

rises so that the float touches the top, the siphon come into operation, and releases the water ; thus all the water in the box is drained out.

2.4. COMPUTATION OF AVERAGE RAINFALL OVER A BASIN

If a basin or catchment area contains more than one raingauge station, the computation of average precipitation or rainfall may be done by the following methods :

1. Arithmetic average method.
2. Thiessen polygon method.
3. Isohyetal method.

1. Arithmetic Average Method

If the rainfall is uniformly distributed in its areal pattern, the simplest method of estimating average rainfall is to compute the arithmetic average of the recorded rainfall values at various stations.

Thus, if $P_1, P_2, P_3, \ldots, P_n$ etc., are the precipitation or rainfall values measured at n gauge stations, we have

$$P_{av} = \frac{P_1 + P_2 + \ldots + P_n}{n} = \frac{\Sigma P}{n} \qquad \ldots(2.1)$$

2. Thiessen Polygon Method

The arithmetic average method is the most approximate method since rainfall varies in intensity and duration from place to place. Hence the rainfall recorded by each rain-gauge station should be weighted according to the area it is assumed to represent.

Thiessen polygon method is a more common method of weighting the rain-gauge observations according to the area. Thiessen polygon method is also called weighted mean method and is more accurate than the arithmetic average method.

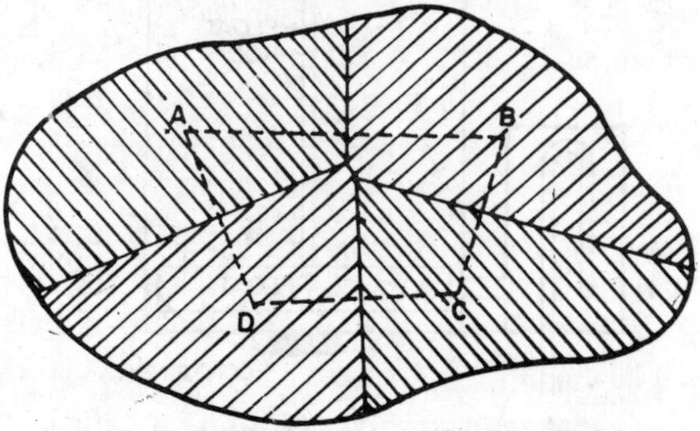

FIG. 2.6. THIESSEN POLYGON METHOD

Procedure

1. Join the adjacent rain-gauge stations A, B, C, D etc., by straight lines.

2. Construct the perpendicular bisectors of each of these lines.

3. A Thiessen network is thus constructed. The polygon formed by the perpendicular bisectors around a station encloses an area which is everywhere closer to that station than to any other station. Find the area of each of these polygons shown hatched in Fig. 2.6.

4. Multiply the area of each Thiessen polygon by the rain-gauge value of the enclosed station.

5. Find the total area (ΣA) of the basin.

6. Compute the average precipitation or rainfall from the equation

$$P_{av} = \frac{A_1 P_1 + A_2 P_2 + \ldots\ldots + A_n P_n}{A_1 + A_2 + \ldots\ldots A_n} = \frac{\Sigma (A \times P)}{\Sigma A} \quad \ldots(2.2)$$

3. Isohyetal Method

The basic assumption in the Thiessen polygon method is that a rain-gauge station best represents the area which is close to it. However, this may not always be valid, specially when the rainfall is controlled by topography or results from intense convection. The isohyetal method is the most elaborate and accurate in such conditions.

An *isohyet* is a line, on a rainfall map of the basin, joining places of equal rainfall readings. An *isohyet map* showing contours of equal rainfall presents a more accurate picture of the rainfall distribution over the basin.

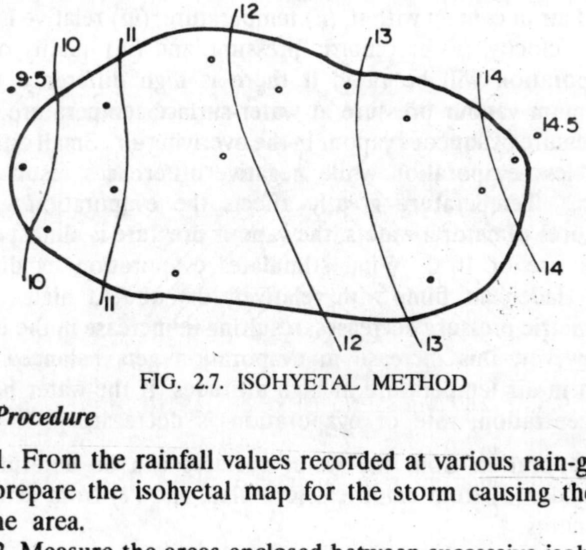

FIG. 2.7. ISOHYETAL METHOD

Procedure

1. From the rainfall values recorded at various rain-gauge stations, prepare the isohyetal map for the storm causing the rainfall over the area.

2. Measure the areas enclosed between successive isohyets with the help of a planimeter.

3. Multiply each of these areas by the average rainfall between the isohyets.

4. The average rainfall is then computed from the expression

$$P_{av} = \frac{\Sigma A \times \left(\frac{P_1 + P_2}{2}\right)}{\Sigma A} \quad \ldots(2.3)$$

2.5. EVAPORATION AND TRANSPIRATION

Evaporation is the transfer of water from the liquid to the vapour state. The rate of evaporation from a water surface is proportional to the difference between the vapour pressure at the surface and the vapour pressure in the overlying air (Dalton's law). *Transpiration* is the process by which plants dissipate water from the surface of their leaves, stalks and trunks in the process of its growth. *Evapo-transpiration* or *consumptive use* of water by a crop is the depth of water consumed by evaporation and transpiration during crop growth, including water consumed by accompanying weed.

Evaporation is one of the most important factor in water resources planning, since it raises the storage requirements of reservoirs and lowers the yield of lakes and ponds.

Evaporation from Water Surfaces

Evaporation from water surfaces (*i.e.* from rivers, lakes, reservoirs etc.) depends upon the following factors : (*i*) vapour pressure of water and air in contact with it, (*ii*) temperature, (*iii*) relative humidity, (*iv*) wind velocity, (*v*) barometric pressure, and (*vi*) quality of water. The evaporation will be rapid if there is high difference between the maximum vapour pressure at water-surface temperature and the actual pressure of aqueous vapour in the overlying air. Small differences result in less evaporation while negative differences result in condensation. Temperature greatly affects the evaporation. At the temperatures of natural waters, the vapour pressure is almost doubled for every rise of 10°C. Wind stimulates evaporation by displacing moisture- laden air films with relatively dry air. At high altitudes, the barometric pressure decreases, resulting in increase in the evaporation. However, this increase in evaporation gets balanced due to reduction in air temperature at high altitudes. If the water has more salt concentration, rate of evaporation is decreased.

Evaporation from the water surfaces can be determined by two methods: (1) from empirical formulae, and (2) from actual field measurements.

Determination of Evaporation from Empirical Formulae

1. *Meyer's Formula*

$$E = C(V - v)\left(1 + \frac{W}{K}\right) \qquad \text{...(2.4 } a)$$

where E = Evaporation loss in cms per month (*i.e.* in 30 days)

V = Maximum vapour pressure, expressed in cms of mercury, at the water surface corresponding to its temperature.

v = Actual vapour pressure, expressed in cms of mercury, at the monthly mean temperature and relative humidity in the atmosphere above.

C = Constant, the value of which may be taken as 1.1 for deep bodies of water and 1.5 for shallow bodies of water.

K = Constant, the value of which may be upto 16.

2. *Rohwer's Formula*

$$E = C'(1.465-0.00732\,p_a)(0.44+0.0732W)(V-v) \quad ...(2.4\,b)$$

where E = Evaporation loss in cms per day (24 hours)

C' = Constant, the value of which may be found out from the available evaporation data; it may be approximately taken equal to 0.75.

p_a = Atmospheric pressure in cms of mercury, at 0°C.

V, v and W have the same meaning as in the Meyer's formula.

Determination of Evaporation from Field Measurements

The evaporation from water surface is generally measured by exposing pans of water to the atmosphere. The water pan may be either installed on the ground near the lake (Fig. 2.8 *a*), or it may be suppported on water surface with the help of floats (Fig. 2.8 *b*). The former is known as *land pan*, while the later is known as a *floating pan.* In each case, the Standard Class *A* pan is a galvanised iron pan of 4′ (1.20 m) diameter and 10″ (25 cm) depth. The land pan is supported on a grid of 5 cm × 10 cm timbers to raise it above the ground to promote air circulation. The floating pan, however, is kept floating on water surface with the help of floats. The change in the water level is measured by a hook gauge provided over a stilling well. Temperature, rainfall, wind speed etc. are also measured. The evaporation measured in the pan in multiplied by a suitable coefficient to get the actual evaporation of the lake surface. This coefficient, known as *pan evaporation coefficient*, varies from 0.6 to 0.8.

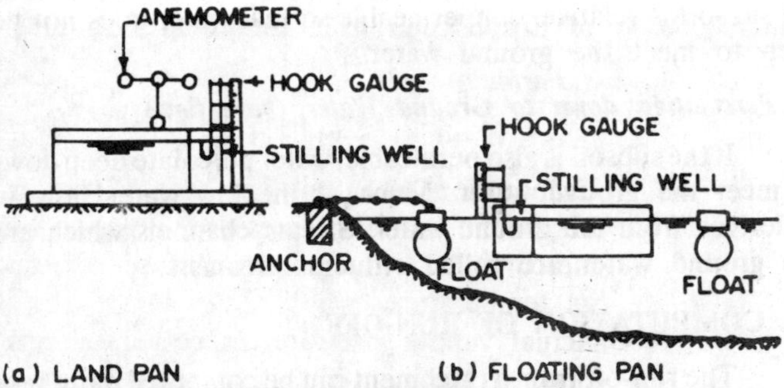

FIG. 2.8. MEASUREMENT OF EVAPORATION

2.6. RUN-OFF

The run-off of a catchment area in any specified period is the total quantity of water draining into a stream or into a reservoir in that period. This can be expressed as (*i*) centimetres of water over a catchment, or (*ii*) the total water in cubic-metres or hectare-metres for a given catchment.

The rainfall is disposed of in the following manner:

1. Basin Recharge.
2. Direct run-off (or simply, run-off).
3. Percolation down to ground water.
4. Evaporation.

1. *Basin Recharge*

The basin recharge consists of the following :

(*i*) Rain intercepted by leaves and stems of vegetation.
(*ii*) Water held up in surface depressions, commonly known as the depression storage.
(*iii*) Soil moisture held as capillary water in pore spaces of soil or as hygroscopic water absorbed on the surface of soil particles.

2. *Direct Run-off*

Direct run-off is that water which reaches the stream shortly after it falls as rain. Direct run-off consist of:

(*i*) Over land flow (a surface run-off).
(*ii*) Inter-flow (Influent stream).

Overland flow is that portion of water which travels across the ground surface to the nearest stream. However, if the soil is permeable, water percolates into it, and when it becomes saturated, flows laterally in the surface soil to a stream channel. The essential condition for inter-flow is that the surface soil is permeable, but the subsoil is relatively impermeable so that water does not percolate deep to meet the ground water.

3. *Percolation down to Ground Water (base flow)*

If the subsoil is also permeable, water percolate deep down-wards to meet the ground water. Much of the low water flow in rivers is derived from the ground water. Stream channels which are below the ground water are called effluent streams.

2.7. COMPUTATION OF RUN-OFF

The run-off from a catchment can be computed daily, monthly or yearly. Following are some of the methods of computing the run-off :

(*a*) By formulae and tables.
(*b*) By Infiltration method.
(*c*) By Unit Hydrograph.
(*d*) By Rational method.

(*a*) RUN-OFF BY FORMULAE AND TABLES

1. Run-off Coefficient

The run-off and the rainfall can be inter-related by run-off coefficient, by the expression

$$R = k\,P \qquad \text{...(2.5)}$$

where R = run-off in cm,
P = rainfall in cm,
k = run-off coefficient.

The run-off coefficient naturally depends upon all the factors which affect the run-off. This method is used only for small water control projects, and should be avoided for the analysis of major storms. The customary values of k are given below :

Area	k
Urban residentials	
Single Houses	... 0.3
Garden apartments	... 0.5
Commerical and Industrial	... 0.9
Forested areas, depending on soil	... 0.05 – 0.2
Parks, farm land, pasture	... 0.05 – 0.3
Asphalt or concrete pavement	... 0.85.

Barlow's Table

T.G. Barlow carried out studies of catchments mostly under 140 sq. km in the United Provinces (U.P.) and gave the following values of k (in percentage) for various classified catchments :

TABLE 2.1 BARLOW'S TABLE

Class	*Description of Catchment*	*Run-off per cent*
A	Flat, cultivated and black cotton soils	10
B	Flat, partly cultivated, various soils	15
C	Average	20
D	Hills and plains with little cultivation	35
E	Very hilly and steep, with hardly any cultivation	45

The above values of run-off percentages are for average monsoon. These are to be multiplied by the following coefficients (Table 2.9) according to the nature of the season.

TABLE 2.2 BARLOW'S COEFFICIENT

Nature of season	Class of Catchment				
	A	B	C	D	E
(1) Light rain, no heavy downpour	0.70	0.8	0.8	0.8	0.8
(2) Average or varying rainfall, no continous downpour	1.0	1.0	1.0	1.0	1.0
(3) Continuous downpour	1.50	1.50	1.60	1.70	1.80

2. Strange's Tables and Curves

W.L. Strange gave tables and curves for run-off resulting from rainfall in the plains of South India. The following tables and graphs are for the former Bombay Presidency. Strange's tables and curves give run-off for daily rainfall, and take into account three types of catchment, (*i.e.* good, average and bad), and three surface conditions, (*i.e.* dry, damp and wet) prior to the rain.

TABLE 2.3 DAILY RUN-OFF ACCORDING TO STRANGE

NOTE : FOR GOOD OR BAD CATCHMENT, ADD OR DEDUCT UPTO 25% OF YIELD.

Daily rainfall in (mm)	Run-off percent and yield when the original state of ground was					
	Dry		Damp		Wet	
	%	Yield in mm	%	Yield in mm	%	Yield in mm
6.25					8	0.5
12.5			6	0.75	12	1.5
18.75			8	1.50	16	3.0
25	3	0.75	11	2.75	18	4.5
31.25	5	1.56	14	4.37	22	6.88
37.5	6	2.25	16	6.00	25	9.37
43.75	8	3.5	19	8.31	30	13.13
50	10	5.0	22	11.00	34	17.00
62.5	15	9.37	29	18.15	43	26.90
75	20	15.00	37	27.75	55	41.25
100	30	30.00	50	50.00	70	70.00

3. Inglis's formula

C.C. Inglis gave the following formulae, derived from data collected from 37 catchments in the Bombay Presidency :

For Ghat areas

$$R = 0.85\,P - 12'' \quad \text{(when } R \text{ and } P \text{ are in inches)} \qquad ...(2.6\ a)$$
$$R = 0.85\,P - 30.5 \quad \text{(when } R \text{ and } P \text{ are in cm)} \qquad ...(2.6)$$

For Non-Ghat areas

$$R = \left(\frac{P-7}{100}\right) \times P \quad \text{(when } R \text{ and } P \text{ are in inches)} \qquad ...(2.7\ a)$$
$$R = \frac{(P-17.8)}{254} \times P \quad \text{(when } R \text{ and } P \text{ are in cm)} \qquad ...(2.7)$$

4. Lacey's formula

$$R = \frac{P}{1 + \frac{120\,F}{PS}} \quad \text{(when } P \text{ and } R \text{ are in inches)} \qquad ..(2.8\ a)$$

and $$R = \frac{P}{1 + \frac{304.8\,F}{PS}} \quad \text{(when } P \text{ and } R \text{ are in cm)} \qquad ...(2.8)$$

In both the expressions,
S = A catchment factor and F = Monsoon duration factor.

Corresponding to the five classes of catchment, defined by Barlow, Lacey gave the following values of the catchment factor (S) :

TABLE 2.4

Class of Catchment	*Value of S*
A	0.25
B	0.60
C	1.00
D	1.70
E	3.45

Lacey also divided the monsoon into three classes, depending upon its duration and gave the following values of monsoon duration factors :

TABLE 2.5

No	*Class of monsoon*	*Monsoon duration factor (F)*
1	Very Short	0.50
2	Standard Length	1.00
3	Very Long	1.50

(5) Khosla's formula

$$R = P - \frac{T - 32}{9.5} \text{ (when } R \text{ and } P \text{ are in inches)} \quad ...(2.9\,a)$$

$$R = P - \frac{T - 32}{3.74} \text{ (when } R \text{ and } P \text{ are in cm)} \quad ...(2.9)$$

where T = mean temperature in °F on the entire catchment.

The temperature introduced in the formula takes into account various factors affecting losses by evaporation, transpiration, sunshine and wind velocity.

(6) Parker's formulae

(*i*) $R = 0.94\,P - 14$ for catchments in the British Isles
(*ii*) $R = 0.94\,P - 16$ for catchments in Germany ...(2.10)
(*iii*) $R = 0.80\,P - 16.5$ for catchments in East U.S.A.

where R and P are in inch units.

(*b*) RUN-OFF BY INFILTRATION METHOD

Infiltration is defined as the movement of water through the soil surface and into the soil. The capacity of any soil to absorb water from rainfall falling continuously at an excessive rate goes on decreasing with time, until a minimum rate of infiltration is reached. At any instant, the *infiltration capacity* of a soil is the maximum rate at which water will enter the soil in a given condition. The *infiltration rate* is the rate at which water actually enters the soil during a storm, and is equal to the infiltration capacity or the rainfall rate, whichever is less. The *infiltration capacity* of soil can be determined experimentally by subjecting an experimental plot to rainfall rates

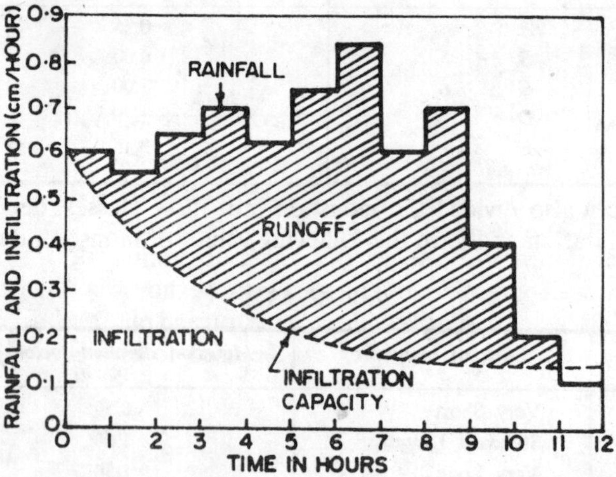

FIG. 2.9. INFILTRATION CAPACITY CURVE AND RUN-OFF

in excess of infiltration capacity, and by measuring the surface run-off (Fig. 2.9). For small areas having uniform infiltration characteristics, the run-off volume can be estimated by subtracting infiltration from the design rainfall.

Infiltration Index

Infiltration index is the average rate of loss such that the volume of rainfall in excess of that rate will be equal to direct run-off. Estimates of run-off volume from large areas, having heterogeneous infiltration and rainfall characteristics, are made by use of *infiltration indices.*

There are two types of infiltration indices.

(*i*) *Average infiltration rate* or *W*-index and (*ii*) ϕ index.

The *W*-index is calculated from the expression

$$W = \frac{P - R}{t_r} \text{ cm/hr} \qquad \text{...(2.11)}$$

where t_r = duration of rainfall in hours.

The ϕ-index is defined as the rate of rainfall above which the rainfall volume equals the run-off volume (Fig. 2.10).

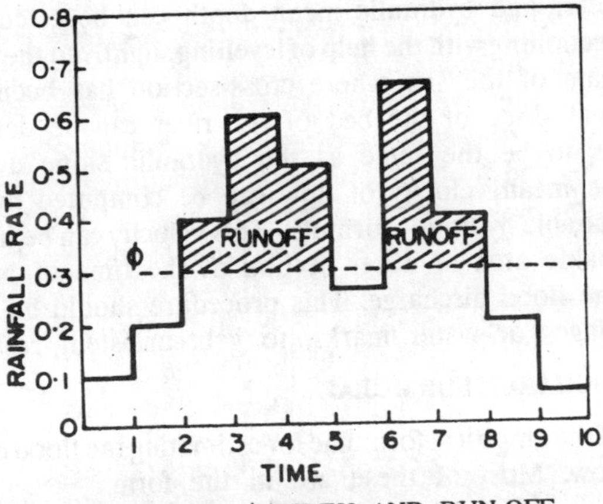

FIG. 2.10. ϕ INDEX AND RUN-OFF

For flood forecasting, appropriate index values must be derived by correlation with those factors which determine the index at any time. In such approach, there seems to be no advantage over the method discussed earlier, in which the run-off and rainfall are correlated. However, the infiltration index can be used to estimate the run-off co-efficient (*k*) from the relation

$$k = \frac{i - W}{i} \qquad \text{...(2.12)}$$

where i = rainfall intensity (cm/hr)

2.8. ESTIMATION OF MAXIMUM RATE OF RUN-OFF OR FLOOD DISCHARGE

The estimation of peak flow or flood can be made by the following methods :

(*a*) By physical indication of past floods.
(*b*) By flood discharge formulae.
(*c*) By flood frequency studies.
(*d*) By unit hydrograph

The method of flood frequency and the method of unit hydrograph can be found in any standard text book on Hydrology.

(*a*) PHYSICAL INDICATION OF PAST FLOODS

Ancient monuments, etc., situated on river banks always bear past flood marks. Old persons in the villages situated on the bank of the river may be contacted to know the maximum water level attained in the past 35 years. The cross-section of the river may be plotted, and the water line corresponding to the highest flood can be drawn on it. From such a cross-section, the water-flow-area, wetted perimeter and hydraulic mean depth can be calculated. By longitudinal sectioning with the help of levelling slightly to the upstream and downstream of the site where cross-section has been plotted, the longitudinal slope of the bed of the river can be determined. Assuming this to be the same as the hydraulic slope during the past flood, the mean velocity of flow can be computed by Chezy's or any other suitable hydraulic formula. This velocity can be multiplied with the probable area of water section at the time of past flood to calculate the flood discharge. This procedure should be repeated at several villages or water marks, to get consistent results.

(*b*) FLOOD DISCHARGE FORMULAE

Some of the empirical formulae for estimating the flood discharge are given below. Most of these are in the form

$$Q = CA^n \qquad ...(2.13)$$

where Q = flood discharge
A = catchment area
n = flood index
C = flood coefficient.

Both C and n depend upon various factors, such as (*i*) size, shape and location of catchment, (*ii*) topography of the catchment, and (*iii*) intensity and duration of rainfall, and distribution pattern of the storm over the basin.

1. Dicken's Formula

In metric unit,

$$Q = CA^{3/4} \quad ...(2.14)$$

where Q = discharge in cumecs
A = area of basin in sq. km.

In F.P.S. units

$$Q_1 = C_1 A_1^{3/4} \quad ...(2.14\ a)$$

where Q_1 = discharge in cusecs
A_1 = area of basin in sq. miles.

The constants C and C_1 depend upon the catchment and may be obtained from Table 2.6.

TABLE 2.6

Region	C	C_1
Northern India	11.4	825
Central India	13.9−19.5	1000 to 1400
Western Ghats	22.2−25	1600 to 1800

2. Ryve's Formula

For Madras catchments

$$Q = CA^{2/3} \quad ...(2.15)$$

(where Q is in cumecs and A in sq. km.)

and

$$Q = C_1 A_1^{2/3} \quad ...(2.15\ a)$$

(where Q_1 is in cusecs and A_1 in sq. miles)

values of C and C_1 may be obtained from Table 2.7

TABLE 2.7

	Location of the Catchment	C	C_1
1.	Areas within 24 km (15 miles) from the coast.	6.75	450
2.	Area within 24 km –161 km (25 to 100 miles) from the coast)	8.75	563
3.	Limited areas near hills	10.1	672

3. Inglis Formula

Inglis formula is applicable for catchments of former Bombay Presidency.

Expressed in metric units,

$$Q = \frac{123A}{\sqrt{A + 10.4}} \approx 123 A^{1/2} \qquad ...(2.16)$$

Expressed in F.P.S. units,

$$Q_1 = \frac{7000 A_1}{\sqrt{A_1 + 4}} \approx 7000 A_1^{1/2} \qquad ...(2.16\ a)$$

4. Nawab Jang Bahadur Formula

For catchments of old Hydrabad State :

Expressed in metric units,

$$Q = CA^{\left(0.993 - \frac{1}{14}\log A\right)} \qquad ...(2.17)$$

where C varies from 48 to 60.

Expressed in F.P.S. units,

$$Q_1 = C_1 A_1^{\left(0.92 - \frac{1}{14}\log A_1\right)} \qquad ...(2.17\ a)$$

where C_1 varies from 1600 to 2000.

5. Fanning's Formula

For American catchments,

In metric units,

$$Q = CA^{5/6} \qquad ...(2.18)$$

where average value of C_1 may be taken equal to 2.54.

In F.P.S. units,

$$Q_1 = C_1 A_1^{5/6} \qquad ...(2.18\ a)$$

where average value of C_1 may be taken equal to 200.

6. Creager's Formula

Applicable for American catchments

Expressed in F.P.S. units,

$$Q_1 = 46\, C_1 A_1^{(0.894 A_1 - 0.048)} \qquad ...(2.19)$$

The constant varies from 30 to 100.

7. Fuller's Formula

This formula takes into account the flood frequency also.

Expressed in metric units,

$$Q_{max} = CA^{0.8} (1 + 0.8 \log T) (1 + 2.67 A^{-0.3}) \quad ...(2.20)$$

where T = Number of years after which such a flood is to reoccur.

Q = Maximum flood (in cumecs) during any part of the day that could occur in T-years.

A = Area of drainage basin, in sq. km.

C = Constant varying from 0.185 to 1.3.

Expressed in F.P.S. units,

$$Q_1 = C_1 A_1^{0.8} (1 + 0.8 \log T)(1 + 2 A_1^{-0.3}) \quad ...(2.20\ a)$$

where C_1 varies from 14 to 98.

PROBLEMS

1. Explain with the help of diagram the hydrologic cycle.

2. What do you understand by precipitation ? Explain various types of precipitation.

3. Explain any one type of automatic raingauge.

4. Describe various methods of computing average rainfall over a basin.

5. What are the methods of computing run-off from a catchment area? Give various formulae stating clearly the area for which each is applicable.

6. Explain various methods of determining flood discharge in a stream.

7. Give various flood discharge formulae applicable for Indian catchments.

3

Surface Sources

3.1. STORAGE RESERVOIRS

Dams are constructed across the rivers and streams to create an artificial lake or reservoir behind it. Dams and reservoirs are the most important and expensive elements in multipurpose river basin development. They require very careful planning, design and operation. A number of problems arise in design, construction and operation of the dams and reservoirs ; for example, selection of site, the relative merits of different types of dams, storage capacity and optimum yield and co-ordinated use of storage for different purposes.

Depending upon the purposes served, reservoirs may be classified as follows : (*i*) storage or conservation reservoirs (*ii*) flood protection reservoirs (*iii*) distribution reservoirs, and (*iv*) multipurpose reservoirs. *Storage reservoirs* are primarily used for water supplies for irrigation, hydroelectric developments, domestic and industrial supplies. A river does not carry the same quantity of water throughout the year. A storage reservoir is constructed to store the excess water during the period of large supplies, and release it gradually as and when it is needed. *Flood protection reservoirs* are those which store water during flood and release it gradually at a safe rate when the flood reduces. *Distribution reservoir* is a small storage reservoir used for water supply in a city or industrial area. A distribution reservoir account for a varying rate of water during the day. Such a distribution reservoir permits the pumping plants and water treatment works etc. to operate at a constant rate. The varying demand rate, exceeding the constant pumping rate is met from the distribution reservoir. A *multipurpose reservoir* is the one which serves more than one purpose.

3.2. INVESTIGATIONS FOR RESERVOIR PLANNING

The following investigations are required for reservoir planning:

1. Engineering surveys 2. Geological investigations.
3. Hydrological investigations.

1. Engineering Surveys

The area at the dam site is surveyed in detail and a contour plan is prepared. From the plan, the following physical characteristics are prepared :

(*a*) Area-elevation curve
(*b*) Storage-elevation curve
(*c*) Map of the area to indicate the land and property to be surveyed
(*d*) Suitable site selection for the dam.

Area-Elevation and Storage-Elevation Curves

Fig. 3.1 shows a typical contour plan. The area A_1, A_2, A_3 enclosed by the successive contours can be determined with a plani-

FIG. 3.1. A TYPICAL CONTOUR PLAN

meter. A close observation of Fig. 3.1 shows that as the value of a contour (*i.e.* elevation) increases, its area increases. A curve, such as curve *AB* shown in Fig. 3.2 may be drawn between elevation and area.

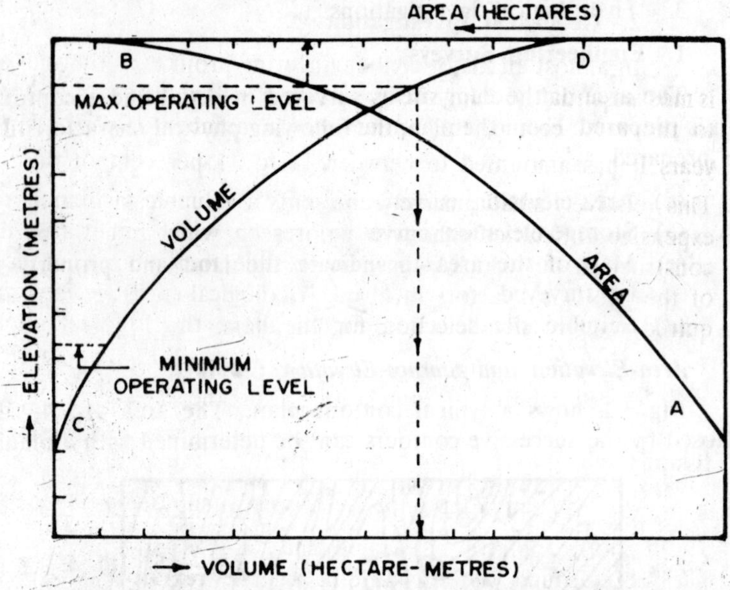

FIG. 3.2. ELEVATION-AREA AND ELEVATION-STORAGE CURVES

The reservoir capacity, or the volume of storage, corresponding to a given water level in the reservoir may be calculated either by trapezoidal formula or by prismoidal formula. Thus, if V is the storage volume and h is the contour interval, the formulae are :

(1) $$V = \Sigma \frac{h}{2}(A_1 + A_2) \text{ .....(Trapezoidal formula)}$$

$$= h\left\{\frac{A_1 + A_n}{2} + A_2 + A_3 + \text{....} A_{n-1}\right\}$$

(2) $$V = \Sigma \frac{h}{2}(A_1 + A_2 + \sqrt{A_1 A_2}) \text{ ...(Cone formula)}$$

(3) $$V = \Sigma \frac{h}{3}(A_1 + 4A_2 + A_3) \text{ ....(Prismoidal formula)}$$

$$= \frac{h}{3}\{(A_1+A_n)+4(A_2+A_4+\text{.....})+2(A_3+A_5+\text{.....})\}$$

where A_n is the area of the contour corresponding to the water surface elevation in the proposed reservoir. The volumes of storage corresponding to various water-surface elevations may be calculated, and a curve, such as *CD* in Fig. 3.2 may be plotted between elevation and storage.

The contour plan (Fig. 3.1) also indicates the water spread corresponding to the reservoir elevation and enables us to determine the area under submergence and the compensation to be paid to the owners.

2. Geological Investigations

In almost all major civil engineering projects, geological advice is most essential. Geological investigations cost very little in comparison to the total cost of the project ; in a number of cases in recent years it has amounted to between $\frac{1}{4}$ to 1 per cent of the whole. This relatively small amount represents a valuable insurance against expensive difficulties, otherwise unforeseen, which might arise during construction but which it may be possible to predict from a study of the geological factors involved. Geological investigations are required to give detailed information about the following items :

1. Water-tightness of reservoir basin.
2. Suitability of foundations for the dam.
3. Geological and structural features, such as folds, faults, fissures etc., of the rocks of the basin.
4. Type and depth of overburden (superficial deposits).
5. Location of permeable and soluble rocks, if any.
6. Ground water conditions in the region.
7. Location of quarry sites for materials required for the dam construction, and quantities available from them.

The geology of the catchment area should also be studied, since it affects the proportions of run-off and percolation. The special requirement for the geology of the reservoir site is that there should be no danger of serious leakage when the ground is under pressure from the full head of water in the reservoir. The geology of the dam site is important from the point of view of secure foundation for the dam. The nature of sub-surface geology should be explored by trial bores or various means of geophysical exploration.

3. Hydrological Investigations

The hydrological investigation is a very important aspect of reservoir planning. The capacity of the irrigation canals and/or the installed capacity of the power houses will depend upon the available supplies from the reservoir. The hydrological investigations may be divided into two heads :

1. Study of run-off pattern at the proposed dam site, to determine the storage capacity corresponding to a given demand, and

2. Determination of the hydrograph of the worst flood, to determine the spillway capacity and design.

3.3. SELECTION OF SITE FOR A RESERVOIR

The final selection of site for a reservoir depends upon the following factors :

(1) The geological conditions of the *catchment area* should be such that percolation losses are minimum, and maximum run-off is obtained.

(2) The reservoir site should be such that quantity of leakage through it is a minimum. Reservoir sites having the presence of highly permeable rocks reduce the water-tighness of the reservoir. Rocks which are not likely to allow passage of much water include shales and slates, schists, gneisses, and crystalline igneous rocks such as granite.

(3) Suitable dam site must exist. The dam should be founded on sound watertight rock base, and percolation below the dam should be minimum. The cost of the dam is often a controlling factor in selection of a site.

(4) The reservoir basin should have narrow opening in the valley so that the length of the dam is less.

(5) The cost of real estate for the reservoir, including road, rail road, dwelling relocation, etc., must be as less as possible.

(6) The topography of the reservoir site should be such that it has adequate capacity without submerging excessive land and other properties.

(7) The site should be such that a deep reservoir is formed. A deep reservoir is preferable to a shallow one because of : (*i*) lower cost of land submerged per unit of capacity, (*ii*) less evaporation losses because of reduction in the water spread area, and (*iii*) less likelihood of weed growth.

(8) The reservoir site should be such that it avoids or excludes water from those tributaries which carry a high percentage of silt in water.

(9) The reservoir site should be such that the water stored in it is suitable for the purpose for which the project is undertaken. The soil and rock mass at the reservoir site must not contain any objectionable soluble minerals and salts.

3.4. STORAGE CAPACITY AND YIELD

Yield. Yield is the amount of water that can be supplied from the reservoir in a specified interval of time. The interval of time chosen for the design varies from a day for small distribution reservoir to a year for large conservation reservoirs. For example, if 25,000 cubic metres of water is supplied from a reservoir in one year, its yield is 25,000 cubic metres/year or 2.5 hectare-metres/year.

Safe yield or firm yield. The maximum quantity of water that can be guaranteed during a critical dry period is known as the safe yield or firm yield.

Secondary yield. Secondary yield is the quantity of water available in excess of safe yield during periods of high flood.

Average yield. The arithmetic average of the firm and the secondary yield over a long period of time is called average yield.

Mass inflow curve

The reservoir capacity corresponding to a specified yield is determined with the help of mass inflow curve and the demand curve. A *mass inflow curve* is a plot between the cumulative inflow in the reservoir with time.

Fig. 3.3 shows a flood hydrograph of inflow for several years Fig. 3.4 shows a mass inflow curve prepared from the flood hydrograph of Fig. 3.3. Taking the starting year (*i.e.,* 1957) as the base, the total quantity of water from 1957 to a time t_1 (say, 1960) that has flown through the river is the volume represented by the hatched area. In the mass curve (Fig. 3.4), the corresponding ordinate at time t_1 (ordinate *AB*) will therefore be equal to the volume of water indicated by the hatched area of Fig. 3.3. Similarly, the ordinates of the mass curve corresponding to other years can be computed from Fig. 3.3 and plotted. A mass curve continuously rises as it

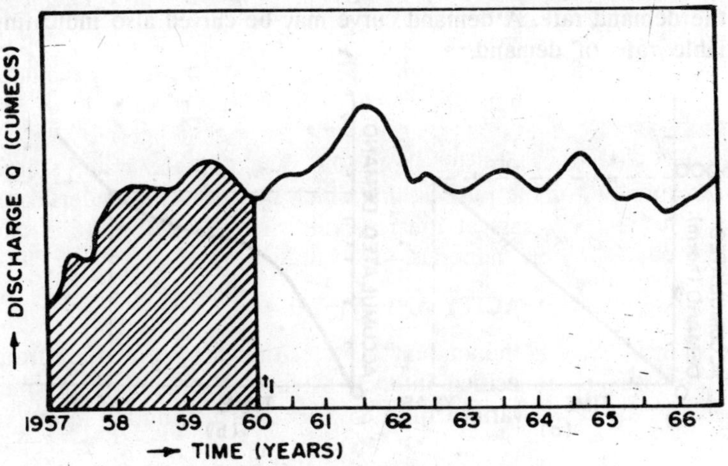

FIG 3.3. FLOOD HYDROGRAPH OF INFLOW

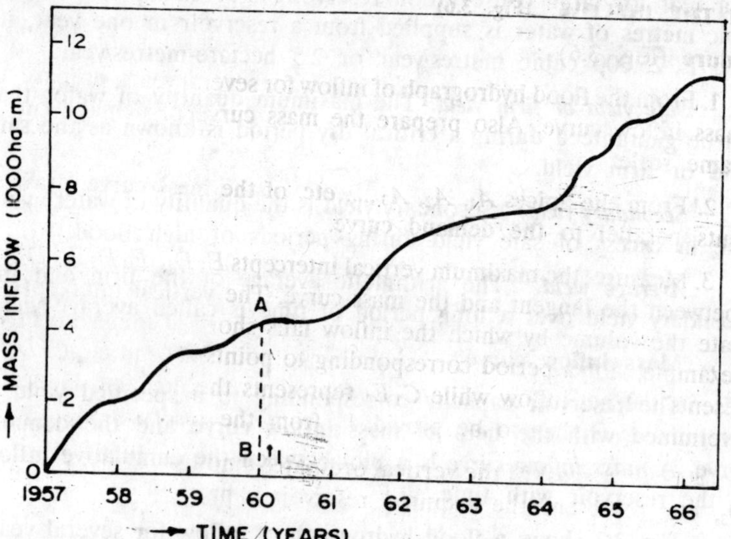

FIG. 3.4. MASS INFLOW CURVE

shows *accumulated inflow.* If there is no inflow during certain period, the mass curve will be horizontal during that period. The mass curve will rise very sharply during the period of high flood. Thus the steepness of the mass curve shows the rate of inflow at that time interval. The hollows on the mass curve show relatively dry periods.

Demand curve. A *demand curve* (Fig. 3.5) is a plot between accumulated demand with time. The demand curve representing a uniform rate of demand is a straight line having the slope equal to the demand rate. A demand curve may be curved also indicating variable rate of demand.

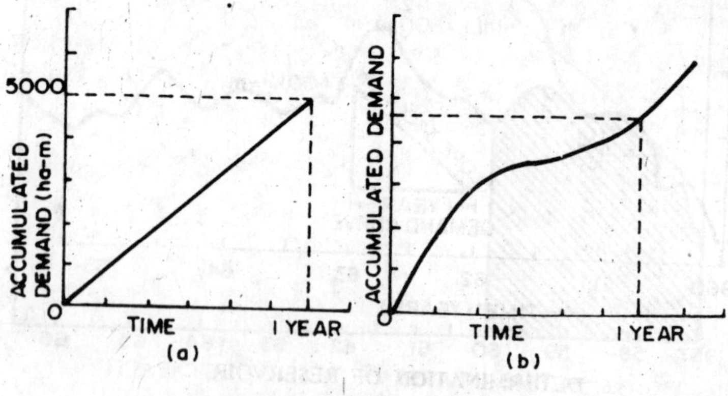

FIG. 3.5. DEMAND CURVE

CALCULATION OF RESERVOIR CAPACITY, FOR A SPECIFIED YIELD FROM THE INFLOW (Fig. 3.6)

Procedure (Fig. 3.6) :

1. From the flood hydrograph of inflow for several years, prepare the mass inflow curve. Also prepare the mass curve of demand on the same scale.

2. From the apices A_1, A_2, A_3 ..... etc. of the mass curve, draw tangents parallel to the demand curve.

3. Measure the maximum vertical intercepts E_1D_1, E_2D_2, E_3D_3, etc. between the tangent and the mass curve. The vertical intercepts indicate the volume by which the inflow falls short of the demand. For example, for a period corresponding to points A_1 and C_1, C_1D_1 represents the net inflow while C_1E_1 represents the demand. Hence the volume E_1D_1 has to be provided from the reservoir storage.

4. The biggest of the vertical ordinates amongst E_1D_1, E_2D_2, E_3D_3, etc. represent the required reservoir capacity.

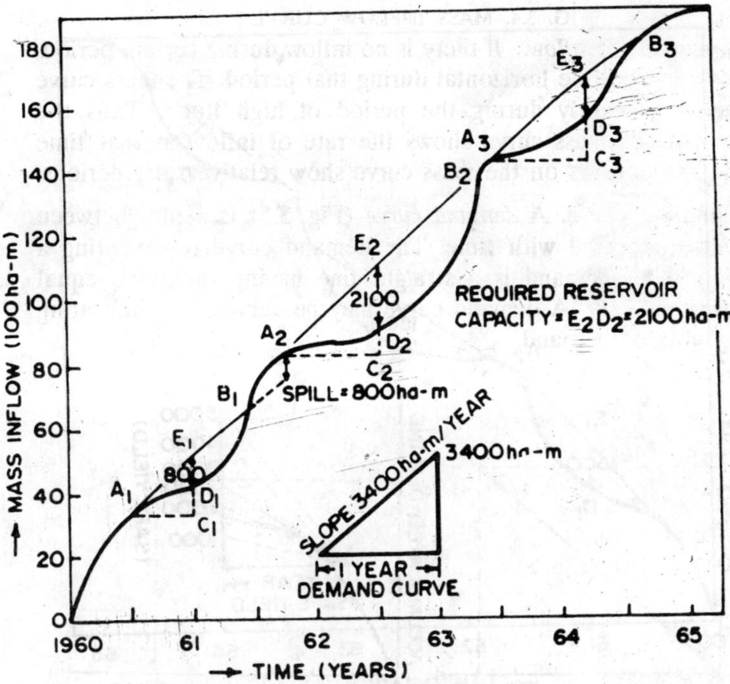

FIG. 3.6. DETERMINATION OF RESERVOIR CAPACITY

It should be noted that the vertical distance between successive tangents represent water wasted over the spillway. The spillway must have sufficient capacity to discharge this flood volume.

Corresponding to the numerical figures indicated in Fig. 3.6, we observe that

(1) The required reservoir capacity is 2100 ha-m.

(2) Assuming the reservoir to be full at A_1, it is depleted to (2100 – 800) = 1300 ha-m at D_1, and is again full at B_1.

(3) Assuming the reservoir to be full at A_2, it is empty at D_2 and is again full at B_2.

(4) The reservoir is full between B_1 and A_2, and the quantity of spill is 800 ha-m.

DETERMINATION OF SAFE YIELD FROM A RESERVOIR OF A GIVEN CAPACITY

The following is the procedure of determining the safe yield from a reservoir of a given storage capacity, with the help of a mass inflow curve :

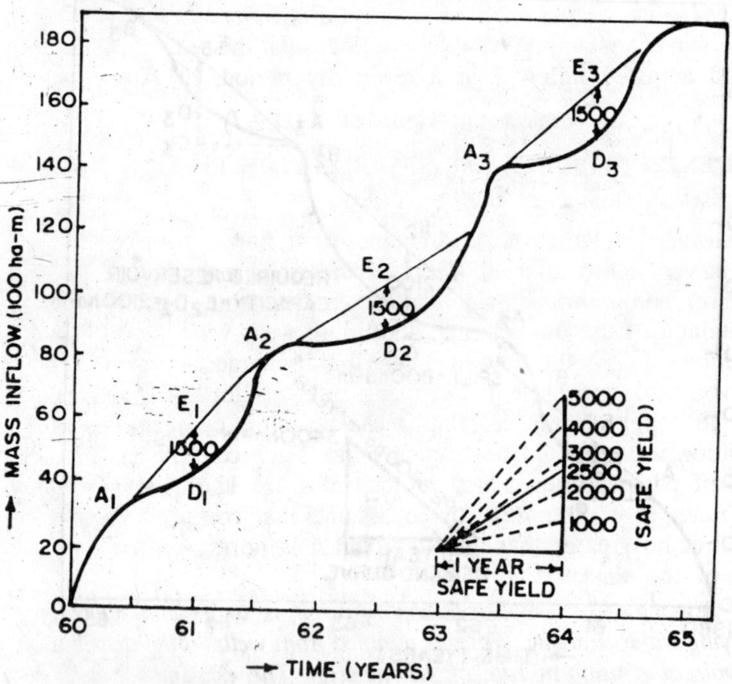

FIG. 3.7. DETERMINATION OF YIELD FROM RESERVOIR OF SPECIFIED CAPACITY

(1) Prepare the mass inflow curve. On the same diagram, draw straight lines, from a common origin, representing demands at various rates, say varying from 0 to 5000 ha-m per year.

(2) From the apices A_1, A_2, A_3, etc. of the mass curve, draw tangents in such a way that their maximum departure from the mass curve does not exceed the specified reservoir capacity. Thus, in Fig. 3.7, the ordinates $E_1 D_1$, $E_2 D_2$, $E_3 D_3$, etc. are all equal to the reservoir capacity (say 1500 ha-m).

(3) Measure the slopes of each of these tangents. The slopes indicate the yield which can be attained in each year from the reservoir of given capacity. The slope of the flattest demand line is the *firm yield.*

ANALYTICAL METHOD FOR COMPUTING RESERVOIR CAPACITY

When the demand is variable, it is preferable to carry out the computations for storage in a tabular form. Example 3.2 gives the procedure for computations. In this method, the variations in precipitation and evaporation data can be easily accounted for. Assuming that the reservoir is full at the beginning of a dry period, the maximum amount of water S that must be withdrawn from storage to maintain a given average draft or demand D equals to the maximum cumulative difference between the draft (or demand) D and adjusted inflow I in a given dry period, or

$$S = \text{maximum value of } \Sigma (D - I)$$

COMPENSATION WATER AND RESIDUAL FLOW RULES

When a dam is constructed across a river or a stream so that storage reservoir is formed behind it, the flow to the downstream side is very much affected. Most of the wells may be situated on the river banks downstream, and hence if a minimum flow is not maintained to the downstream side, these wells will dry up. *Compensation water* is the flow that must be discharged below a storage reservoir to compensate the riparian interests for the water taken away to supply. While calculating the reservoir capacity, the compensation water should be kept in mind. Each country has its own rules of compensation water ; such rules are known as *residual flow rules.* It is often realistic to assume that compensation water need not be greater than the flow which is normally exceeded for 90% of the time.

Example 3.1. *A distribution reservoir is to be constructed for supplying water to a city. Water is pumped from wells to the distribution reservoir at a uniform rate of 2.5 cumecs. The estimated hourly requirements for the maximum day are tabulated below. Estimate the capacity of the distribution reservoir.*

TABLE 3.1

Time (hours)	*01*	*02*	*03*	*04*	*05*	*06*	*07*	*08*	*09*	*10*	*11*	*12*
Demand (million litres per hour)	*3.2*	*3.6*	*3.4*	*3.9*	*4.6*	*6.9*	*8.8*	*9.2*	*9.8*	*9.2*	*9.0*	*8.9*
Time (hours)	*13*	*14*	*15*	*16*	*17*	*18*	*19*	*20*	*21*	*22*	*23*	*24*
Demand (million litres per hour)	*8.3*	*8.0*	*7.8*	*7.8*	*8.5*	*9.2*	*9.0*	*8.8*	*6.1*	*5.2*	*4.8*	*4.0*

Solution :

The average pumping rate is determined by the total demand by 24. Hence average pumping rate $= \frac{168}{24} = 7$ million litres/hour.

TABLE 3.2

Hours ending	*Demand (million litres)*	*Pumping rate (million litres/hour)*	*Requirements from storage (million litres)*
01	3.2	7	0
02	3.6	7	0
03	3.4	7	0
04	3.9	7	0
05	4.6	7	0
06	6.9	7	0
07	8.8	7	1.8
08	9.2	7	2.2
09	9.8	7	2.8
10	9.2	7	2.2
11	9.0	7	2.0
12	8.9	7	1.9
13	8.3	7	1.3
14	8.0	7	1.0
15	7.8	7	0.8
16	7.8	7	0.8
17	8.5	7	1.5
18	9.2	7	2.2
19	9.0	7	2.0
20	8.8	7	1.8
21	6.1	7	0
22	5.2	7	0
23	4.8	7	0
24	4.0	7	0
sum :	168.0	168.0	24.3

The hourly demand as well as the water drawn from the reservoir tabulated in Table 3.2. The required reservoir capacity will be equal to the sum of the hourly requirements from storage. This comes out to be **24.3 million litres.**

Example 3.2. *Table 3.3 gives the monthly inflows during the critical low water period at the site of a proposed dam, the corresponding monthly evaporations and precipitations at a nearby station, and the estimated monthly demand for water. Prior water rights require the release of full natural flow or 10 hectare per month, whichever is least. Assume that 30 per cent of the rainfall on the land area to be flooded by the reservoir has reached the stream in the past. Using a net increased pool area of 500 hectares, find the required useful storage. Use a pan evaporation coefficient of 0.72.*

TABLE 3.3

Month	*Inflow at the proposed site (hectare-m)*	*Pan evaporation (cm)*	*Precipitation (cm)*	*Demand (ha-m)*
January	8.6	2.2	0.8	14.5
February	2.2	2.3	1.2	15.8
March	1.8	3.1	0.0	16.2
April	0.0	8.6	0.0	16.8
May	0.0	12.8	0.0	17.5
June	13.5	15.6	4.8	18.0
July	280.6	12.3	12.2	18.0
August	510.2	10.6	18.6	17.0
September	136.0	10.0	8.6	16.5
October	52.5	8.2	1.5	16.0
November	20.6	5.8	0.0	15.8
December	12.3	3.0	0.0	15.0

Solution :

Computations are arranged in Table 3.4 in which the first five columns are the same as given in Table 3.3. Column 6 gives the downstream requirements due to water rights, and the entries in this column are equal to the river flow or 10 ha-m, whichever is the minimum.

Column 7 gives the evaporation over the whole of the reservoir area. Its value E is calculated from the relation :

$$E = \frac{\text{Reservoir area} \times \text{Pan Evaporation}}{100} \times \text{Pan evap. coefficient}$$

$$= \frac{500 \times \text{column (3)}}{100} \times 0.72 \text{ hectare-metres}$$

$$= 3.6 \times \text{column (3) hectare-metres.}$$

TABLE 3.4

Month	Inflow (ha-m)	Pan. Evap. (cm)	Precipitation (cm)	Demand (ha-m) (D)	D/S requirements (ha-m)	Evaporation E (ha-m) =3.6×col. 3	Precipitation P (ha-m) = 3.6× col. 4	Adjusted Inflow I (ha-m) = col. 2−col. 6 − col. 7+col. 8	Water reqd. from storage S (ha-m) = col. 5 − col. 9
(1)	(2)	(3)	(4)	(5)	(6)	(7)	(8)	(9)	(10)
Jan.	8.6	2.2	0.8	14.5	8.6	7.92	2.8	− 5.12	19.62
Feb.	2.2	2.3	1.2	15.8	2.2	8.28	4.2	− 4.08	19.88
Mar.	1.8	3.1	0.0	16.2	1.8	11.16	0.0	− 11.16	27.36
April	0.0	8.6	0.0	16.8	0.0	30.96	0.0	− 30.96	47.76
May	0.0	12.8	0.0	17.5	0.0	46.08	0.0	− 46.08	53.58
June	13.5	15.6	4.8	18.0	10.0	56.16	16.8	− 35.86	53.86
July	280.6	12.3	12.2	18.0	10.0	44.28	42.7	+ 269.02	Nill
Aug.	510.2	10.6	18.6	17.0	10.0	38.16	65.1	+ 527.14	Nill
Sept.	136.0	10.0	8.6	16.5	10.0	36.00	30.1	+ 120.10	Nill
Oct.	52.5	8.2	1.5	16.0	10.0	29.52	5.25	+ 18.23	Nill
Nov.	20.6	5.8	0.0	15.8	10.0	20.88	0.0	− 10.28	26.08
Dec.	12.3	3.0	0.0	15.0	10.0	10.80	0.0	− 8.50	23.50
Sum	1038.3	94.5	47.7	197.1	82.6	340.20	166.95	782.45	281.64

Column 8 gives the precipitaion in hectare metres, falling over the reservoir area. Since 30% of the precipitation is already reaching and has been included in the inflow (column 2), only 70% of the precipitation has been included in the computation. Thus the precipitation P is calculated from the relation

$$P = \frac{\text{Reservoir area} \times \text{column (4)}}{100} \times 0.7$$
$$= \frac{500 \times \text{column (4)}}{100} \times 0.7$$
$$= 3.5 \times \text{column (4) hectare-metres.}$$

Column 9 gives the *adjusted inflow* (I) computed from the relation

$$I = \text{column (2)} - \text{column (6)} - \text{column (7)} + \text{column (8)}$$

Column (10)* gives the water required from storage (*i.e.* S) and is computed from the following relation :

$$S = D - I$$

or Column (10) = column (5) − column (9)

In the above relation, only *positive values* are to be included. When column (9) (*i.e.* adjusted inflow) is more than column (5), zero values are to be written in column (10) indicating that no water is required from the storage since demand is much less than the adjusted inflow. The required storage is the sum of the monthly increments of demand in excess of stream flow. The required storage capacity in this case works out to be **281.64 ha-m.**

Example 3.3. *Table 3.5 gives the details about the average seasonal discharges of a river for 12 years. Determine the storage capacity required to maintain a flow of 475 cumecs throughout the year.*

TABLE 3.5

Year	*16th June to 30th Sept. (cumecs)*	*1st Oct. to 31st March (cumecs)*	*1st April to 15th June (cumecs)*
1960–61	1050	300	50
1961–62	3000	250	40
1962–63	3500	370	90
1963–64	2000	150	120
1964–65	1200	350	65
1965–66	1400	400	100
1966–67	3600	200	80
1967–68	3000	150	120
1968–69	700	210	50
1969–70	800	120	80
1970–71	2400	320	120
1971–72	3200	280	80

Solution :

Let the periods from 16th June to 30th Sept., 1st Oct. to 31st March and 1st April to 15th June be designated as *M, W* and *S*. The data for preparing mass inflow curve are tabulated in Table 3.6. The mass inflow curve is shown in Fig. 3.8.

TABLE 3.6 DATA FOR MASS–INFLOW CURVE

S.No.	Year	Seasons	No. of days	Av. discharge (cumecs)	Volume (million ha-m)	Cumulative volume (million ha-m)
1	1960–61	M	107	1050	0.9707	0.9707
		W	182	300	0.4717	1.4424
		S	76	50	0.0328	1.4752
2	1961–62	M	107	3000	2.7734	4.2486
		W	182	250	0.3931	4.6417
		S	76	40	0.0263	4.6680
3	1962–63	M	107	3500	3.2357	7.9037
		W	182	370	0.5818	8.4855
		S	76	90	0.0591	8.5446
4	1963–64	M	107	2000	1.8490	10.3936
		W	182	150	0.2359	10.6295
		S	76	120	0.0788	10.7083
5	1964–65	M	107	1200	1.1094	11.8177
		W	182	350	0.5504	12.3681
		S	76	65	0.0427	12.4108
6	1965–66	M	107	1400	1.2943	13.7051
		W	182	400	0.6290	14.3341
		S	76	100	0.0657	14.3998
7	1966–67	M	107	3600	3.3281	17.7279
		W	182	200	0.3145	18.0424
		S	76	80	0.0525	18.0949
8	1967–68	M	107	3000	2.7734	20.8673
		W	182	150	0.2359	21.1032
		S	76	120	0.0788	21.1820
9	1968–69	M	107	700	0.6471	21.8291
		W	182	210	0.3302	22.1593
		S	76	50	0.0328	22.1921
10	1969–70	M	107	800	0.7396	22.9317
		W	182	120	0.1887	23.1204
		S	76	80	0.0525	23.1729
11	1970–71	M	107	2400	2.2188	25.3917
		W	182	320	0.5032	25.8949
		S	76	120	0.0788	25.9737
12	1971–72	M	107	3200	2.9583	28.9320
		W	182	280	0.4403	29.3723
		S	76	80	0.0525	29.4248

$$\text{Yearly demand} = 475 \times 365 \text{ cumec-days}$$
$$= 475 \times 365 \times 8.64 = 1497960 \text{ ha-m}$$
$$\approx 1.5 \text{ million ha-m.}$$

In Fig. 3.8, tangents T_1, T_2, T_3, T_4, T_5 etc. are drawn to the curve at the apexes, and parallel to the demand curve which has a slope

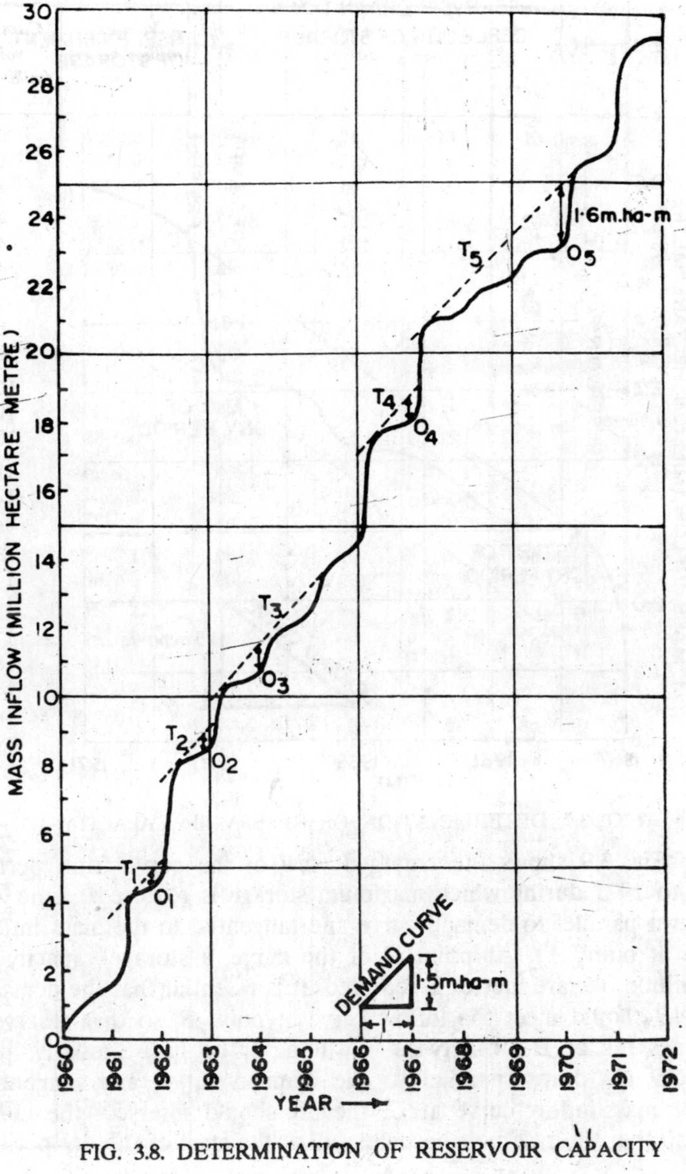

FIG. 3.8. DETERMINATION OF RESERVOIR CAPACITY

of 1.5 m ha-m in 1 year. The ordinates $O_1, O_2, ...O_5$ indicate the deficiencies during the dry periods, assuming that the reservoir was full at the beginning of the period. The maximum of these ordinates (*i.e.* O_5 = **1.6 m ha-m**) gives the desired reservoir capacity.

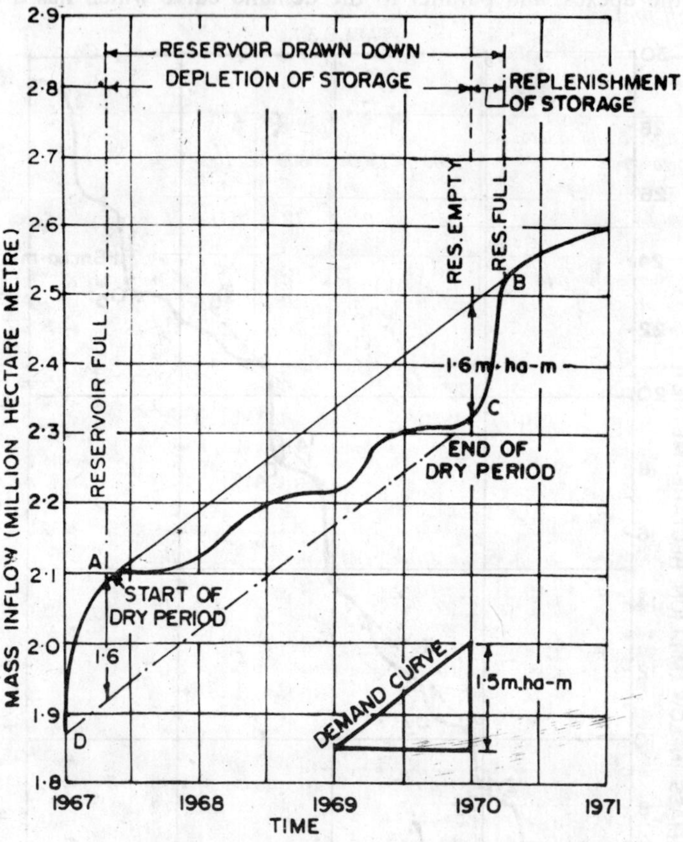

FIG. 3.9. DETERMINATION OF RESERVOIR CAPACITY

Fig. 3.9 shows the enlarged view of the curve from period 1967 to 1971 during which maximum storage is required. Line *AB* is drawn parallel to demand curve and tangential to the mass inflow curve at point *A*. At point *C* of the curve, a storage capacity of 1.6 million hectare metres is required. It is essential that the demand line *AB* should meet the inflow curve at point *B*, so that reservoir becomes full at *B* ; otherwise it will never be full. Similarly, if a line *CD* is drawn parallel to the demand curve, and tangential to the mass-inflow curve at *C*, then it should intersect the curve at *D* so that the reservoir becomes full at the start of the dry period.

Example 3.4 *Draw the mass inflow diagram and compute the storage needed for an impounding reservoir, for a constant draft of 300 million litres per month, with the following recorded mean monthly runoff values :*

TABLE 3.7.

Order of the month	*1*	*2*	*3*	*4*	*5*	*6*	*7*
Observed monthy mean run off (million litres)	*1200*	*1500*	*600*	*100*	*80*	*30*	*10*
Order of the month	***8***	***9***	***10***	***11***	***12***	***13***	***14***
Observed monthly mean runoff (million litres)	*0*	*20*	*60*	*800*	*1000*	*1200*	*700*

Solution.

The computations are shown in Table 3.8, where Q denotes the runoff (or discharge) in the stream and D denotes the draft required, per month.

TABLE 3.8. COMPUTATION FOR STORAGE

Order of month	*Run off* Q	*Draft* D	*Cumulative run off* ΣQ	*Deficiency* $(D - Q)$	*Cumulative Deficiency* $\Sigma (D - Q)$	*Reservoir state*
(1)	*(2)*	*(3)*	*(4)* = Σ (2)	*(5)* = (3) − (2)	*(6)* = Σ (5)	*(7)*
1	1200	300	1200	− 900	0 (2400)	
2	1500	300	2700	− 1200	0 (1500)	
3	600	300	3300	− 300	0 (300) ↑	Res. Full at the beginning of dry period.
4	100	300	3400	200	200	
5	80	300	3480	220	420	
6	30	300	3510	270	690	
7	10	300	3520	290	980	
8	0	300	3520	300	1280	
9	20	300	3540	280	1560	
10	60	300	3600	240	1800*	* Max. deficiency
11	800	300	4400	− 500	1300	
12	1000	300	5400	− 700	600	
13	1200	300	6600	− 900	0 (300)	Res. Full
14	700	300	7300	− 400	0 (700)	

Columns (1), (2) and (3) are based on given data. Column (4) gives cumulative runoff or inflow (ΣQ), which is equal to Σ (2). Column (5) is obtained by subtracting column (2) from col. (3); the negative values denote that the reservoir is full. When the deficiency changes from negative value to positive value, it marks the beginning of the dry period. Column (6) is equal to Σ (5), in which the negative values are not included until the beginning of dry period. The *cumulative surplus* calculated backwards from the beginning of dry period is shown in brackets in col. (6). From col. (6) of the Table, the max. deficiency comes out to be **1800 million litres**, which is the required storage.

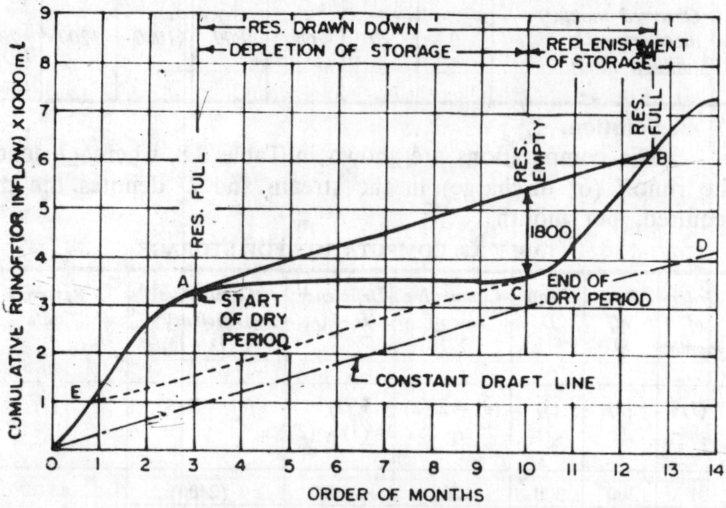

FIG. 3.10. DETERMINATION OF RESERVOIR CAPACITY.

The graphical solution is shown in Fig. 3.10 in which the curve *OACB* shows the cumulative runoff (or inflow). Line *OD* is the constant draft line which has a slope of 300 Ml/month. Point *A* denotes the start of dry period. A line *AB* is drawn parallel to the draft line *OD*, and tangent to the curve at *A*. Point *C* corresponds to the max. deficiency, equal to the vertical intercept between the tangent to the curve, and its value comes out to be equal to 1800 million litres. Hence the required storage capacity of the reservoir is 1800 million litres. It is to be noted that at point *B*, where the tangent meets the curve, the reservoir again becomes full. Similarly, if a tangent *CE* is drawn to the curve at *C*, parallel to the draft line, it should intersect the run off curve if the reservoir is to be full at the beginning of the dry period.

3.5. DAMS

A dam is a hydraulic structure constructed across a river to store water on its upstream side. It is an impervious or fairly impervious barrier put across a natural stream so that a reservoir is formed. Due to the construction of the dam, water level in the river at its upstream side is very much increased, and a large area may be submerged depending upon the water spread of the reservoir so formed. Dams may be classified into different categories, depending upon the purpose or basis of the classification. Table 3.9 gives a summary of various types of dams.

TABLE 3.9 TYPES OF DAMS

Basis of classification	*Types*	*Common examples*
(*a*) Classification according to use	(*i*) Storage dam	Gravity dam, earth dam, rockfill dam, Arch dam etc.
	(*ii*) Diversion dam	Weir, barrage
	(*iii*) Detention dam	Dike, water spreading dam, debris dam
(*b*) Classification by hydraulic design	(*i*) Overflow dam	Spillway
	(*ii*) Non-overflow type	Gravity dam, earth dam, rockfill dam
(*c*) Classification by materials	(*i*) Rigid dam	Gravity dam, arch dam, buttress dam, steel dam, timber dam
	(*ii*) Non-rigid dams	Earth dam, rockfill dam

A *storage dam* is the most common type of dam, and is constructed to impound water to its upstream side during periods of excess supply in the river. A *diversion dam* simply raises water level slightly in the river, and thus provides head for carrying or diverting water into ditches, canals, or other conveyance systems to the place of use. A *detention dam* is constructed to store water during floods and release it gradually at a safe rate, when the flood recedes. A *non-overflow dam* is the one in which the top of the dam is kept at a higher elevation than the maximum expected high water level. An *overflow dam*,, however, is designed to carry surplus discharge (including floods) over its crest. *Rigid dams* are those which are constructed of rigid materials such as masonry, concrete, steel or timber. Rigid dams may be further classified as follows : (*i*) solid masonry or concrete gravity dam (*ii*) arched masonry or concrete dam (*iii*) concrete buttress dam (*iv*) steel dam and (*v*) timber dam. Non-rigid dams are those which are constructed of non-rigid materials such as earth and rockfill.

The most common types of non-rigid dams are : (*i*) earth dam (*ii*) rockfill dam (*ii*) combined earth and rockfill dam.

1. GRAVITY DAMS

A *gravity dam* is the one in which the external forces (such as water pressure, wave pressure, silt pressure, uplift pressure etc.) are resisted by the weight of the dam itself. Thus the forces disturbing the stability of the dam are resisted by the gravity forces of the mass of the dam. A *gravity dam* may be constructed either of masonry or of concrete. *Masonry gravity dams* are now-a-days constructed of only small heights. All major and important gravity dams are now constructed of concrete only. A gravity dam may be either straight or curved in plan.

Fig. 3.11 shows a gravity dam, which may be subjected to the following forces : (*i*) water pressure *P*, (*ii*) weight of the dam, *W*, (*iii*) uplift pressure, *U*, (*iv*) pressure due to earth quake, P_e, (*v*) ice pressure, P_i, (*vi*) wave pressure, P_w, and (*vii*) silt pressure P_s. A moderate size gravity dam may have a drainage/inspection gallery. Most of the gravity dams are solid so that no bending stress is introduced at any point. Gravity dams are particularly suited across gorges with very steep side slopes where earth dams might slip. Where good foundations are available, gravity dams can be built upto any height. The highest dam in the world are of gravity type.

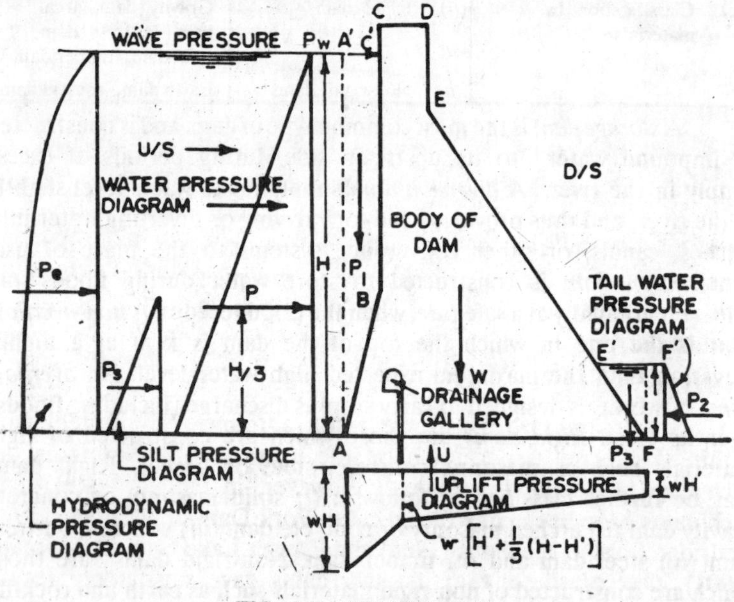

FIG. 3.11. GRAVITY DAM

2. ARCH DAMS

An arch dam (Fig. 3.12) is a dam curved in plan and carries a major part of its water load horizontally to the abutments by arch action. This part of water load depends primarily upon the amount of

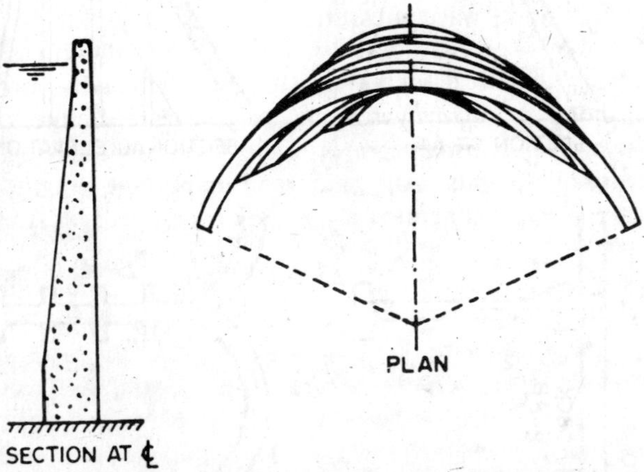

FIG. 3.12. ARCH DAM

curvature. The balance of the water load is transferred to the foundation by cantilever action. The thrust developed by the water load carried by arch action, essentially requires strong side walls of the canyon to resist the arch forces. The weight of arch dam is not counted on to assist materially in the resistance of external loads. For this reason, uplift on the base is not an important design factor.

3. BUTTRESS DAMS

A buttress dam (Fig. 3.13) consists of a number of buttresses or piers, dividing the space to be dammed into a number of spans. To hold up water and retain the water between these buttresses, panels are constructed of horizontal arches or flat slabs. When the panels consist of arches, it is known as *muliple arches type buttress dam.* If the panels consist of flat slab, it is known as *deck type buttress dam.*

4. STEEL DAMS

Steel dams are constructed with a framework of steel with a thin skin plate as deck slab on the upstream side. In India, no such dam has been constructed. However, in United States three such dams have been constructed : Ash Fork Dam in Arizona (1898), Redridge Dam in Michigan (1905) and Hauser Lake Dam in Montana (1901). Out of these, the first two dams gave satisfactory results

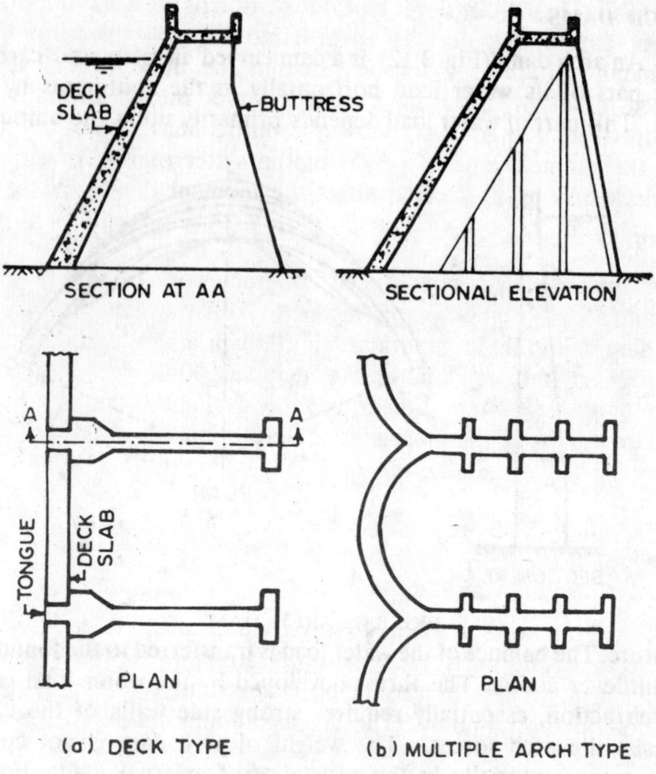

FIG. 3.13. BUTTRESS DAMS

while the third dam failed only after one year of service. The failure was mainly due to undermining of the foundation by leakage through or under the steel sheet pile.

Steel dams (Fig. 314) are generally of two types : (*i*) *direct strutted type* and (*ii*) *cantilever type.* In the direct strutted type, the load on the deck plate is carried directly to the foundations through

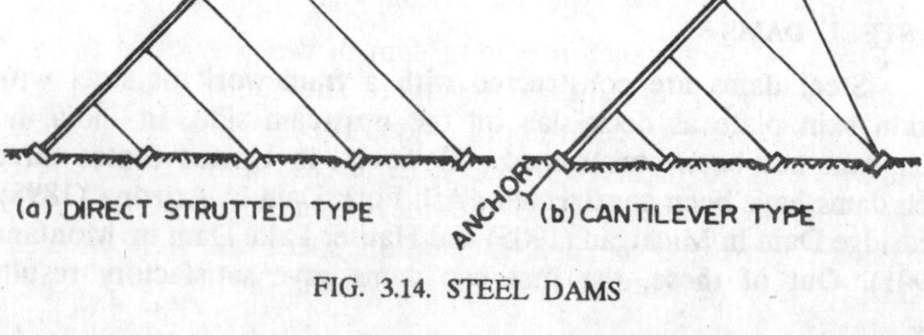

FIG. 3.14. STEEL DAMS

inclined struts. In the cantilever type, the section of the bent supporting the upper part of the deck is formed into a cantilever truss. This arrangement introduces a tensile force in the deck girders which is resisted either by anchoring the deck girder into the foundation at the upstream toe or by framing the entire bent rigidly together so that the moment of the weight of the water on the lower part of the deck may be utilized to offset the moment of the cantilever.

5. TIMBER DAMS

A timber dam is constructed of framework of timber struts and beams, with timber plank facing to resist water pressure. A timber dam is an ideal temporary dam, though a well designed, constructed and maintained timber dam may last 30-40 years. They are suitable to places where timber can be available in plenty.

Timber dams are normally found to be of three types :

1. A-frame type (Fig. 3.15).
2. Rock-filled crib type (Fig. 3.16)
3. Beaver type.

1. A frame type timber dam

Fig. 3.15 shows a typical A-frame type timber dam. It consists of five component parts : (*a*) sills (*b*) struts (*c*) wales (*d*) studs and (*e*) lagging. The sills should be fastened to the ledge rock by wedge bolts or anchor bolts. The lagging should not be of less than 5 cm thickness.

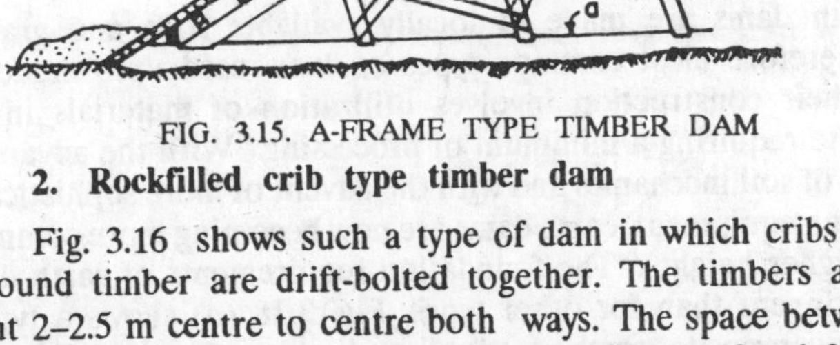

FIG. 3.15. A-FRAME TYPE TIMBER DAM

2. Rockfilled crib type timber dam

Fig. 3.16 shows such a type of dam in which cribs of square or round timber are drift-bolted together. The timbers are spaced about 2–2.5 m centre to centre both ways. The space between them is filled with rock fragments or boulders. In the case of rock foundation, the bottom cribs are pinned to the rock foundation. If, however, the dam is constructed on earth foundation, sheet piling is provided both at the u/s as well as d/s side as shown in Fig. 3.16.

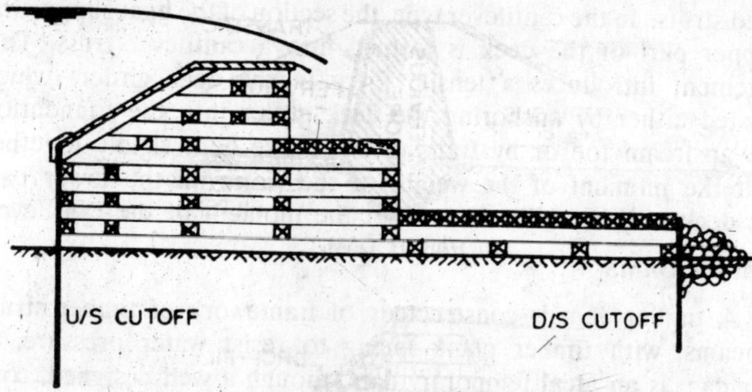

FIG. 3.16. ROCKFILLED TIMBER CRIB DAM.

3. Beaver type timber dam

Fig. 3.17 shows Beaver type timber dam which is the lowest in cost if plenty of timber is available. This is used only for low height and u/s slope of the dam is not kept steeper than 1 in 2. It consists of a number of timbers the butts of which point downstream, and between the butts are placed spacer logs which are drift-pinned. Fill of earth or sand is placed over the plank deck for its stability.

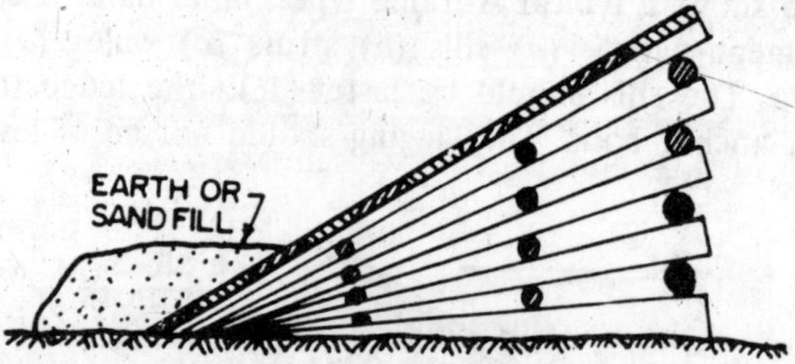

FIG. 3.17. BEAVER TYPE TIMBER DAM

6. EARTH DAMS AND ROCKFILL DAMS

Earth dams are made of locally available soils and gravels, and are therefore most common types of dams used upto moderate heights. Their construction involves utilization of materials in the natural state requiring a minimum of processing. With the advancing knowledge of soil mechanics and with the advent of more sophisticated earth moving equipment, earth dams are now becoming more common, even for higher heights. The foundation requirements of earth dams are less stringent than for other types. Fig. 3.18 (*a*) shows a typical section of composite earth dam.

A rockfill dam is an embankment which uses variable sizes of rock to provide stability and an impervious membrane to provide water tightness. In modern practice, the rockfill dam has four fun-

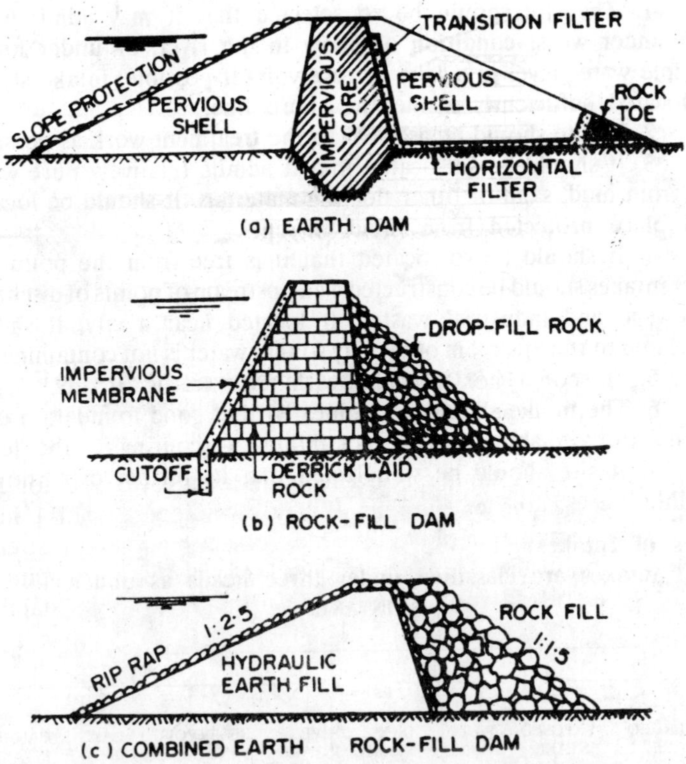

FIG. 3.18. EARTH AND ROCKFILL DAMS

damental parts : (1) dumped rockfill at the downstream, (2) upstream rubble cushion of laid-up stone bonding into the dumped rock, (3) upstream impervious facing resting on rubble cushion, and (4) upstream cutoff to check under-seepage. Fig. 3.18(*b*) shows a typical section. Fig. 3.18 (*c*) shows a combined earth and rockfill dam in which the usual derrick laid rockfill is replaced by the hydraulic earthfill and no upstream impervious membrane is provided.

3.6. INTAKES

Intakes are the structures used for admitting water from the surface sources (*i.e.*, river, reservoir or lake), and conveying it further to the treatment plant. Generally, an intake is a masonry or concrete structure with an aim of providing relatively clean water, free from pollution, sand and objectionable floating material.

Selection of site

The following points should be considered in selecting a suitable site for the intake structure :

1. The site should be so selected that it may admit water even under worst condition of flow in the river, or under lowest possible water level in a lake or reservoir. If possible, intake should be located sufficiently inside the shore line.

2. Its site should be as near to the treatment work as possible.

3. It should be so located that it admits relatively pure water free from mud, sand or other floating materials. It should be located at a place protected from rapid currents.

4. It should be so located that it is free from the pollution. River intakes should be constructed well upstream of points of discharge of sewage and industrial wastes. If located near a city, it should be located to the upstream of the city so that water is not contaminated.

5. It should not interfere with river traffic, if any.

6. The intake should be so located that good foundation conditions are available and the possibility of scouring is the least.

7. It site should be so selected that its further expansion is possible.

Types of Intakes

Intakes are classified under three heads as under :

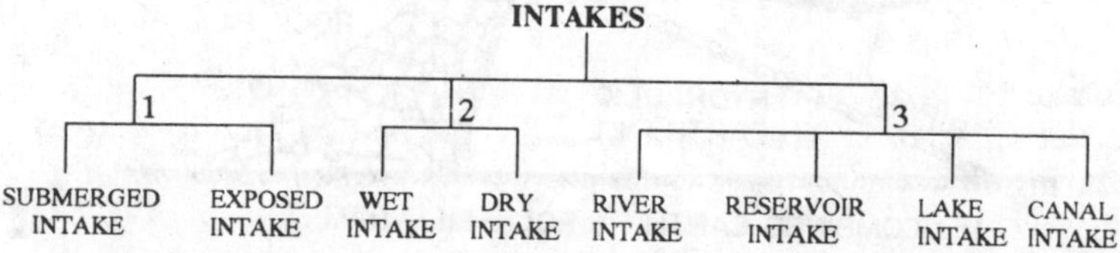

Submerged intake is the one which is constructed entirely under water. Such an intake (Fig. 3.23) is commonly used to obtain supply from a lake. An *exposed intake* is in the form of a well or tower constructed near the bank of a river, or in some cases even away from the river banks. Exposed intakes are more common due to ease in its operation (Fig. 3.19). A *wet intake* is that type of intake tower in which the water level is practically the same as the water level of the sources of supply. Such an intake is sometimes known as jack well (Fig. 3.20) and is most commonly used. In the case of *dry intake,* however, there is no water in the water tower (Fig. 3.21 *c*). Water enter enters through entry port directly into the conveying pipes. The dry tower is simply used for the operation of valves etc.

River Intakes

A river intake is located to the upstream of the city so that pollution is minimized. They are either located sufficiently inside the river so that demands of water are met with in all the seasons of the year, or they may be located near the river bank where a

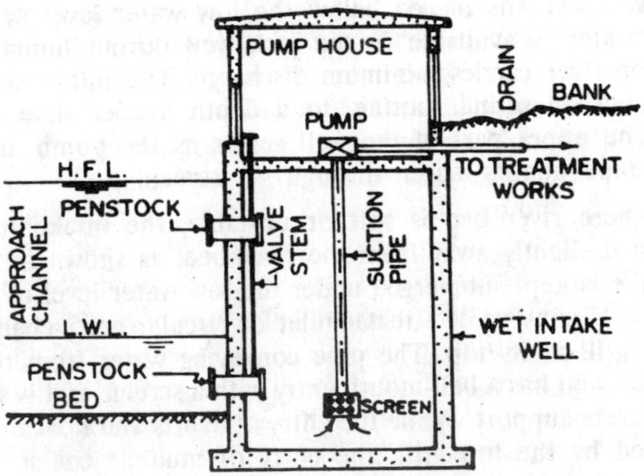

FIG. 3.19. RIVER INTAKE.

sufficient depth of water is available. Sometimes, an approach channel is constructed and water is led to the intake tower. If the water level in the river is low, a weir may be constructed across it to raise the water level and divert it to the intake tower.

Fig. 3.19 shows a wet type intake well founded on river bed. The intake tower permits entry of water through several entry ports located at various levels to cope with the fluctuations in the water level during different seasons. These entry ports are sometimes known as *penstocks* and are provided with suitably designed screens to exclude debris and floating material from entry. The entry ports contain valves which can be operated from the upper part of the well.

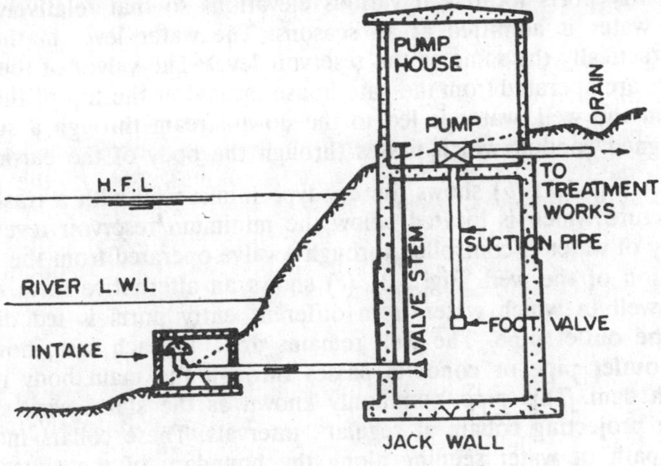

FIG. 3.20. RIVER INTAKE.

The lowest entry is placed below the low water level of the river so that water is available in the jack well during summer season also when river carries minimum discharge. The intake well should be founded on sound footing, to a depth deeper than the scour depth. The upper part of the well serves as the pump house. The suction pipe admits water through a screen.

Where river bed is soft or unstable, the intake tower may be founded slightly away from the river bed, as shown in Fig. 3.20. The intake is kept submerged under the low water level of the river. It essentially consists of a rectangular or circular entry chamber with a strong grill at its top. The pipe conveying water from the intake to the jack well has a bell-mouth entry with a screen, and is supported on a concrete support. While the entry of debris and floating material is checked by the top grill, the entry of mud or coarse sand etc. is checked by the screen provided at the bell-mouth entry. Water enters to the jack well through a valve which can be controlled from the pump house.

Reservoir Intake

When the flow in the river is not guaranteed throughout the year, a dam is constructed across it to store water in the reservoir so formed. The reservoir intakes are practically similar to the river intake, except that these are located near the upstream face of the dam where maximum depth of water is available. Their design depends upon the type of dam. Fig. 3.21 (*a*) shows a typical intake for an earth dam with several entry ports. The intake is constructed near the toe of the dam. The access to the intake tower is provided through a foot bridge. Water may enter the well through a number of entry ports located at various elevations so that relatively clear top water is admitted at all seasons. The water level in the well is practically the same as the reservoir level. The valves of the entry ports are operated from the gate house located at the top of the well. From the well, water is led to the down-stream through a suitably designed conduit which passes through the body of the earth dam.

Fig. 3.21 (*b*) shows the dry type intake well with a trash-rack structure which is located below the minimum reservoir level. The entry of water is controlled through a valve operated from the upper portion of the well. Fig. 3.21 (*c*) shows an alternative form of the dry well in which water from different entry ports is led directly to the outlet pipe. The well remains dry. In each case, however, the outlet pipe or conduits passes through the main body of the earth dam. This pipe, commonly known as the sluice way, should have projecting collars at regular intervals. These collars increase the path of water seeping along the boundary of the sluice way. The length of the seepage path should be more so that no damage

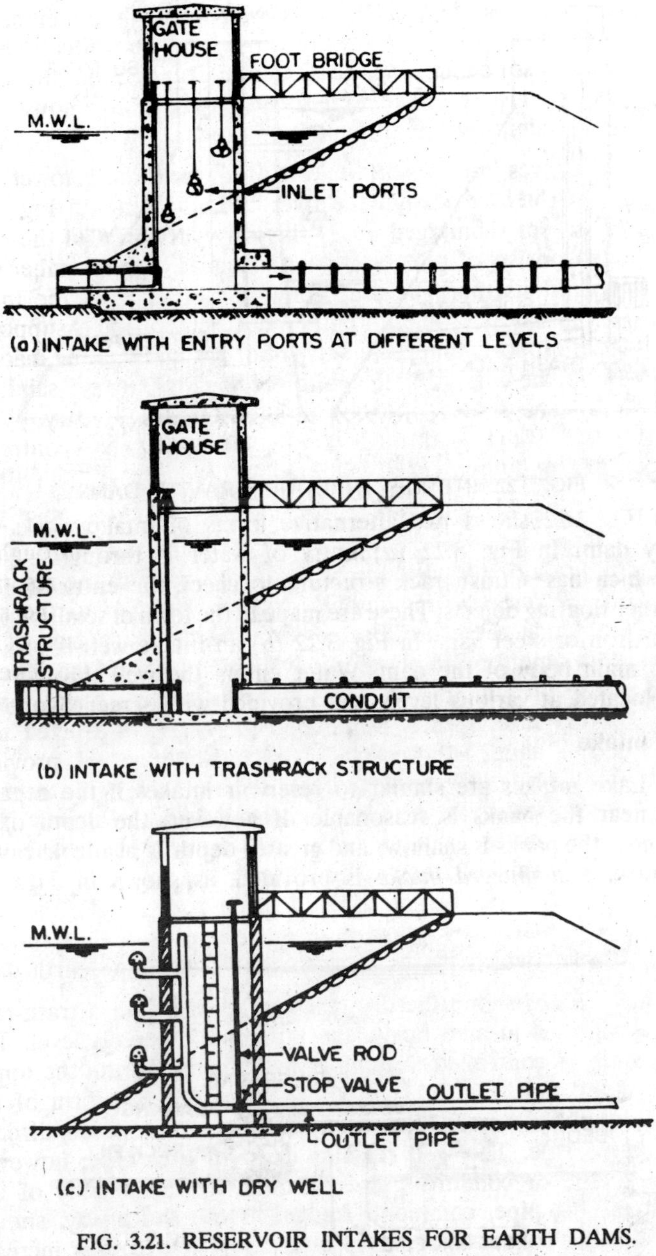

FIG. 3.21. RESERVOIR INTAKES FOR EARTH DAMS.

is done to the embankment, since many failures of earth dams have been reported due to seepage of water around sluice way.

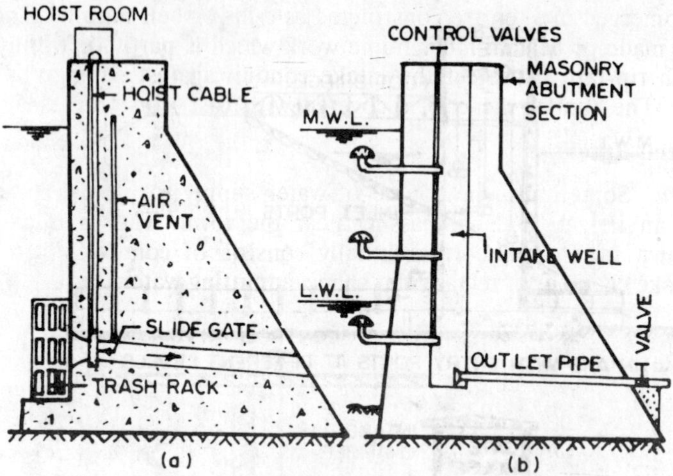

FIG. 3.22. INTAKES THROUGH GRAVITY DAMS.

Fig. 3.22 shows two alternative forms of intake works in a gravity dam. In Fig. 3.22 (*a*) entry of water is through a single port which has a trash rack structure to check the entry of debris and other floating objects. These are made in the form of semipolygonal grid of iron or steel bars. In Fig. 3.22 (*b*) an intake well is provided in the main body of the dam. Water enters the well through inlet ports located at various levels and provided with screened opening.

Lake Intake

Lake intakes are similar to reservoir intakes if the depth of water near the banks is reasonable. If however, the depth of the water near the banks is shallow, and greater depth is available only at its centre, a *submerged intake* is provided, as shown in Fig. 3.23.

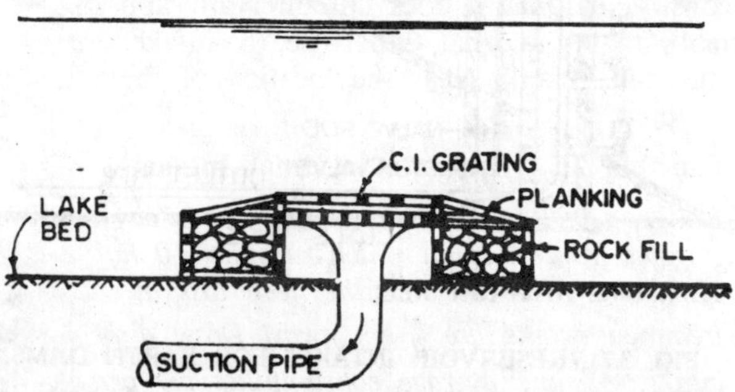

FIG. 3.23. SUBMERGED INTAKE CRIB.

Submerged intakes are constructed as cribs or bell mouths. The cribs are made of heavy timber frame work which is partly or wholly filled with rip-rap to protect the intake conduit against damage by waves etc. The top of the crib is covered with cast iron or mesh grating.

Canal Intake

Sometimes, the source of water supply to a small town may be an irrigation canal passing near the town. The canal intake is shown in Fig. 3.24. It essentially consists of concrete or masonry intake chamber of rectangular shape, admitting water through a coarse

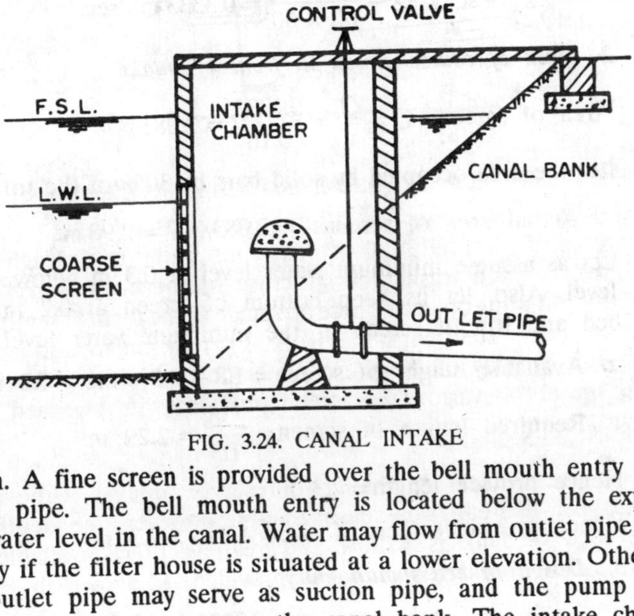

FIG. 3.24. CANAL INTAKE

screen. A fine screen is provided over the bell mouth entry of the outlet pipe. The bell mouth entry is located below the expected low water level in the canal. Water may flow from outlet pipe under gravity if the filter house is situated at a lower elevation. Otherwise, the outlet pipe may serve as suction pipe, and the pump house may be located on or near the canal bank. The intake chamber is so constructed that it does not offer any appreciable resistance to normal flow in the canal. Otherwise, the intake chamber is located inside the canal bank. Near the location of the intake work, the canal is lined.

Example 3.5. Design of a canal Intake

Design a bell mouth canal intake for a city of 80000 persons drawing water from a canal which runs for 10 hours a day with a depth of 1.8 m. Also calculate the head loss in the intake conduit if the treatment works are $\frac{1}{2}$ *km away. Draw a neat sketch of the canal intake. Assume average consumption per person = 150 l/day. Assume the velocity through the screens and bell mouth to be less than 16 cm/s and 32 cm/s respectively.*

Solution

1. ***Discharge through intake***

Daily discharge = $150 \times 80000 = 12000000$ l / day.

Since the canal runs only for 10 hours per day,

$$\text{Intake load/hour} = \frac{12000000}{10} = 1200000 \text{ l / hr}$$

$$= 1200 \text{ m}^3\text{/hr}$$

$$\therefore \qquad Q = \frac{1200}{60 \times 60} = 0.3333 \text{ m}^3\text{/sec.}$$

2. ***Area of coarse screen in front of intake***

$$\text{Area of screen, } A_s = \frac{Q}{v} = \frac{0.3333}{0.16} = 2.083 \text{ m}^2$$

Let the area occupied by solid bars be 30% of the total area.

$\therefore$ Actual area of screens $= 2.083/0.7 = 2.98 \text{ m}^2$

Let as assume minimum water level at 0.3 m below normal water level. Also, let us keep bottom of screen at 0.2 m above canal bed and top of screen at the minimum water level.

$\therefore$ Available height of screen $= 1.8 - 0.3 - 0.2 = 1.3$ m

$$\therefore \text{ Required length of screen} = \frac{2.98}{1.3} = 2.29 \text{ m}$$

Hence provide length = 2.3 m

Hence provide screen of size = 1.3 m × 2.3 m

3. ***Design of bell mouth entry***

$$\text{Area of bell mouth,} \qquad A_b = \frac{0.3333}{0.32} = 1.042 \text{ m}^2$$

$$\therefore \qquad \text{Dia. } d_b = \sqrt{\frac{1.042 \times 4}{\pi}} = 1.15 \text{ m.}$$

Hence provide bell mouth of 1.2 m diameter.

4. ***Design of intake conduit***

Let us assume a velocity of 1.5 m/sec in the conduit.

$$\therefore \text{ Dia. of intake conduit, } D = \sqrt{\left(\frac{0.3333}{1.5}\right) \times \frac{4}{\pi}} = 0.532 \text{ m}$$

However, provide 0.5 m dia. conduit, so that actual velocity of flow is

$$V = \frac{0.3333 \times 4}{\pi\,(0.5)^2} \approx 1.7 \text{ m/sec.}$$

For head loss through the conduit, consider Eq. 16.10 :

$$V = 0.849\, C\, R^{0.63}\, S^{0.54}$$

Take $C = 130$ for cast iron pipes

Also, $R = D/4 = 0.5/4 = 0.125$ m.

Hence slope S of the energy line is given by

$$1.7 = 0.849\,(130)\,(0.125)^{0.63}\,.\,S^{0.54}$$

or $$1.7 = 29.779\, S^{0.54}$$

From which $$S = \left(\frac{1.7}{29.779}\right)^{1/0.54} = 4.98 \times 10^{-3}$$

But $$S = H_L / L$$

$\therefore$ Hence loss $H_L = S.L = 4.98 \times 10^{-3}\left(\frac{1}{2} \times 1000\right) = \mathbf{2.49\ m}$

For the arrangement of various components of the canal intake, refer Fig. 3.24.

PROBLEMS

1. Describe in brief various investigations required for reservoir planning.

2. What are the factors on which the selection of the site of a reservoir depend ?

3. What do you understand by mass inflow curve and how it is prepared?

4. What do you understand by demand curve ? Explain the method of calculating reservoir capacity for a specified yield, from the mass inflow curve.

5. Explain how do you determine the safe yield from a reservoir of a given capacity.

6. Classify various types of dams. Discuss in brief merits and demerits of various types of dams.

7. Discuss the physical factors that govern the selection of type of dam.

8. What are intakes ? What are the important considerations which govern the selection of site of an intake.

9. Describe the river intake and reservoir intake.

10. Describe the working of a canal intake.

11. Describe, with the help of sketches, a reservoir intake for an earth dam.

4

Ground Water : Wells

4.1. INTRODUCTION

Ground water hydrology is the science of the occurrence, distribution and movement of water below the surface of earth. The largest available source of the fresh water lies underground. The total ground water potential is estimated to be one-third the capacity of oceans.

The main source of ground water is precipitation. A portion of rain falling on the earth's surface infiltrates into ground, travels down and when checked by impervious layer to travel further down, forms ground water. The ground water reservoir consists of water held in voids within a geologic stratum. Other sources of ground water include water from deep in the earth which is carried upward in intrusive rocks and water which is trapped in sedimentary rocks during their formation. The quantities of such waters are small and they are often so highly mineralised as to be unsuited for use. Water bearing formations of the earth's crust act as conduits for transmission and as revervoirs for storage of ground water.

The discharge from ground water occurs in two ways : (1) natural way (2) artificial way. The natural discharge occurs as flow in lakes, reservoirs, rivers oceans and springs. Pumpage from wells constitutes the major artificial discharge of ground water.

4.2. TYPES OF AQUIFERS

Aquifers are mainly of two types :

1. Unconfined aquifer.
2. Confined aquifer (artesian aquifer).

Unconfined Aquifer

Unconfined aquifer, or *water-table aquifer* is the one in which a water table serves as the upper surface of the zone of saturation. It is also sometimes known as the *free, phreatic or non-artesian* aquifer. In such an aquifer, the water table varies in undulating form and in slope. Rises and falls in the water table correspond to changes in the volume of water in storage within the unconfined aquifer.

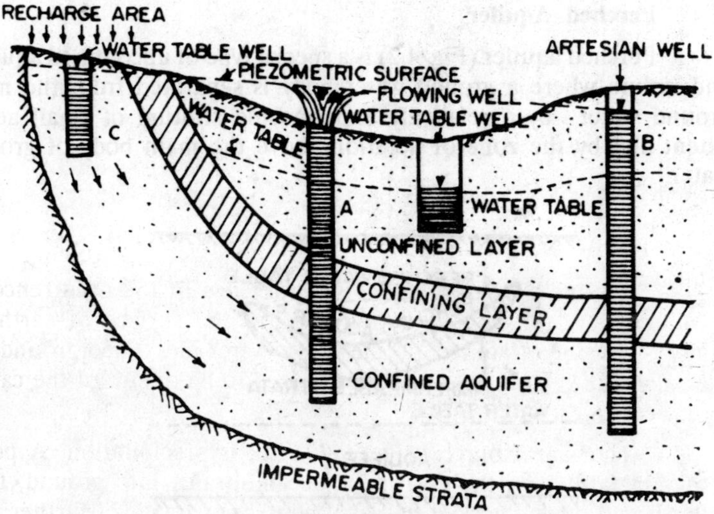

FIG. 4.1. CONFINED AND UNCONFINED AQUIFERS.

Confined aquifer or artesian aquifer

Confined aquifer or artesian aquifer is the one in which ground water is confined under pressure greater than atmospheric by overlying, relatively impermeable strata. Artesian aquifers are analogous to pipelines. The static pressure at a point within the artesian aquifer is equivalent to the elevation of the water table in the recharge area less the loss in head through the aquifer to the point under consideration. In a well penetrating such an aquifer, the water level will rise to the level of the local static pressure or artesian head. Water enters a confined aquifer in an area where the confining bed rises to the surface or *ends underground* and the aquifer becomes unconfined. Artesian aquifers usually have relatively small recharge areas as compared with unconfined aquifers. When water is withdrawn from an artesian well, a local depression of the piezometric surface results. This decrease in pressure permits a slight expansion of the water and in some cases a compaction of the aquifer.

Flowing well and Artesian well

When a well penetrates a confined aquifer, water rises in the well to the level of local static pressure or artesian head. If this artesian pressure is sufficient to raise the water above the ground level, a *flowing well* occurs, such as well *A* in Fig. 4.1. If however, the water level in such a well is below the ground level, but is above the local water table, it is known as the *artesian well,* such as well *B* in Fig. 4.1.

Perched Aquifer

Perched aquifer (Fig. 4.2) is a special type of unconfined aquifer, and occurs where a ground water body is separated from the main ground water by a relatively impermeable stratum of small aerial extent and by the zone of aeration above the main body of ground water.

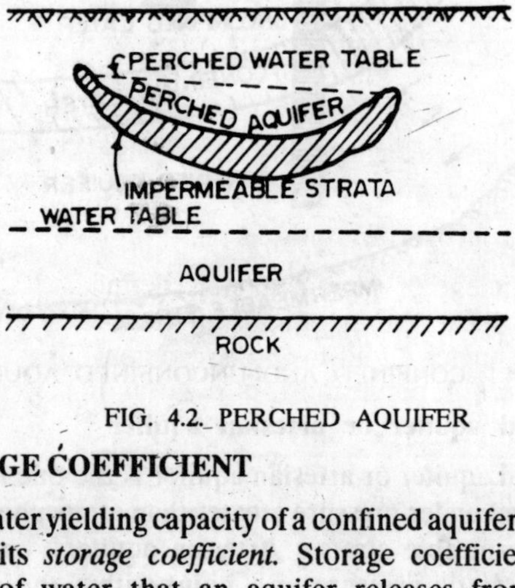

FIG. 4.2. PERCHED AQUIFER

4.3. STORAGE COEFFICIENT

The water yielding capacity of a confined aquifer can be expressed in terms of its *storage coefficient.* Storage coefficient is defined as the volume of water that an aquifer releases from or takes into storage per unit surface area of aquifer per unit change in the component of head normal to that surface.

Let us consider a vertical column of unit area (one metre × one metre) extending through a confined aquifer (Fig. 4.3). Then the storage coefficient, *S,* is the volume of water, in cubic metres, released from the equifer when the piezometric surface declines by one metre. In most of the confined aquifers, the value of storage coefficient ranges between 0.00005 to 0.005. Its value can be determined from pumping tests of wells penetrating fully into confined aquifer.

In an unconfined aquifer, when the water table is lowered by one metre, the water from 1 metre height of the vertical column

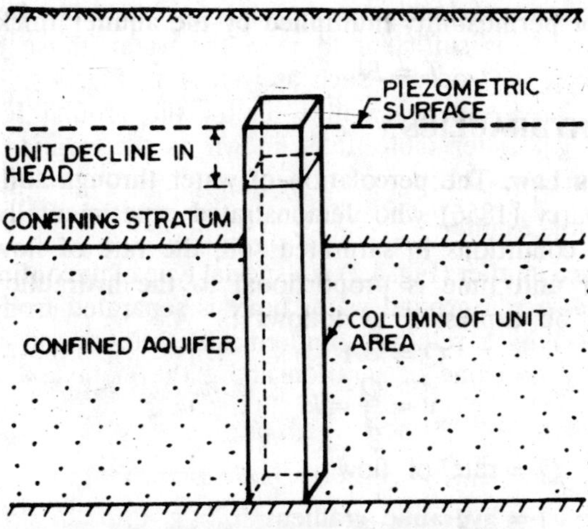

FIG. 4.3. STORAGE COEFFICIENT

of unit area drains freely under gravity. Thus, storage coefficient for an unconfined aquifer corresponds to its specific yield.

Coefficient of permeability (*k*)

The coefficient of permeability is defined as the velocity of flow which will occur through the total cross-sectional area of the soil (or aquifer) under a unit hydraulic gradient. Some typical values of the coefficient of permeability are given in the table below :

TABLE 4.1

Soil type	*Coefficient of permeability cm/sec*
Clean gravel	1.0 and greater
Clean sand (coarse)	1.0–0.01
Sand (mixture)	0.01–0.005
Fine sand	0.05–0.001
Silty sand	0.002–0.0001
Silt	0.0005–0.00001
Clay	0.000001 and smaller

Coefficient of Transmissibility (*T*)

Coefficient of transmissibility is defined as the rate of flow of water (in m^3 / day or gallons/day) through a vertical strip of aquifer of unit width (1 m or 1 ft) and extending the full saturation height under unit hydraulic gradient, at a temperature of 60° F.

Thus, the coefficient of transmissibility T equals to the field coefficient of permeability multiplied by the aquifer thickness (B):

$$T = Bk.$$

4.4. WELL HYDRAULICS

Darcy's Law. The percolation of water through soil was first studied by Darcy (1856) who demonstrated experimentally that for laminar flow conditions in saturated soil, the rate of flow or the discharge per unit time is proportional to the hydraulic gradient, and it could be expressed as follows :

$$Q = kiA \qquad ...(4.1)$$

or

$$v = \frac{Q}{A} = ki \qquad ...(4.2)$$

where Q = rate of flow

i = hydraulic gradient

k = Darcy's coefficient of permeability

A = total cross-sectional area of soil mass perpendicular to the direction of flow

v = flow velocity.

Eq. 4.2 demonstrates the linear dependency between the hydraulic gradient and the discharge velocity. However, it in no way describes the state of affairs within an individual pore. Strictly speaking, Darcy's law represents the statistical macroscopic equivalent of the Navier-Stokes equation of motion for the viscous flow of ground water.

Darcy's law is valid only for laminar flow. Because of very small pore dimensions in fine grained soils, a laminar flow should exist, but in coarse grained soils, turbulent flow may be expected under certain conditions. It has been borne out by experiments that the limits of validity of Darcy's law may be fixed with respect to particle size, velocity of flow and hydraulic gradient. Fancher, Lewis and Barnes demonstrated that flow through sands remains laminar and the Darcy's law valid so long as the Reynold's number, expressed in the form below is equal to or less than unity :

$$\frac{\rho v d}{\mu} \leq 1 \qquad ...(4.3)$$

where ρ = mass density

μ = dynamic viscosity

d = diameter or particle size

v = velocity of flow.

For the ground water flow occurring in nature, the law is generally within its validity limits. But in rock aquifers, in unconsolidated aquifers with the steep hydraulic gradients, or in those containing large diameter solution openings, Darcy's law may not be applicable. Also, the flow in the immediate vicinity of wells have steep hydraulic gradient and the Darcy's law is not applicable in the immediate vicinity of the well.

STEADY RADIAL FLOW TO A WELL : DUPUIT'S THEORY

When a well is penetrated into an extensive homogeneous aquifer, the water table initially remains horizontal in the well. When the well is pumped, water is removed from the aquifer and the water table or piezometric surface, depending upon the type of the acquifer, is lowered resulting in a circular depression in the water table or the piezometric surface. This depression is called the *cone of depression* or the *drawdown curve.* At any point, away from the well, the *drawdown* is the vertical distance by which the water table or the piezometric surface is lowered. The analysis of such radial flow towards a well was originally proposed by Dupuit in 1863 and later modified by Thiem (1906). For the sake of analysis, we shall take two cases : (1) well in unconfined aquifer, and (2) well fully penetrating a confined aquifer.

(1) Unconfined Aquifer

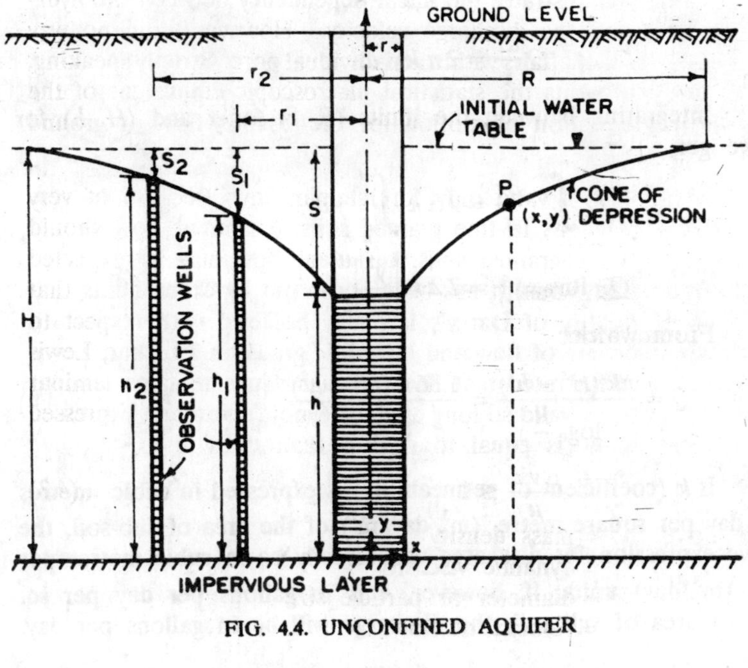

FIG. 4.4. UNCONFINED AQUIFER

Fig. 4.4 shows a well penetrating an unconfined or free aquifer to its full depth.

Let r = radius of the well

H = thickness of the aquifer, measured from the impermeable layer to the initial level of water table

s = drawdown at the well

h = depth of water in the well, measured above the impermeable layer.

Considering the origin of co-ordinates at point O at the centre of the well at its bottom, let the co-ordinates of any point P on the drawdown curve be (x, y).

Then, from Darcy's law

$$Q = k A_x i_x$$

where A_x = area of cross-section of the saturated part of the aquifer at P

$= (2\pi x) \times (y) = 2\pi x y$

i_x = hydraulic gradient at $P = \frac{dy}{dx}$

Hence $$Q = k(2\pi x y)\frac{dy}{dx}.$$

$$\therefore \quad Q\frac{dx}{x} = 2\pi k y . dy.$$

Integrating between the limits (R, r) for x and (H, h) for y, we get

$$Q\int_r^R \frac{dx}{x} = 2\pi k \int_h^H y\,dy$$

$$\therefore \quad Q(\log_e x)_r^R = 2\pi k\left[\frac{y^2}{2}\right]_h^H$$

From which

$$Q = \frac{\pi k(H^2 - h^2)}{\log_e \frac{R}{r}} = \frac{1.36\,k(H^2 - h^2)}{\log_{10}\frac{R}{r}} \quad ...(4.4)$$

If k (coefficient of permeability) is expressed in cubic metres per day per square metre ($m^3/day/m^2$) of the area of sub-soil, the above expression for discharge will directly be in cubic metres per day (m^3/day) units. If, however, k is in gallons per day per sq. foot of area of sub soil, the discharge will be in gallons per day.

In the above expression, R, commonly known as *radius of zero drawdown*, is the radius, measured from the centre of the well to a point where the drawdown curve meets the original water table tangentially. *In practice, the selection of the radius of influence R is approximate and arbitrary, but the variation in Q is small for a wide range of R.* Suggested values of R fall in the range of 100 to 300 metres.

Alternatively, R may be computed from the following approximate expression given by Sichardt :

$$R = 3000\, s \sqrt{k}$$

where R and s are in metres, and k is in m/sec.

If there are two observation wells at radial distance r_1 and r_2 $(r_2 > r_1)$, and if the depths of water in them are h_1 and h_2 respectively, Eq. 4.4 can also be expressed in the form :

$$Q = \frac{\pi k (h_2^2 - h_1^2)}{\log_e \frac{r_2}{r_1}} \quad \text{...(4.4 } a\text{)}$$

or

$$Q = \frac{1.36\, k\, (h_2^2 - h_1^2)}{\log_{10} \frac{r_2}{r_1}} \quad \text{...(4.4 } b\text{)}$$

If the drawdown (s) is measured at the well, we have

$$s = H - h$$

and

$$H = s + h, \text{ or } H + h = (s + 2h)$$

Then, from Eq. 4.4,

$$Q = \frac{\pi k\, (H - h)\, (H + h)}{\log_e \frac{R}{r}} = \frac{\pi\, ks\, (s + 2h)}{\log_e \frac{R}{r}}$$

or

$$Q = \frac{\pi ks\, (s + 2L)}{\log_e \frac{R}{r}} = \frac{2\pi ks\, (L + s/2)}{\log_e \frac{R}{r}}$$

where $h = L =$ Length of the strainer

or

$$Q = \frac{2.72\, ks\, (L + s/2)}{\log_{10} \frac{R}{r}} \quad \text{...(4.5)}$$

Assumptions and Limitations of Dupuit's Theory

Dupuit's theory of flow for unconfined aquifer is based on the following assumptions :

1. The velocity of flow is proportional to the tangent of the hydraulic gradient instead of sine.

2. The flow is horizontal and uniform everywhere in the vertical section.

3. Aquifer is homogeneous, isotropic and of infinite aerial extent.

4. The well penetrates and receives water from the entire thickness of the aquifer.

5. The co-efficient of transmissibility is constant at all places and at all times.

6. Natural ground water regime affecting an aquifer remains constant with time.

7. Flow is laminar and Darcy's law is applicable.

Out of these, assumptions (1), (2) and (7) are of particular importance. The flow is not horizontal, especially near the well. Also, the piezometric surface attains greater slope as it approaches the well boundary, with the result that assumption 1 is an approximation. Due to these reasons, the parabolic form of piezometric surface computed from the Dupuit's theory deviates from the observed surface. This deviation is large at the well face, resulting in the formation of seepage face. In addition to these, the velocity near the well increases and the flow no longer remains laminar. Thus, Darcy's law equation is not valid near the well face.

2. Confined Aquifer

Fig. 4.5 shows a well fully penetrating a confined or artesian aquifer. Let (x, y) be the coordinates of any point P on the drawdown curve, measured with respect to the origin O. Then, from Darcy's law, flow crossing a vertical plane through P is given by

$$Q = k\, i_x A_x$$

where A_x = cross-sectional area of flow, measured at P $= 2\pi x b$

b = thickness of confined aquifer

i_x = hydraulic gradient at $P = \frac{dy}{dx}$

$$\therefore \quad Q = k\left(\frac{dy}{dx}\right)(2\pi x b)$$

$$\therefore \quad Q\frac{dx}{x} = 2\pi k b . dy.$$

Integrating between the limits (R, r) for x and (H, h) for y, we get

$$Q\int_r^R \frac{dx}{x} = 2\pi k b \int_h^H dy$$

$$\therefore \quad Q\left[\log_e x\right]_r^R = 2\pi k b \left[y\right]_h^H$$

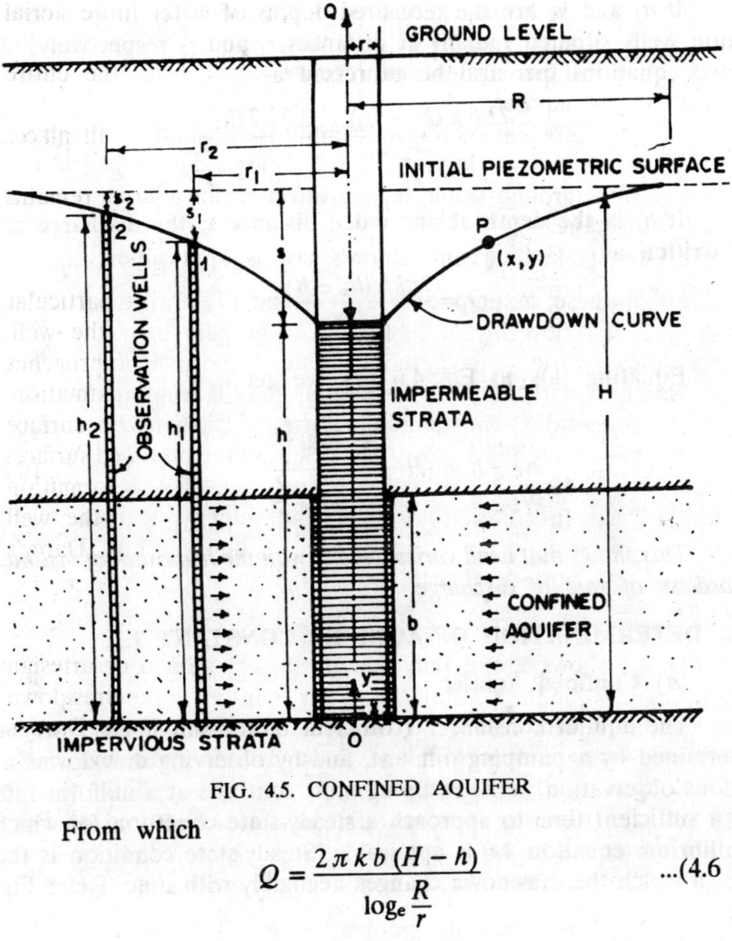

FIG. 4.5. CONFINED AQUIFER

From which

$$Q = \frac{2\pi k b (H - h)}{\log_e \frac{R}{r}} \quad \text{...(4.6 } a)$$

$$= \frac{2.72\, bk (H - h)}{\log_{10} \frac{R}{r}} \quad \text{...(4.6 } b)$$

$$= \frac{2\pi b k s}{\log_e \frac{R}{r}} = \frac{2.72\, b k s}{\log_{10} \frac{R}{r}} \quad \text{...(4.6 } c)$$

$$= \frac{2.72 T s}{\log_{10} \frac{R}{r}} \quad \text{...(4.6 } d)$$

where T = coefficient of transmissibility $= bk$

s = drawdown at the well.

Eq. 4.6 (a) is known as the *equilibrium equation* or Thiem equation.

If h_1 and h_2 are the measured depths of water in two obser vation wells situated radially at distances r_1 and r_2 respectively, the above equations can also be expressed as

$$Q = \frac{2.72\, b\, k\, (h_2 - h_1)}{\log_{10} \frac{r_2}{r_1}} = \frac{2.72\, T(h_2 - h_1)}{\log_{10} \frac{r_2}{r_1}} \qquad ...(4.6\ e)$$

If h_x is the depth at any radial distance x, the discharge can be written as

$$Q = \frac{2\pi kb\, (h_x - h)}{\log_e \frac{r_x}{r}}$$

Equating this to Eq. 4.6 (*a*), we get

$$h_x - h = (H - h) \frac{\log_e \frac{r_x}{r}}{\log_e \frac{R}{r}} \qquad ...(4.7)$$

This shows that head varies linearly with the logarithm of distance regardless of rate of discharge.

4.5. DETERMINATION OF AQUIFER CONSTANT *T*

(*a*) Confined Aquifer

The aquifer constant T (cofficient of transmissibility) can be determined by a pumping out test, and by observing drawdowns at various observation wells. Pumping must continue at a uniform rate for a sufficient time to approach a steady-state condition for which equilibrium equation 4.6 is applicable. Steady-state condition is the one in which the drawdown changes negligibly with time. Refer Fig. 4.5.

Let s_1 = drawdown in well 1 = $(H - h_1)$

s_2 = drawdown in well 2 = $(H - h_2)$.

$\therefore\ h_2 - h_1 = (H - s_2) - (H - s_1) = s_1 - s_2.$

Then from equation 4.6 (*d*),

$$Q = \frac{2.72T\, (h_2 - h_1)}{\log_{10} \frac{r_2}{r_1}} = \frac{2.72T\, (s_1 - s_2)}{\log_{10} \frac{r_2}{r_1}}$$

$$\therefore \quad T = \frac{Q}{2.72\, (s_1 - s_2)} \log_{10} \frac{r_2}{r_1} \qquad ...(4.8)$$

Choosing $r_2 = 10\, r_1$, we find $\log_{10} \frac{r_2}{r_1} = 1$.

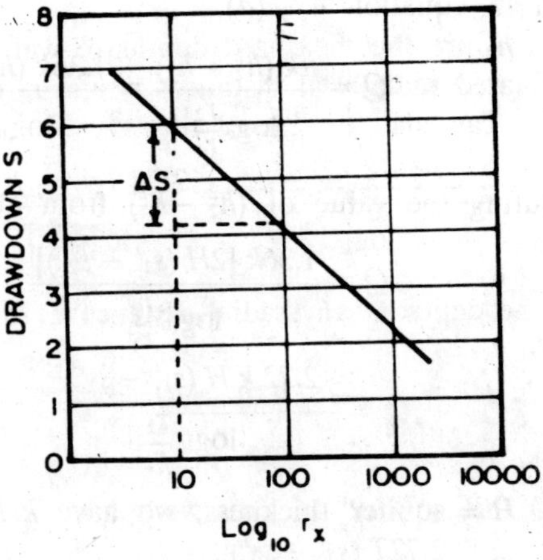

FIG. 4.6.

$$\text{Hence}\quad T = \frac{Q}{2.72\,(s_1 - s_2)} = \frac{Q}{2.72\,\Delta s} \qquad \text{...(4.9)}$$

where Δs = difference in drawdowns at the two wells so selected that $r_2 = 10\,r_1$

The method, therefore, consists in observing drawdowns s_1, s_2, ... , s_x at certain observation wells distant $r_1, r_2, ... r_x$, etc., and plotting a graph between s_x as ordinate and $\log_{10} r_x$ as abscissa, thus getting a straight line as shown in Fig. 4.6. From the graph, Δs can be obtained for one log cycle of distance and can be substituted in equation 4.9 to get T.

(*b*) **Unconfined Aquifer**

The above method of determination of T can also be extended to an unconfined aquifer.

Refer Fig. 4.7.

$$h_1 = H - s_1$$

$$h_2 = H - s_2$$

$$\therefore \quad h_2^2 - h_1^2 = (H - s_2)^2 - (H - s_1)^2$$

$$= 2H\left[\left(s_1 - \frac{s_1^2}{2H}\right) - \left(s_2 - \frac{s_2^2}{2H}\right)\right]$$

$$= 2H\,(s_1' - s_2')$$

where s_1' and s_2' are *modified drawdowns* given by

$$s_1' = s_1 - \frac{s_1^2}{2H}$$

and

$$s_2' = s_2 - \frac{s_2^2}{2H}$$

Now from equation 4.4 (*a*),

$$Q = \frac{\pi k\,(h_2^2 - h_1^2)}{\log_e \dfrac{r_2}{r_1}} = \frac{1.36k\,(h_2^2 - h_1^2)}{\log_{10} \dfrac{r_2}{r_1}}$$

Substituting the value of $(h_2^2 - h_1^2)$ from above, we get

$$Q = \frac{1.36k\,[2H\,(s_1' - s_2')]}{\log_{10} \dfrac{r_2}{r_1}}$$

$$= \frac{2.72\,k\,H\,(s_1' - s_2')}{\log_{10} \dfrac{r_2}{r_1}}$$

Since H = aquifer thickness, we have $kH = T$

$$\therefore \quad Q = \frac{2.72T\,(s_1' - s_2')}{\log_{10} \dfrac{r_2}{r_1}}$$

From which

$$T = \frac{Q}{2.72\,(s_1' - s_2')} \log_e \frac{r_2}{r_1} \qquad ...(4.10)$$

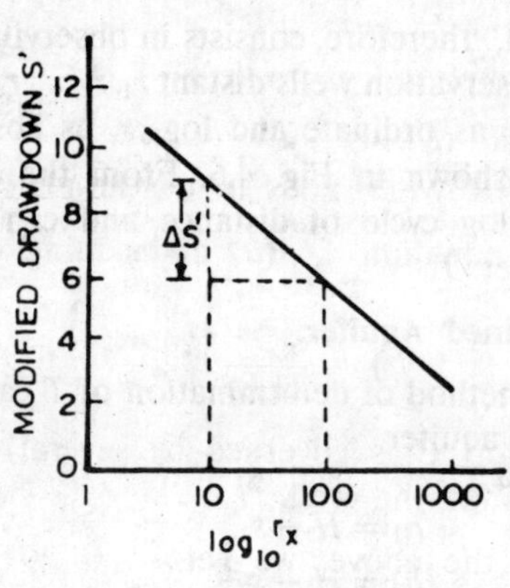

FIG. 4.7.

It should be noted that equations 4.8 and 4.10 are identical. Choosing $r_2 = 10\,r_1$ as before, we get

$$T = \frac{Q}{2.72\,\Delta s'} \qquad ...(4.11)$$

Thus, the observed drawdowns are corrected or modified, and a graph is plotted between the modified drawdown (s'') and logarithm of distance of observation wells from the discharge well. Measuring $\Delta s'$ for one log cycle of distance, we get T from equation 4.11.

4.6. CHARACTERISTIC WELL LOSSES : SPECIFIC CAPACITY OF WELL

Well loss. When water is pumped out of a well, the total drawdown caused includes not only that of the logarithmic drawdown curve at the well face, but also drawdown caused by flow through well screen and axial movement within the well. The later drawdown is called *'well loss'*. Since turbulent flow generally occurs near the well face, this loss may be taken to be proportional to Q^n.

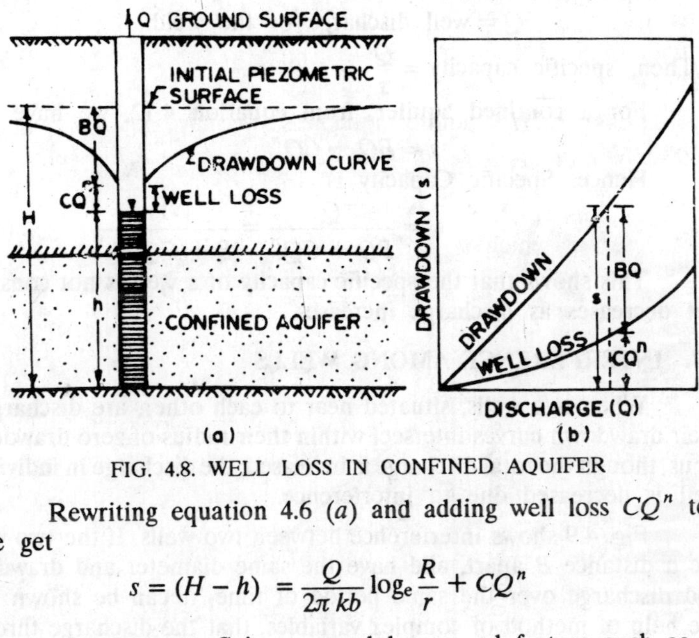

FIG. 4.8. WELL LOSS IN CONFINED AQUIFER

Rewriting equation 4.6 (*a*) and adding well loss CQ^n to it, we get

$$s = (H - h) = \frac{Q}{2\pi kb} \log_e \frac{R}{r} + CQ^n$$

where the constant C is governed by several factors such as well radius, condition of the well.

Rewriting the above, we get

$$s = BQ + CQ^n \qquad \ldots(4.12)$$

where $B = \dfrac{\log_e \dfrac{R}{r}}{2\pi kb}$

BQ = aquifer loss

CQ^n = well loss.

Fig. 4.8. (*a*) illustrates that the total drawdown consists of the sum of the aquifer loss (BQ) and the well loss (CQ^n). Fig. 4.8 (*b*) shows the variations of drawdown and well loss with the discharge.

For a screen which is not clogged or encrusted and whose size is compatible to the surrounding porous media, the portion of the well loss caused by water entering the well is small in comparison with the portion resulting from axial movement inside the well to the pump intake.

Specific Capacity

The specific capacity of a well is the measure of the effectiveness of the well, and is defined as the yield of the well per unit drawdown.

Thus, if s = drawdown

Q = well discharge or the yield.

Then, specific capacity $= \dfrac{Q}{s}$

For a confined aquifer, from equation 4.12, we have

$$s = BQ + CQ^n$$

Hence, Specific Capacity is

$$= \frac{Q}{BQ + CQ^n} = \frac{1}{B + CQ^{n-1}} \qquad \text{...(4.13)}$$

This shows that the specific capacity of a well is not constant, but decreases as discharge increases.

4.7. INTERFERENCE AMONG WELLS

When two wells, situated near to each other, are discharging, their drawdown curves intersect within their radius of zero drawdown. Thus, though the total discharge is increased, the discharge in individual well is decreased due to interference.

Fig. 4.9 shows interference between two wells. If the two wells are a distance B apart, and have the same diameter and drawdown and discharge over the same period of time, it can be shown with the help of method of complex variables, that the discharge through each well is given by

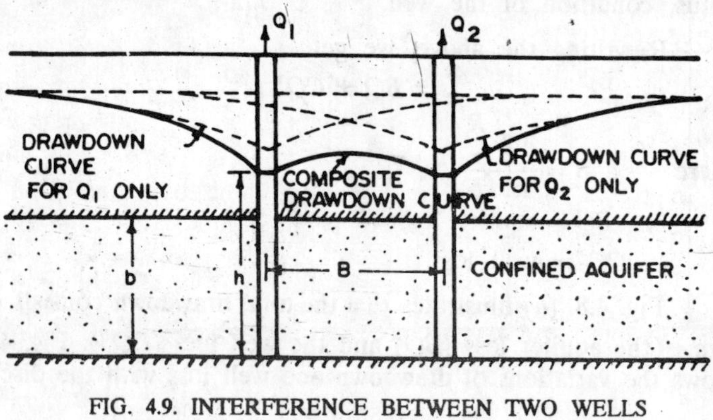

FIG. 4.9. INTERFERENCE BETWEEN TWO WELLS

$$Q_1 = Q_2 = \frac{2\pi kb\,(H - h)}{\log_e \dfrac{R^2}{rB}} \qquad ...(4.14)$$

where R is the radius of area of influence ($R >> B$).

If there were only one well, then the discharge, under the same drawdown, would have been, from equation 4.6 (a),

$$Q = \frac{2\pi kb\,(H - h)}{\log_e \dfrac{R}{r}} \qquad ...(4.6\ a)$$

Since $\qquad R \gg B, \ \dfrac{H^2}{rB} > \dfrac{R}{r}$

Hence $\qquad Q > Q_1.$

Thus, discharge in each well decreases due to the interference.

Similarly, if there are three wells forming an equilateral triangle a distance B on a side, and if all the three wells have the same characteristics,

$$Q_1 = Q_2 = Q_3 = \frac{2\pi kb\,(H - h)}{\log_e \dfrac{R^3}{rB^2}} \qquad ...(4.15)$$

4.8. FULLY PENETRATING ARTESIAN-GRAVITY WELL

Sometimes, in an artesian well (*i.e.*, a well in a confined aquifer) high pumping rates may lower the water at the well face to a level below the top of the confined aquifer, as shown in Fig. 4.10. In such a case, the flow pattern close to the well is similar to that

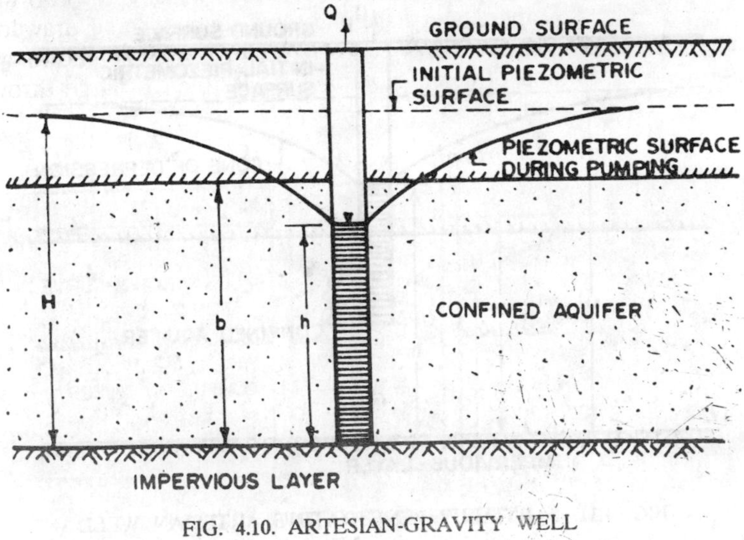

FIG. 4.10. ARTESIAN-GRAVITY WELL

for a gravity well (*i.e.* a well in unconfined aquifer) whereas at distances farther from the well, the flow is artesian. This type of well is known as a *combined artesian-gravity well.*

Fig. 4.10 shows an artesian gravity well. The flow from such a well can be computed from the following expression developed by Muskat :

$$Q = \frac{\pi k\,(2bH - b^2 - h^2)}{\log_e \dfrac{R}{r}} \qquad \text{...(4.16 } a\text{)}$$

4.9. PARTIALLY PENETRATING ARTESIAN WELL

A partially penetrating artesian well is the one in which the well screen does not penetrate to the full depth of the confined aquifer. The pattern of flow in the aquifer in the vicinity of such a well deviates from that for a fully penetrating well. In practice, we often encounter such wells that extend only part way through the water bearing strata.

Fig. 4.11 shows a partially penetrating artesian well in which the strainer length b_1 is less than the aquifer thickness b. The discharge Q_p from such a well can be computed from the following equation:

$$Q_p = \frac{2\pi kb\,(H - h)}{\log_e \dfrac{R}{r}}\, G = Q \,.\, G \qquad \text{...(4.16 } b\text{)}$$

where Q_p = discharge for the partially penetrating well.

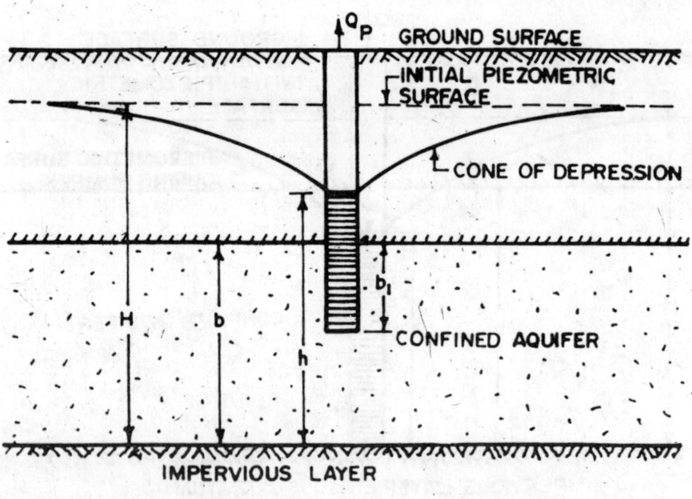

FIG. 4.11. PARTIALLY PENETRATING ARTESIAN WELL

Q = discharge for a fully penetrating well for the same drawdown $(H - h)$.

G = correction factor for partial penetration

$= \frac{Q_p}{Q}$.

A reasonable estimation of the correction factor G can be obtained from the following expression developed by Kozeny :

$$G = \frac{b_1}{b}\left(1 + 7\sqrt{\frac{r}{2b_1}}\cos\frac{\pi b_1}{2b}\right) \quad ...(4.17)$$

4.10. SPHERICAL FLOW IN A WELL

Fig. 4.12 shows a special case of partially penetrating well, where the well just penetrates the top surface of a semi-infinite porous medium. Here $b_1 = 0$, and equation 4.16 does not apply because the flow towards the well is purely spherical. The discharge Q_s from such a well can be computed from the expression

$$Q_s = 2\pi k r (H - h) \quad ...(4.18)$$

For the case of simple radial flow in a fully penetrating well, the discharge Q is given by equation 4.6 (a) :

$$Q = \frac{2\pi kb\,(H - h)}{\log_e \frac{R}{r}} \quad ...(4.6\ a)$$

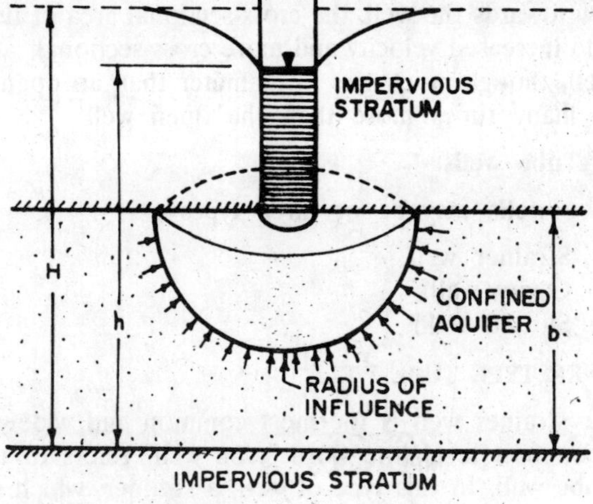

FIG. 4.12. SPHERICAL FLOW IN A WELL

$$\therefore \quad \frac{Q_s}{Q} = \frac{r}{b} \log_e \frac{R}{r} = 2.303 \frac{r}{b} \log_{10} \frac{R}{r} \quad \text{...(4.19)}$$

As a numerical example,

Let $r = 8 \text{ cm} = 0.08 \text{ m}$

$\frac{R}{r} = 1000$

b = thickness of aquifer = 16 m.

$$\therefore \quad \frac{Q_s}{Q} = 2.303 \times \frac{0.08}{16} \log_{10} 1000 \approx \frac{1}{30}.$$

This shows that the spherical flow is very much less efficient than the radial flow.

4.11. TUBE WELLS

A *water well* is a hole or shaft, usually vertical, excavated in the earth for bringing ground water to the surface. Wells can be mainly divided into two classes.

1. Dug wells or open wells.
2. Bored or drilled wells or tube wells.

An *open well* is comparatively of bigger diameter and is suitable for discharges upto 0.005 cumec. This is because the cross-sectional area of flow is less in the open well, and the water can be withdrawn safely only at the critical velocity for the soil. A *tube well* is a long pipe sunk into the ground with a strainer which allows water to pass through but prevents sand from coming in. Because of the strainer, high velocity of flow can be permitted without danger of soil particles being carried away with water. Also because of the radial flow towards the well, the cross-sectional area of flow is more. Due to the increased velocity and more cross-sectional area of flow, a tube well, though much less in diameter than an open well, gives discharge many times more than the open well.

Types of Tube wells

Tube wells may be of three types :

1. Strainer well.
2. Cavity well.
3. Slotted well.

1. STRAINER TYPE TUBE WELL

The strainer well is the most common and widely used tube well. In common term, the word "tube well" refers to the strainer type o. tube well. In this type of well, a strainer, which is a special type of wire mesh, is wrapped round the main tube of the well. The main pipe contains bigger holes or slots than the openings of the strainer. The total area of the opening of the tube is kept equal to the openings of the strainer so that the velocity of flow does not change. Due to fineness of the openings of the strainer, a higher operational velocity of water can be permitted. Little annular space is left between the strainer and the pipe so that the open area of pipe perforations is not reduced. The mesh size of the strainer is generally kept equal to D_{60} to D_{70} of the surrounding soil.

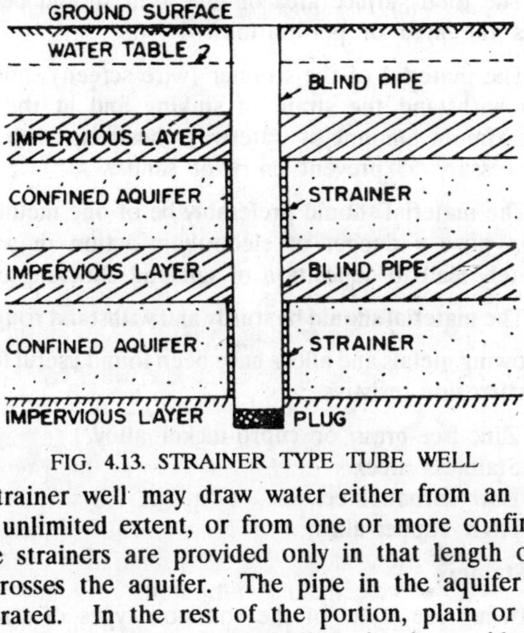

FIG. 4.13. STRAINER TYPE TUBE WELL

A strainer well may draw water either from an unconfined aquifer of unlimited extent, or from one or more confined aquifer layers. The strainers are provided only in that length of the pipe where it crosses the aquifer. The pipe in the aquifer portion is kept perforated. In the rest of the portion, plain or blind pipe is provided. At the bottom, a short blind pipe is provided to permit settlement of any sand if passed through the strainer. The well is generally plugged at the bottom.

Abyssinian tube well is a special type of strainer tube well, in which the diameter of pipe is kept $1\frac{1}{2}''$ and the strainer is provided only for a length of about 4 to 5 feet.

Design criteria for strainer type tube well

The following points should be noted for the design of a strainer type of tube well :

(1) The wire screen should not be in contact with the slotted tubes otherwise a large part of the opening will be covered by the wire screen.

(2) The area of the opening in the wire screen should be equal to the area of waterway in the perforated or slotted tube to permit no change in the velocity of incoming water particles. If there is any change, sand will be deposited in the annular space.

(3) The design should permit less velocity through soil than the exit critical velocity which is 1.25 cm/sec.

(4) The discharging velocity should vary from 1 m/sec to 2 m/sec.

(5) The total surface area of tube well should be more than three times the area of perforations.

(6) The material of the strainer (wire screen) should be such that it can withstand the strain of sinking and at the same time can give maximum amount of waterway consistent with fineness of opening necessary to prevent entry of sand.

(7) The material should preferably be of one metal. A bimetal construction pose a danger of electrolytic action in water, which may ultimately lead to deposition of salt and chocking of openings.

(8) The material should be sturdy and withstand rough handling.

Following metals and alloys have been found useful for strainers. They are corrosion-resistive :

(*i*) Zinc-free-brass or cupro-nickel alloy.
(*ii*) Stainless steel.
(*iii*) Low carbon steel.
(*iv*) High copper-alloy.

Types of Stainers

Following are some of the common types of strainers used in tube wells :

(*i*) Cook strainer.
(*ii*) Tej strainer.
(*iii*) Brownlie strainer.
(*iv*) Ashford strainer.
(*v*) Leggett strainer
(*vi*) Phoenix strainer.
(*vii*) Layne and Bowler strainer

(1) ***Cook strainer*** :

This is a very costly strainer of American patent. It is made up of solid drawn brass tube slotted with wedge-shaped horizontal slots. The slots are made with a slot cutting machine from inside the tube. The slots are wide inside and narrower outside, as shown in Fig. 4.14. The gauge of slots depends on the coarseness of sand, and varies from 0.15 to 0.4 mm.

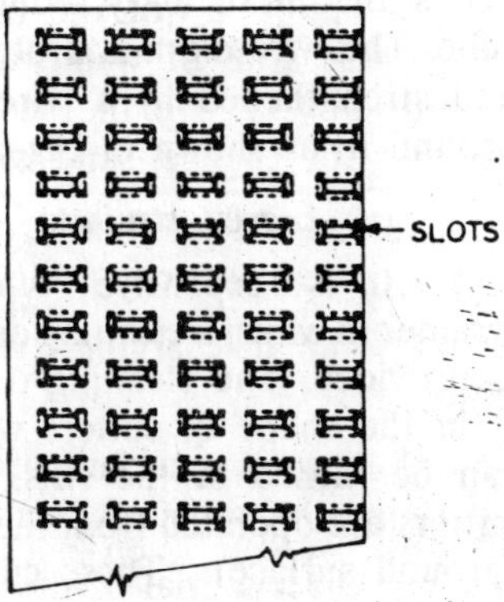

FIG. 4.14. COOK STRAINER

(ii) ***Tej strainer***

It is similar to cook strainer, but is manufactured in India. It consists of a brass tube constructed of a brass sheet bent round to form the tube, the vertical joint being brazed. The slots are cut in the sheet before it is bent. The strainer is generally manufactured from 7.5 cm diameter upwards, and is made in $2\frac{1}{2}$ metre lengths.

The individual lengths of the strainer are then joined together by means of screwed collars of brass.

(iii) ***Brownlie strainer***

The Brownlie strainer is made of a polygonal convoluted steel plate having perforations. A wire mesh surrounds the steel tube, as shown in Fig. 4.15. The mesh consists of heavy parallel copper wires woven with copper ribbons. Since the wire mesh is slightly away from the perforated tube, it is known as the best type of strainer.

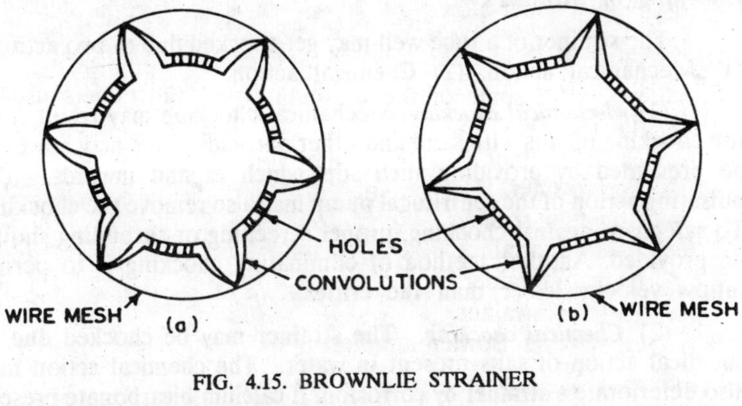

FIG. 4.15. BROWNLIE STRAINER

(iv) ***Ashford strainer***

This is very delicate strainer and consists of perforated tube with a wire round it over which a wire mesh is soldered. The wire keeps the mesh away from the tube. The wire mesh is protected and strengthened by a wire net around it, as shown in Fig. 4.16.

(*v*) ***Leggett strainer***

It is expensive type of strainer in which a cleaning device is provided. The cleaning device is in the shape of cutters which can be turned in the slits. The cutters are operated from the top (ground surface). These cutters clean the strainer clogged by the solid matter.

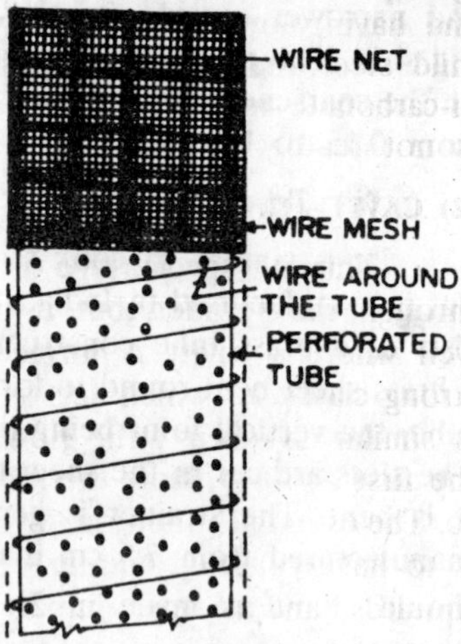

FIG. 4.16. ASHFORD STRAINER

(*vi*) ***Phoenix strainer***

It is a mild steel tube in which the openings are made by cutting slits from inside. The tube is cadmium plated to keep it free from danger of chocking and corrosion caused by chemical action.

(*vii*) ***Layne and Bowler strainer***

It is a robust type strainer manufactured in America. It consists of wedge-shape steel wire wound to suitable pitch round a slotted or perforated steel or wrought iron pipe. The joint of the strainer pipes are made by screwed collars.

Chocking of Strainers

The strainer of a tube well may get chocked due to two actions: (1) Mechanical action, (2) Chemical action.

(1) ***Mechanical chocking.*** Mechanical chocking may result from the chocking of slits with sand and other particles. This may however, be prevented by providing such slits which expand inwards. The pulsating action of the centrifugal pump may also remove the chocking. To safeguard against chocking, proper screening or shrouding should be provided. Another method of eliminating chocking is to permit inflow velocity lesser than the critical.

(2) ***Chemical chocking.*** The strainer may be chocked due to chemical action of salts present in water. The chemical action may also deteriorate a strainer by corrosion. If calcium bicarbonate present in water exceed by an amount of 15 parts per million parts of water, carbon dioxide is released when pressure is reduced due to pumping and calcium carbonate is precipitated on the strainer. The cumulative action of such precipitation reduces the yield. The chemical chocking

by deposition of carbonates is reduced by providing large slit area and having low inflow velocity. Similarly, sodium salts may attack mild steel and cast iron strainer pipes causing chocking. Sodium bi-carbonate may attack copper to a certain extent, though brass is not readily attacked.

(2) CAVITY TYPE TUBE WELL

This is a special type of tube well in which water is not drawn through the strainer, but it is drawn through the bottom of the well where a cavity is formed. The tube well pipe penetrates a strong clay layer which acts as strong roof. Thus, a cavity tube well is similar to deep well. However, a deep well draws water from the first aquifer below the *mota* while the cavity well need not do so. The essential condition for a cavity tube well to function efficiently is to have confined aquifer of good specific yield, and the aquifer should have a strong impervious material above it. In the initial stage of pumping with the help of centrifugal pump or an air lift pump, fine sand comes with water and consequently a hollow or cavity is formed. As the spherical surface area of cavity increases outwards, the radial critical velocity decreases, the sand particles stop entering the well. At this stage, an equilibrium in the cavity formation is established and clean water continues to enter the well on further pumping at the same constant discharge. After the formation of the cavity, the velocity of entry of water at the bottom of the pipe is lesser than the critical.

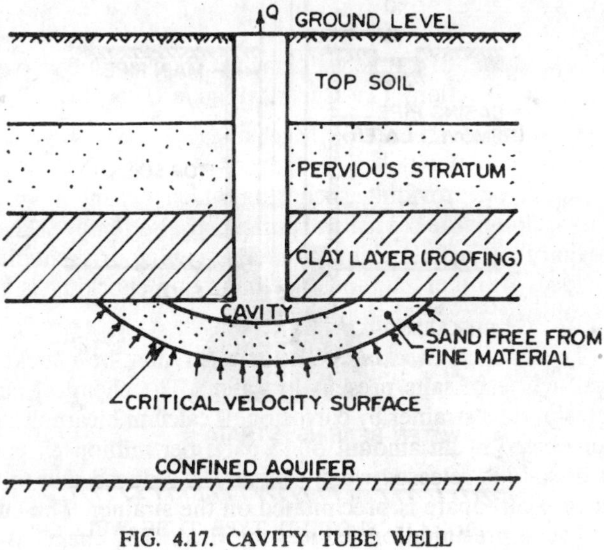

FIG. 4.17. CAVITY TUBE WELL

The essential difference between a strainer tube well and a cavity tube well lies in the fact that in the former case, due to the provision of a strainer, the flow is *radial*, while in the later case the flow is spherical. In the strainer tube well, the area of flow is increased by increasing the length of the strainer pipe (in an unconfined aquifer of infinite extent) while in the case of cavity type, the area of flow is increased by enlarging the size of the cavity. The cavity, formed due to a particular discharge Q_1, increases in size if an increased discharge Q is pumped out.

(3) SLOTTED TYPE TUBE WELL

A slotted tube well is resorted to under two circumstances (1) *sufficient* depth of water bearing stratum is not available even upto a depth of 75 to 100 m, so that strainer type tube well cannot be used and (2) suitable strong roof is not available so that a cavity well cannot be formed. In such a circumstance, a slotted well is used and it is made to penetrate to some depth in the water bearing strata. A slotted tube well essentially consists of a slotted tube penetrating the confined aquifer. The size of slots may be 25 mm × 3 mm at 10 to 12 mm spacings. In order to prevent the fine particles entering the pipe, it is shrouded with a mixture of gravel and bajri (coarse sand).

First of all, a casing pipe of 36 cm diameter is lowered, and soil is excavated out, and the water bearing strata is penetrated

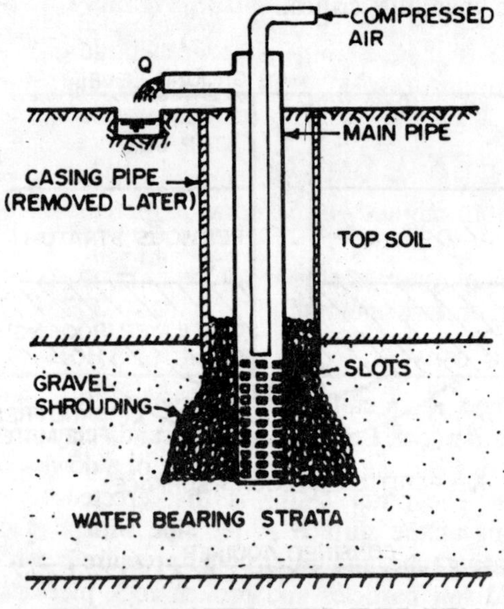

FIG. 4.18. SLOTTED TYPE TUBE WELL

by a depth of about 5 m length. The perforated pipe, sometimes known as the education pipe, of 15 cm diameter is then lowered, the slotted portion being only 5 m long and the rest of the length being of plain pipe. Gravel is then poured from the top, upto about 3 to 4 m higher than the top level of perforated portion of the pipe. The casing pipe is then withdrawn 5 cm at a time and the well is developed with the help of compressed air pumped into the education pipe. Finally, when the casing pipe is fully withdrawn, the annular space between the casing pipe and the education pipe is suitably plugged. By developing the well with the help of compressed air, the sand surrounding the gravel filter is freed of finer particles and the chances of getting the filter chocked are reduced. Due to the provision of gravel shrouding, a larger area of radial flow is obtained. There are two essential differences between a strainer tube well and a slotted tube well : (1) in the strainer tube well, the strainer pipes are surrounded by wire mesh to prevent the fine particles from entering the well, while in the slotted tube well the gravel shrouding serves this purpose, (2) a strainer tube well may have several alternative lengths of strainer pipes and plain pipes, while a slotted tube well has the slotted pipe length only at its bottom. Thus, a strainer tube well draws water from several aquifers sandwiched between impervious layers, while a slotted tube well draws water only from one pervious stratum which has sufficient water bearing capacity.

4.12. METHODS FOR DRILLING TUBE WELLS

For installing a tube well in the ground, so as to penetrate the required stratum, a hole, slightly of larger diameter than the diameter of the strainer pipe, is bored. The most common methods for boring a well are :

1. Wash boring or water-jet boring method.
2. Cable tool method (also known as percussion or standard method)
3. Hydraulic rotary method.
4. Reverse rotary method.

1. WASH BORING OR WATER JET BORING METHOD

This method is suitable at places where the well is to be sunk into formation consisting of gravel, sand, clay or other soft deposits. The boring is done by cutting action of a downward-directed stream of water. The outer casing is first erected in position in a suitable pit dug at the surface. A jet pipe with a nozzle is then lowered in the casing tube, and water under pressure is forced through it. The dislodged soil particles and broken rock pieces form slurry with water and are lifted up through the annular space between

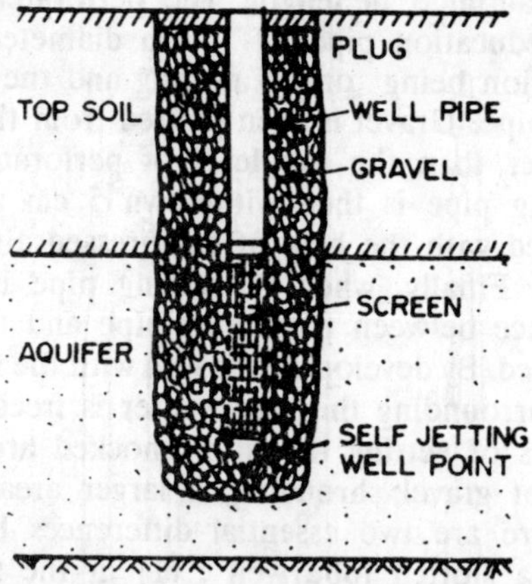

FIG. 4.19. SELF-JETTING WELL POINT

the casing and jet pipe by the returning water in the upward direction. The casing pipe having shoes at the bottom is kept rotated slowly, and is thus lowered. In penetrating clays and hard pans, various types of jetting drill bits are fitted to the drill pipe which is raised and lowered sharply, causing the bit to shatter the formation. When the casing pipe has penetrated to a sufficient depth into the aquifer, the well pipe attached with screen etc. is lowered in the casing pipe. The outer casing is then pulled. Sometimes, a *self-jetting well point* is used. In this, the casing pipe is not used, but instead, a tube of brass screen ending into a jetting nozzle is screwed to the main well pipe (Fig. 4.19). As the jetting action progresses, the well pipe goes on sinking. An annular space round the well pipe is automatically created due to the upward motion of water carrying the dislodged particles. When the well pipe has been sunk to the desired depth, jetting is stopped and the annular space is packed with gravel.

2. CABLE TOOL METHOD (PERCUSSION METHOD)

Cable tool method, also sometimes known as the Percussion method or Standard method, is used for drilling deep wells through consolidated rock materials. In this method, a standard well drilling rig consists of a mast, a multiline hoist, a walking beam and an engine—all assembled and mounted on a truck for easy portability. A string of percussion tools consists of (Fig. 4.20) : a rope socket, a set of jars, a drill stem and a drilling bit—the total weight of these amounting to several thousand kilograms. The drilling bits are manufactured in 1 to 3 metre lengths, and may weigh upto 1500 kg. A pit is dug at the site where the well is to be drilled. A casing pipe, with a drive shoe is inserted in the pit. The string

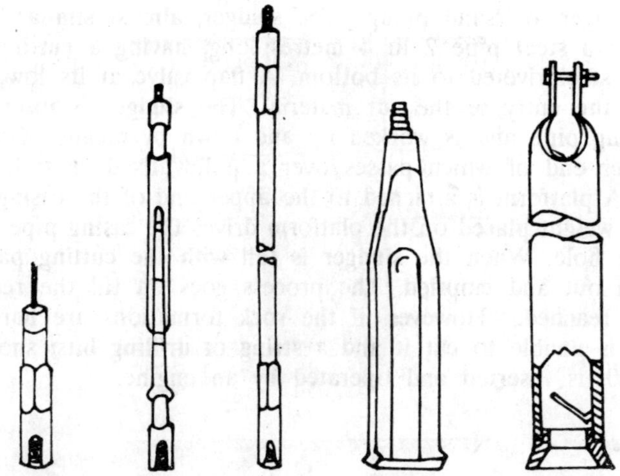

FIG. 4.20. PERCUSSION DRILLING TOOLS

of drilling tools is inserted in the first length of the casing pipe. Drilling is then accomplished by regular lifting and dropping of the string of tools mechanically. During drilling, the tools make 40 to 60 strokes per minute ranging from 40 cm to 1 m in length. The drilling line is kept continuously rotated so that the drilling bit will form a round hole. After the bit has cut 1 to $1\frac{1}{2}$ metre through the formation, the string of tools are taken out and a bailer (Fig. 4.20) is inserted in the hole to remove the drill cuttings. The bailer essentially consists of a pipe like section with a valve at the bottom. When the bailer is inserted in the hole, the valve is automatically opened by the upward movement of the cuttings. The valve, however, prevents the cutting from moving in the downward direction and thus escaping during lifting. When the bailer is full, it is lifted up to the surface and emptied. The length of the bailer varies with its diameter, and may range from 3 to 12 metres. When the cuttings have been taken out, the string of tools are again inserted and blow given to break the formation by impact. If no water is encountered in hole, water is added from the surface to form a paste with the cuttings. The casing is driven down by means of drive clamps fastened to the drill stem. The up and down motion of the tools striking the top of the casing, protected by a drive head, sinks the casing. The individual subsequent lengths of casing are joined by threaded or welded joints.

In soft and fissured rock formations, manual labour may be used (Fig. 4.21). In this case, the boring is done with the help of a 'sludger' or 'sand pump'. The sludger, almost similar to the bailer, is a steel pipe 2 to 4 metres long, having a cutting shoe of hard steel riveted to its bottom. A flap valve at its lower end permits the entry of the cut material. The sludger is inserted in the casing pipe and is worked up and down by means of a rope the other end of which passes over a pulley fixed centrally to a tripod. A platform is attached to the upper end of the casing pipe, and the weight placed on the platform drives the casing pipe slowly into the hole. When the sludger is full with the cutting paste, it is taken out and emptied. The process goes on till the required level is reached. However, if the rock formations are hard, the sludger is unable to cut it and a string of drilling bits, shown in Fig. 4.20, is inserted and operated by an engine.

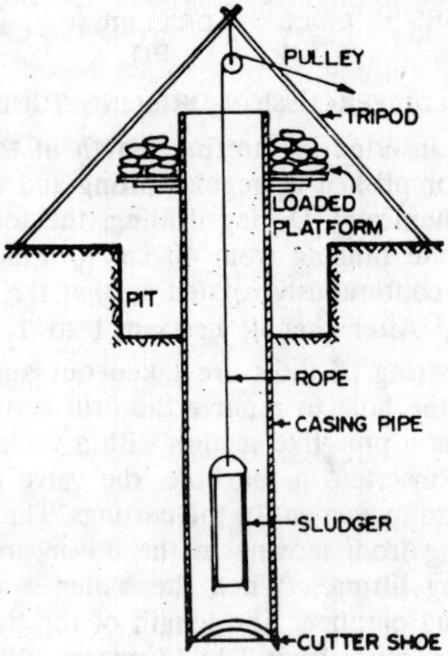

FIG. 4.21. PERCUSSION BORING BY MANUAL LABOUR.

In both the methods, the record of the material collected by bailer or sludger is kept. A bore log is then plotted to know the depths of various formations. The well pipe with strainers at determined levels of aquifer is then lowered to the desired depth. The well is then shrouded, and the casing pipe is taken out when the shrouding is done in steps. Wells by cable-tool method have been drilled in diameters upto 30 cm and to depths as great as 1600 metres.

3. HYDRAULIC ROTARY METHOD

Hydraulic rotary method, sometimes known as *rotary boring method*, is used for drilling large bores in unconsolidated strata. This is the fastest method and has been used for well upto 45 cm diameter (upto 150 cm with a reamer), and for depth over 1600 m. Oil wells over 7000 metre deep have been drilled by this method.

In this method, the boring is done with the help of a drilling bit attached at the end of a string of hollow pipes (Fig. 4.22). A mixture of clay and water, known as *drilling mud*, is continuously circulated through the drill shaft in the hole. Material loosened by the bit is carried upward in the hole by the rising mud. Ordinarily, no casing is required since the drilling mud forms a clay lining and supports the walls of the hole.

The drill bits have hollow shanks and one or more centrally located orifices for jetting the mud into the bottom of the hole. The drill rod, made of heavy pipe, carries the drill bit at one end, and is screwed to a square section known as *kelly*. A rotating table, which fits closely around the kelly, rotates the drill rod which slides downward as the hole deepens. The rising drilling mud carrying rock fragments is taken to the settling basin where the cuttings settle. The mud is recirculated to the hole. To maintain the required consis-

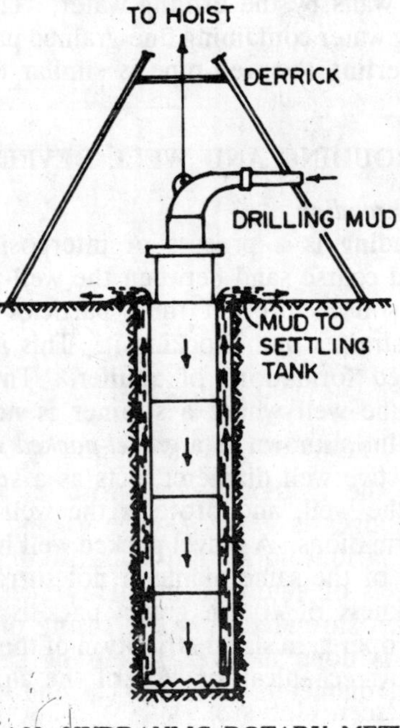

FIG. 4.22. HYDRAULIC ROTARY METHOD

tency, clay and water is added to the circulating mud from time to time. A complete boring record is maintained to know the type of formations at various depths. When desired level is reached, the drill rod etc. are taken out and the well pipe containing strainer pipes at appropriate locations (opposite aquifers) is lowered. Since the well walls are coated with clay, it should be washed to get more discharge. Back washing is done by lowering the drill pipe and bit in the well pipe, and forcing water containing calgon (sodium-hexa-meta phosphate). Calgon has the property of dispersing clay colloids. A collar, of the size of well-pipe is attached to the drill rod just above the bit. This forces the water through the strainer causing washing action on the clay wall. At the same time, the drill rod is plunged up and down causing surging action. When washing in the bottom is done, the bit is raised through some distance and the operation is repeated.

4. REVERSE ROTARY METHOD

Reverse rotary method, similar to the hydraulic rotary method, is very much used in Europe. In this method, the cuttings are used for this purpose. A mixture of water and fine grained material is circulated in the hole. The procedure is essentially a suction dredging method. The walls of the hole during drilling are supported by hydrostatic pressure acting against the film of fine-grained material deposited on the walls by the drilling water. The method of recirculation of drilling water containing fine-grained particles and cleaning the well after inserting the well-pipe is similar to that of hydraulic rotary methods.

4.13. WELL SHROUDING AND WELL DEVELOPMENT

(*a*) *Well Shrouding*

Well shrouding is a process of interposing coarse material such as gravel and coarse sand between the well-pipe (strainer pipe) and the aquifer soil to prevent finer particles of soil coming in contact with the strainer and chocking it. This is essential in sandy and unconsolidated formations of aquifer. This is also essential in slotted type tube well where a strainer is not used. Such tube well is also sometimes known as a *gravel-packed well*. The shrouding increases the effective well diameter, acts as a strainer to keep fine material out of the well, and protects the well-pipe from caving of surrounding formations. A gravel packed well has a greater specific capacity than one of the same diameter not surrounded by a gravel. A minimum thickness of 40 cm gravel pack is necessary to make it effective. The proper gain size distribution of the shrouding material depends upon the mechanical analysis of the aquifer and upon the perforations or screen slot size.

The amount of shrouding material per 30 cm length of the casing pipe can be calculated accurately before hand. In the beginning, the shrouding material for 60 cm length is shovelled in from the top between the tube-pipe and casing. The casing is raised by 30 cm with the help of jack. Then the quantity for each 30 cm is added and the casing pipe is withdrawn 30 cm at a time till the strainer is covered. The shrouding material is sometimes placed through small pipes or pilot holes around the tube-pipe which feed the material down into position. A further refinement is the use of bladeless pumps for pumping the shrouding material into place.

(*b*) **Well Development**

Well development is the process of removing fine material from the aquifer formation surrounding the strainer pipe, and is aimed at (*i*) increasing the specific capacity of the well, (*ii*) preventing sand flowing in, and (*iii*) obtaining maximum economic well life. The actual yield of the well can be known only after well development. Thus, it also helps in determining the required characteristics of the pump and power unit to be installed. Depending upon formation characteristics of the aquifer, a well may be developed by one of the following methods.

1. Development by pumping.
2. Development by surging.
3. Development by compressed air.
4. Development by back washing.
5. Development by dry ice.

1. Development by pumping

In this method, a variable speed pump is used. The method is based on the principle that irregular and non-continuous pumping agitates the fine material surrounding the well so that it can be carried into the well and pumped out. Initially, the pump is started with a very low discharge. The fine particles start coming. This low speed is maintained till clear water comes. The discharge is then increased in steps until maximum discharge or well capacity is reached. The pump is then stopped and levels permitted to increase till it comes to normal. The pump is then again started and the procedure repeated, till no fine particles come.

2. Development by Surging

In this method, surging effect is created by up and down movement of a hollow surge block or a bailer. Calgon (sodium-hexametaphosphate) is added to water, so that it acts as dispersing agent for fine grained particles. When the surge block is moved up, it sucks water in. When it is moved down, it forces water-calgon solution back in the formation. Further upward motion bring with it fine material. The surge block is connected to a string of hollow pipe from which the water charged with fine particles is pumped out continuously. The procedure is repeated by increasing the speed of surging till clear water comes out.

3. Development by Compressed Air.

In this method, the development is done with the help of an air compressor, a discharge pipe and an air pipe. The air pipe is put into the discharge pipe and is lowered into the well tube, till the assembly reaches near the bottom of the strainer-pipe section. The lower end of the air pipe is kept emerging out of the discharge pipe by a small length. The air entry to the air pipe is first closed and the compressor is then started till a pressure of 6 to 10 kg/cm^2 is built up. The air is then suddenly made to enter the pipe, at this pressure, with the help of suitable quick-opening valve. This sudden entry of air into well creates a powerful surge within the well causing loosening of fine material surrounding the perforations. When the pressure decreases, water enters the well bringing the loosened particles with it. The continuous air injection creates an air lift pump, and the water carrying fine particles is pumped out. The process is repeated till clear water comes. The pipe assembly is then lifted up, and the surging is again created. The operation is repeated at intervals along the screen section till the well is fully developed.

4. Development by Back Washing

In this method, in addition to the compressor, a discharge pipe and an air pipe, and additional small air pipe is used. The well is sealed at its top so that is becomes air-tight. The discharge pipe and air pipe assembly is lowered in the well, as in the previous method, but the end of the air pipe is kept inside the discharge pipe. A small air pipe is fitted at the top of the air-tight cover and is provided with a three-way cock. With the help of the three-way cock, air can be admitted to the well either through the long air pipe (but inside the discharge pipe) or through the long air pipe fitted at the top. Air is first made to enter the long air pipe. This forces air and water out of the well through the discharge pipe. When clear water comes the valve is closed, and water level is allowed to increase in the well. The valve is then turned to the other side so that air enters through the discharge pipe and at the same time agitates the fine particles surrounding the well. Calgon is often added to water. When air starts escaping from the discharge pipe, the valve is turned so that air enters the long air pipe, and the assembly works as an air-lift pump and the water is pumped out. The procedure is repeated till clear water comes and the well is fully developed.

5. Development by dry ice (solid sodium dioxide)

In this method, well is developed with the help of two chemicals : hydrochloric acid and solid sodium dioxide (known as dry ice). First of all, hydrochloric acid is poured into the well. The well is capped at the top and compressed air is forced into the well. The pressure of the compressed air forces the chemical into the formation. The cap is then removed and blocks of dry ice are dropped into the well. The sublimation releases gaseous carbon dioxide, and a high pressure of this gas is built up in the well. On releasing the pressure the muddy water is forced up in the form of a jet and is automatically thrown out of the well. Explosion of mud and water extending 40 metres into the air from a well in Utah (U.S.A) was observed when the well was developed with dry ice.

4.14. OPEN WELLS

As stated earlier, an open well is essentially of a bigger diameter than of a tube well, and derives its water only from one pervious stratum. Since a tube well, in general, may derive water from more than one aquifer formation, it has greater depth than an open well. The economically feasible depth of an open well is limited to 30 metres below the ground surface. In a lined open well, the entry of water is from the bottom and *not* from the sides.

An open well is classified as :

(*i*) Shallow well. (*ii*) Deep well.

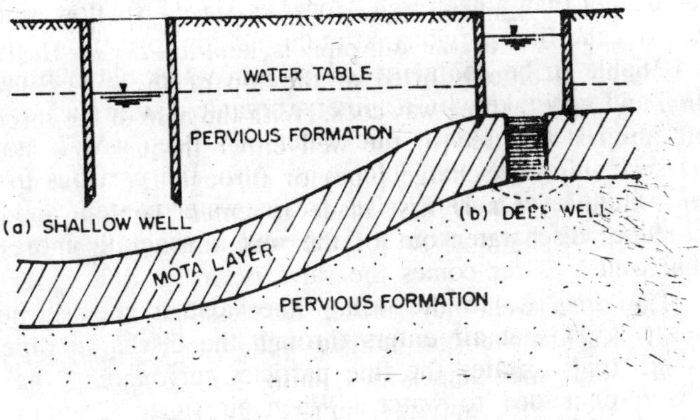

FIG. 4.23. SHALLOW AND DEEP WELLS

The nomenclature of shallow and deep well has nothing to do with the actual depth of the well. A *deep well* is a well which is supported on a *mota* layer and draws its water supply, through a hole bored in it, from the pervious formation below the *mota* layer.

A *shallow well,* on the other end, penetrates the pervious stratum only and draws its water supply through it. The term *mota* layer also sometimes known as *matbarwa* or *nagasan,* refers to a layer of clay, cemented sand, kankar or any other hard material. The *mota* layer gives structural support to the open well, and is found throughout the Indo-Gangetic plain. These mota layers may either be continuous, or may be localised and may be found in different thicknesses and depths at different places.

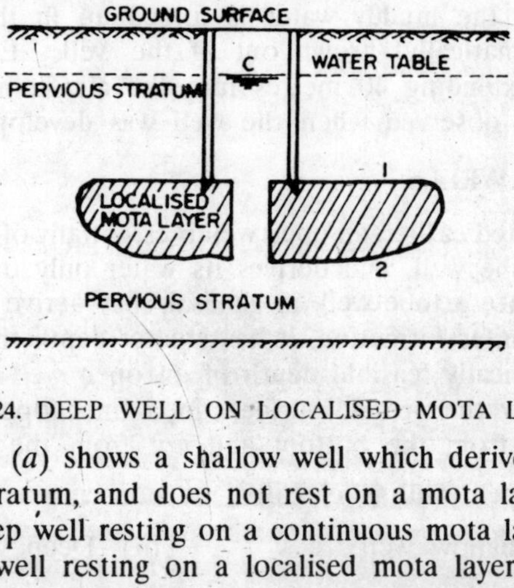

FIG. 4.24. DEEP WELL ON LOCALISED MOTA LAYER

Fig. 4.23 (*a*) shows a shallow well which derives water from the pervious stratum, and does not rest on a mota layer. Fig. 4.23 (*b*) shows a deep well resting on a continuous mota layer. Fig. 4.24 shows a deep well resting on a localised mota layer and deriving its water from the second pervious stratum. Actually, a shallow well can be deeper than a deep well. However, since a shallow well draws water from the first pervious stratum (*i.e.* top formation), the water in it is liable to be contaminated by rain water percolating in the vicinity, and may take with it mineral organic matter such as decomposing animals and plants. The water in a deep well is not liable to get such impurities and infections. Also, the pervious formation below a mota layer normally has greater water content and specific yield. Hence discharge from a deep well is generally more than a shallow well.

The open well may further be classified as :

(*i*) Kachha well or unlined well.

(*ii*) Well with impervious lining.

(*iii*) Well with pervious lining.

Kachha well. A kachha well is a temporary well of a very shallow depth. It is suitable only in hard formations the walls of which can stand vertically. They are suitable only when the water table is very near the ground surface. Such wells often collapse after some time, and are dangerous.

Well with impervious or pucca lining. This is the most common type of open well, and is suitable for all types of formations. Once constructed, it becomes a permanent source of water supply. Impervious lining for an open well in sandy formations is most essential to gives structural stability to the well.

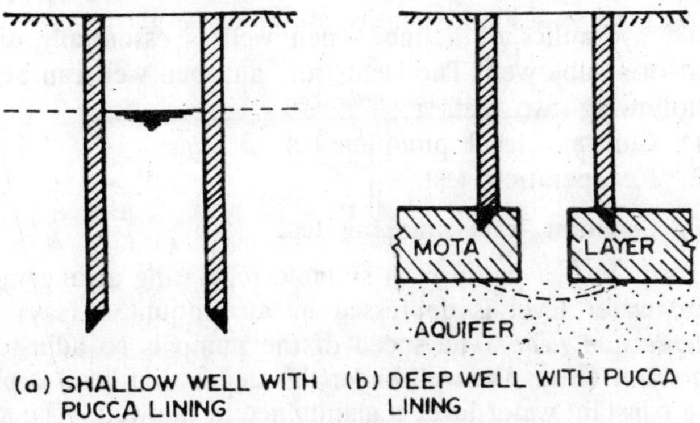

FIG. 4.25. WELLS WITH PUCCA LINING

The thickness of impervious lining (steining) varies from 30 to 60 cm and may be either in brick masonry or in stone masonry. The linings carry well curbs under them. Well curbs may be constructed of either wood, iron or reinforced concrete. In a pucca well, the flow is not radial. Water enters only from the bottom and, after a virtual cavity has been formed at the bottom, the flow is spherical.

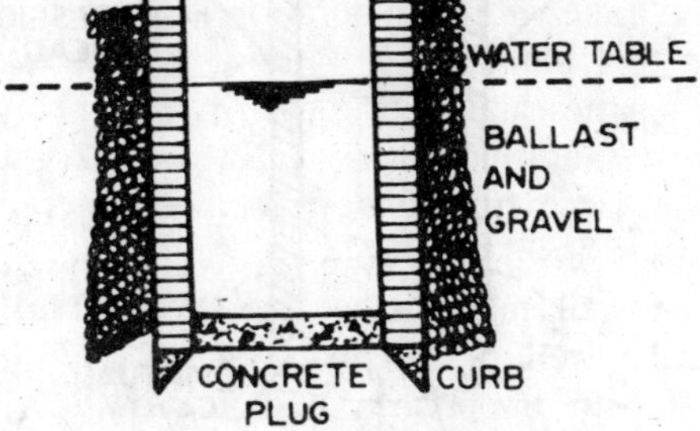

FIG. 4.26. WELL WITH PERVIOUS LINING.

Well with pervious lining. Such types of wells are suitable in coarse formations. The lining consists of dry bricks or stones with no mortar or binding material. Due to this, water enters from the sides, and flow is therefore radial. Such wells are generally plugged at the bottom. If there is no plug at the bottom, the flow is a combination of radial and spherical pattern.

If a well with pervious lining is constructed in a sandy formation, brick ballast upto 20 mm size and gravel is placed behind the lining to prevent sand coming into the well along with the flowing water.

4.15. YIELD OF AN OPEN WELL

The hydraulics of a lined open well is essentially different from that of a tube well. The yield from an open well can be found by the following two tests :

(1) Constant level pumping test.
(2) Recuperation test.

(1) Constant level pumping test

In this test, a pump with suitable regulating arrangement is used. The water level is depressed by an amount h (say) known as the *depression head*. The speed of the pump is so adjusted that whatever water enters the well under this depression head is pumped out and a constant water level is maintained in the well. The amount of water pumped out is measured with the help of a V-notch, or any other arrangement, in a given amount of time for which the pump speed was requlated to a constant value. The quantity pumped out in one hour gives the yield of the well per hour.

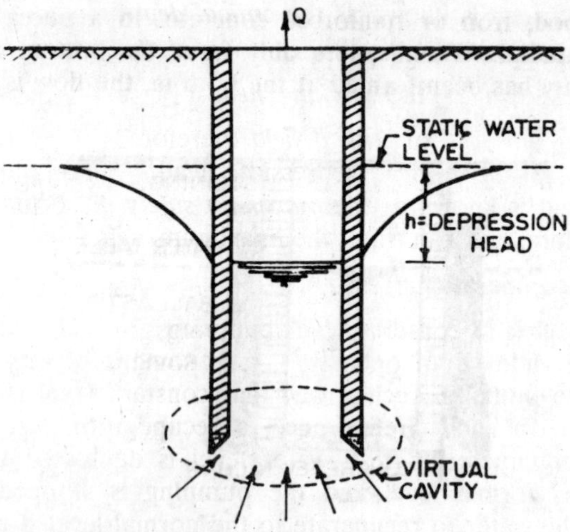

FIG. 4.27. CONSTANT LEVEL PUMPING TEST

The formula for discharge, in cumecs, from an open well with impervious lining may be written as :

$$Q = A \times v$$

or $$Q = A \times C \times h \text{ cumecs} \quad ...(4.20)$$

where Q = discharge, in cubic metres second

A = cross-sectional area of flow into the well as its base, in m^2

v = mean velocity of water percolating into the well, in metres/sec

h = depression head in metres

C = percolation intensity coefficient. This is a constant of the formation around the well. Its value is greater for coarser soil and smaller for finer soils. (m/sec under unit head)

The above formula can also be derived from Darcy's law as under :

$$Q = k.A.i = k.A.\frac{h}{L} = \frac{k}{L}.Ah$$

or $$Q = C.Ah \quad ...(4.20)$$

Due to cavity formation, the area A is taken to be equal to $\frac{4}{3}$ times the actual cross-sectional area of the bottom of the well.

From the above expression, it is clear that the discharge increases with the percolation head h. However, the percolation head cannot be increased beyond a certain critical value because otherwise the percolation velocity will be exceeded and the soil particles will be disturbed and dislodged. The critical value of h at which the velocity is critical is known as the *critical depression head.* Normally, the depression head is kept equal to $\frac{1}{2}$ of the critical head ; such a head is known as the *working head.*

Maximum yield or *critical yield* therefore will be obtained corresponding to the critical depression head. The yield under the working head is known as the *maximum safe yield.* From a pumping test, therefore, we can find the maximum safe yield.

2. Recuperation Test

Though the constant level pumping gives an accurate value of the safe yield of an open well, it is sometimes very difficult to regulate the pump in such a way that constant level is maintained in the well. In such circumstance, a recuperation test is resorted to. In the recuperation test, water level is depressed to any level below the normal level and the pumping is stopped. The time taken for the water to recuperate to the normal level is noted. From the data, the discharge from the well can be calculated as under (Fig. 4.28) :

Let aa = static water level in the well, before the pumping started

bb = water level in the well when the pumping stopped

h_1 = depression head in the well when the pumping stopped (metes)

cc = water level in the well at a time T after the pumping stopped

h_2 = depression head in the well at a time T after the pumping stopped (metres)

h = depression head in the well at a time t after the pumping stopped (metres)

dh = decrease in depression head in a time dt

t, T = times in hours.

Thus, in a times t, reckoned from the instant of stopping the pump, the water level recuperates by $(h_1 - h)$ metres. In a time dt after this, the head recuperates by a value dh metres.

$\therefore$ Volume of water entering the well, when the head recuperates by dh is $dV = a\, dh$ ...(1)

where A = cross-sectional area of well at its bottom.

Again, if Q is the rate of discharge in the well at the time t_1 under the depression head h, the volume of water entering the well in a time dt hours is given by

$$dV = Q.\, dt$$

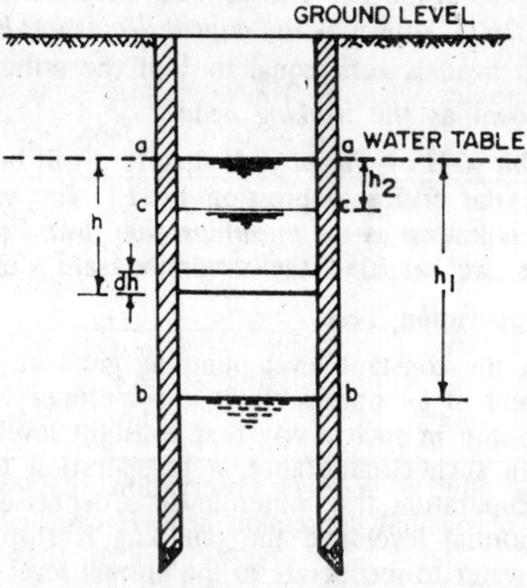

FIG. 4.28. RECUPERATION TEST

But $$Q \propto h$$

or $$Q = Kh \qquad ...(2)$$

$\therefore$ $$dV = Khdt \qquad ...(3)$$

where K is a constant depending upon the soil at the base of the well through which water enters.

Equating (1) and (3), we get

$$Khdt = -Adh \qquad ...(4)$$

The minus sign indicates that h decreases as time t increases. Integrating the above between the limits :

$$t = 0 \text{ when } h = h_1$$

$$t = T \text{ when } h = h_2$$

we get $$\frac{K}{A}\int_0^T dt = -\int_{h_1}^{h_2} \frac{dh}{h}$$

or $$\frac{K}{A}\int_0^T dt = \int_{h_2}^{h_1} \frac{dh}{h}$$

From which

$$\frac{K}{A}T = [\log_e h]_{h_2}^{h_1}$$

$$\therefore \qquad \frac{K}{A} = \frac{1}{T}\log_e \frac{h_1}{h_2} = \frac{2.303}{T}\log_{10}\frac{h_1}{h_2} \qquad ...(4.21)$$

Thus knowing the value of h_1, h_2 and T from a recuperation test, the quantity K/A can be calculated. K/A is known as the *specific yield* or specific capacity of an open well, in cubic metres per hour per sq. metre of the area through which water percolates under one metre depression head. In the absence of the recuperation test, the following rough values of K/A specified by Marriot can be adopted.

TABLE 4.2.

Type of soil	$\frac{K}{A}$ *Cubic metres per hour, sq. metre of area under unit depression head*
Clay	0.25
Fine sand	0.5
Coarse sand	1.00

Knowing the value of K/A by observation, the discharge Q from a well under a constant depression head H can be calculated as under :

$$Q = KH \qquad \text{(from 2)}$$

or $$Q = \left(\frac{K}{A}\right).A.H \qquad ...(4.22)$$

or $$Q = \frac{2.303}{T}\left(\log_{10}\frac{h_1}{h_2}\right)AH \text{ cubic metre/hour} \qquad ...(4.23)$$

It should be noted that the time T in the above expression is in hours. If T is substituted in seconds, Q will be in cumecs. The discharge Q will be the maximum yield of the well if H corresponds to the maximum depression head. If H is the average depression head, Q will be the average yield.

4.16. SELECTION OF SUITABLE SITE FOR A TUBE WELL

The site where a tube well is to be located must be carefully selected. Improper or careless selection of the site have resulted in tube wells having very low yield or without any yield during summer season. Examples of such tube wells are present in Mahi area where some tube wells did not discharge any water during summers and had to be shifted.

The following points should be considered for the selection of the site :

(1) Tube well should be selected at a site where large underground reservoir exists. A judgment of such a place can be made on the following considerations :

(*a*) The place will be usually dark and green in summer also– indicating a high water table.

(*b*) Certain type of worms and insects such as gnats hover in a column round the spot where there is likelihood of sub-soil water.

(*c*) By a study of geological strata by actual boring.

(2) If the geological explorations indicate ridges and depressions of impermeable strata inside the ground, the tube well should be located where there is depression in the valley (Fig. 4.29).

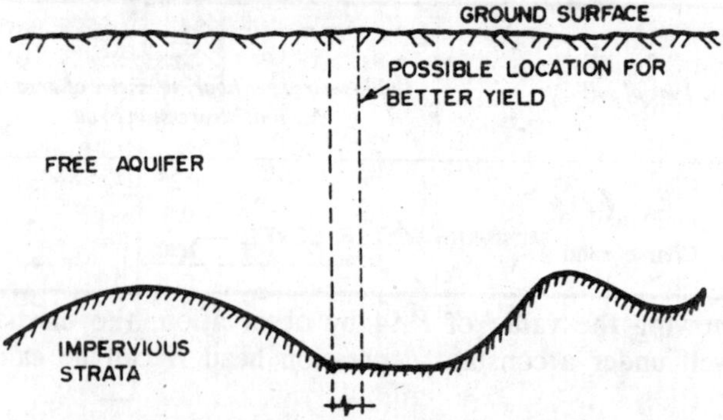

FIG. 4.29.

(3) The area should have an access for the availability of a cheap electric supply so that motor driven pumps can be fitted with tube wells without an exhorbitant cost. Diesel driven sets are not as efficient as those by electric power.

(4) The area should have a well distributed and uniform demand for irrigation throughout the year. If the demand is not uniform, tube well will be idling for some time or has to work inefficiently and thus the cost for overall irrigation will increase.

(5) The area around the tube well should have an intensive cultivation and the tube well should be located centrally so as to reduce the length of guls and thereby transient losses. In any case the tube well should have a good command.

(6) The water available from the tube well should be tested to find out the irrigation quality of water. If the water is found to contain harmful salts, the site may have to be changed in spite of various advantages.

4.17. SECTION OF A TUBE WELL

Fig. 4.30 shows the section of a tube well with a centrifugal pump.

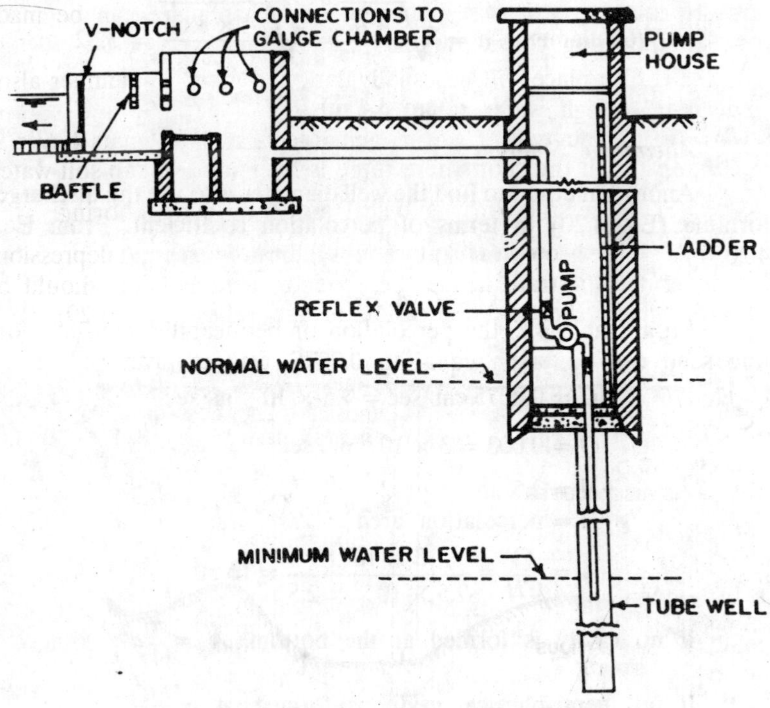

FIG. 4.30. SECTION OF A TUBE WELL

ILLUSTRATIVE EXAMPLES

Example 4.1. *Design an open well in fine sand to give a discharge of 0.003 cumecs when worked under a depression head of 2.5 metres.*

Solution :

Required discharge

$$Q = 0.003 \text{ cumecs}$$
$$= 0.003 \times 3600 = 10.8 \text{ m}^3/\text{hour}$$

From Eq. 4.22,

$$Q = \left(\frac{K}{A}\right) A.H \qquad ...(1)$$

For fine sand, $\frac{K}{A} = 0.5 \text{ m}^3/\text{hour per m}^2$ of area, under unit depression head

$$H = \text{depression head} = 2.5 \text{ m}$$

Substituting values in (1), we get

$$10.8 = 0.5 \times A \times 2.5$$

$$\therefore \quad A = \frac{10.8}{0.5 \times 2.5} \text{ m}^2 = 8.64 \text{ m}^2.$$

$$\therefore \text{ Well diameter } d = \sqrt{\frac{4A}{\pi}} = \sqrt{\frac{4 \times 8.64}{\pi}} = 3.32 \text{ m.}$$

(say 3.4 m)

Alternative Solution

Another method to find the well diameter is to use the discharge formula (Eq. 4.20) in terms of percolation coefficient. From Eq. 4.20,

$$Q = C.A.H. \qquad ...(2)$$

From Table 4.2, the percolation or permeability constant for fine sand can be taken equal to 0.0075 cm/sec (average).

$$C = 0.0075 \text{ cm/sec} = 7.5 \times 10^{-5} \text{ m/sec}$$

$$Q = 0.003 = 3 \times 10^{-3} \text{ m}^3/\text{sec}$$

$$H = 2.5 \text{ m}$$

$$A = \text{percolation area}$$

$$\therefore \quad A = \frac{Q}{CH} = \frac{3 \times 10^{-3}}{7.5 \times 10^{-5} \times 2.5} = 16 \text{ m}^2$$

If no cavity is formed at the bottom, $A = \frac{\pi}{4} d^2$

If full hemispherical cavity is formed, $A = \frac{\pi d^2}{2}$

Usually the area of cavity is between the two values.

$$\therefore \qquad A = \frac{1}{2}\left(\frac{\pi}{4}d^2 + \frac{\pi d^2}{2}\right) = \frac{3}{8}\pi d^2 \approx \frac{1}{3}\pi d^2$$

Thus, the area of percolation at the well bottom may be taken equal to $\frac{4}{3}$ times the cross-sectional area of well bottom

$$\left(i.e., \frac{\pi}{3}d^2 = \frac{4}{3} \times \frac{\pi}{4}d^2\right)$$

$$\therefore \qquad \frac{\pi}{3}d^2 = 16$$

$$\therefore \qquad d = \sqrt{\frac{16 \times 3}{\pi}} = \mathbf{3.9\ m.}$$

Example 4.2. *During a recuperation test, the water in an open well was depressed, by pumping, by 2.5 metres and it recuperated 1.8 metres in 80 minutes. Find (a) yield from a well of 4 m diameter under a depression head of 3 metres, (b) the diameter of the well to yield 8 litres/second under a depression head of 2 metres.*

Solution :

From Eq. 4.21, the specific yield is given by

$$\frac{K}{A} = \frac{2.303}{T}\log_{10}\frac{h_1}{h_2}$$

where T = time in hours $= \frac{80}{60} = \frac{4}{3}$ hours

$h_1 = 2.5$ m

$h_2 = 2.5 - 1.8 = 0.7$ m

$$\therefore \qquad \frac{K}{A} = \frac{2.303 \times 3}{4}\log_{10}\frac{2.5}{0.7} = 0.955$$

(*a*) Yield from the well of diameter 4 m :

$$A = \frac{\pi}{4}(4)^2 = 12.56\ \text{m}^2$$

$$Q = \left(\frac{K}{A}\right)A.H$$

$$= 0.955 \times 12.56 \times 3 = 36\ \text{m}^3/\text{hour}$$

$$= \mathbf{10\ lit/sec.}$$

(*b*) Yield = 8 lit/sec = 28.8 m^3/hour

(since 1 lit/sec = 3.6 m^3/hour)

$$\therefore \qquad Q = \left(\frac{K}{A}\right)A.H.$$

$$\therefore \quad A = \frac{Q}{H} \times \left(\frac{A}{K}\right) = \frac{28.8}{2} \times \frac{1}{0.955}$$

$$= 15.05 \text{ m}^2$$

$$\therefore \quad d = 4.37 \text{ m} \approx \mathbf{4.4\ m} \text{ (say).}$$

Example 4.3. *A tube well of 30 cm diameter penetrates fully in an artesian aquifer. The strainer length is 15 m. Calculate the yield from the well under a drawdown of 3 m. The aquifer consists of sand of effective size of 0.2 mm having coefficient of permeability equal to 50 m/day. Assume radius of drawdown equal to 150 metres.*

Solution :

From Eq. 4.6 c, the discharge is given by

$$Q = \frac{2.72\, bks}{\log_{10} \frac{R}{r}}$$

Here b = thickness of aquifer
= length of strainer = 15 m
s = drawdown = 3 m
k = coefficient of permeability= 50 m/day
R = 150 m
r = 15 cm = 0.15 m

$$\therefore \quad Q = \frac{2.72 \times 15 \times 50 \times 3}{\log_{10} \frac{150}{0.15}} \text{ m}^3/\text{day}$$

$$= 2040 \text{ m}^3/\text{day} = 85 \text{ m}^3/\text{hour}$$

$$= \mathbf{23.6\ lit/sec.}$$

Example 4.4. *A tube well penetrates fully an unconfined aquifer. Calculate the discharge from the tube well under the following conditions:*

Diameter of the well = *30 cm*

Drawdown = *2 m*

Effective length of the strainer under the above drawdown
= *10 m*

Coefficient of permeability of aquifer
= *0.05* cm/sec

Radius of zero drawdown
= *300 m*

Solution :

From Eq. 4.5, discharge is given by

$$Q = \frac{2.72\, ks\left(L + \frac{s}{2}\right)}{\log_{10} \frac{R}{r}}$$

where $k = 0.05 \text{ cm/sec} = \frac{0.05}{100} \text{ m/sec}$

$= 5 \times 10^{-4}$ m/sec

s = drawdown = 2 m

L = effective length of strainer = 10 m

R = 300 m.

r = 0.15 m.

$$\therefore \quad Q = \frac{2.72 \times 5 \times 10^{-4} \times 2\,(10 + 1)}{\log_{10} \frac{300}{0.15}} \text{ m}^3\text{/sec}$$

$= 9.05 \times 10^{-3}$ m^3/sec $\approx$ **0.009 cumecs**

= **9.05 lit/sec.**

Example 4.5. *Design a tube well for the following data*

(*i*) *Yield required* = *0.08 cumecs*
(*ii*) *Thickness of confined aquifer* = *30 m*
(*iii*) *Radius of circle of influence* = *300 m*
(*iv*) *Permeability coefficient* = *60 m/day*
(*v*) *Drawdown* = *5 m.*

Solution:

From Eq. 4.6 *c*, we have

$$Q = \frac{2.72\, bks}{\log_{10} \frac{R}{r}}$$

Here $Q = 0.08$ cumecs $= 0.08$ m^3/sec

$b = 30$ m

$s = 5$ m

$k = 60 \text{ m/day} = \frac{60}{3600 \times 24} \text{ m/sec}$

$R = 300$

$r = ?$

Hence

$$0.08 = \frac{2.72 \times 30 \times 5}{\log_{10} \frac{R}{r}} \times \frac{60}{3600 \times 24}$$

$$\therefore \qquad \log_{10}\frac{R}{r} = \frac{2.72 \times 30 \times 5 \times 60}{3600 \times 24 \times 0.08} = 3.54$$

$$\therefore \qquad \frac{R}{r} = 3481$$

$$\therefore \qquad r = \frac{R}{3481} = \frac{300}{3481} = 0.0862 \text{ m.}$$

Hence adopt well of **9 cm** diameter.

Example 4.6. *An artesian tube well has a diameter of 20 cm. The thickness of aquifer is 30 m and its permeability is 36 m/day. Find its yield under a drawdown of 4 m at the well face. Use radius of influence as recommended by Sichardt.*

Solution :

From Eq. 4.6 c, we have

$$Q = \frac{2.72\, b\, k\, s}{\log_{10}\dfrac{R}{r}}$$

Here $b = 30$ m

$$k = 36 \text{ m/day} \left(= \frac{36}{24 \times 3600} \text{ m/sec} \right)$$

$$s = 4 \text{ m}$$

$$r = 0.10 \text{ m}$$

$$R = 3000\, s \sqrt{k}, \text{ from Sichardt formula}$$

$$= 3000 \times 4 \sqrt{\frac{36}{24 \times 3600}} = 245 \text{ m.}$$

$$\therefore \qquad Q = \frac{2.72 \times 30 \times 36 \times 4}{\log_{10}\dfrac{245}{0.1}}$$

$$= 3470 \text{ m}^3\text{/day} = \mathbf{144.5\ m^3/hour}$$

$$= \mathbf{40.2\ lit/sec} \text{ (or } \mathbf{0.0402\ cumecs}\text{).}$$

Example 4.7. *A well penetrates fully a 10 m thick water bearing stratum of medium sand having coefficient of permeability of 0.005 m/sec. The well radius is 10 cm and is to be worked under a drawdown of 4 m at the well face. Calculate the discharge from the well. What will be the percentage increase in the discharge if the radius of the well is doubled ?*

Take R = 300 m in each case.

Solution :

From Eq. 4.6 *c*, we have

$$Q = \frac{2.72\, bks}{\log_{10} \frac{R}{r}}$$

$$= \frac{2.72 \times 10 \times 0.005 \times 4}{\log_{10} \frac{300}{r}} = \frac{0.544}{\log_{10} \frac{300}{r}}$$

(*i*) In the first case, $r = 10$ cm $= 0.1$ m.

$$\therefore \qquad Q_1 = \frac{0.544}{\log_{10} \frac{300}{0.1}} = \frac{0.544}{3.477} = \mathbf{0.1565} \text{ cumecs.}$$

(*ii*) In the second case, $r = 20$ cm $= 0.2$ m.

$$\therefore \qquad Q_2 = \frac{0.544}{\log_{10} \frac{300}{0.2}} = \frac{0.544}{3.126} = 0.1715 \text{ cumecs.}$$

$$\therefore \ \% \text{ increase in discharge} = \frac{0.1715 - 0.1565}{0.1565} \times 100$$

$$= \mathbf{9.6\ \%}$$

Thus the size of the well has very little influence on the discharge. By doubling the well diameter, the discharge has increased by 9.6% only.

Example 4.8. *In order to determine the field permeability of a free aquifer, pumping out test was performed and following observations were made :*

Diameter of well = *20 cm*

Discharge from the well = *240* m^3*/hour*

R.L. of original water surface, before pumping started = *240.5 m*

R.L. of water in the well at constant pumping= *235.6* m

R.L. of the impervious layer = *210 m*

R.L. of water in observation well = *239.8 m*

Radial distance of observation well from the tube well = *50 m.*

Calculate k. Also calculate (i) the error in k if observations are not taken in the observation well, and radius of influence is assumed to be 300 m. (ii) actual radius of influence based on the observations of observation well.

Solution : (Refer Fig. 4.7)

(*a*) From Eq. 4.4 *b*, we have

$$Q = \frac{1.36\,k\,(h_2^2 - h_1^2)}{\log_{10}\dfrac{r_2}{r_1}}$$

Using suffix 1 for the well face, and suffix 2 for the observation well, we have

$$r_1 = 10 \text{ cm} = 0.1 \text{ m}$$
$$r_2 = 50 \text{ m}$$
$$h_1 = 235.6 - 210 = 25.6 \text{ m}$$
$$h_2 = 239.8 - 210 = 29.8 \text{ m}$$

$$\therefore \quad (h_2 + h_1) = 29.8 + 25.6 = 55.4 \text{ m}$$

$$(h_2 - h_1) = 29.8 - 25.6 = 4.2 \text{ m}$$

$$Q = 240 \text{ m}^3/\text{hour}$$

or
$$240 = \frac{1.36 \times k\,(55.4 \times 4.2)}{\log_{10}\dfrac{50}{0.1}}$$

$$k = \frac{240}{1.36 \times 55.4 \times 4.2}\log_{10} 500$$
$$= 2.045 \text{ m/hour} \qquad ...(a)$$
$$= \mathbf{49.15 \text{ m/day.}}$$

(*b*) If *k* is computed on the basis of assumed radius of influence, we have, from Eq. 4.4.

$$Q = \frac{1.36\,k\,(H^2 - h^2)}{\log_{10}\dfrac{R}{r}}$$

Here
$$\frac{R}{r} = \frac{300}{0.1} = 3000$$

$$H = 240.5 - 210 = 30.5 \text{ m}$$
$$h = 235.6 - 210 = 25.6 \text{ m}$$

$$\therefore \quad \left.\begin{aligned} H + h &= 30.5 + 25.6 = 56.1 \text{ m} \\ H - h &= 30.5 - 25.6 = 4.9 \text{ m} \end{aligned}\right] \qquad ...(b)$$

$$240 = \frac{1.36\,k\,(56.1)\,(4.9)}{\log_{10} 3000}$$

$$\therefore \quad k = \frac{240}{1.36 \times 56.1 \times 4.9}\log_{10} 3000$$
$$= 2.23 \text{ m/hour} = 53.57 \text{ m/day}$$

$$\therefore \quad \%\text{ error} = \frac{53.57 - 49.15}{49.15} \times 100 = \mathbf{9\ \%}.$$

(*c*) Calculation of actual radius of influence :

From equation 4.4 :

$$Q = \frac{1.36\,k\,(H^2 - h^2)}{\log_{10} \frac{R}{r}}$$

where $k = 2.045$ m/hour [From (*a*)]

$H + h = 56.1$ m

$H - h = 4.9$ m [From (*b*)]

$Q = 240\ \text{m}^3/\text{hour}.$

$$\therefore \quad 240 = \frac{1.36 \times 2.045\,(56.1)\,(4.9)}{\log_{10} \frac{R}{r}}$$

$$\therefore \quad \log_{10} \frac{R}{r} = \frac{1.36 \times 2.045 \times 56.1 \times 4.9}{240} = 3.185$$

$$\therefore \quad \frac{R}{r} = 1531$$

$$\therefore \quad R = 1531 \times r = 1531 \times 0.1$$

or $\quad R \approx \mathbf{153\ m}$

Example 4.9. *Two tube wells, each of 20 cm diameter are spaced at 100 m distance. Both the wells penetrate fully a confined aquifer of 12 m thickness. Calculate the discharge if only one well is discharging under a depression head of 3 m. What will be the percentage decrease in discharge of the well if both the wells are discharging under the depression head of 3 m ? Take radius of influence for each well equal to 250 metres, and coefficient of permeability of the aquifer as 60 metres/day.*

Solution : (Refer Fig. 4.9)

In the first case, when only one well is discharging, the discharge is given by equation 4.6 *c*

$$Q = \frac{2.72\,bks}{\log_{10} \frac{R}{r}}$$

where $b = 12$ m

$s = 3$ m

$k = 60$ m/day

$$\log_{10} \frac{R}{r} = \log_{10} \frac{250}{0.1} = 3.398$$

$$\therefore \quad Q = \frac{2.72 \times 12 \times 60 \times 3}{3.398}\ \text{m}^3/\text{day}$$

$$= 1728\ \text{m}^3/\text{day} = \mathbf{72.0\ m^3/hour.}$$

When both the wells are discharging, the discharge from each well is given by Eq. 4.14 :

$$Q_1 = Q_2 = \frac{2\pi kb\,(H-h)}{\log_e \frac{R^2}{rB}} = \frac{2.72\,kbs}{\log_{10} \frac{R^2}{rB}}$$

$$= \frac{2.72 \times 60 \times 12 \times 3}{\log_{10}\left(\frac{250 \times 250}{0.1 \times 100}\right)} = \frac{2.72 \times 60 \times 12 \times 3}{3.796}$$

$$= 1545 \text{ m}^3/\text{day} = \mathbf{64.4\ m^3/hour}$$

$\therefore$ % decrease in the discharge

$$= \frac{72.0 - 64.4}{72.0} \times 100 = \mathbf{10.55\%}$$

Example 4.10. *A gravity well has a diameter of 60 cm. The depth of water in the well is 40 metres before pumping is started. When pumping is being done at the rate of 2000 litres per minute, the drawdown in a well 10 metres away is 4 metres and in another well 20 metres away is 2 metres. Determine (a) radius of zero drawdown (b) coefficient of permeability (c) drawdown in the well (d) specific capacity of the well (e) maximum rate at which water can be pumped from the well.*

Solution :

(*a*) For a tube well in unconfined aquifer, we have from Eq. 4.4 *b*

$$Q = \frac{1.36\,k\,(H^2 - h^2)}{\log_{10} \frac{R}{r}}$$

Here, we have $H = 40$; At $r = 10$ m, $h = 40 - 4 = 36$ m

At $r = 20$ m, $h = 40 - 2 = 38$ m.

Applying the above equation at these two locations, we have

$$\frac{40^2 - 36^2}{\log_{10} \frac{R}{10}} = \frac{40^2 - 38^2}{\log_{10} \frac{R}{20}}$$

or $$\log_{10} \frac{R}{20} = \frac{40^2 - 38^2}{40^2 - 36^2} . \log_{10} \frac{R}{10} = 0.5132 \log_{10} \frac{R}{10}$$

or $$\log_{10} \frac{R}{20} = \log_{10} \left(\frac{R}{10}\right)^{0.5132}$$

or $$\frac{R}{20} = \frac{R^{0.5132}}{3.2599}$$

$\therefore$ $$R^{1-0.5132} = \frac{20}{3.2599} = 6.1352$$

$$\therefore \quad R = (6.1352)^{\frac{1}{0.4868}} = (6.1352)^{2.0542}$$

$$\therefore \quad R = \mathbf{41.52\ m}$$

(*b*) $Q = 2000$ litres/min.

$= 2\ m^3/min.$

(since 1 litre/min. $= 0.001\ m^3/min.$)

$R = 41.52$ m ; $H = 40$ m,

$h = 36$ m ; $r = 10$ m

$$\therefore \quad Q = 2 = \frac{1.36\,k\,(40^2 - 36^2)}{\log_{10}\dfrac{41.52}{10}}$$

$$\therefore \quad k = \frac{2}{1.36\,(40^2 - 36^2)} \log_{10}\frac{41.52}{10}$$

$= 0.003$ m/min.

$= \mathbf{4.31\ m/day.}$

(*c*) Depth of water in the well is given by

$$2 = \frac{1.36 \times 0.003\,(40^2 - H_0^2)}{\log_{10}\dfrac{41.52}{0.30}}$$

$$\therefore \quad 1600 - H_0^2 = \frac{2}{1.36 \times 0.003} \times \log_{10}\frac{41.52}{0.30}$$

$= 1049.58$

$$\therefore \quad H_0 = 23.46 \text{ m.}$$

Hence drawdown at the well

$= 40 - 23.46$

$= \mathbf{16.54\ m}$

(*d*) ***Specific capacity*** is defined as the discharge per unit drawdown. Let it be designated by S_c. It is not constant, but decreases as the discharge increases (see §4.6). Let us assume that the yield is directly proportional to the drawdown or to the radius of zero-drawdown.

$$\therefore \quad Q \propto R$$

or $Q = C.R$ where C is a constant.

For a given data,

$$C = \frac{Q}{R} = \frac{2}{41.52} = 0.04817.$$

$$\therefore \qquad R = \frac{Q}{C} = \frac{Q}{0.04817} \text{ in general.}$$

Now corresponding to drawdown of 1 m, discharge $Q = S_c$. Hence radius of zero drawdown

$$= \frac{S_c}{C} = \frac{S_c}{0.04817}$$

Hence $$Q = S_c = \frac{1.36\,k\,(40^2 - 39^2)}{\log_{10}\dfrac{R}{r}}$$

$$= \frac{1.36 \times 0.003\,(40^2 - 39^2)}{\log_{10}\left(\dfrac{S_c}{0.04817 \times 0.3}\right)}$$

or $$S_c \log_{10}(69.2\,S_c) = 0.3223.$$

Solving this by trial and error, we get

$$S_c = 0.258 \text{ m}^3/\text{min}.$$

Hence, specific capacity of the well is **0.258 m³/min/m** depression head.

(*e*) maximum rate of discharge Q_m will be obtained when drawdown in the well is equal to H, *i.e.*, when H_0 = zero.

$$\therefore \qquad Q_m = \frac{1.36 \times 0.003\,(40^2 - 0^2)}{\log_{10}\dfrac{Q_m}{0.04817 \times 0.3}}$$

or $$Q_m \log_{10} 69.2\,Q_m = 6.528$$

Solving this by trial and error, we get

$$\mathbf{Q_m = 2.85\ m^3/min.}$$

4.18. UNSTEADY FLOW

The analysis of flow towards wells discussed in the previous articles is based on the assumption that steady state of flow is developed immediately after pumping is started. Actually, the cone of depression fluctuates with time. The gradual approach of cone of depression towards a steady state is produced primarily by the removal of water from storage as cone deepens. Hence a *storage co-efficient* comes into play. The storage co-efficient is a dimensionless constant of the aquifer and may be interpreted as the amount of water in storage *released from storage* from a column of aquifer of unit cross-section under a unit decline of head.

Equations developed for unsteady or transient well flow normally show how the drawdown s of the piezometric surface or water table is related to the time of pumping the well. In Fig. 4.31,

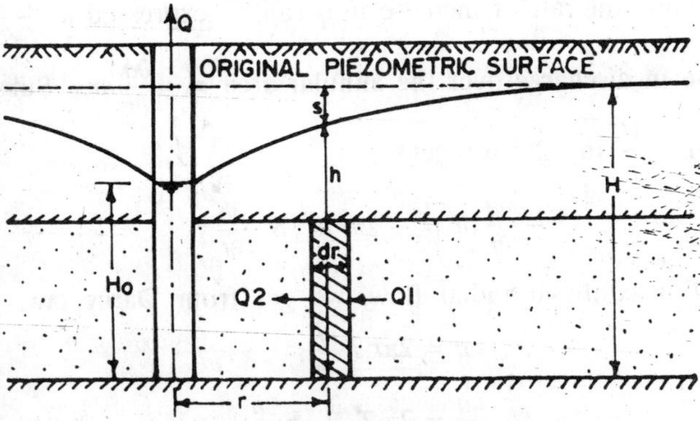

FIG. 4.31. TRANSIENT FLOW

consider an annular cylinder of thickness *dr* and radius *r*. Due to unsteady flow conditions, water will be released from storage of this elementary cylinder. According to definition of storage co-efficient *S*, the rate $\frac{\partial V}{\partial t}$ at which a certain volume *V* is released from storage over an aquifer of area *A* is given by

$$\frac{\partial V}{\partial t} = -\frac{\partial h}{\partial t} . SA \qquad ...(1)$$

where V = volume of water released per horizontal area A of aquifer.

h = height of piezometric surface or water tabl above lower boundary of aquifer.

S = storage co-efficient.

A = area of the aquifer to which $\frac{\partial V}{dt}$ applies.

t = time.

Since h decreases with time t, minus sign with $\partial h/\partial t$ has been used. For the elementary cylinder of the aquifer, $A = 2\pi r\, dr$. Hence

$$\frac{\partial V}{\partial t} = -2\pi r\, dr . S . \frac{\partial h}{\partial t} \qquad ...(2)$$

Let Q_1 = discharge entering the outer face of the cylinder.

Q_2 = discharge leaving the inner face of the cylinder.

$\therefore$ Change in discharge

$$\delta q = Q_1 - Q_2.$$

Since the rate of increase in q can be expressed as $-\dfrac{\partial q}{\partial r}$, the increase in discharge over the annular area $= -\dfrac{\partial q}{\partial r}\,dr$. Substituting this for $\dfrac{\partial V}{\partial t}$ in (2), we get

$$-\frac{\partial q}{\partial r}\,dr = -2\pi r\,dr\,.\,S\,.\frac{\partial h}{\partial t} \qquad ...(3)$$

For confined radial flow, we get from Darcy law,

$$q = 2\pi r\,T\frac{\partial h}{\partial r}$$

$$\therefore \qquad \frac{\partial q}{\partial r} = 2\pi\,T\frac{\partial}{\partial r}\left(r\frac{\partial h}{\partial r}\right)$$

$$= 2\pi\,T\left(\frac{\partial h}{\partial r} + r\frac{\partial^2 h}{\partial r^2}\right) \qquad ...(4)$$

Substituting in (3), we get

$$2\pi\,T\left(\frac{\partial h}{\partial r} + r\frac{\partial^2 h}{\partial r^2}\right) = 2\pi r\,.\,S\frac{\partial h}{\partial t}$$

which simplifies to

$$\frac{\partial^2 h}{\partial r^2} + \frac{1}{r}\frac{\partial h}{\partial r} = \frac{S}{T}\frac{\partial h}{\partial t} \qquad ...(4.24)$$

This is the basic equation for unsteady flow towards the well. Theis (1935) obtained a solution for this equation based on the analogy between ground water flow and heat conduction, by assuming that the well is replaced by mathematical sink of constant strength such that $h = H$ before pumping begins and that $h \to H$ as $r \to \infty$ after pumping begins $(t \geq 0)$. The solution is

$$H - h = \frac{Q}{4\pi\,T}\int_{u = r^2 S/4Tt}^{\infty}\frac{e^{-u}}{u}\,du \qquad ...(4.25\ a)$$

or

$$s = \frac{Q}{4\pi\,T}W(u) \qquad ...(4.25)$$

where

$$W(u) = \int_u^{\infty}\frac{e^{-u}}{u}\,du = \text{well function.}$$

Eq. 4.25 is known as the *non-equilibrium equation* or *Theis equation*. The integral in the above equation is a function of the lower limit, and is known as exponential integral. It can be expressed as a convergent series so that Eq. 4.25 can also be expressed as

$$s = \frac{Q}{4\pi\,T}\left[-0.5772 - \log_e u + u - \frac{u^2}{2\,.\,2\,!} + \frac{u^3}{3\,.\,3\,!} - .....\right] \qquad ...(4.25\ b)$$

TABLE 4.3. VALUES OF WELL FUNCTION W(u)
($u = N \times 10^n$)

$N =$ ↓ / $u =$ →	$N \times 10^{-15}$	$N \times 10^{-14}$	$N \times 10^{-13}$	$N \times 10^{-12}$	$N \times 10^{-11}$	$N \times 10^{-10}$	$N \times 10^{-9}$	$N \times 10^{-8}$
1	33.96	31.66	29.36	27.05	24.75	22.45	20.15	17.84
2	33.27	30.97	28.66	26.36	24.06	21.76	19.45	17.15
3	32.86	30.56	28.26	25.96	23.65	21.35	19.05	16.75
4	32.57	30.27	27.97	25.67	23.36	21.06	18.76	16.46
5	32.35	30.05	27.75	25.44	23.14	20.84	18.54	16.23
6	32.17	29.87	27.50	25.26	22.96	20.66	18.35	16.05
7	32.02	29.71	27.41	25.11	22.81	20.50	18.20	15.90
8	31.88	29.58	27.28	24.97	22.67	20.37	18.07	15.76
9	31.76	29.46	27.16	24.86	22.55	20.25	17.95	15.65

$N =$ ↓ / $u =$ →	$N \times 10^{-7}$	$N \times 10^{-6}$	$N \times 10^{-5}$	$N \times 10^{-4}$	$N \times 10^{-3}$	$N \times 10^{-2}$	$N \times 10^{-1}$	N
1	15.54	13.24	10.94	8.633	6.332	4.038	1.823	0.2194
2	14.85	12.55	10.24	7.940	5.639	3.355	1.223	0.04891
3	14.44	12.14	9.837	7.535	5.235	2.959	0.906	0.01305
4	14.15	11.85	9.550	7.247	4.948	2.681	0.702	0.003779
5	13.93	11.63	9.326	7.024	4.726	2.468	0.560	0.001148
6	13.75	11.45	9.144	6.842	4.545	2.295	0.454	0.000360
7	13.60	11.29	8.990	6.688	4.392	2.151	0.374	0.000116
8	13.46	11.16	8.856	6.555	4.259	2.027	0.311	0.0000377
9	13.34	11.04	8.739	6.437	4.142	1.919	0.260	0.0000125

Wenzel tabulated the values of $W(u)$ for various values of u ranging from 10^{-15} to 9.9. Table 4.3 gives the values of $W(u)$ for u ranging from 10^{-15} to 9.

From the table, when

$u = 5$, $\quad W(u) = 0.001148$

and when $u = 5 \times 10^{-3}$, $\quad W(u) = 4.726$.

The values of *formation constants* S and T can be found by measuring drawdowns in observation wells when the well under study is pumped at a constant rate of discharge Q. There are several methods of determining S and T, but we will discuss here following three methods :

(*a*) Theis method,
(*b*) Jacob's method,
(*c*) Chow's method.

(*a*) THEIS METHOD

Theis proposed a curve-fitting method for finding formation constants S and T from a pump out test. From Eq. 4.25, we observe that

$$s = \left[\frac{Q}{4\pi T}\right] W(u) \text{ or } \log W(u) = \left[\log \frac{4\pi T}{Q}\right] + \log s \quad \text{....(4.26 } a)$$

and

$$\frac{r^2}{t} = \left[\frac{4T}{S}\right] u \text{ or } \log u = \left[\log \frac{S}{4T}\right] + \log \frac{r^2}{t} \quad \text{...(4.26 } b)$$

If a constant withdrawal rate (Q) is maintained, the bracketed portions of the above two equations are constant for a given pumping test. It is to be noted that s is related to r^2/t in a manner that is similar to the relation of $W(u)$ to u. Hence if a plot or *data curve* is made between s and r^2/t on logarithmic co-ordinates tracing paper (Fig. 4.32 *a*) to the same scale as the *type curve* $W(u)$ versus u (Fig. 4.32 *b*), the data curve will be similar to the type curve.

Procedure :

(*i*) In the *observation well* situated at a radial distance r from the main well, observe s and t.

(*ii*) Plot s versus r^2/t on a log-log tracing paper (Fig. 4.25 *a*).

(*iii*) Plot $W(u)$ versus u on log-log graph paper (Fig. 4.32 *b*).

(*iv*) Keep Fig. 4.32 *a* on Fig. 4.32 *b*, and adjust it in such a way that when the co-ordinate axes are held parallel, the data

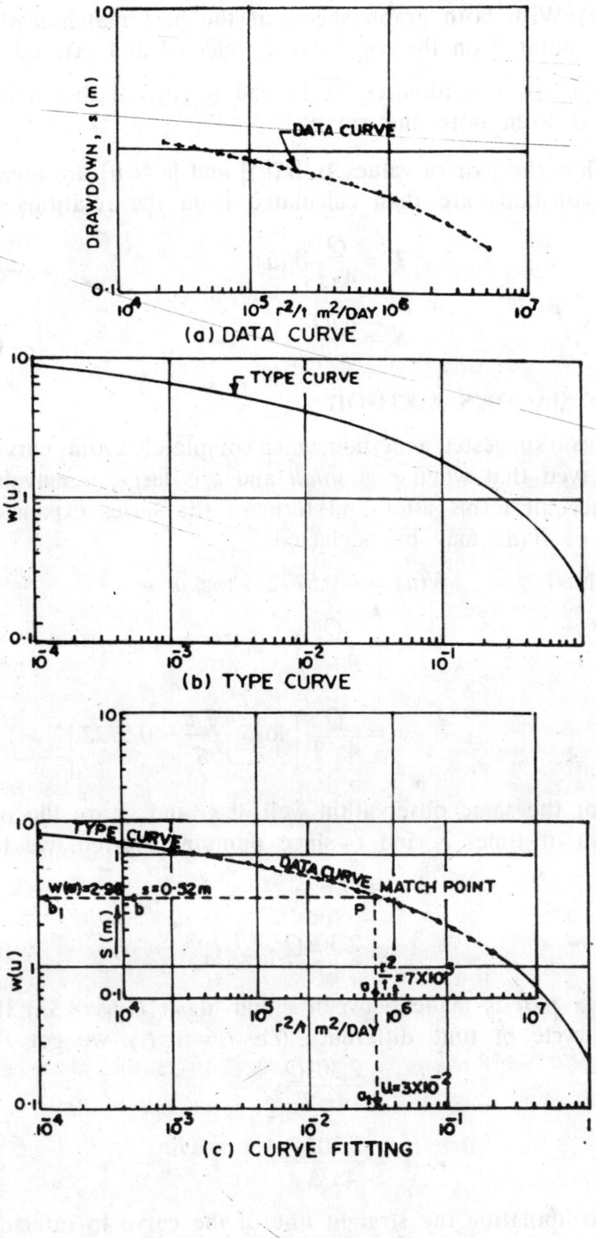

FIG. 4.32. THEIS METHOD.

curve is oriented in a position which represents the best fit of the field data to the type curve.

(*v*) With both graph sheets at the best match position, an arbitrary point P on the top curve is selected and pricked through.

(*vi*) The co-ordinates (a, b) and (a_1, b_1) of the match point are noted from both the curves.

Thus the pair of values $[s, W(u)]$ and $[r^2/t, u]$ are known. The aquifer constants are then calculated from the relations :

$$T = \frac{Q}{4\pi s} W(u) \qquad \text{...(4.27 a)}$$

and

$$S = \frac{4uT}{r^2/t} \qquad \text{...(4.27 b)}$$

(*b*) **JACOB'S METHOD**

Jacob suggested a method which completely avoids curve fitting. He observed that when r is *small* and t is large, u may be small and hence all terms after 2nd term of the series expansion (Eq. 4.25 *b*) of $W(u)$ may be neglected.

Thus $$W(u) = -0.5772 - \log_e u$$

$$\therefore \quad s = \frac{Q}{4\pi T}[-0.5772 - \log_e u]$$

or

$$s = \frac{Q}{4\pi T}\left[\log_e \frac{4Tt}{r^2 S} - 0.5772\right] \qquad \text{...(4.28)}$$

For the same observation well, if s_1 and s_2 are the observed drawdown at times t_1 and t_2 since pumping started, we have

$$\Delta s = s_2 - s_1$$

$$= \frac{2.303Q}{4\pi T} \log_{10} \frac{t_2}{t_1} \qquad \text{...(4.29)}$$

If a plot is made between s and $\log_{10} t$ (Fig. 4.33) then for one log cycle of time difference (*i.e.* $t_2 = 10\,t_1$), we get

$$\Delta s = \frac{2.303Q}{4\pi T}$$

or

$$T = \frac{2.303Q}{4\pi \Delta s} \qquad \text{...(4.30)}$$

Extrapolating the straight line of the curve to intersect with the zero drawdown axis permits the calculation for S. Let t_0 be the time (since pumping started), at which the drawdown in observation well is still zero, we have from equation 4.28 :

$$s = 0 = \frac{Q}{4\pi T}\left[\frac{4Tt_0}{r^2 S} - 0.5772\right]$$

or $$\log_e \frac{4Tt_0}{r^2 S} = 0.5772$$

or $$\frac{4Tt_0}{r^2 S} = e^{0.5772}$$

which gives $$S = \frac{2.25\, Tt_0}{r^2} \qquad \text{...(4.31)}$$

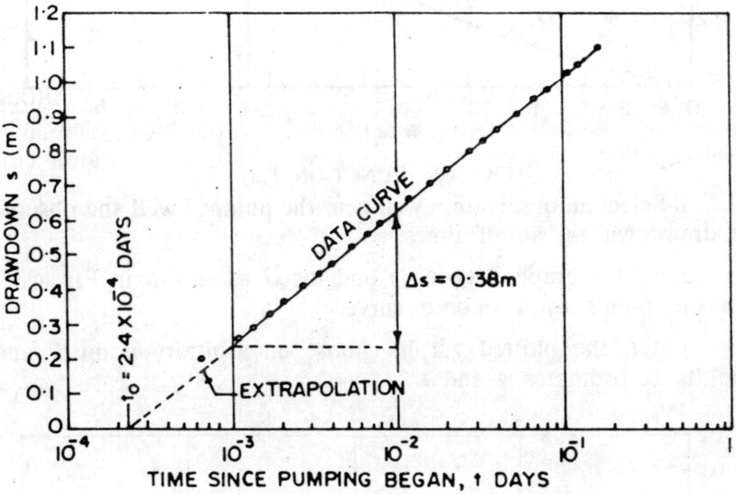

FIG. 4.33. JACOB'S METHOD

Thus storage coefficient S can be calculated from Eq. 4.31. *The straight line approximation for this method should be restricted to values of u less than about 0.01 to avoid large error.*

(c) CHOW'S METHOD

Chow gave a method of solution which avoids curve fitting. He introduced a function $F(u)$ given by the relation :

$$F(u) = \frac{W(u)}{2.303} \qquad \text{...(4.32)}$$

The relation between $F(u)$, $W(u)$ and u is shown in Fig. 4.34 with the u scale on the curve. The function $F(u)$ is introduced to relate $W(u)$ and u to a certain combination of s and t. The procedure is as follows :

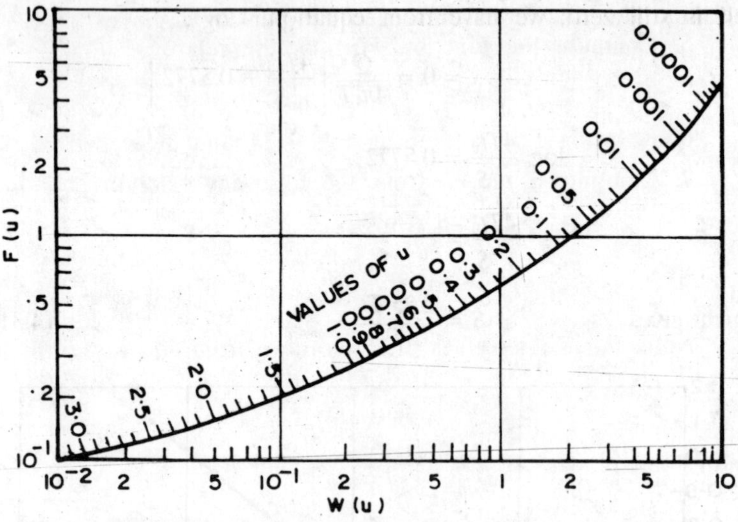

FIG. 4.34. FUNCTION $F(u)$

1. Select an observation well near the pumped well and observe the drawdown (s) at all times.

2. Plot a graph between s and $\log_{10} t$ as shown in Fig. 4.35. Join the points by a smooth curve.

3. On the plotted graph, choose an arbitrary point P and note its co-ordinates s and t.

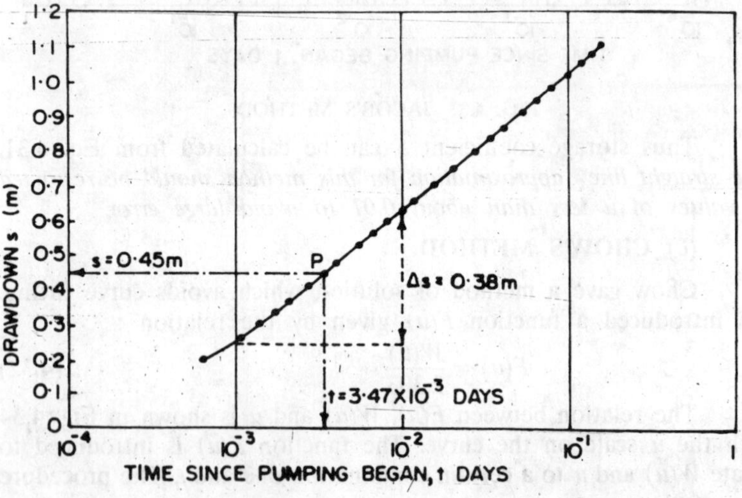

FIG. 4.35. CHOW'S METHOD

4. Draw the tangent to the curve at the chosen point P and determine the drawdown difference Δs per log cycle of time.

5. Compute the function $F(u)$ by the relation

$$F(u) = \frac{s}{\Delta s} \qquad ...(4.33)$$

6. Knowing $F(u)$ and using Fig. 4.34, find u and $W(u)$.

7. Compute S and T from the following relations :

$$T = \frac{Q}{4\pi s} W(u)$$

and

$$S = \frac{4uTt}{r^2}$$

Note. For small values of u, we have from Jacob's method:

$$\Delta s = \frac{2.303Q}{4\pi T} \log_{10} \frac{t_2}{t_1}$$

$$= \frac{2.303Q}{4\pi T} \quad (\text{when } t_2 = 10\, t_1)$$

Also,

$$s = \frac{Q}{4\pi T} W(u)$$

$$\therefore \quad F(u) = \frac{s}{\Delta s} = \frac{W(u)}{2.303} \qquad ...(4.32\ a)$$

For $F(u) < 2$, the above equation can safely be used. For $F(u) > 2, u$ becomes large. Hence Eq. 4.32 should be used.

Example 4.11. Theis Method. *A well fully penetrating a confined aquifer is pumped at a uniform rate of 2500 litres per minute. The drawdowns in an observation well situated at 60 m away are given in Table 4.4. Using Theis method, determine the formation constants of the aquifer (Adapted from data from U.S. Geological Survey).*

TABLE 4.4.

Time since pumping began t (min)	***Drawdown s (m)***	***t (min)***	***s (m)***	***t (min)***	***s (m)***
0	0.0	12	0.59	150	1.03
1.0	0.20	14	0.62	180	1.05
1.5	0.26	18	0.66	210	1.08
2.0	0.30	24	0.71	240	1.10
2.5	0.33	30	0.75		
3	0.36	40	0.80		
4	0.41	50	0.83		
5	0.45	60	0.86		
6	0.48	80	0.91		
8	0.53	100	0.95		
10	0.56	120	0.98		

Solution :

The values of t, s and r^2/t are tabuted in Table 4.5.

TABLE 4.5

COMPUTATIONS FOR DATA CURVE (r = 60 m)

Time t		r^2/t	s	Time t		r^2/t	s
min.	*days*	*(m^2/day)*	*(m)*	*min.*	*days*	*(m^2/day)*	*(m)*
0	0	∞	0.0	18	1.25×10^{-2}	2.88×10^5	0.66
1.0	6.94×10^{-4}	5.18×10^6	0.20	24	1.67×10^{-2}	2.16×10^5	0.71
1.5	1.04×10^{-3}	3.46×10^6	0.26	30	2.08×10^{-2}	1.73×10^5	0.75
2.0	1.39×10^{-3}	2.59×10^6	0.30	40	2.78×10^{-2}	1.30×10^5	0.80
2.5	1.74×10^{-3}	2.07×10^6	0.33	50	3.47×10^{-2}	1.04×10^5	0.83
3.0	2.08×10^{-3}	1.73×10^6	0.36	60	4.17×10^{-2}	8.64×10^4	0.86
4	2.78×10^{-3}	1.30×10^6	0.41	80	5.56×10^{-2}	6.48×10^4	0.91
5	3.47×10^{-3}	1.04×10^6	0.45	100	6.94×10^{-2}	5.8×10^4	0.95
6	4.17×10^{-3}	8.64×10^5	0.48	120	8.33×10^{-2}	4.32×10^4	0.98
8	5.56×10^{-3}	6.48×10^5	0.53	150	1.04×10^{-1}	3.46×10^4	1.03
10	6.94×10^{-3}	5.18×10^5	0.56	180	1.25×10^{-1}	2.88×10^4	1.05
12	8.33×10^{-3}	4.32×10^5	0.59	210	1.46×10^{-1}	2.47×10^4	1.08
14	9.72×10^{-3}	3.70×10^5	0.62	240	1.67×10^{-1}	2.16×10^4	1.10

A plot is made between s and r^2/t as shown in Fig. 4.32 (*a*) on a log-log tracing paper. The type curve [$W(u)$ versus u] is shown in Fig. 4.32 (*b*), prepared from data of Table 4.3.

The data curve of Fig. 4.32 (*a*) is superimposed on the type curve and properly oriented to get the best possible fit. A match point P is pricked (Fig. 4.32 *c*) and its co-ordinates are found as follows :

$$s = 0.52 \text{ m} ; \qquad W(u) = 2.96$$

$$\frac{r^2}{t} = 7 \times 10^5 ; \qquad u = 3 \times 10^{-2}$$

Discharge $Q = 2500$ litres/min. $= 2.5$ m^3/min.

$$= 3600 \text{ m}^3\text{/day}$$

$$\therefore \quad T = \frac{Q}{4\pi s} W(u) = \frac{3600}{4 \times \pi \times 0.52} \times 2.96$$

$$= \mathbf{1631\ m^3/day/m}$$

and
$$S = \frac{4uT}{r^2/t} = \frac{4 \times 3 \times 10^{-2}}{7 \times 10^5} \times 1631 = \mathbf{0.00028}$$

Example 4.12 Jacob's Method. *Solve example 4.11 by Jacob's method.*

Solution:

Fig. 4.33 shows the plot between s and t. Most of the points lie on a straight line. The drawdown Δs for one log cycle of time difference comes out to be 0.38 m. Hence from Eq. 4.30,

$$T = \frac{2.303Q}{4\pi \, \Delta s} = \frac{2.303 \times 3600}{4\,\pi \times 0.38}$$

$$= \mathbf{1736\ m^3/day/m.}$$

Extending the straight line so as to cut the time axis, we get

$$t_0 = 2.4 \times 10^{-4} \text{ days.}$$

$$\therefore \quad S = \frac{2.25Tt_0}{r^2} = \frac{2.25 \times 1736 \times 2.4 \times 10^{-4}}{(60)^2}$$

$$= \mathbf{0.00026}$$

Example 4.13. Chow's Method. *Solve example 4.11 by Chow's method.*

Solution :

The plot between s and t is shown in Fig. 4.35. Choose any point P whose co-ordinates are $s = 0.45$ m and $t = 3.47 \times 10^{-3}$ day.

Since the curve is a straight line, the tangent to the curve at P will coincide with the straight line itself. For one log cycle of time, we get

$$\Delta s = 0.38$$

$$\therefore \quad F(u) = \frac{s}{\Delta s} = \frac{0.45}{0.38} = 1.184$$

From Fig. 4.34, corresponding to $F(u) = 1.184$, we get

$$W(u) = 2.7 \text{ and } u \approx 0.04$$

Alternatively, from Eq. 4.32 (*a*),

$$W(u) = 2.303\,F(u) = 2.303 \times 1.184 = 2.727,$$

and from Table 4.2, $u = 0.035$

Hence
$$T = \frac{Q}{4\pi s} W(u) = \frac{3600}{4\pi \times 0.45} \times 2.727$$
$$= \mathbf{1736\ m^3/day/m}$$

$$S = \frac{4utT}{r^2} = \frac{4 \times 0.035 \times 3.47 \times 10^{-3}}{3600} \times 1736$$
$$= \mathbf{0.000234}.$$

4.19. OTHER SOURCES OF UNDERGROUND WATER

The ground water can be tapped with the help of the following sources :

1. Wells
2. Springs
3. Infiltration galleries
4. Karez.

Out of these, wells have been discussed at length in the previous articles. We shall now discuss in brief the other sources of ground water.

SPRINGS

Ground water which flow out at the surface constitutes a spring. A pervious layer, sandwiched between two impervious layers gives rise to a natural spring. A spring indicates the outcropping of the water table. Springs are of the following types :

1. Stratum springs.
2. Valley springs.
3. Fault springs.
4. Springs from solution channels.
5. Artesian spring (geyser).
6. Deep seated springs.

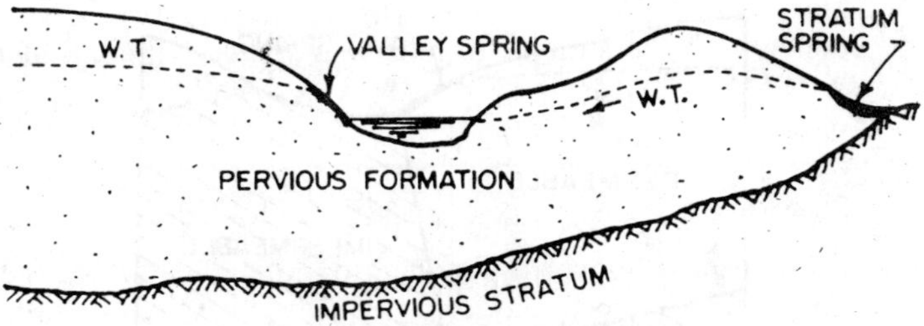

FIG. 4.36. SPRINGS

Stratum spring is formed when the downward passage of ground water in a permeable deposit is hindered by an underlying impervious layer. The ground water emerges as a spring at the contact of the formations, and hence such spring is also known as *contact spring* (Fig. 4.36). Another kind of stratum spring may arise at the junction of two formations where permeable beds dip beneath an impermeable cover. This may be called an *overflow spring*, since the ground water literally overflows at the edge of the impervious stratum (Fig. 4.37).

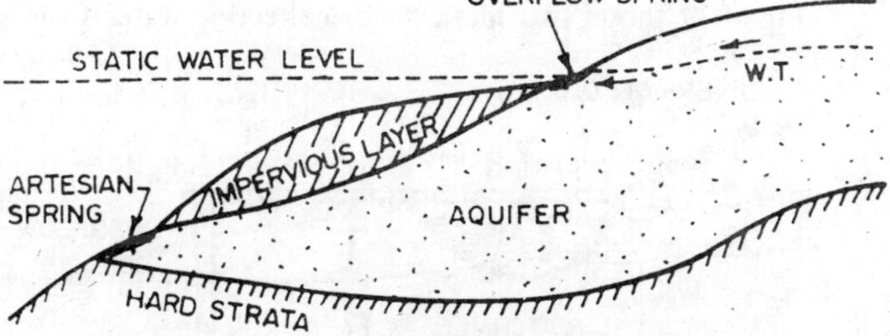

FIG. 4.37. SPRINGS

A *valley spring* is due to the emergence of the water table where the topography falls below its level such as that shown in Fig. 4.36. Water flows out at the surface and feeds the surface stream.

When a permeable formation is brought against an impermeable one by faulting, ground water in the permeable bed may escape along the fault line and appear at the surface as a *fault spring* (Fig. 4.38).

In limestone districts where there is much underground water in circulation, streams flow along bedding planes in channels enlarged by solutions, and may emerge as springs in valleys, to feed surface stream.

Artesian spring arises where water is confined in permeable rock under artesian head beneath a cover of impervious layer (Fig.

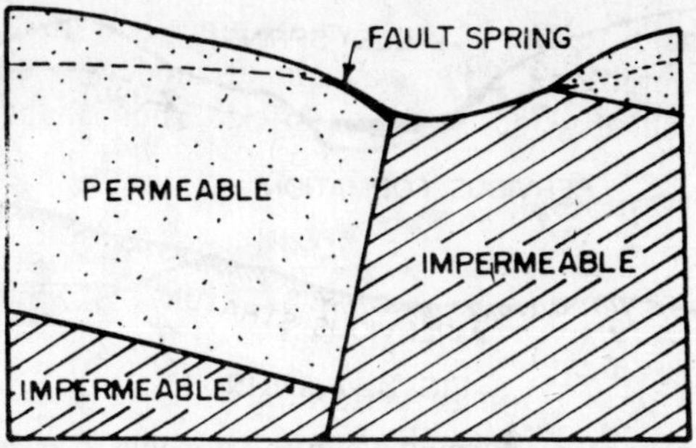

FIG. 4.38 FAULT SPRING

4.37). If this cover is locally broken by fracturing or if the artesian aquifer outcrops, the artesian water below emerges as an artesian spring. Artesian spring, also known as *Geysers,* are of infrequent occurrence.

The *deep-seated springs* are associated with vulcanicity and deep fractures, and yield juvenile water and not the rainwater.

Fig. 4.39 shows two methods of collecting water from springs.

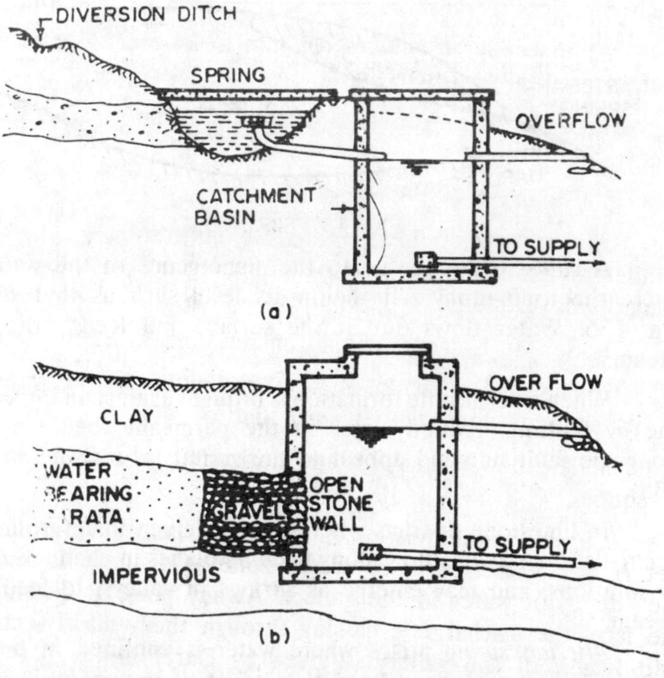

FIG. 4.39. COLLECTION OF WATER FROM SPRINGS

Infiltration Galleries

It is a small tunnel in the form of pipes laid under the ground to tap up underground flow available at moderate depth. These tunnels could be of wood, stone or bricks. Their shape is generally

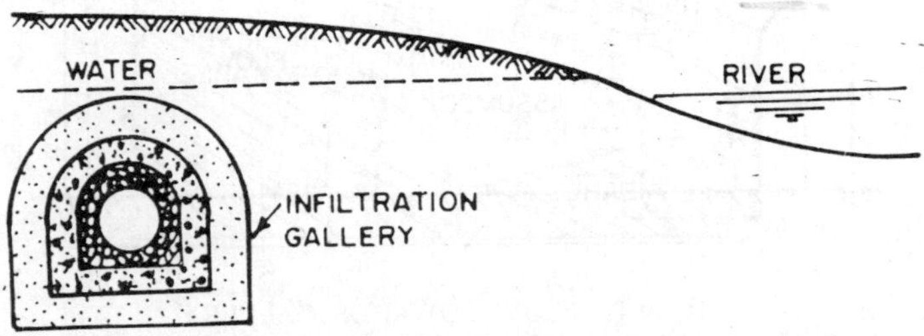

FIG. 4.40. INFILTRATION GALLERY

circular, and they are covered by graded filters all around. The ground water can enter the tunnel through side and bottom. These galleries are often constructed parallel to a river bank to collect the inflow of water to river. Fig. 4.40 shows a typical infiltration gallery surrounded by a three layer graded filter consisting of :

(*i*) 16 cm ballast (25 mm to 50 mm size) around the pipe which has perforations.

(*ii*) 16 cm large pebbles (12 mm to 24 mm) as intermediate layer.

(*iii*) 24 cm fine pebbles (3 mm to 10 mm size) as the outer layer.

Flow into Infiltration Gallery

The analysis of flow towards an infiltration gallery running parallel to a stream etc, can be done on the basis of following assumptions proposed by Dupuit :

1. Soil is isotropic, homogeneous and incompressible.

2. The tangent of the angle of inclination of the water table, *i.e.* its slope is equal to its sine, and

3. The flow is uniform and horizontal throughout the depth of aquifer.

Consider an infiltration gallery at a distance L from the source. Let H be the height of saturated zone at distance L and H_0 be the depth of water in the gallery. At any point P distant x from the face, the discharge q passing through the vertical section, per unit length of the gallery is given by Darcy law :

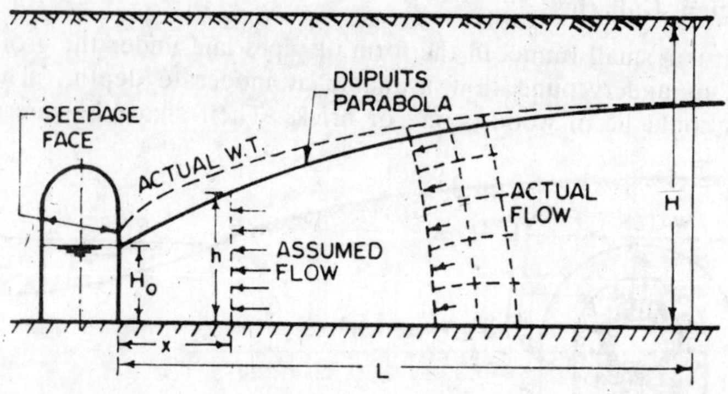

FIG. 4.41. FLOW TOWARDS GALLERY

$$q = k\frac{dh}{dx}(h \times 1) \qquad ...(4.32)$$

$$\therefore \quad q\int_0^L dx = k\int_{H_0}^{H} h\,dh$$

$$\therefore \quad q = k\frac{H^2 - H_0^2}{2L} \qquad ...(4.33)$$

If however, the gallery receives discharge from both the sides, the above discharge will be doubled. Eq. 4.32 traces a parabolic water table which departs significantly from the actual water table shown by dotted line.

Example 4.14. *An infiltration gallery is situated at a distance of 15 m from a diffusion ditch which penetrates fully the pervious formation of depth 6 m. If the depth of the water in the infiltration gallery is 2 m above the bottom of the formation, compute the discharge per metre of the gallery. Take k = 0.5 cm/sec.*

Solution :

$$k = 0.5 \text{ cm/sec} = 0.005 \text{ m/sec.}$$

$$L = 15 \text{ m}$$

$$H = 6 \text{ m}$$

$$H_0 = 2 \text{ m}$$

$$\therefore \quad q = k\frac{H^2 - H_0^2}{2L} = 0.005\,\frac{6^2 - 2^2}{2 \times 15}$$

$$= 0.00533 \text{ m}^3/\text{sec}$$

$$= 460.8 \text{ m}^3/\text{day/m}$$

KAREZ

A karez is an underground tunnel driven into the hill-side to tap water from the underground springs. The tunnel has certain bed inclination. Water from the karez is taken in an open channel.

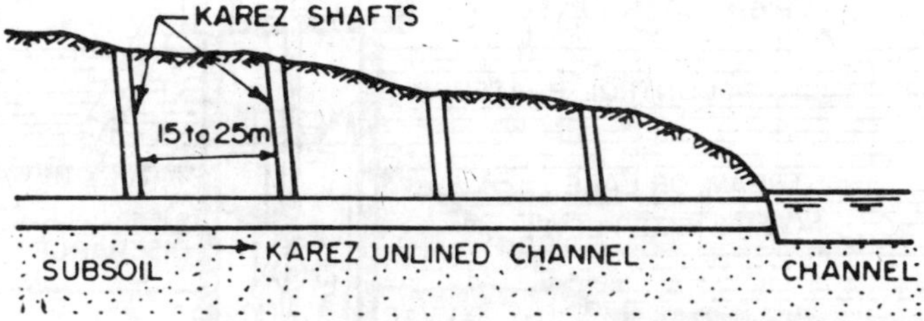

FIG. 4.42. KAREZ

Such tunnels are in use in Baluchistan and North West Frontier Province (West Pakistan).

4.20. RADIAL COLLECTOR WELLS

Radial collector wells or *horizontal wells* are becoming very popular these days. The system consists of a vertical central cylinder of monolithic concrete of about 5 m in diameter which is sunk into the aquifer by excavating the inside earth material. When the desired level is reached, it is sealed at its bottom with a thick concrete plug. The concreted cylinder has precast ports at its bottom end. Through these ports, a number of radial collectors, 15 to 20 cm in diameters are jacked horizontally into the water bearing formations. These form radial pattern of horizontal pipes as shown in Fig. 4.43. Perforated pipes with proper screens are later inserted in these horizontal holes. The length of radial collector may be upto 60 m. Patents for this type of well were obtained by L. Ranney (for Ranney Method Water supplies Inc. Columbus, Ohio) and H. Fehlmann (for Grundwasserbauten A.G. Berne, Switzerland). In the Ranney well, slotted pipe is placed directly in the collector hole. In the Fehlmann well, a blank pipe is installed first, and then the perforated pipe is placed inside and the blank casing pipe is removed. During the development of the well, fines are removed and natural gravel packs are formed around the collector wells. The inner end of each collector well is fitted with a sluice valve which can be operated from the pump house above.

Collector wells extract relatively large supplies of groundwater from valley fills and other alluvial deposits of high permeability and ample rate of recharge. The entry velocity is very low. Hence it prevents the danger of clogging and incrustation due to carbonate

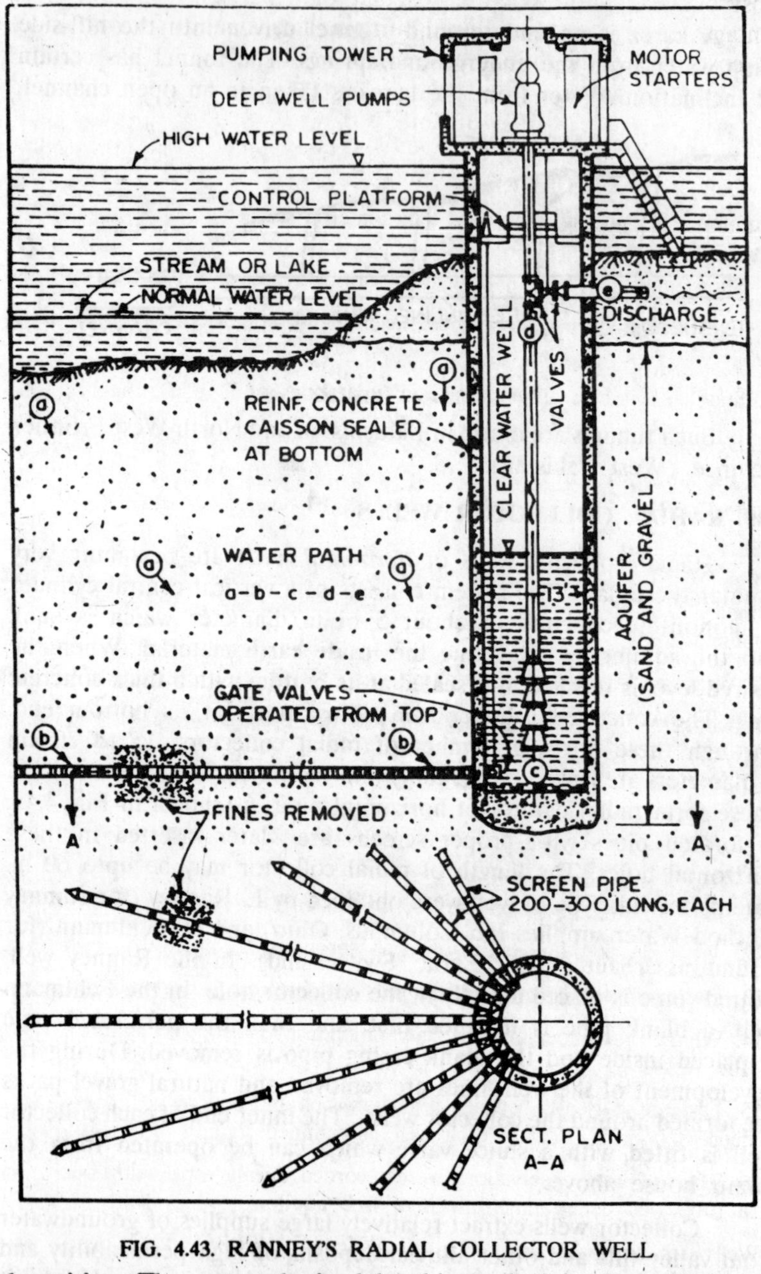

FIG. 4.43. RANNEY'S RADIAL COLLECTOR WELL

deposition. The water obtained by this method is clean, fresh and biologically pure. It normally does not require any treatment. The

initial cost of a collector well exceeds that of a vertical well, but the advantages of large yields, reduced pumping heads and low maintenances costs offset the initial cost.

PROBLEMS

1. Define the following terms :

Aquifer, aquiclude, specific yield ; piezometric surface ; water table; perched uquifer.

2. Show, with the help of sketches, various types of wells.

3. Explain the terms 'storage coefficient' and 'coefficient of transmissibility'.

4. State and discuss assumptions and limitations of Dupuit's theory.

5. Derive an expression for discharge from a well in an unconfined aquifer. The well fully penetrates it.

6. Derive an expression for discharge from a well fully penetrating a confined aquifer.

7. Explain the method determining the coefficient of transmissibility of a confined aquifer by pumping out test. How can this method be extended for unconfined aquifer ?

8. Write notes on the following :

(*i*) well loss (*ii*) specific capacity of well (*iii*) spherical flow in well (*iv*) interference among wells.

9. Describe various types of tube wells.

10. Describe, with the help of sketches, some of the common types of strainers used in tube wells.

11. Explain the Percussion method of drilling a tube well.

12. What do you understand by well shrouding ?

13. Describe in brief various methods of developing a tube well.

14. What do you understand by recuperation test ? Derive the equations used in the test.

15. Distinguish clearly between a shallow well and a deep well. How does a deep well differ from a tube well in confined aquifer ?

16. The following observations were recorded during a pumping out test on a tube well penetrating fully in a free aquifer :

Well diameter : 25 cm

Discharge from the well : 300 m^3/hour

R.L. of original water surface, before pumping started = 122.0 m.

R.L. of water in the well at constant pumping = 171.1 m

R.L. of water in the observation well = 121.3 m.

R.L. of impervious layer = 92.0 m

Radial distance of observation well from the tube well = 50 m

Determine : (*a*) the field permeability coefficient of the free aquifer, and (*b*) radius of zero drawdown. [**Ans.** (*a*) 60.7 m/day (*b*) 157 m]

17. Design a tube well for the following data :

(*i*) Yield required = 0.2 cumec
(*ii*) Thickness of confined aquifer = 40 m
(*iii*) Radius of circle of influence = 300 m
(*iv*) Permeability coefficient = 80 m/day
(*v*) Drawdown = 6 m [**Ans.** 28 cm, or say 30 cm]

18. During a recuperation test, the water in an open well was depressed by pumping by 2 m and it recuperated 1.5 m in 1 hour. Estimate the yield from a well of 2 m diameter under a depression head of 2 m situated in the same area. Derive the expression your use. [**Ans.** 8.7 m^3/hour]

19. A tube well penetrates fully a 8 m thick water bearing stratum (confined) of medium sand having coefficient of permeability of 0.004 m/sec. The well radius is 15 cm and is to be worked under a drawdown of 3 m at the well face. Calculate the discharge from the well. What will be percentage increase in the discharge if the radius of the well is doubled ? Take radius of zero drawdown equal to 400 m in each case.

[**Ans.** (*i*) 275 m^3/hour (*ii*) 9.6%]

20. Design an open well in fine sand to give a discharge of 0.005 cumecs when worked under a depression head of 3 metres. Take the value of the specific yield for fine sand as 0.5 m^3/hour per square metre of area, under unit depression head. [**Ans.** Dia. 3.9 m]

5

Water Demand and Quantity

5.1. INTRODUCTION

Before designing a proper water works project, it is essential to determine the quantity of water that is required daily. This involves the determination of the following items.

1. ***Population determination.*** Determination of population is one of the most important factors in the planning, if the project has to serve the community for a certain design period. Normally, a design period of 20 to 40 years is selected. What will be the population at the end of the design period, is the basic question. This can be achieved by using various methods for population forecast.

2. ***Rate of demand.*** The water consumption in a city may be conveniently divided into the following categories : (*i*) domestic (*ii*) trade (*iii*) agricultural (*iv*) public and (*v*) losses. The total consumption of water depends upon several factors, such as climatic condition, cost of water, living standards of the inhabitants, pressure in the pipelines, type of supply etc. The total quantity of water required divided by the total population gives *per capita* water demand. The accurate measurement of consumption is often very difficult because standards of supply and maintenance vary widely.

5.2. DESIGN PERIOD

Generally, water supply projects are designed for a design period of 20 to 40 years, after their completion. The time lay between the design and completion should not be more than 2 years. In some specific components of the project, the design period may be modified. Different segments of water treatment and distribution systems may be approximately designed for differing periods of time using differing capacity criteria, so that expenditure far ahead of utility is avoided. Table 5.1 gives the design periods far various components of a water supply project.

TABLE 5.1
DESIGN PERIODS FOR PROJECT COMPONENTS

	Component	*Design period (years)*
1	Storage by dams	50
2.	Infiltration works	30
3.	Pump sets	
	(*i*) All prime movers except electric motors	30
	(*ii*) Electric motors and pumps	15
4.	Water treatment units	15
5.	Pipe connections to the several treatment units and other small appurtenances	30
6	Raw water and clear water conveying mains	30
7.	Clear water reservoirs at the head works, balancing tanks and service reservoirs (over head or ground level)	15
8.	Distribution system	30

5.3. POPULATION FORECAST

The data about the present population of a city under question can always be obtained from the records of the municipality or civic body. However, a water supply project is designed to cater the needs of the community upto the end the design period which may extend upto 2 to 4 decades, before the project is abandoned or enlarged by reason of inadequacy. There are several methods for population forecast, but the judgment must be exercised by the engineer as to which method is most applicable for a particular location. The increase in population of city depends upon several factors such as living conditions of the city and its environs, industrial potential, state of development, location with respect to road and rail links, climatic conditions etc. The entire population of a city may not be evenly distributed, due to variations in the land use pattern and available facilities etc. The *population density,* indicating the number of persons per unit area, and the distribution of population should also be studied for efficient design of the distribution system.

Following are some of the important methods of population forecasts or population projections :

1. Arithmetical increase method.
2. Geometrical increase method.
3. Incremental increase method.
4. Decreased rate of growth method.
5. Graphical extension method.
6. Graphical comparison method.
7. Zoning method or master plan method.
8. Ratio and correlation method.
9. Growth composition analysis method.

1. Arithmetical Increase Method

This is the most simple method of population forecast, though it generally gives lower results. In this method, the increase in population from decade to decade is assumed constant. Mathematically, this hypothesis may be expressed as

$$\frac{dP}{dt} = K \qquad ...(5.1\ a)$$

where $\frac{dP}{dt}$ is the rate of change of population and K is a constant. From the census data of past 3 or 4 decades, the increase in population for each decade is found, and from that an *average* increment is found. For each successive future decade, this average increment is added. The future population P_n after n decades is thus given by

$$P_n = P + nI \qquad ...(5.1)$$

where P_n = future population at the end of n decades
P = present population,
I = average increment for a decade.

This method should be used for forecasting population of those large cities, which have reached their saturation population.

2. Geometrical Increase Method or Uniform Percentage Growth Method

In this method, it is assumed that the *percentage increase* in population from decade to decade is constant. From the population data of previous three or four decades, the percentage increase in population is found and its average is found. If I_g is the average percentage increase per decade, or r_g is the increase per decade expressed as ratio, the population P_n after n decades is given by

$$P_n = P\left(1 + \frac{I_g}{100}\right)^n = P(1 + r_g)^n \qquad ...(5.2)$$

Eq. 5.2 can be derived very easily as under :

Let P be the present population and P_1 be the population after one decade.

Then, $$P_1 = P + \frac{I_g}{100}P = P\left(1 + \frac{I_g}{100}\right)^1 \qquad ...(i)$$

Similarly, population P_2 after two decades is

$$P_2 = P_1 + \frac{I_g}{100}P_1 = P_1\left(1+\frac{I_g}{100}\right)^1 = P\left(1+\frac{I_g}{100}\right)^2 \qquad ...(ii)$$

Hence $$P_n = P\left(1 + \frac{I_g}{100}\right)^n$$

While the arithmetical average method is analogous to the 'simple interest method', this method is analogous to the compulation of income by the 'compound interest method'. This method gives *higher results* since the percent increase never remains constant but, instead, decreases when the growth of the city reaches to saturation.

The value of r_g can be found from the expression

$$r_g = \left(\frac{P_n}{P}\right)^{1/n} - 1 \qquad ...(5.2\ a)$$

Alternatively, r_g can be determined by computing the *average* of growth rates of several known decades of the past :

$$r = \frac{\text{increase in population}}{\text{original population}} \text{ for each decade.}$$

Knowing $r_1, r_2 ..... r_n$ for each decade, the average value r_g can be found either by *arithmatic average method* or by *geometric average method* :

(*i*) By arithmatic average method :

$$r_g = \frac{r_1 + r_2 + \ldots\ldots r_n}{n} \qquad ...(5.2\ b)$$

(*b*) By geometric average method

$$r_g = \left(r_1 . r_2 \ldots\ldots r_n\right)^{1/n} \qquad ...(5.2\ c)$$

The field engineers use the arithmatic average method for computing r_g (or I_g) since it gives slightly *higher* (and hence safer) values. However, the Manual on water supply and treatment recommends to use the *geometric mean method.*

3. Incremental Increase Method

This method combines both the arithmetic average method and the geometrical average method. From the census data for the past several decades, the *actual increase* in each decade is first found. Then the *increment* in *increase for each decade* is found. From these, an *average increment of the increases* (known as incremental increase) is found. The population in the next decade is found by adding to the present population the average increase plus the *average incremental increase* per decade. The process is repeated for the second future decade, and so on. Thus the future population at the end of n decades is given by :

$$P_n = P + nI + \frac{n(n+1)}{2} r \qquad ...(5.3)$$

where P = present population
I = average increase per decade
r = average *incremental increase*
n = number of decades.

Eq. 5.3 can be easily derived as under :

Let P be at the present population. The population P_1 after one decade will be

$$P_1 = P + I + 1r \qquad ...(i)$$

Similarly, population P_2 after 2 decades is

$$P_2 = P_1 + I + 2r = P + 2I + 3r = P + 2I + \frac{2(2+1)}{2} r$$

Population P_3 after 3 decades is

$$P_3 = P_2 + I + 3r = P + 3I + 6r = P + 3I + \frac{3(3+1)}{2} r$$

Hence, population P_n after n decades is

$$P_n = P + nI + \frac{n(n+1)}{2} r$$

4. Decreased Rate of Growth Method or Logistic Method

It is found that the rate of increase of population never remains constant, but varies. Fig. 5.1 shows a plot between the population P and the time T for a developing city. The population of a city will grow until it reaches a saturation population which is established by limit of economic opportunity. All populations thus grow according to the *logistic* or *S-curve.* The curve *ABC* (Fig. 5.1 *a*) starts with a low rate of growth, followed by a high rate and then at a progressively lower rate to the saturation population. Thus in Fig. 5.1 (*a*) part *AB* has geometric increase while there is first order increase from *B* to *C*. From *D* to *E*, near point of inflection, there is straight line increase. The curve *abc* is the *first derivative curve* indicating the rate of growth.

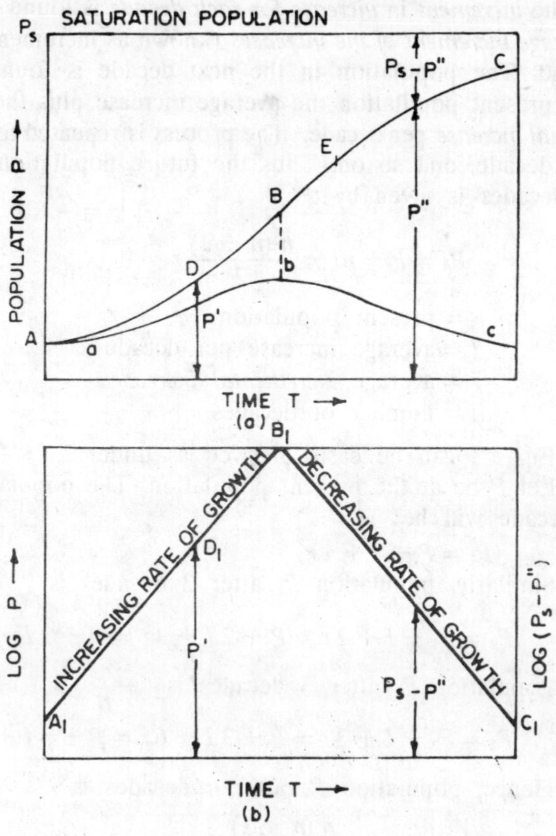

FIG. 5.1. INCREASE IN POPULATION WITH TIME : LOGISTIC CURVE

It is seen that in the part *bc* of the curve, the rate of increase *decreases*. Fig 5.1 (*b* shows the same plot in which the population is plotted on log scale. It is clear that for the part $A_1 B_1$, we have increasing rate of growth while for the part $B_1 C_1$, there is decreasing rate of growth of population. Thus, as the city becomes large, a decreased rate of growth may be expected. This factor should be taken into account while computing future population, as illustrated in Example 5.1.

Logistic curve analysis

The logistic curve used in modelling population trends has S-shape, as shown in Fig. 5.1 (*a*). The Gompertz curve and the logistic curve are both used in establishing long term population trends of large population centres. Both of these curves are S–shaped and have upper and lower asymptotes. According to P.F. Verhulst, the logistic curve can be represented by the equation

$$\log_e \left(\frac{P_s - P}{P}\right) - \log\left(\frac{P_s - P_0}{P_0}\right) = - K P_s \cdot t \qquad ...(5.4)$$

where P_s = saturation population

P_0 = population at starting point A

P = population at any time t from origin A.

K = constants

$$\therefore \quad \log_e \left[\left(\frac{P_s - P}{P}\right) \times \left(\frac{P_0}{P_s - P_0}\right)\right] = - K P_s t$$

or
$$\frac{P_s - P}{P} \times \frac{P_0}{P_s - P_0} = \log_e^{-1} (- K P_s \cdot t)$$

or
$$\frac{P_s}{P} = 1 + \left(\frac{P_s - P_0}{P_0}\right) \log_e^{-1} (- K P_s t)$$

$$\therefore \quad P = \frac{P_s}{1 + \dfrac{P_s - P_0}{P_0} \log_e^{-1} (- K P_s \cdot t)} \qquad ...(5.5)$$

Selecting $\dfrac{P_s - P_0}{P_0} = m$ and $- K \cdot P_s = n$, where m and n are constants, we get

$$\therefore \quad P = \frac{P_s}{1 + m \log_e^{-1} (nt)} \qquad ...(5.6)$$

If three pairs of characteristic values P_0, P_1, and P_2 at time $t = t_0$, $t = t_1$ and $t = t_2 = 2t$, are selected from the useful range of census population data, the values of P_s, m and n can be found from the following simultaneous equations :

$$P_s = \frac{2 P_0 P_1 P_2 - P_1^2 (P_0 + P_2)}{P_0 P_2 - P_1^2} \qquad ...(5.7\ a)$$

$$m = \frac{P_s - P_0}{P_0} \qquad ...(5.7\ b)$$

$$n = \frac{1}{t_1} \log_e \left[\frac{P_0 (P_s - P_1)}{P_1 (P_s - P_0)}\right] \qquad ...(5.7\ c)$$

Eq. 5.6 can also atternatively expressed as :

$$P = \frac{P_s}{1 + e^{a + bt}} \qquad ...(5.8)$$

P_s, a and b may be determined from *three successive census populations* and the Eqs :

$$P_s = \frac{2\,P_0\,P_1\,P_2 - P_1^2\,(P_0 + P_2)}{P_0\,P_2 - P_1^2} \quad \text{...(5.9 a)}$$

$$a = \log_e \frac{P_s - P_0}{P_0} \quad \text{...(5.9 b)}$$

$$b = \frac{1}{n} \log_e \frac{P_0\,(P_s - P_1)}{P_1\,(P_s - P_0)} \quad \text{...(5.9 c)}$$

where n is the time interval between successive censuses. The values of P_s, a and b so obtained may be substituted in Eq. 5.8 to estimate the population for any period t beyond the base year corresponding to P_0. Eq 5.8 in more useful for computation with the help of electronic calculators. See example 5.4 for illustration.

5. Graphical Extension Method

In this method, a curve is drawn between the population P and time T, with the help of census data of previous few decades, so that the shape of the population curve is obtained – upto the present period. The curve is then carefully extended from the present

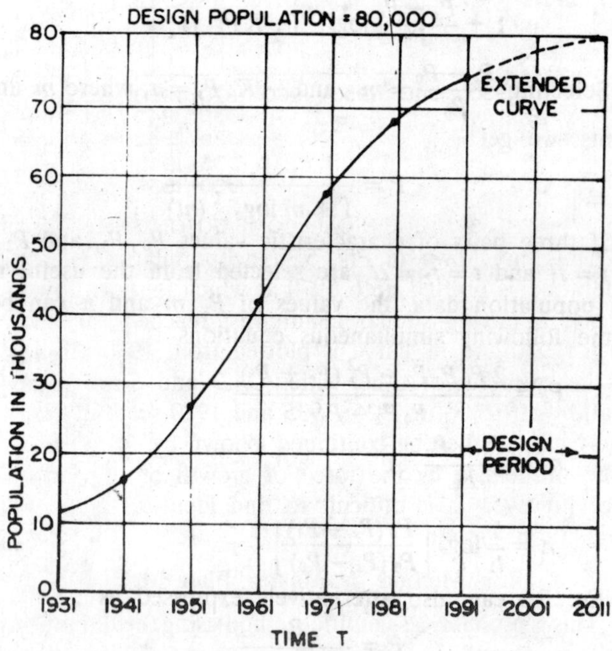

FIG. 5.2. GRAPHICAL EXTENSION METHOD.

to the future decades. From the extended part of the curve, the population at the end of any future decade is approximately determined.

6. Graphical Comparison Method

This method is a variation of the previous method. It assumes that the city under consideration will develop as similar cities developed in the past. The method consist of plotting curves of cities that, one or more decades ago, had reached the present population of the city under consideration.

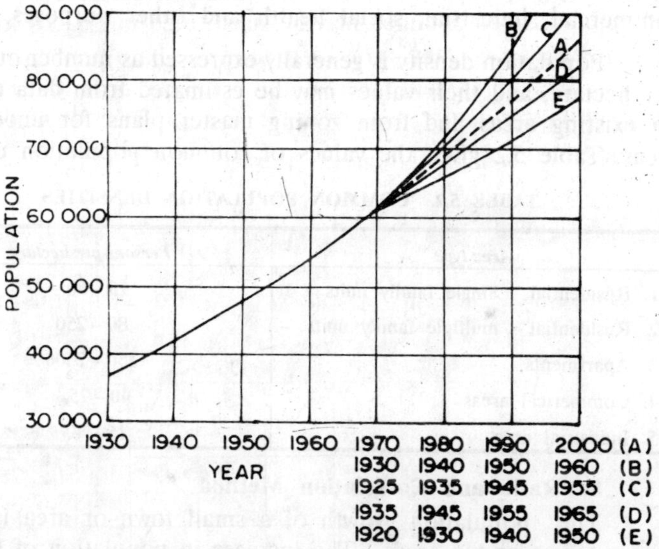

FIG. 5.3. GRAPHICAL COMPARISON METHOD

Thus, as shown in Fig. 5.3, the population of city *A* under consideration is plotted upto 1970 at which its population is 62,000. The city *B* having similar conditions, reached the population of 62000 in 1930 and its curve is plotted from 1930 onwards. Similar curves are plotted for other cities *C*, *D* and *E* which reached the population of 62000 in 1925, 1935 and 1920 respectively. The curve of city *A* can be then be continued (shown by dotted line), allowing it to be influenced by the rate of growth of the larger cities. In practice however, is is difficult to find identical cities with respect to population growth.

7. Zoning Method or Master Plan Method

This is probably a scientific method using the limitations imposed by the town planner in the increase in *density of population* of various parts of the city. For this, a master plan of the city is prepared, dividing it into various zones such as industrial, commercial, residential and other zones. Each zone Is allowed to develop as per master plan only. The future population of each zone, when fully developed can be easily found. For example, sector *A* of a *residential zone* has 1000 plots. Allowing 5 persons per plot, the population of this sector, when fully developed, will be 1000 × 5= 5000 persons. Similarly, the development of each zone can be estimated. This method is more advantageous because of the fact that the total water requirement of the city depends not only for domestic purposes, but also for commercial, industrial, social health and other purposes,

Population density is generally expressed as number of persons per hectare, and their values may be estimated from data collected on existing areas and from zoning master plans for undeveloped areas. Table 5.2 gives the values of common population densities.

TABLE 5.2. COMMON POPULATION DENSITIES

Area type	*Persons per hectare*
1. Residential - Single family units	15 – 80
2. Residential - multiple family units.	80 – 250
3. Apartments	250 – 2500
4. Commerical areas	40 – 75
5. Industrial area	15 – 40

8. Ratio and Correlation Method

The population growth of a small town or area is related to big towns or big areas. The increase in population of big cities bear a direct relationship to the population of the whole state or country. In this method, the local to national (or state) population ratio is determined in the previous two to four decades. Depending upon conditions or other factors, even changing ratio may be adopted. These ratios may be used in predicting the future population. This method takes into account the regional and national factors affecting population growth. This method is useful for only those areas whose population growth in the past is fairly consistent with that of state or nation.

9. Growth Composition Analysis Method

The change in population of a city is due to three reasons: (*i*) birth, (*ii*) death, and (*iii*) migration from villages or other towns. The population forecast may be made by proper analysis of these three factors. The difference between birth rate and death rate gives the *natural increase* in the population. Thus,

$$P_n = P + \text{Natural increase} + \text{Migration.}$$

The estimated natural increase is given by the following expression :

$$\text{Natural increase} = T(I_B P - I_D P) \quad ...(5.10)$$

where T = design (forecast) period.
P = present population.
I_B = average birth rate per year.
I_D = average death rate per year.

5.4. FACTORS AFFECTING POPULATION GROWTH

The population growth of a city depends upon following factors. These factors affect considerably the estimated population.

1. *Economic factors.* Such as development of new industries, discovery of oil or other minerals in the vicinity of the city.

2. *Development programmes.* Development of projects of national importance, such as river valley projects etc.

3. *Social facilities.* Educational, medical, recreational and other social facilities.

4. *Communication links.* Connection of the town with other big cities, and also to the mandies of agricultural products.

5. *Tourism.* Tourist facilities, religious places or historical buildings.

6. *Community life.* Living habits, social customs, and general education in the community.

7. *Unforeseen factors.* Earthquakes, floods, epidemics, frequent famines etc.

5.5. DETERMINATION OF POPULATION FOR INTER-CENSAL AND POST-CENSAL YEARS

Sometimes, it may be required to determine the population for the intermediate portion of a censal period, from the available data. This can be done with the help of arithmetical increase method and the geometrical increase method. Let dP be the increase in population during a time period dT.

If $\frac{dP}{dT} = K_A =$ constant, then the growth will be arithmetic.

If $\frac{dP}{dT} = K_G \cdot P$, where K_G is the proportionality factor, then the growth will be geometric. The values of the factors K_A and K_G can be determined from the following expressions :

$$K_A = \frac{P_L - P_E}{T_L - T_E} \quad ...(5.11)$$

and
$$K_G = \frac{\log_e P_L - \log_e P_E}{T_L - T_E} \quad ...(5.12)$$

where P_L = population at the *last* census at data T_L

P_E = population at the *earlier* census at date T_E.

Now if the population P_M is the desired mid-year population at a date T_M, its value is given by the following expressions.

Arithmetical Increase Method

For inter censal period :

$$P_M = P_E + K_A \,.\, (T_M - T_E) \quad ...(5.13\ a)$$

or
$$P_M = P_E + \frac{T_M - T_E}{T_L - T_E}(P_L - P_E) \quad ...(5.13)$$

For post-censal period,

$$P_M = P_L + K_A\,(T_M - T_L)$$

or
$$P_M = P_L + \frac{T_M - T_L}{T_L - T_E}(P_L - P_E) \quad ...(5.14)$$

Geometrical Increase Method

For inter-censal period,

$$\log_e P_M = \log_e P_E + K_G\,(T_M - T_E) \quad ...(5.15\ a)$$

or
$$\log_{10} P_M = \log_{10} P_E + \frac{T_M - T_E}{T_L - T_E}(\log_{10} P_L - \log_{10} P_E) \quad ...(5.15)$$

For post-censal period :

$$\log_e P_M = \log_e P_L + K_G\,(T_M - T_L) \quad ...(5.16\ a)$$

or
$$\log_{10} P_M = \log_{10} P_L + \frac{T_M - T_L}{T_L - T_E}(\log_{10} P_L - \log_{10} P_E) \quad ...(5.16)$$

Example 5.1. *The following is the population data of a city, available from past census records. Determine the population of the city in 2011 by (a) arithmetical increase method (b) geometrical increase method (c) incremental increase, method (d) graphical method (e) decreased rate of growth method.*

Year	1931	1941	1951	1961	1971	1981	1991
Population (P)	12000	16500	26800	41500	57500	68000	74100

Solution :

The computations about increment, % increment and incremental increase per decade are arranged in Table 5.3 below :

TABLE 5.3.

Year 1	*Population* 2	*Increment per decade* 3	*% increment per decade* 4	*Incremental increase* 5	*Decrease in % increment* 6
1931	12000				
		4500	37.50		
1941	16500			+ 5800	
		10300	62.42		
1951	26800			+ 4400	7.57
		14700	54.85		
1961	41500			+ 1300	16.30
		16000	38.55		
1971	57500			− 5500	20.29
		10,500	18.26		
1981	68000			− 4400	9.29
		6100	8.97		
1991	74100				
	Total	62,100	220.55	+ 1600	53.45
	Average	$\frac{62,100}{6}$ = 10350	$\frac{220.55}{6}$ = 36.76	$\frac{1600}{5}$ = 320	$\frac{53.45}{4}$ = 13.36

In the above table, percentage increase for the first decade (1931 to 1941)

$$= \frac{16500 - 12000}{12000} \times 100 = \frac{4500}{12000} \times 100 = 37.5\%.$$

Similarly, % increment for other decades have been calculated.

1. *Arithmetical Increase Method*

$$P_n = P + nI \quad \text{(Eq. 5.1)}$$

where P = population in 1991 = 74,100

n = number of decades $= \frac{1991 - 1971}{10} = 2$

I = average increase per decade = 10350 (from Table 5.3)

$$\therefore \quad P_n = 74100 + 2 \times 10350$$
$$= \mathbf{94800.}$$

2. *Geometrical Increase Method*

$$P_n = P\left(1 + \frac{I_g}{100}\right)^n \qquad \text{...(5.2)}$$

Here, I_g = average per cent increase per decade = 36.76% (from Table 5.3)

$$\therefore \quad P_n = 74100\left(1+\frac{36.76}{100}\right)^2$$

$$= \mathbf{1{,}38{,}590}.$$

The above computations are based on the value of I_g computed by arithmatic average method. If, however, geometric average method is used, as recommended by the Manual, we have

$$I_g = \left(I_{g1}\,.\,I_{g2}\,.\,.\,.\,.\,.\,I_{gn}\right)^{1/n}$$

$$= (37.50\times62.42\times54.85\times38.55\times18.26\times8.97)^{1/6}$$

$$= 30.54 \text{ (against a value of 36.76)}$$

$$\therefore \quad P_n = 74100\left(1+\frac{30.54}{100}\right)^2 = \mathbf{126272}$$

3. *Incremental Increase Method*

$$P_n = P + nI + \frac{n\,(n+1)}{2}r \qquad ...(5.3)$$

where, $I = 10{,}350$

and r = average incremental increase
= 320 (from Table 5.3)

$$\therefore \quad P_n = 74100 + 2\times10350 + \frac{2\,(2+1)}{2}\times320$$

$$= \mathbf{95760}.$$

4. *Graphical Extension Method*

Fig. 5.2 shows the plot between the population and the time. The dotted portion of the curve is the extended part from 1991 to 2011, following closely its trend. From the extended part, the population at the end of 2011 = 80,000.

5. *Decreased Rate of Growth Method*

Column 6 of Table 5.3 gives the decrease in the per cent increment found in column 4. In the initial portion of the census records, there is no decrease in the percent increment, and hence this period has not been included in the computations. The total decrease in percent increment for four decades comes out to be 53.45, giving an *average rate of decrease in the percentage growth*

$$= \frac{53.45}{4} = 13.36\%$$

In column 4, the average increment rate per decade was found to be 36.76%, but due to decrease in the rate of growth, this figure will be modified as under :

Year	*Average increment per decade*	*Average rate of decrease in the increment*	*Net increment rate (%)*
2001	*36.76*	*13.36*	*23.40*
2011	*23.40*	*13.36*	*10.04*

Hence the population at the end of each decade will be as under :

2001 74100 + 23.40% of 74100 = 91439

2011 91439 + 10.04% of 91439 = **100619**

Example 5.2. *In a town, it has been decided to provide 200 litres per head per day in the 21st century.*

Estimate the domestic water requirements of this town in the year AD 2000 by projecting the population of the town by the incremental increase method, from the data given below :

Year	*Population*
1940	2,50,000
1950	4,80,500
1960	5,50,300
1970	6,38,600
1980	6,95,200

Solution : The computations about increase in population per decade and incremental increase are done in Table 5.4 :

TABLE 5.4

Year	*Population*	*Increment per decade*	*Incremental Increase*
1940	250000		
		230500	
1950	480500		(–) 160700
		69800	
1960	550300		(+) 18500
		88300	
1970	638600		(–) 31700
		56600	
1980	695200		
	Total	445200	(–) 173900
	Average	$\frac{445200}{4} = 111300$	$(-)\frac{173900}{3} = (-)\ 57967$

Here, $I = 111300$ and $r = -57967$

Expected population in the year 2000 *AD* (*i.e.* after 2 decades) is

$$P_n = P + n\,I + \frac{n\,(n+1)}{2}\,r, \text{ where } n = 2$$

$$\therefore \quad P_{2000} = 695200 + 2\,(111300) + \frac{2\,(2+1)}{2}\,[\,-57967\,]$$

$$= 695200 + 222600 - 173901 = 743899$$

Hence water requirement in 2000 *AD* @ 200 litres/head/day

$$= 743899 \times 200 \approx 148.8 \times 10^6 \text{ litres/day}$$

$$= \mathbf{148.8} \text{ million litres/day}$$

Example 5.3. *The population of city in successive decennial census is given as 41500 and 57500. Assuming the census date as 10th April, find the mid year population as 10th July for (a) 3rd inter-censal year, and (b) 6th post-censal year by the arithmetical increase method and the geometrical increase method.*

Solution :

(a) For 3rd inter-censal year

$$T_M - T_E = 3 + (10^{th} \text{ July} - 10^{th} \text{ April})$$

$$= 3.25$$

$$T_L - T_E = 10 \text{ years}$$

$$\therefore \quad \frac{T_M - T_E}{T_L - T_E} = \frac{3.25}{10} = 0.325$$

Arithmetical Increase	*Geometrical Increase*
$P_L = 57500$	$\log_{10} P_L = 4.7597$
$P_E = 41500$	$\log_{10} P_E = 4.6181$
$P_L - P_E = 16000$	$\log_{10} P_L - \log_{10} P_E = 0.1416$
$0.325\,(P_L - P_E) = 5200$	$0.325\,(\log_{10} P_L - \log_{10} P_E) = 0.0460$
$\therefore P_M = 41500 + 5200$	$\log_{10} P_M = 4.6181 + 0.0460$
$= \mathbf{46700.}$	$= 4.6641$
	$\therefore P_M = \mathbf{46140.}$

(b) For 6th post-censal year

$$T_M - T_L = 6.25\,; \qquad T_L - T_E = 10 \text{ years}$$

$$\therefore \quad \frac{T_M - T_L}{T_L - T_E} = \frac{6.25}{10} = 0.625$$

Arithmetical Increase	*Geometrical Increase*
$P_L = 57500$	$\log_{10} P_L = 4.7597$
$P_E = 41500$	$\log_{10} P_E = 4.6181$
$P_L - P_E = 16000$	$\log_{10} P_L - \log_{10} P_E = 0.1416$
$0.625\,(P_L - P_E)$	$0.625\,(\log_{10} P_L - \log_{10} P_E)$
$= 10000$	$= 0.0885$
$\therefore \quad P_M = 57500 + 10000$	$\log_{10} P_M = 4.7597 + 0.0885$
$= \mathbf{67500}$	$= 4.8482$
	$\therefore \quad P_M = \mathbf{70500}$

Note. The geometric estimates are higher for post-censal years and lower for inter-censal years.

Example 5.4. *A city has following recorded population :*

1951	*50000*
1971	*110000*
1991	*160000*

Estimate : (a) the saturation population, and (b) expected population in 2011.

Solution :

Here $n = 20$ years

$$P_0 = 50{,}000\,;\ P_1 = 1{,}10{,}000\,;\ P_2 = 1{,}60{,}000$$

Hence from Eqs. 5.9 (*a*), (*b*) and (*c*)

$$P_s = \frac{2\times 50000\times 110000\times 160000 - (110000)^2(50000+160000)}{50000\times 160000 - (110000)^2}$$

$$\approx \mathbf{190488}$$

$$a = \log_e \frac{190488 - 50000}{50000} = 1.033$$

$$b = \frac{1}{20}\log_e \frac{50000\,(190488 - 110000)}{110000\,(190488 - 50000)} = -\,0.0673$$

Hence the equation of the logistic curve is

$$P = \frac{190488}{1 + e^{1.033 - 0.0673\,t}}$$

In 2011, $t = 2011 - 1951 = 60$ years.

$$\therefore \quad P_{2001} = \frac{190488}{1 + e^{1.033 - 0.0673 \times 60}}$$

$$= \frac{190488}{1 + 0.04954} \approx \mathbf{181500}$$

5.6. WATER DEMAND

An average person may consume no more than 5 to 8 litres a day in liquid and solid foods, including 3 to 6 litres in the form of water, milk and other beverages. However, the per capita consumption of water drawn from public supply is quite large. Total water requirements may be divided into the following five categories:

1. Residential or domestic use.
2. Institutional use.
3. Public or civic use.
4. Industrial use.
5. Water system losses.

1. Residential or domestic use

The residential or domestic use includes water requirements for drinking, cooking, bathing, washing of clothes, utensils and house, and flushing of water closets. Provision is sometimes made for domestic animals. IS : 1172–1957 recommends a per capita water consumption of 135 litres per day. Table 5.5 gives the break up of water requirements for domestic purposes, which forms about 50% of the total water requirements per head per day, for all the five categories mentioned above. Table 5.6 gives the water requirements for domestic animals. It should be noted that water required for lawn sprinkling and for residential gardens is over and above the values given in Table 5.5.

TABLE 5.5.
WATER REQUIREMENTS FOR DOMESTIC PURPOSES

S.No.	*Description*	*Amount of water in litres per head per day*
1	Bathing	55
2	Washing of clothes	20
3	Flushing of W.C.	30
4	Washing the house	10
5	Washing of utensils	10
6	Cooking	5
7	Drinking	5
	Total	**135 litres**

TABLE 5.6.
CONSUMPTION OF WATER FOR DOMESTIC ANIMALS AND LIVE-STOCKS

S.No.	Animals	Water consumption in litres per animal per day
1	Cow and buffalo	40 to 60
2	Horse	40 to 50
3	Dog	8 to 12
4	Sheep or goat	5 to 10

The Manual on water supply and treatment, prepared by the Ministry of Urban Development (MUD) New Delhi recommends the following rates in litres per capita per day (lcpd) for domestic and non-domestic needs (Table 5.7) :

TABLE 5.7.
WATER FOR DOMESTIC AND NON-DOMESTIC NEEDS

Description	Amount of water (lcpd)
1. For communities with population upto 20,000	
(a) Water supply through stand post	40 (min.)
(b) Water supply through house service connection	70 to 100
2. For communities with population 20,000 to 100,000	100 to 150
3. For communities with population above 100,000	150 to 200

2. Institutional Use

The manual on water supply and Treatment recommends the values of water requirements for *institutional needs* as given in Table 5.7.

3. Public or Civic use

Water required for public or civic uses may be for the following purposes : (*i*) Road washing, (*ii*) Sanitation, (*iii*) Public parks, and (*iv*) Fire fighting. For road washing in the municipality area, a provision of 5 litres per head per day is made. Similarly, for sanitary purposes, such as cleaning public sanitary blocks, flushing sewer systems etc., a provision of 3 to 5 litres per head per day may be made. Water required for maintaining public parks etc. may be 2 to 3 litres per square metre per day.

TABLE 5.8
WATER FOR INSTITUTIONAL NEEDS

	Institution	*Water requirement (litres per heal per day)*
1.	Hospitals (including laundry)	
	(a) No. of beds exceeding 100	450 (per bed)
	(b) No. of beds not exceeding 100	340 (per bed)
2.	Hotels	180 (per bed)
3.	Hostels	135
4.	Nurse's homes and medical quarters	135
5.	boarding schools/colleges	135
6.	Restaurants	70 (per seat)
7.	Air ports and sea ports	70
8.	Junction stations and intermediate stations where mail and express stoppage (both railways and bus stations) is provided)	70
9.	Terminal stations	45
10.	Intermediate stations (excluding mail and express stops)	45 (could be reduced to 25 where bathing facilites are not provides)
11.	Day schools /colleges	45
12.	Offices	45
13.	Factories	45 (could be reduced to 30 where no bathing rooms are required to be provided)
14.	Cinema, concert halls and theatres	15

Fire demand

Water required for fire fighting is usually known as *fire demand.* It is treated as a function of population and may be computed from the following formulae :

1 ***Kuichling's formula***

$$Q = 3182 \sqrt{P} \qquad \text{...(5.17)}$$

where Q = quantity of water in litres per minute
P = Population in thousands

2. ***Buston's formula***

$$Q = 5663 \sqrt{P} \qquad \text{...(5.18)}$$

3. ***Freeman's formula***

$$Q = 1136\left(\frac{P}{5} + 10\right) \qquad ...(5.19)$$

and $F = 2.8\sqrt{P}$ ...(5.19 *a*)

where F = number of simultaneous fire streams

4. ***National Board of Fire Underwriters formula***

$$Q = 4637\sqrt{P}\,(1 - 0.01\sqrt{P}) \qquad ...(5.20)$$

Though the total demand of water for extinguishing fire is usually very small, the rate of consumption is very high. It depends upon bulk, congestion and fire resistance of buildings. The minimum limit of fire demand is the amount and rate of supply that would extinguish the largest probable fire in the city. Fire hydrants of 15 to 20 cm diameter are normally provided on all street corners, and at suitable intermediate points. These are generally connected to water supply mains. When fire occurs, pumps installed on fire brigade trucks are rushed to the site and connected to fire hydrants from where they throw jet of water under very high pressure. The pressure varies between 1 to 2 kg/cm^2 (0.1 to 0.2 N/mm^2). For a fire of moderate nature, three streams each of 1100 litres per minute are required. For a city of one lac population, the fire flow, required by National Board of Fire Underwriters (now known as 'American Insurance Association) comes out to be 40,500 litres/minute. Assuming an average total consumption of 150 litres per capita per day, total water requirement comes out to be 150 lac litres per day or 10,417 litres/minute. Thus the rate of flow required for fire demand, (*i.e.* 40,500 litres/minute) is very much higher than the total flow rate required for other purposes. However, the provision for fire stream is made only for 3 to 5 hours fire flow. The total quantity of water calculated on yearly basis is usually very small because fire breaks out only few times in a year. Thus, for the city of population of 1 lac, if fire breaks out 12 times a year, each of a duration of 4 hours, the per capita water demand

$$= \frac{\text{Yearly requirement of water}}{365 \times \text{Population}}$$

$$= \frac{(10417 \times 60 \times 4)\,12}{365 \times 100000} = 0.82 \text{ litres.}$$

Thus for Indian conditions, a provision of 1 litre per head per day will be sufficient for fire fighting. Each fire hydrant provided for this purpose has three streams, and each stream discharges 1100 litres per head per day.

The above formulae for water demand do not take into account the frequency of fire that may occur. It may be determined from the following expression :

$$Q = \frac{4360\,T^{0.275}}{(t+12)^{0.757}} \text{ litres/minute} \qquad ...(5.21)$$

where t = duration of fire in minutes

T = period of occurrence of fire, in years.

The recommended minimum values for the above formula are:

$t = 30$ minutes and $T = 1$ year

The manual on water supply and Treatment by MUD recommends that a provision in kilo litres per day based on Eq. 5.22, where P is the population in thousands may be adopted for communities larger than 50,000 :

$$Q = 100\sqrt{P} \qquad ...(5.22)$$

It is desirable that one third of the fire fighting requirements form the part of service storage. Thus, for a population of 100,000, Q will be $= 100\sqrt{100} = 100$ kilo litres per day.

4. Industrial use

The presence of industries in or near the city has great impact on the water demand. The quantity of water required depends upon the type of industry. For a city with moderate factories, a provision of 20 to 25 percent of per capita consumption may be made for this purpose. The fore cast for this demand will be based on nature and magnitude of each industry and the potential for its expansion. Table 5.9 gives data about the needs of some industries.

TABLE 5.9. INDUSTRIAL NEED.

Industry	*Unit of production*	*Water requirement in kilolitres per unit*
1. Automobile	Vehicle	40
2. Distillary	Kilolitre (proof alcohol)	122–170
3. Fertilizer	Tonne	80–200
4. Leather	100 kg (tonne)	4
5. Paper	Toone	200–400
6. Special quality paper	Toone	400–1000
7. Straw board	Tonne	70–100
8. Petroleum refinery	Tonne (crude)	1.5–2.0
9. Steel	Tonne	200–250
10. Sugar	Tonne (cane crushed)	1–2
11. Textile	100 kg (goods)	8–14

5. Water System Losses

Losses from a water distribution system consists of (*i*) leakage and over-flow from service reservoirs, (*ii*) leakage from main and service pipe connections, (*iii*) leakage and losses on consumer's premises when they get un-metered house-hold supplies (*iv*) under-registration of supply meters, and (*v*) large leakage or wastage from public taps. Losses in the supply lines are mainly due to defective pipe joints, cracked pipes, and loose valves and fittings. In the case of a well maintained, and fully metered water distribution system, the losses may hardly exceed 20% of the total consumption. In a system where the supply is partly metered (for domestic connections) and partly un-metered (for municipal taps), the losses may be upto 50% of the total supply.

Example 5.5. *Compute the 'fire demand' for a city having population of 1,40,000 using various formulae.*

Solution. P = population in thousands = 140.

1. *Kuichling's formula*

$$Q = 3182\sqrt{P} = 3182\sqrt{140} = \mathbf{37650\ litres/min.}$$
$$= \mathbf{0.627\ cumecs.}$$

2. *Buston's formula*

$$Q = 5663\sqrt{P} = 5663\sqrt{140} = \mathbf{67000\ litres/min.}$$
$$= \mathbf{1.117\ cumecs.}$$

3. *Freeman's formula*

$$Q = 1136\left(\frac{P}{5} + 10\right) = 1136\left(\frac{140}{5} + 10\right)$$

$$= \mathbf{43168}\ \text{litres/min} = \mathbf{0.719}\ \text{cumecs.}$$

No. of fire streams,

$$F = 2.8\sqrt{P} = 2.8\sqrt{140} \approx \mathbf{33.}$$

4. *National Board of Fire Underwriter's formula*

$$Q = 4637\sqrt{140}\,(1 - 0.01\sqrt{140})$$
$$= \mathbf{48{,}374\ litres/min. = 0.806\ cumecs.}$$

5.7. FACTORS AFFECTING RATE OF DEMAND

The average per capita water consumption varies widely in Indian cities, the demand figures varying from 80 litres to 160 litres. Such variation depends upon a number of factors summarised below:

1 *Size and type of community*. The fluctuations in demand depends upon the size of the city. In a large city, the fluctuations

in demand may be narrow. In a small city, per capita water consumption is to be expected to be small because there are only limited uses for water in a small town. The fluctuation in use in an industrial community is normally much smaller than in a residential community.

2. ***Standard of living.*** The higher the standard of living is, the higher the demand and greater the variation in demand.

3. ***Climatic conditions.*** If the community is located in hot climate, water use will be increased by bathing, lawn sprinkling and use in parks and recreation fields. Similarly, in extreme cold climates, water may be wasted at the faucets to prevent freezing of pipes, resulting in increased consumption. Extremes of heat and cold cause variations in demand.

4. ***Quality of water.*** Poor-quality water results in a reduction in use.

5. ***Pressure in the supply.*** High pressure results in increased use while low pressure results in decreased use.

6. ***System of supply.*** Water supply may either be continuous, (*i.e.* for all the 24 hours) or intermittent. Generally, intermittent supply will reduce rate of demand.

7. ***Sewerage.*** Use of water increases when sewers are installed in an area previously without them.

8. ***Metering.*** Use of water decreases when the supplies are metered.

9. ***Water rates.*** An increase in water rates will reduce consumption in metered areas.

10. ***Age of community.*** Older, more stable communities use less water than rapidly developing communities where new homes are being constructed and owners are planting new lawns.

11. ***Lawn sprinkling.*** Enforcement of lawn sprinkling regulations can reduce peak demands significantly.

5.8. VARIATIONS IN RATE OF DEMAND

The average daily per capita consumption is obtained by dividing the quantity of water supplied during the year by the number of days in the year and the number of persons served. This per capita consumption or demand varies not only from year to year and from season to season, but more important from day to day and hour to hour. These variations are expressed as percentage of the annual average daily consumption. Some common values are as under :

1. ***Maximum seasonal consumption.*** 130% of annual average daily rate of demand.

2. ***Maximum monthly consumption.*** 140% of annual average daily rate of demand.

3. ***Maximum daily consumption.*** 180% of annual average daily consumption.

4. ***Maximum hourly consumption.*** 150% of average for the day.

A convenient formula for estimating consumption is given by R.O. Goodrich :

$$p = 180\, t^{-0.10} \quad \text{...(5.23)}$$

where, p = percentage of annual average consumption for time t days from $\frac{2}{24}$ to 360.

Substituting $t = 1$, we get $p = 180$, *i.e.*, the consumption for maximum is 180% of the average.

Substituting $t = 30$ days, we get

$$p = 180\,(30)^{-0.10} = 128\%.$$

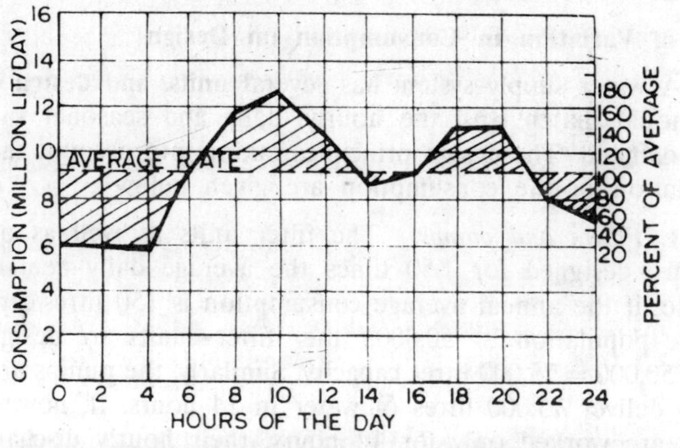

FIG. 5.4. VARIATION IN CONSUMPTION THROUGHOUT THE DAY.

Fig. 5.4 shows the variation in consumption throughout the day. In most of the Indian cities, the peak demand occurs in the morning and evening. During the night hours, the consumption is below the average. The average consumption is shown by the dotted lines. A term *absolute maximum hourly demand* is used to indicate the consumption of maximum hour on maximum day of maximum month of maximum season. Let the annual average consumption in a city be 150 litres per capita per day. The rate of consumption on maximum day will be = 150×1.8 =270 litres. Maximum hourly consumption will be $= \frac{150}{24} \times 1.8 \times 1.5 = 16.9$ litres/hour. The absolute maximum hourly demand (adopting the monthly and seasonal factors suggested above) will be $= \frac{150}{24} \times 1.30 \times 1.40 \times 1.80 \times 1.50 = \frac{150}{24} \times 4.914 = 30.7$ litres/hour (against an annual average of $\frac{150}{24} = 6.25$ litres hour).

In order to cope with the hourly variations in the demand, either the pumps may be run at variable speed (which is difficult and cumbersome) or else the pumps may be run at average speed and store the water during the period of less consumption. The second alternative is generally followed. If the pumps are run for all the 24 hours, the rate of delivery by the pumps will be equal to average demand. If they are run only for t hours in a day, the rate of pumping will be $\frac{24}{t}$ times the average consumption. The excess water during the slack demand period is stored in a service reservoir, to be consumed during the periods of peak demand.

Effect of Variation in Consumption on Design

A water supply system has several units, and design of each unit should match with the hourly, daily and seasonal variations in the demand. The design principles taking into account the effect of variation in the consumption are given below :

1. ***Filters and pumps.*** The filter units as well as pumping units are designed for 1.50 times the average daily demand. For example, if the annual average consumption is 150 litres/capita/day, and the population is 50,000, the filter units are designed for $1.50 \times 50{,}000 = 75{,}000$ litres capacity. Similarly, the pumps are designed to deliver 75,000 litres of water in 24 hours. If, however, the pumps are worked only for 12 hours, their hourly discharge will e $\frac{75000}{12} = 6250$ litres per hour.

2. ***Distribution mains.*** Distribution mains are designed for the maximum hourly demand of the maximum day. Adopting the factors suggested above, the multiplying factor for the supply will be $= 1.8 \times 1.5 = 2.7$.

3. ***Sedimentation tanks and water reservoirs.*** The sedimentation tanks and the clean water reservoirs are designed for the average daily rate of consumption.

PROBLEMS

1. Explain in brief different methods used for prediction of future population of a city, with reference to the design of a water supply system.

2. What do you understand by 'per capita demand' of water ? How is it determined ?

3. Discuss various factors that affect the rate of demand.

4. Explain in brief various expressions used to determine the 'fire demand'.

5. Explain in brief various factors that affect population growth.

6. Write a note on variations in rate of demand. Explain clearly how you take into account these variations in the design of various units.

7. The following data shows the variation in population of a town from 1922 to 1972. Estimate the population of the city in the year 2002. Use various methods :

Year	1922	1932	1942	1952	1962	1972
Population	72,000	85,000	1,10,500	1,44,000	1,84,000	2,21,000

[**Ans.** (*i*) Arithmetical increase method : 3,10,400
(*ii*) Geometrical increase method : 4,35,000
(*iii*) Incremental increase method : 3,46,400

8. In two periods, each of 20 years, a city population grew from 30000 to 172000 to 292000. Find (*a*) saturation population (*b*) the coefficients of logistic equation and (*c*) expected population in the next 20 years.

[**Ans.** (*a*) 3,12,800 (*b*) $a = 2.243$ $b = -0.122$ (*c*) 310800

6

Quality of Water

6.1. INTRODUCTION

The primary objective of water treatment and purification is to collect water from best available source and subject it to processing which will ensure water of good physical quality, free from unpleasant taste or odour and containing nothing which might be detrimental to health. It is not possible to find absolutely pure water in nature. Chemically pure water is the one which contains two parts of hydrogen and one part of oxygen. Water drops just emitting out of the clouds may be considered as chemically pure water, but as the drops fall down they absorb some gases and other matter. It is however not essential to have chemically pure water. Sometimes presence of some minerals give good taste to water. Supplies of water may be obtained by drawing directly upon rivers, lakes, or springs. Depending largely upon the sources from which they are derived, the various waters so obtained may differ greatly in purity and suitability for the purposes for which they are required. Underground water may contain much matter dissolved from state through which it passes. Shallow wells are much more prone to bacterial pollution than are deep wells. Streams and rivers collect impurities from the earth and through the discharge of sewage and industrial effluents and these are carried to ponds, lakes and reservoirs. In the past, there have been many cases of recurring epidemics of enteric fevers traceable to drinking water supplies.

The following are important requirements of water for *domestic use* :

1. It should be colourless, and sparkling clear. It must be free from solids in suspension and must not deposit a sediment on standing.
2. It should be of good taste, free from odour.
3. It should be reasonably soft.
4. It should be plentiful and cheap.
5. It should be free from disease producing bacteria or organisms.
6. It should be free from objectionable dissolved gases, such as sulphuretted hydrogen. It should, however, have sufficient quantity of dissolved oxygen.
7. It should be free from harmful salts.
8. It should be free from objectionable minerals, such as iron, manganese, lead, arsenic and other poisonous metals.
9. It should be free from radio-active substance such as radium, stronsium etc.
10. It should be reasonably free from phenolic compounds, chlorides, fluoride and iodine.
11. It should not lead to scale formation and should be non-corrosive.

The above requirements are only qualitative. The quantitative requirements are given in the subsequent articles.

6.2. COMMON IMPURITIES IN WATER AND THEIR EFFECT

Impurities in water are classified into three heads :

(*i*) Suspended impurities.

(*ii*) Dissolved impurities.

(*iii*) Colloidal impurities.

Suspended impurities are those which normally remain in suspension. They are microscopic and make water turbid. Dissolved impurities are not visible, but they are large since water is good solvent. They cause bad taste, hardness and alkalinity. Sometimes, they are harmful. Colloidal impurities are electrically charged. Due to this, the colloidal particles, usually very small in size, remain in constant motion and do not settle. Table 6.1 gives summary of various suspended and dissolved impurities.

TABLE 6.1. SUSPENDED AND DISSOLVED IMPURITIES

Type	*Constituents*		*Effect*
1 **Suspended Impurities**	(*a*) Bacteria		Some cause diseases
	(*b*) Algae, Protozoa		Odour, colour, turbidity
	(*c*) Silts		Murkiness or turbidity
2 **Dissolved Impurities**	(*a*) *Salts*		Alkalinity
	(*i*) Calcium and Magnesium	Bicarbonate	Alkalinity, hardness
		Carbonate	Hardness
		Sulphate	Hardness, corrosion
		Chloride	
	(*ii*) Sodium	Bicarbonate	Alkalinity, softening effect
		Carbonate	Alkalinity, softening effect
		Sulphate	Foaming in boilers
		Fluoride	Dental flurosis or mottled enamel
		Chloride	Taste
	(*b*) *Metals and Compounds*		
	(*i*) Iron oxide		Taste red colour, corrosiveness, hardness
	(*ii*) Manganese		Black or brown colour
	(*iii*) Lead		Cumulative poisoning
	(*iv*) Arsenic		Toxicity, poisoning
	(*v*) Barium		Toxic effect on heart, nerves
	(*vi*) Cadmium		Toxic, illness
	(*viii*) Cyanide		Fatal
	(*viii*) Boron		Affect central nervous system
	(*ix*) Selenium		Highly toxic to animals, fish
	(*x*) Silver		Discolouration of skin; eyes
	(*xi*) Nitrates		Blue baby conditions; infant poisoning; colour; acidity
	(*c*) *Vegetable dyes*		
	(*d*) *Gases*	Oxygen	Corrosiveness to metals
		Carbon dioxide	Acidity, corrosiveness
		Hydrogen sulphide	Odour, acidity, corrosiveness

6.3. QUALITY OF SOURCE

Natural water is available from the following sources :

(*i*) Pure water

(*ii*) Rain water

(*iii*) Surface runoff water

(*iv*) River water.

(*v*) Lakes and pond water, swamp water.

(*vi*) Ground water

Fig. 6.1 Shows the characteristic properties of natural water in hydrologic cycle.

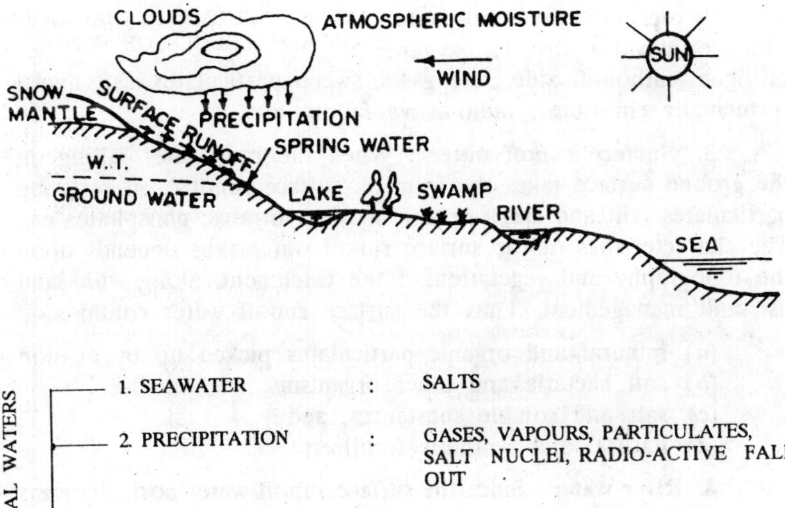

CHARACTERISTICS OF NATURAL WATERS

1. SEAWATER	:	SALTS
2. PRECIPITATION	:	GASES, VAPOURS, PARTICULATES, SALT NUCLEI, RADIO-ACTIVE FALL OUT
3. SURFACE RUNOFF	:	PARTICULATES, ORGANIC MATTER, NITRATES, PHOSPHATES, BIOCIDES
4. GROUND WATER AND SPRING WATER	:	CARBONATES, CHLORIDES AND SULPHATES OF CALCIUM AND MAGNESIUM; IRON AND MANGENESE ; SO_2 ; H_2S
5. LAKE WATER	:	ALGAE, ODOURS, TASTES
6. SWAMP WATER	:	COLOUR, ODOURS, TASTES
7. RIVER WATER	:	PARTICULATES ; ORGANIC MATTER WASTE WATERS

FIG. 6.1. CHARACTERISTIC PROPERTIES OF NATURAL WATERS IN HYDROLOGIC CYCLE

1. Pure water : Chemically pure water is a combination of two elements : hydrogen and oxygen, with a chemical symbol H_2O and chemical name as *hydrogen monoxide.* It is not possible to find absolutely pure water in nature. Also, it is not economically feasible to manufacture pure water in this form. However, pure water is entirely free from all the three types of impurities mentioned in § 6.1.

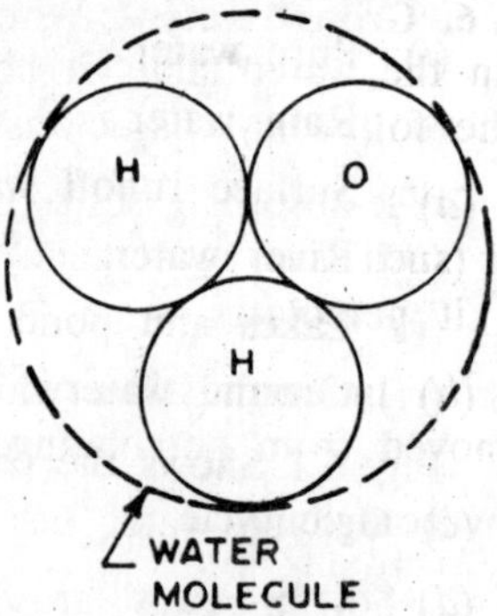

FIG. 6.2. SCHEMATIC DIAGRAM OF A MOLECULE OF WATER

2. Rain water : When precipitation take place, the rain water, falling through atmosphere absorbs various gases and vapours which are normally present in the atmosphere. Thus, rain water absorbs oxygen, nitrogen, carbon-dioxide, rare gases, sweep particulates, salt nuclei (principally chlorides), radio-active fallout etc.

3. Surface runoff water : When the rain water falling on the ground surface takes the form of 'surface runoff', it picks up particulates (silt and clay), organic matter, nitrates, phosphates etc. The characteristics of the surface runoff water thus depends upon the topography and vegetation of the catchment, along with land use and management. Thus the surface runoff water contains

(*a*) mineral and organic particulates picked up by erosion
(*b*) soil bacteria and other organisms
(*c*) salt and soluble substances, and
(*d*) natural and synthetic fertilizers.

4. River water : Since the surface runoff water normally joins the rivers and streams, the characteristics of river water are practically the same as those of surface runoff water. However, if the river receives the municipal and industrial waste waters, additional pollution parameters are added to it.

5. Lake and Pond water : When water is ponded in lakes and reservoirs, algae and similar other organisms grow on the surface, giving rise to odors, tastes and colour. Decaying vegetation further intensify these elements. In addition to this, thermal stratification takes place in reservoirs resulting in

(*i*) low dissolved oxygen
(*ii*) dissolution of iron and manganese
(*iii*) production of H_2S

(*iv*) increase in CO_2 and
(*v*) reduction in pH of water.

6. Ground water : When rain water infiltrates into the ground to join the water table, it becomes 'ground water'. Ground water has the following characteristics :

(*a*) It absorbs gases of decomposition and degradable organic matter (such as H_2S, methane) within the pores of soil mantle through which it percolates.

(*b*) In earth strata which in rich is organic matter, oxygen is removed from percolating water and CO_2 is added.

(*c*) Ground water has lower pH value.

(*d*) Soil minerals are dissolved in ground water. Carbonates, sulphates and chlorides are added, resulting in hardness.

(*e*) Ground water may contain iron and manganese in soluble form.

Effect of Impoundment on water quality

When water is impounded in the reservoirs, specially for storage purposes, its characteristics are changed. The impoundment may produce both beneficial as well as detrimental effects on water quality, and these effects are summarised in Table 6.2.

Beneficial effects

1. Reduction in turbidity at upper levels, because of reduced velocities and long detention time.
2. Reduction in hardness caused by algal carbon dioxide and subsequent precipitation of calcium carbonate
3. Organic oxidation.
4. Reduction in B.O.D. caused by bio-degradation during storage.
5. Colour reduction due to settling silts.
6. Reduction in coliform density due to natural die off during storage.

Detrimental Effects.

1. Lower atmospheric reaeration caused by reduced velocities and increased depth.
2. Increased algal blooms with resultant taste and odour problems, in addition to the aesthetic aspect.
3. Back up of pollutants present in the receiving water.

4. Thermal stratification resulting in low D.O., dissolution of iron and manganese, production of H_2S, increase in CO_2 and reduction in pH.

TABLE 6.2
INFLUENCE OF IMPOUNDMENT ON WATER QUALITY

Benefits	***Detriments***
1. Turbidity reduction	1. Less mixing and less reaeration.
2. Hardness reduction	2. Algal blooms : bad taste and odour
3. Organic oxidation	3. Back up of pollutants
4. B.O.D. reduction	4. Thermal stratification
5. Colour reduction	(*i*) Low dissolved oxygen
6. Coliform reduction	(*ii*) Iron and mangenese dissolution
	(*iii*) Hydrogen sulphide production
	(*iv*) Increase in CO_2
	(*v*) Reduction in pH.

Thermal stratification

Thermal stratification is the term which is applied to the variation in the *temperature* of the impounded water with depth. Fig 6.3 shows thermal stratification in a reservoir, during summer. The entire depth can be divided into three zones: (*i*) Epilimnion (*ii*) Mesolimnion (or thermocline) and (*iii*) Hypolimnion.

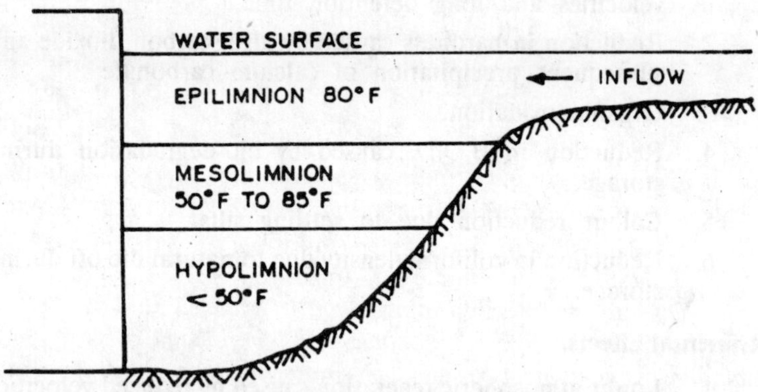

FIG. 6.3. SUMMER STRATIFICATION.

(*i*) ***Epilimnion*** : This is the top zone, 6 to 15 m deep according to depth of reservoir. The water in this zone is of good quality, having high dissolved oxygen.

*(ii) **Mesolimnion (or Thermocline)***. It is a zone, 3 to 6 m deep, having rapidly changing temperature at the rate of 1° C or more per 1 m change in depth. The primary disadvantage of this zone is that the oxygen content is depleted through microbial activity. Since there is no opportunity of reaeration, anearobic conditions are set in with an increase in iron, carbon dioxide, manganese, H_2S taste and odour. All these are potentially detrimental to water quality.

*(iii) **Hypolimnion*** : This is the bottom most zone which is cold. It has practically no dissolved oxygen. It has very high dissolved iron and manganese. Anaerobic conditions exist in this zone. Also, there is an increase in CO_2 and decrease in pH.

6.4. WATER ANALYSIS

Various types of impurities present in water can be determined by water analysis. This analysis is done both for raw water as well as treated or purified water. The examination of raw water will enable us to determine the outline or processes of water purification. Water analysis of purified water is done to know whether the degree of purification has reached the required standards or not. Following are the purposes of water analysis of raw water and purified water:

1. To classify the water with respect to general level of mineral constituents.
2. To determine the degree of clarity and ascertain the nature of matter in suspension.
3. To determine the chemical and bacteriological pollution of water.
4. To determine the presence/absence of an excess of any particular constituent affecting potable quality and general use.
5. To determine the level of organic impurities.
6. To set the outlines of purification process and specify various stages in it.
7. To ascertain whether purification of water has reached the required standards or not.

Thus to summarise, examination of water is used to classify, prescribe treatment, control treatment and purification processes and maintain public supplies of an appropriate standard of organic quality, clarity and palatability.

The examination of water may be divided into three classes:

1. Physical Examination.
2. Chemical examination.
3. Microbiological examination.

Sampling. The proper collection of water sample entails the most careful consideration of all relevant circumstances, as regards choice of sampling point, method of collection and frequency of testing. Frequency of testing directly depends upon the variability of quality of raw water and to the hazards of pollution. For deep underground water, where chances of pollution are less, the testing may be less frequent. In the case of impounded water, where chemical and physical data are intimately related to biological investigation, seasonal factors and pollutional aspects determine how often a river water is to be examined. The samples should be collected in clean and sterilized bottles having stoppers. The water and the bottle should not be touched while collecting the sample.

6.5. PHYSICAL EXAMINATION

These consist of (*i*) Colour test (*ii*) taste and odour test (*iii*) temperature test, and (*iv*) turbidity test.

1. COLOUR TEST

Water gets its colour due to leaves, peat, logs and other organic matter either in true solution or in colloidal suspension. Colour can be measured against various standards or scales such as Hazen or Platinic chloride scale, Burgess scale or cobalt scale using a tintometer. The apparent colour of water due to turbidity is first removed by centrifuging. This is achieved by placing the sample of water in a centrifuge and rotating it at a very high speed so that the vortex caused due to centrifuge action spills out suspended matter. The sample is then ready for determination of its true colour, using a tintometer. The instrument has an eye-piece with two holes. A the slide of *standard coloured water* is seen through one hole, while the slide of the water to be tested is seen through the other hole. A number of slides of standard colour in water are kept ready for comparison. The intensity of colour in water is measured in terms of arbitrary *unit of colour* on the cobalt scale. The unit of colour is that produced by one milligram of platinum cobalt in one litre of distilled water. For public water supply, the number on cobalt scale should not exceed 20, and should preferably be less than 10. According to IS 3025-1964, if 1.245 g of potassium chloroplatinate and 1.0 g of cobalt chloride are dissolved in distilled water and diluted to 1 litre, it has a colour of 500 Hazen units.

2. TASTE AND ODOUR TESTS

Taste and odour tests are of prime importance in assessing the palatability of water. They may be there due to the presence of mineral salts, tarry substances, industrial wastes, domestic sewage, decomposing organic matter, certain types of microscopic organisms or chemical compounds such as phenols etc. However, taste and odour are the most difficult physical characteristics to measure in any numerical sense because of personal factors related to taste and odour, atmospheric conditions of impurity, temperature and humidity. Odour is meaured in terms of *thresh hold odour number.* This is numerically equal to the volume of sample of water in cubic centimetres required to be added to 100 c.c. of odour free fresh water when the mixure just starts giving the typical smell in the diluted sample. For example, if 6 c.c. of water added to 100 c.c. of fresh water does not give any odour, while if more than 6 c.c. of water when added to 100 c.c. of water gives odour in the diluted sample, the thresh hold number will be 6. For public water supply, thresh hold number should not be more than 3. The test is conducted at 20°*C*. *Alternatively,* the odour test may be carried out by inhaling through two tubes of an instrument called *Osmoscope.* It consists of two 10 mm dia. glass tubes 30 cm long kept about 8 mm apart. The lower end of each tube is plain and is held slightly above water surface while the upper beveled end can be inserted in the nostril. Diluted water samples in the ratio of 1:2, 1:4, 1:8, 1:16 etc. are prepared. One tube is kept in a flask containing one of the above diluted water sample, while the other tube is kept in the flask containing original water sample. The mixure with greatest dilution is tested first, and then others with lesser and lesser dilution. The mixure giving first detectable odour will be taken as *thresh hold odour,* while the number of additions of fresh water in such mixure is known as *odour-intensity* or pO *value.* Following table gives the odour interpretation of the pO values :

pO value	*Meaning*
0	No perceptible odour.
1	Very faint odour, detectable only to trained observer.
2	Faint odour, detectable to average person only if he is told.
3	Distinct odour, readily detectable by most people.
4	Very distinct odour.
5	Strong and intense odour.
6	Extremely strong odour.

3. TEMPERATURE TEST

The temperature of water to be supplied should be between 10°C to 20°C. Temperature higher than 25°C is considered objectionable. It can be measured with ordinary thermometers graduated in

0.1°*C*, range 0 to 50°. In case of large bodies of surface water, where it is required to record temperature at a certain depth, *'broken capillary thermometer'* is used. At depths greater than 15 m, a *thermo-couple* may be used. Temperature measurements are sometimes important to identify the magnitude of density, viscosity, vapour pressure and surface tension of water. Other factors such as saturation values of solids and gases dissolved in it, B.O.D. values and the biological activities are dependent on temperature of water.

4. TURBIDITY TEST

Turbidity is imparted by the colloidal matter present in water. The colloidal matter may be clay and loam or microscopic organisms. It is a measure of the resistance of water to the passage of light through it. Hence turbidity is estimated against standard suspensions of a siliceous material such as Fuller's earth, the silica scale being based on standardization by photometric means. The *standard unit* of turbidity is the turbidity produced by one part of Fuller's earth in a million parts of distilled water. Turbidity is expressed in terms of parts of suspended matter per million parts of water by weight, abbreviated as p.p.m. One unit or 1 p.p.m. is equivalent to 1 mg per litre. The permissible turbidity of domestic water may be between 5 to 10 p.p.m.

The following are common methods of measuring turbidity of water : (*i*) by turbidity rod (*ii*) by Jackson's turbidimeter (*iii*) by Baylis turbidimeter (*iv*) by Nephelometers.

(*i*) **Turbidity Rod.** Turbidity rod is used for measuring turbidity of water in the field. It consists of a graduated alumininum rod, about 20.3 cm in length, at the upper end of which is attached a graduated non-stretchable tape of about 12.2 cm long. At the lower end of the aluminium rod, a screw containing a platinum needle (of 1 mm diameter and 2.5 cm length) and a nickel ring is inserted. The graduated tape has a mark at its top end specifying the position of eye during the test. In order to find the turbidity, the lower end of the rod is gradually immersed in water whose turbidity is to determined. Eye is kept constantly at the marked position and the platinum needle is watched. The rod is moved slowly in water till the platinum needle just disappears from the vision due to turbidity of water. The reading of the graduated tape near the water surface directly gives turbidity in p.p.m. The rod gives only rough value of the turbidity of water.

(*ii*) **Jackson's Turbidimeter.** This is a laboratory apparatus which is used to measure turbidity when it is more than 25 p.p.m., and preferably when it is more than 100 p.p.m. It consists of a

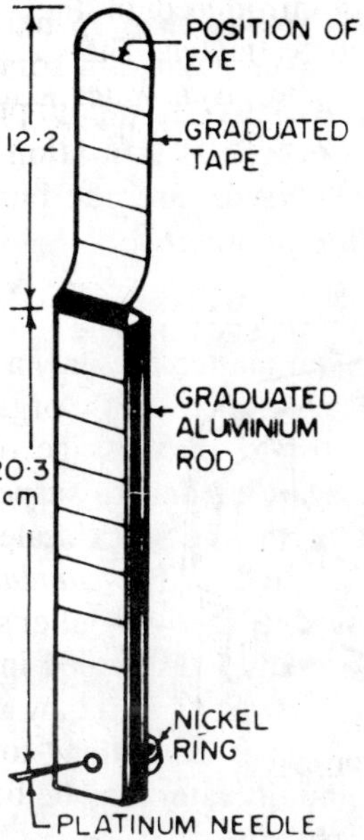

FIG. 6.4. TURBIDITY ROD

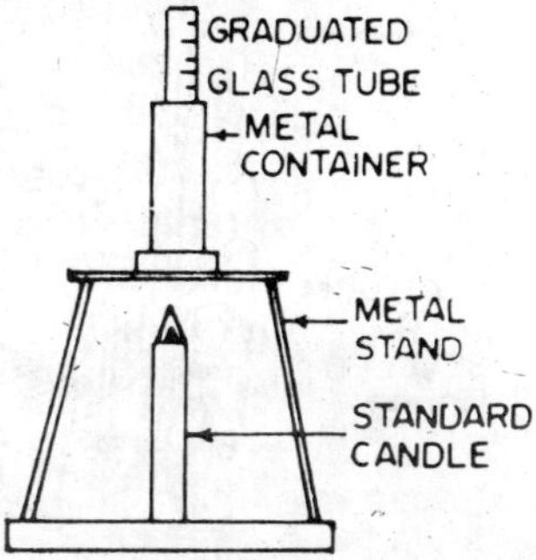

FIG. 6.5. JACKSON'S TURBIDIMETER

metal stand holding a metal container and a graduated glass tube in it, as shown in Fig. 6.5. A standard candle is placed below the stand. The water sample is poured in the sample and the image of the flame of the standard candle is seen through the turbid water in the glass tube. The level of water in the glass tube is gradually increased till the image of the flame ceases to be seen. The height of water column, measured in the graduated glass tube provides the measure of the turbidity of the water. The longer the light path, the lower the turbidity; a light path of 21.5 cm corresponds to 100 JTU (Jackson Turbidity Unit) while light path of 10.8 cm corresponds to 200 JTU where 1 JTU = 1 p.p.m.

(*iii*) **Baylis Turbidimeter**. This is a very accurate instrument and is preferred when turbidity of the sample is less than 5 units. It consists of a galvansied iron box in which two glass tubes are kept at one end and a 250 watt bulb with reflector is placed at the other end as shown in Fig. 6.6. One tube contains standard solution of known turbidity while in the other tube the water sample is kept. The tube is held firmly in a platform with beveled holes at its bottom end. The tubes are surrounded on all its four sides by blue cobalt plates, and at its bottom by a white opal glass plate. Because of blue cobalt plates, blue light is cast in both the tubes, and a comparison is made. If the light differs, another tube containing standard solution of different turbidity is introduced in the place of the first one till the colour in both the tubes match. The standard solution at this stage give the turbidity of the given water sample. The turbidity is expressed either as p.p.m. or BTU (Baylis' Turbidity Unit), both being equivalent.

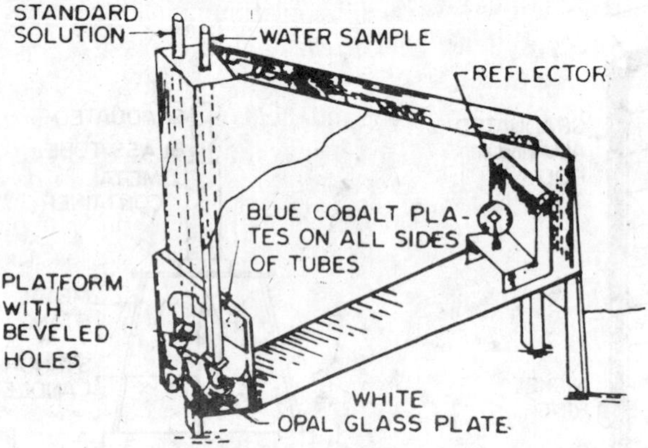

FIG. 6.6. BAYLIS' TURBIDIMETER

(iv) Nephelometers : Nephelometers are the *modern commerical turbidimeters* in which the light intensity is measured at right anlges to the incident ray in contrast to the conventional turbidimeters (described earlier) in which the light intensity is measured after the same passes straight through the turbid water. To achieve this, nephelometers use photometers. A photometer measures the intensity of light passing through the turbid water, after the same is scattered at right angles to the incident ray. The value of turbidity measured by a nephelometer is generally expressed as NTU (Naphelometric turbidity unit). Since formazin polymer is used as reference turbidity standard suspension, naphelometric turbidity units are also sometimes referred to as FTU (Formazin turbidity unit).

6.6. CHEMICAL EXAMINATION

Chemical tests are carried out to determine the following :

(*i*) Total solids. (*ii*) Chlorides. (*iii*) Hardness. (*iv*) pH value. (*v*) Metals and other chemical substances. (*vi*) Nitrogen and its compounds. (*vii*) Dissolved gases.

1. TOTAL SOLIDS

In this test, the suspended, dissolved and colloidal solids are determined separately and then added together to get total solids.

High solid contents indicate either contamination or presence of excessive mineral matters. The amount of total solids should preferably be less than 500 p.p.m. In order to determine the suspended solids, water is filtered through fine filter paper. The material retained on the filter is dried and weighed. In order to determine dissolved and colloidal solids, the filtered sample of water is evaporated and the dry residue left is weighed. A platinum dish is used for this purpose. In order to determine *'fixed solids'* (*i.e.* total solids exclusive of organic matter), the total solids obtained by filtering and evaporating are kept in the platinum dish and held over a Bunsen flame till every part of solids is raised to a bright red heat. The organic matter is burnt off in this process and only inorganic matter is left behind. The dish is cooled, and weight of the matter remaining behind gives the amount of 'fixed solids' which can be expressed in p.p.m.

2. CHLORIDES

The test is primarily carried out to determine the content of sodium chloride in water. Rain water carries small amount of chloride derived from the ocean. Natural surface waters derive chlorides from oil-field wastes, salt water, mineral deposits etc., and the normal contents of chlorides in these may be of the order of 10 to 20 p.p.m. Greater amounts arise either from the natural occurrence of chlorides in the geological strata concerned, or in the case of rivers, from the entry of sewage effluents and wastes. Chlorides are estimated by titration with standard silver nitrate solution using potassium chromate as indicator. For this, 50 c.c. of water sample is taken by a pipette, in a porcelain dish, and two or three drops of potassium chromate solution is added to the water in the dish. The water sample is then titrated with standard solution of silver nitrate.

3. HARDNESS

The hardness of water is due to presence of certain salts, such as carbonates, bi-carbonates, chlorides and sulphates, of calcium and magnesium dissolved in it. Calcium and magnesium salts are soap destroying and are therefore the prime constituents concerned with hardness. *Total hardness* is composed of two components – temporary hardness, and permanent hardness. *Temporary hardness,* also known as carbonate hardness, is due to the presence of carbonates and bicarbonates of calcium and magnesium. It is easily removed by either boiling the water and by adding lime to water. The *permanent hardness,* also called 'non-carbonate hardness', is due to presence of sulphates, chlorides and nitrates of calcium and magnesium. It cannot be removed by boiling and requires special methods of water softening. Hardness is expressed either in p.p.m. or in terms of *degrees of hardness*. Water is said to have one degree of hardness when its soap destroying power is equivalent to the effect of 14.25 milligrams of calcium carbonate in one litre of water (or 1 gram of calcium carbonate in one imperial gallon). Thus 1 degree of hardness is equivalent to 1 p.p.m.

Table 6.3 gives the nature of water and their scale of hardness in degrees. Table 6.4 gives the hardness in p.p.m. and the corresponding nature of water.

TABLE 6.3.

S. No.	*Degree of hardness*	*Nature of water*
1	1	Extremely soft water
2	2	Very soft water
3	3	Soft water
4	6	Reasonably soft water
5	7	Reasonably hard water
6	9	Hard water
7	11	Very hard water
8	15	Excessive hard water
9	17	Too hard to use.

TABLE 6.4.

Class	1	2	3	4
Hardness p.p.m.	0–55	56–100	101–200	201–500
Class of water	Soft	Slightly hard	Moderately hard	Very hard

There are three methods of determining total hardness of water (*i*) Clark's method, (*ii*) Hehner's method. (*iii*) Versenate method.

(i) Clark's method. This method is based on the premise that hardness-producing substances react with soap and form insoluble compounds before latter is produced. Hence total hardness is found by determining the standard soap solution required to obtain a permanent lather with the water sample of known volume with constant shaking. This method has become obsolete.

(ii) Hehner's method. In this method the temporary hardness is determined by titration with a standard solution of sulphuric acid, using methyl orange as indicator. To determine the permanent hardness, standard sodium carbonate solution is added to the water sample and evaporated to dryness. The amount of sodium carbonate in excess over that required to convert the sulphates and chlorides into carbonates gives the permanent hardness.

(*iii*) ***Versenate method.*** In this method, hardness is determined by titrating against Di-Ethylene diamine tetracitic acid (EDTA) solution, using Erio chrome Block T as indicator at pH 8.5 to 11.

For potable water, hardness of water should be more than 5 degrees but less than 8 degrees. Excessive hardness is objectionable because (*i*) great deal of soap is required for washing clothes etc., (*ii*) scale is formed in boilers and hot water heating systems and (*iii*) corrosion and incrustation of pipe lines and plumbing fixures is promoted.

4. HYDROGEN ION CONCENTRATION (pH VALUE)

The hydrogen-ion concentration or pH value of water is a measure of degree of acidity or alkalinity of water. In water, some of the molecules (H_2O) are dissociated into two ions – the hydroxyl OH^-) ions or anions which are negatively charged and the hydrogen H^+) ions or cations which are positively charged. The concentration of the two *free* or *uncombined* components is extremely small, but it is governed by the following *law of mass action* in physical chemistry

$$\frac{\text{Concentration of cations} \times \text{Concentration of anions}}{\text{Concentration of undissociated molecules}} = \text{Constant}$$

Since dissociation of water is only very slight, the concentration of undissociated molecules (*i.e.* the denominator) may be taken as 100% or unity. Hence we have :

$$(H^+) \times (OH^-) = \text{Ionisation Constant} \qquad ...(6.1)$$

When concentrations in the above equation are expressed in terms of gram ions per litre, this constant is 10^{-14} at 21°C.

$$(H^+) \times (OH^-) = 10^{-14} \qquad ....(6.1\ a)$$

For absolutely pure water, concentrations of hydroxyl and hydrogen ions is equal, and at 21°C, its value is 10^{-7} each. Thus,

$$(10^{-7} \text{ gm H–ion}) \times (10^{-7} \text{ gm of OH–ion}) = 10^{-14} \qquad ...(6.1\ b)$$

From Eq. 6.1 (*a*), if the concentration of H ions is known, the concentration of OH ions can be easily found. For example, if concentration of H ions is 10^{-6}, concentration of OH ions will be 10^{-8} so that Eq. 6.1 (*a*) is satisfied. Water becomes *acidic* when concentration of H ions is *increased*, and *alkaline* when concentration of H ions is decreased. Because of very low value of H-ion (or OH-ion) concentration, this figure is inconvenient to be used, and instead a term pH value has been introduced. pH value is the logarithm of the reciprocal of hydrogen ion concentration.

Thus $$\text{pH} = -\log_{10}(\text{H}) = \log_{10}\frac{1}{(\text{H})} \qquad \text{...(6.2)}$$

Similarly, pOH value of water by definition can be expressed as

$$\text{pOH} = -\log_{10}[\text{OH}] = \frac{1}{\log_{10}(\text{OH})} \qquad \text{...(6.2 } a\text{)}$$

Hence from Eq. 6.1 (*a*)

$$\log_{10}(\text{pH} + \text{pOH}) = \log_{10} 14$$

or $$\text{pH} + \text{pOH} = 14 \qquad \text{...(6.2 } b\text{)}$$

For pure water $$(\text{H}) = 10^{-7}$$

$$\therefore \qquad \text{pH} = \log_{10}\frac{1}{10^{-7}} = 7$$

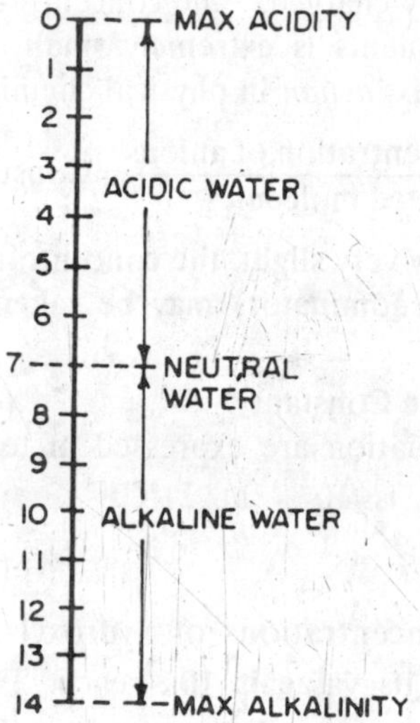

FIG. 6.7. pH SCALE

Water becomes acidic when hydrogen ion concentration increases (say, from 10^{-7} to 10^{-5}), in that case, the pH value is decreased from 7 to 5. Fig. 6.7 gives the pH scale indicating the range of acidity or alkalinity of water. For water with maximum acidity, pH value is zero, while for water with maximum alkalinity, pH value is 14.

Acidic water causes tuberculation while alkaline water causes incrustation. For potable waters, the pH value should be between 6 to 9, and preferably between 7 and 8.5.

Measurement of pH value. pH measurement is important since it provides means of classifying and of corelating other characteristics or behaviours such as corossive acitivity or other inter-related factors controlling biological function in a body of water. Its knowledge is also helpful in controlling softening and coagulation processes in water treatment. There are two methods of determining pH value of water : (*i*) Colourimetric method, and (*ii*) Electrometric method.

1. Colourimetric method

This is probably the most single method in which some *indicator* is added to the water sample, and the colour of the solution is compared with the standard colours of known pH value. These standard colours may be in the form of coloured liquids in glass tubes, coloured glass discs or coloured charts supplied by the manufacturer, with each indicator. For acidic range, the common indicators are thymol blue (acid range), bromophenol blue, methyl red and chlorphenol red. For alkaline range, the common indicators are thymol blue (alkaline range), phonolphalein red and tolyl red.

2. Electrometric method

In this method, a pH meter is used. The specimen of water is kept in a beaker and the two electrodes of the insrtument are dipped in water connected to a dry cell. The current passing through the circuit indicates the pH value which is read directly on the dial. The electrodes commonly used are made of calomel, glass or quinhydrone. In some of the latest pH meters, the pH value is displayed electronically on the dial, upto second decimal place of the value.

Example 6.1. *The measured pH values of incoming and outgoing waters at a water treatment plant are 7.3 and 8.5 respectively. Determine the average pH of water, assuming linear variation of pH with time.*

Solution

For incoming water, $(pH)_1 = 7.3 = \log_{10} \frac{1}{H_1^+}$

For outgoing water, $(pH)_2 = 8.5 = \log \frac{1}{H_2^+}$

$$\therefore \quad H_1^+ = 10^{-7.3} \text{ and } H_2^+ = 10^{-8.5}$$

$$\therefore \quad H_{av}^+ = \frac{H_1^+ + H_2^+}{2} = \frac{10^{-7.3} + 10^{-8.5}}{2}$$

$$= 10^{-8.5}\left[\frac{(10)^{1.2} + 1}{2}\right] = 10^{-8.5}\left[\frac{15.85 + 1}{2}\right]$$

$$= 8.42 \times 10^{-8.5}$$

$$\therefore \quad (pH)_{av} = \log_{10} \frac{1}{8.42 \times 10^{-8.5}} \approx 7.575$$

Example 6.2. *Compute pH and pOH values of freshly distilled water at 21°C.*

Solution

From Eq. 6.1 (*a*),

$$[H^+] \times [OH^-] = 10^{-14} \text{ at } 21°C.$$

In freshly distilled water, concentration of both the ions will be the same.

$$\therefore \quad [H^+] = 10^{-7} \text{ and } [OH^-] = 10^{-7}$$

$$\therefore \quad pH = -\log_{10}[H] = -\log_{10}[10^{-7}] = \mathbf{7}$$

Also, $\quad pOH = -\log_{10}[OH] = -\log_{10}[10^{-7}] = \mathbf{7}$

Example 6.3. *A waste water from a factory having pH = 10 contains KOH only. Find out the total quantity of KOH per day if the waste water discharge is 80 m³/day.*

Solution.

From Eq. 6.2 (*b*), we have

$$pH + pOH = 14$$

$$\therefore \quad pOH = 14 - pH = 14 - 10 = 4$$

$$\therefore \quad [OH] = 10^{-4} \text{ moles/litre}$$

(Note that a 'mole' is the molecular weight in g.)

Now, molecular weight of KOH is equal to (39+16+1) = 56 g.

$\therefore$ KOH in g/litre $= 56 \times 10^{-3}$

Now waste water discharge $= 80 \text{ m}^3/\text{day} = 80 \times 10^3$ litres/day

$$\therefore \quad \text{Quantity KOH} = (80 \times 10^3) \times (56 \times 10^{-3}) \text{ g/day}$$
$$= 4480 \text{ g/day} = \mathbf{4.48 \ kg/day}$$

Example 6.4. *Find out the pH of the mixture of the following two solutions :*

Solution A : volume = 500 ml ; pH = 7

Solution B : volume = 500 ml ; pH = 5

Solution. Since the volume of each solution is equal, the morality of mixture will be half in 1000 ml.

Now, $\quad [H^+]$ of $A = 10^{-7}$ mole/litre

and $\quad [H^+]$ of $B = 10^{-5}$ mole/litre

$$[H^+]_{mix} = (\tfrac{1}{2} \times 10^{-7}) + (\tfrac{1}{2} \times 10^{-5})$$

$$= (0.5 \times 10^{-7}) + (50 \times 10^{-7})$$

$$= 50.5 \times 10^{-7} \text{ mole/litre.}$$

$$\therefore \quad [pH]_{mix} = -\log_{10}[H^+]$$

$$= -\log_{10}[50.5 \times 10^{-7}] = -\log_{10} 50.5 - \log_{10} 10^{-7}$$
$$= -1.703 + 7 \approx 5.297 \approx \mathbf{5.3}$$

Example 6.5. *Find out the pH of a mixture formed by mixing the following two solutions :*

Solution A : volume 300 ml ; pH = 7
Solution B : volume 700 ml ; pH = 5

Solution

$$[H^+]_A = 10^{-7} \text{ moles/litre}$$

and $$[H^+]_B = 10^{-5} \text{ moles/litre}$$

For solution A

$$\text{Concentration of } [H^+] \text{ in } 300 \text{ ml} = 10^{-7} \times \frac{300}{1000}$$

$$\therefore \quad [H^+]_A = 3 \times 10^{-8} \text{ moles/litre.}$$

For solution B

$$\text{Concentration of } [H^+] \text{ in } 700 \text{ ml} = 10^{-5} \times \frac{700}{1000}$$

$$\therefore \quad [H^+]_B = 7 \times 10^{-6} \text{ moles/litre.}$$

$$\therefore \text{ Concentration of } [H^+]_{mix} = [H^+]_A + [H^+]_B$$
$$= [3 \times 10^{-8}] + [7 \times 10^{-6}] \text{ moles/litres}$$
$$= [0.03 \times 10^{-6}] + [7 \times 10^{-6}] \text{ moles/litres.}$$
$$= 7.03 \times 10^{-6} \text{ moles/litre}$$

$$\therefore \quad pH = -\log_{10}(7.03 \times 10^{-6}) = -0.847 + 6$$
$$= \mathbf{5.153}$$

5. METALS AND OTHER CHEMICAL SUBSTANCES

(*i*) Iron and Manganese

Iron in water causes hardness, bad tastes, discolouration of clothes and plumbing fixures and incrustation in water mains. Iron in water is frequently accompanied by heavy growths of *crenothrix* which exaggerates the staining, pipe clogging and other troubles. A water which contains iron is termed ferryginous. Estimation of iron is a simple calorimetric procedure and thiocyanate or thioglycollic acid are most usually employed for the purpose. Manganese imparts a brownish or purplish colour to water and laundered goods (when oxidised) and stains plumbing fixures. It also impairs tastes of coffee and tea. Manganese is estimated by matching the pink colour produced on oxidation to permanganate. Sodium paraperiodate or ammonium persulphate is used to colour both the water sample and manganese colour standards.

(*ii*) **Lead and Arsenic**

Lead is a well known cumulative poison. Lead taken in body by either brief or prolonged exposures can seriously injure health or cause death. Under normal conditions, the concentration of lead should be less than 0.05 mg/l. The presence of lead can be detected by six drops of sulphuric acid in a glass of water, which will yield white precipitate. The newer organic reagents have provided in dithizone another method for estimation of lead in addition to the older determination as sulphide. The toxicity of arsenic is well known. Chronic arsenic poisoning is difficult to diagnose, and can be quite disabling prior to detection. Arsenic can be removed from water in ion exchange equipment using activated alumina or bone char. Arsenic can be determined in trace amounts by heteropoly blue calorimetric method and the Gutzeit method of comparing stains produced on treated paper strips.

(*iii*) **Fluorides and Iodides**

If present in small concentrations, both fluoride and iodide may be useful to human beings. Iodine upto 1 p.p.m. keep away goiter while fluoride upto 1.2 p.p.m. will prevent dental caries of the children. Excess concentrations (more than 3 p.p.m.) can cause dental fluorosis or mottled enamel in children. Fluorides may be estimated calorimetrically by developing a colour with zirconium – alizarin reagent. Iodide can be determined in water supplies by utilizing its ability to catalyze the reduction of ceric ions by arsenious acid, the effect being proportional to the amount of iodide present.

(*iv*) **Barium and Boron**

Barium can have serious toxic effects on the heart, blood vessels, and nerves. It occurs naturally in some mineral springs as the carbonate salt. Ingestion of large amounts of boron can affect the central nervous system, and protracted ingestion may cause *borism.* Its concentrations can be determined by calorimetric and titrimetric methods. In the presence of boron, a solution of carmine or carminic acid in concentrated sulphuric acid changes from bright red to a bluish red or blue, depening on the concentration of boron present.

(*v*) **Cadmium and Hexavalent Chromium**

Cadminum is highly toxic. It can become a water contaminant through careless discharge of electroplating plant washes or from galvanizing operations in which cadmium is a contaminant. Cadmium ions from an intense red compound with diphenyl-thiocarbzone (dithizone) which can be extracted into carbon tetrachloride and used for the chlorimetric determination of cadmium. Chromium, when inhaled can be cause of cancer in man, though the results of ingestion are not known. The likely sources of hexavalent chromium in water could be pollution by wastes from chromium plating shops. Soluble hexavalent chromium in the absence of interfering amounts of substances such as molytodenum vanadium and mercury may be determined colorimetrically by reaction with diphenyl-carbazide in acid solution.

(*vi*) **Sodium and Potassium**

Biological regulatory mechanism in the body of a normal human maintains a constant sodium content in the body despite wide variations in sodium intake. Excess quantities are excreted, primarily in the urine. However, some patients of heart, kidney or lever diseases are unable to rid themselves of sodium resulting in a condition known as *edema* in which large amounts of sodium-containing fluid accumulate. The concentration of sodium can be determined either by the flame photometric method by the grammetric method. Sodium and potassium may however by estimated together and expressed as sodium chloride. In special circumstances, sodium may be separately determined as sodium zinc uranyl acetate. Potassium may be estimated as the perchlorate which is sparingly soluble in organic solvents or as dipotassium cobalt-nitrite.

(*vii*) **Phenols**

Water receives phenols from trade wastes, washing from tarmac road, gas liquours and creosoted surfaces. Trace phenol concentrations of the order of 0.001 ppm can impart an objectionable taste to a water following marginal chlorination. The removal of phenolic tastes from a water supply offers a serious challenge at the treatment plant. Phenols can be determined by amino antipyring method or by Gibbs colorimetric method.

6. NITROGEN AND ITS COMPOUNDS

Nitrogen can be present in water in the following four different forms :

(*i*) Ammoniacal Nitrogen (free and saline ammonia)
(*ii*) Albuminoid Nitrogen.
(*iii*) Nitrite
(*iv*) Nitrate.

(*i*) **Ammoniacal Nitrogen**

The occurrence of *free ammonia* indicates the direct inclusion of organic matter, particularly those arising from the excrement (urine) of animal and human species. Surface water may also get polluted from ammonia from discharges of gas industries. Underground waters drawn from strata overlaid with clay may sometimes suffer deoxygenation and comparatively large quantities of free ammonia can arise from reduction of nitrate. Free ammonia is estimated in the laboratory by distillation of mildly alkalized water, the ammonia in the distillate being measured from the coloration produced with Nessler's reagent.

(*ii*) Albuminoid Nitrogen

The albuminoid nitrogen is normally derived from the animal and plant life normal to the aquatic environment. Its determination gives an approximate indication of the quantity of proteinaceous nitrogen present in water. Its presence gives an indication of organic pollution in a water supply. The quantites of albuminoid nitrogen is determined by adding alkaline solution of potassium permanganate or sulphuric acid to the water sample and then boiling the same when it liberates ammonia gas.

(*iii*) Nitrite

Nitrite, a stage in the nitrogen cycle, occurs in water as an intermediate in an oxidation or reduction process. Nitrate can arise either directly from biological disruption of organic matter via ammonium compounds or by the reduction of nitrate effected by biological environs favouring an anaerobiosis. In raw surface supplies the trace amounts of nitrite indicates presence of pollution. In order to determine, it, the nitrite in the sample is treated with sulphanilic acid ; the diazo salt so formed couples with an amine to give a red colour which is matched against that produced with a series of standard nitrite solutions.

(*iv*) Nitrate

Nitrate constitutes the final stage in the oxidation of nitrogen compounds, and normally reaches important concentrations in the final stages of biologic oxidation. The nitrate contained in pure well waters derived from an extensive catchment is largely the result of biological activity in the surface layers of the soil, enhanced by cultivation and the application of manures. When the nitrate is in excessive amounts, it contributes to the illness known as infant methemoglobinemia. Nitrate is measured either by reduction to ammonia or by matching the colours produced with phenoldisulphonic acid.

7. DISSOLVED GASES

(*i*) Dissolved Oxygen (DO)

Surface waters get oxygen dissolved either from atmosphere or due to activities of algae and other tiny plant life in water. Its content in surface water is dependent upon the amount and character of the unstable organic matter in the water. The amount of oxygen that water can hold depends upon temperature – it can hold smaller oxygen at higher temperature. Adequate dissolved oxygen is necessary for the life of fish and other aquatic organisms. The dissolved oxygen concentration may also be associated with corrosivity of water, photosynthetic activity and septicity.

The dissolved oxygen can be determined by using various modifications of the basic Winkler method or by the polarographic method.

(*ii*) **Bio-chemical oxygen demand (B.O.D.)**

The bio-chemical oxygen demand determination is a measure of the oxygen required to oxidize the organic matter in a sample, through the action of micro-organisms contained in the sample. Unpolluted waters should have less than 5 ppm of BOD found from an incubation period of five days and a temperature of 20°C. Higher values of BOD lead to the suspicion regarding the quality of water. The detailed procedure for determination of BOD has been discussed in Vol. 2.

(*iii*) **Carbon-di-oxide**

Carbon di-oxide gets dissolved in water from the atmosphere from decomposing organic matter at the earth's surface or from underground sources. The carbon di-oxide content of water may contribute significantly to some corrosive situations. The amount of carbon di-oxide can be reduced either by aeration or by addition of alkali. Its quantity in water is estimated using alkali to pH 8.0 – 8.5, using phenolphthalein as indicator.

(*iv*) **Hydrogen Sulphide**

It is mostly found in ground water and may be produced either by reduction of sulphate by inorganic process or by decomposition of organic matter through biological agencies and sulphate reducing bacteria. It gives disaggreeable rotten-egg odour. It can be reduced by aeration. Its estimation is made by titration with iodine.

6.7. MICRO-ORGANISMS IN WATER

A contaminated water may contain a host of micro-organisms, due to which water borne diseases may spread if water is not properly treated before it is supplied for domestic use. A micro-organism that requires living tissue to grow is called a *pathogen* and is a parasite which must depend upon host organism as a proper environment for growth and reproduction. Pathogenic micro-organisms are harmful to man, while the majority of other micro-organisms are beneficial to man. *Micro-organisms can be either plants or animals.*

When the assemblages of aquatic organisms drift more or less passively with waves and current, they constitute the *plankton.* If, however, they change their position or location due to their own effort, they are called *nekton.* If they maintain their stability due to surface tension forces, they are referred as *neuston.* The various micro-organisms found in water may be broadly classified under three categories:

1. Aquatic plants.
2. Aquatic animals.
3. Aquatic molds, bacteria and viruses.

1. AQUATIC PLANTS

Under this category, the following are included :

(i) *Spermophyta* : Water weeds
(ii) *Bryophyta* : Mosses and liverworts
(iii) *Pteridophyta* : Ferns and horsetails
(iv) *Thallophyta* : Algae.

Out of these, water weeds and algae are of practical importance to a water works engineer. If favourable conditions of light, temperature and nutrition exist, *water weeds* grow in dense form and infect lakes, ponds, reservoirs and canals. They very much reduce the carrying capacity of canals, apart from furnishing shelter and food to other micro-organisms. Their decay during autumn and add organic matter in water.

Algae are considered as simple, photosynthetic plants with unicellular organs of reproduction. They are, thus, the organisms that are self-nourishing by deriving energy from simple inorganic substances (*i.e.* gases and salts dissolved in water) with the aid of sunlight. In this process, they produce oxygen. In fresh water, algae are generally microscopic in size, while salt water algae may be several hundred metres in length. They do not possess a root system, and in majority of cases, they are non-motile. Their reproduction may be simply by a sexual fission or by the most complex sexual processes. Algae impart objectionable taste and odour to water and may trouble the water works engineer by clogging the filter beds. They may also cause turbidity in the water and an apparent colour, when present in large numbers.

2. AQUATIC ANIMALS

They include the following :

(i) *Vertebrata* : Fish and amphibians
(ii) *Mollusca* : Mussels, snails, slugs, limplets, cocklets
(iii) *Arthopoda* : Crustacea, insects, spiders, mites

(*iv*) *Worms* : Aquatic earth worms, thread worms, rotifera
(*v*) *Metazoa* : Hydra, polyzoa etc.
(*vi*) *Protozoa* : Endameba histolytica etc.

Vertebrata are the aquatic animals which possess a backbone or spinal column. Fish and amphibians are placed under this phylum. When algae growth is controlled by chemicals, or when natural water is polluted by the discharge containing organic matter, insecticides and other organic waste, the vertebrates either die or their growth is retarded. *Mollusca* sometimes transmit intestinal infections. *Arthopods* such as crustaceans, insects, spiders, mites possess leg-like appendages, and foresake the water environment when they emerge as adults.

Worms

Occasionally, water consumers complain of small worms in the water. Worms may be classified as :

(*i*) *Nematodes* or round worms,

(*ii*) Flat worms (*platyhelminthes*), and

(*iii*) *Rotifers*

Flat worms contain two groups : tapeworms (cestodes) and flukes (trematodes). The tapeworm is a segmented worm consisting of a common head and a series of segments, one behind the other, each segment being capable of developing another worm. Worms may be either parasitic or free living forms. The hook worms, flukes, tape worms, ascarids etc. are parasitic forms. *Rotifers* take their name from the wreath of cilia surrounding of the mouth of the organism the beating of which gives the illusion of rotation. They are typical multicellular and microscopical animals, which are usually free swimming. It has the ability of pulling in tremendous quantities of liquid in search of food. They live in fresh water and are able to withstand being dried up by forming cysts, in which form they are probably dispersed by the wind.

Metazoa

Metazoa, other than worms, are of three types :

(*i*) *Porifera*,

(*ii*) *Polyzoa,* and

(*iii*) *Coelenterata.*

Porifera or sponges are agglomerations of animal cells in gelatinous masses. The *polyzoa* (or bryozoa) are microscopic animals which form moss-like or coral like masses attached to the interior of pipes or to other submerged surfaces. The growth of polyzoa can cause serious trouble by reducing the bore of the pipe, clogging meters, strainers and screens and increasing of frictional losses. *Coelenterates,* such as *hydra* occur in large numbers in water reservoirs and on slow sand filters.

Protozoa

Protozoa are essentially unicellular (single-celled) animals that reproduce by binary fission. They are the lowest and simplest forms of animal life. From the protozoa, animal organisms in water rise in the scale of complexity in structure and life processes. They differ from bacteria by having at least one well-defined nucleus. They are mostly planktonic and motive forms. One form of protozoa – *Endemeba histolytica* causes amoebic dysentery. It forms cysts which are excreted in the bowel discharges of infected persons and which will live for long periods in water. The protozoa are bacteria eaters and survive in dilute organic wastes by easting bacteria and thus destroy the pathogens. When the concentration of organic waste is sufficiently great, the protozoa can utilize the soluble organic compounds for their food. They form scum or unsightly deposits on the porcelain utensils and bathroom fittings. They are removed by standard water treatment – coagulation followed by filtration. They are however resistant to chlorination, though break point chlorination with a free chlorine residual of 2 p.p.m. kills them in a contact period of 30 minutes.

3. AQUATIC MOLDS, BACTERIA AND VIRUSES

Strictly speaking, molds (fungi), bacteria and viruses come under the category of aquatic plant, but because of their special importance in water works planning, these have been kept in a separate category.

Fungi

Fungi are multicellular, non-photosynthetic plants capable of growing in low-moisture and low pH environments. Their optimal temperature lies between 20 to 30°C. They flourish over a wide range of pH (4 to 10) and themselves modify the pH by producing organic acids and ammonia as well. The reproductive stage of fungi is a spore. Spores are transmitted long distances in the air by wind currents. Absence of suitable environment prevents most fungi spores from germinating. The fungi are divided into five classes :

1. Myxomycetes : slime fungi.
2. Phycomycetes : algae fungi.

3. Ascomycetes : sac fungi.

4. Basidiomycetes : rust smuts and mushrooms.

5. Fungi imperfecti : miscellaneous.

Fungi are 5 to 10 microns wide. Occasionally, they find conditions favourable in water mains and grow luxuriantly. Large numbers die and the decomposition of their bodies cause disagreeable tastes and odours. *Actinomycetes* which are related to both bacteria and fungi, are also responsible for earthy, muddy and musty odours in water. Fungi can be controlled by treatment of water with chlorine or copper sulphate.

Viruses

Viruses are infectious agents of both plant and animal cells. They are ultramicroscopic, obligate introcellular parasites that manifest their presence by destruction or impairment of host cells. They can pass through an ultramicrosopic filter, and they fall in the size range of 10 to 500 milli-microns. Because of their small size, viruses lack the biochemical systems needed for normal metabolic cell functions and are essentially units organised solely for self-replication. A typical virus particle consists of an outer protein coat enclosing a core of nuclei acid. One group of viruses, the *bacteriophages* are infectious agents of bacteria and are parasitic to bacteria. It initiates infection by attaching itself by its tail to the wall of a bacterial cell. They play significant roles in epidemiological studies of waterborne diseases. Out of several other forms of viruses, *adenoviruses* are associated with upper respiratory infections in children. *Enteroviruses* are found in gastro-intestinal tract and faeces of man and many lower animals. *Enteric viruses* include *coxsackie viruses,* infectious *hepatitis viruses, poliviruses, reoviruses* etc.

Bacteria

Bacteria are single-celled microorganisms with rigid cell walls that take in soluble food and convert it to new cells. Bacteria may be classified in the following ways :

(*i*) According to shape

(*a*) Round, ovoid or spherical bacteria called *cocci.* They are called *micrococcus* (singly), *diplococcus* (in pairs), *streptococcus* (in chains) or *staphylococcus* (in irregular bunches).

(*b*) Straight, rod like cells with square or rounded ends (called *bacilli*).

(*c*) Curved rods comma shaped (called *vibrio*), or helical or spiral shaped (called *spirilla*).

The true bacteria (*eubacteria*) vary in diameter from 0.5 to 1.5 μ, and are less than 10 μ long.

(*ii*) **According to type**

(*a*) *Saprophytic bacteria.* They are beneficial to man and obtain organic matter in solution from dead and decaying tissue in plants or animals.

(*b*) *Parasitic bacteria.* They live and multiply on or within the body of a living organism of a higher type.

(*c*) *Pathogenic bacteria.* They are capable of causing disease within the living organisms on which they subsist.

(*d*) *Non-pathogenic bacteria.* These do not cause disease.

(*iii*) **According to oxygen need**

(*a*) *Aerobic bacteria.* Bacteria that need oxygen to live are called aerobic.

(*b*) *Anaerobic bacteria.* Bacteria that survive in the absence of oxygen are called anaerobic.

(*c*) *Facultative bacteria.* Bacteria that can live and multiply with or without oxygen are called facultative bacteria.

(*iv*) **According to temperature at which they flourish**

(*a*) *Psychrophilic bacteria* : Those which can survive between 10 to 20°C.

(*b*) *Mesophilic bacteria* : Those which can survive between 20° to 40°C.

(*c*) *Thermophilic bacteria* : Those which can survive between 40 to 65°C.

Bacteria are less sensitive to cold than heat. At low temperatures, they are essentially dormant and can survive for long periods of time. Some bacteria tolerate low pH while others tolerate high pH, the optimum being between 6.5 and 7.5.

(*v*) **According to origin of occurrence** (applicable for bacteria found in water).

(*a*) *Natural water bacteria* : Chromogenic and non-chromogenic bacteria, Fluorescent bacteria, cocci etc. These are widely distributed in water which is their natural environment.

(*b*) *Bacteria from soil* : Sporing and non-sporing organisms.

(*c*) *Bacteria from sewage and animal excrement* : Coliform group of bacteria ; disease producing types such as typhoid and paratyphoid bacteria, intestinal organisms such as the coli-aerogences group, the faecal streptococci and anaerobic spore-bearing bacilli.

(*vi*) Sub-classification of bacteria from sewage and animal excrement

The bacteria of excretal origin may be of three types :

(*a*) The coli-aerogenes group bacteria.

(*b*) Clostridium welchii.

(*c*) Faecal streptococci.

(a) Coli-aerogenes group (coliform group)

The intestines of men and warm blooded animals contain certain harmless bacteria which are excreted with faeces. They are numerous in sewage, and will be present in water which has been contaminated by sewage or excreta. Formerly these organism were known as *colon bacilli,* but at present they are known as *coliforms.* The number of coliform organisms in human faeces is estimated to lie between 10^{11} and 10^{13} per capita daily. The *Coliform group* includes all of the aerobic and facultative anaerobic, non-sporeforming, Gram-negative rod shaped bacteria that ferment lactose (milk sugar) with the production, at 35°C, of gas within 48 hours. The coliform group is composed of two important species, *Escherichia coli (E-coli)* and *Aerobacteria aerogenes.* They are of advantage as indicators of water contamination. Since organisms of coliform group can normally be expected to live longer in water than disease bacteria, a negative test for gasformers indicates that the water is safe.

B-coli (Bacterium coli) and E-coli (Escherichia coli)

The terms B-coli has been used and sometimes is still used to include the whole *coli-aerogenes group.* It should be noted that coliform group of bacteria includes not only organisms originating in the intestinal tract of warm blooded animals (faecal coli, principally *Escherichia coli* or *E-coli*) but also organisms from soil or vegetation (principally, *Aerobacter aerogenes*). A special committee of Members of the Society of General Microbiology and the Society of Applied Bacteriology have proposed that the term *coliform* should refer to *Gram-negative rod resembling Bacterium coli (B-coli) in morphology and staining reactions but not necessarily in cultural and biochemical characteristics*. Such a definition of coliforms includes organisms of the enteric, food poisoning, dysentery groups, plant pathogens and feeble fermenters of lactose. The organisms originating in the intestinal tract of warm blooded animals are called *Escherichia-coli (E-coli),* formerly known as *Bact. coli type I.* Other members of coli-aerogenes group (such as *Bact. aerogenes, Bact. Cloacae* and the *Intermediates*) are found predominantly in soil and vegetation, particularly when the latter is decomposing.

All evidence has so far shown that the *E-coli* (or Bact. co[l]i type I) is the most frequent type present in human and anim[al] intestine ; apart from excretal contamination, it is rarely found outside the animal body. The presence of *E-coli* in a water supply therefore indicates pollution by faecal contamination. Other coli-form bacteria will usually be present as well, but if *E-coli* are absent than the inference is that pollution primarily arises from soil or vegetation.

(*b*) **Clostridium Welchii**

This micro-organism, first described by Welch and Nuttal in 1892, is a Gram-positive rod, having elliptical sub-terminal spores. The organism is found in cultivated soil, sewage and polluted water. Intestines are its natural habitat where it causes no harm but assists digestion.

(*c*) **Faecal streptococci**

Faecal streptococci are found in human intestines. It is a Gram-positive coccus occurring in pairs or short chains. It grows in presence of bile salt and is capable of growing at temperatures as high as 45°C. However, they are not as numerous as *E-coli* in all normal cases. Hence the test for faecal streptococci offers no advantage over the *E-coli test* except in cases of doubt.

6.8. MICROBIOLOGICAL EXAMINATION OF WATER

Microbiological examination of water includes both bacteriological and biological examinations. *Bacteriological examination* of a sample of water is aimed at determining its fitness for use for human consumption, and for use in industries such as food processing and dairy. The water, used should be free from faecal or sewage contamination because the micro-organisms causing water-borne diseases (such as typhoid, cholera and dysentery etc.) are excreted in the faeces of individuals suffering from these diseases. *Biological examination* of water is aimed at determining the presence of the microscopic organisms, other than bacteria, such as algae, moulds or fungi, yeasts, protozoa, rotifers etc. many of which affect the quality of water for drinking and industrial uses.

The following are the *purposes of bacteriological examination of water* :

(*i*) To detect and assess the degree of excremental pollution in the sources of supply.

(*ii*) To assess the amount of treatment required to render a source of supply safe for consumption.

(*iii*) To ascertain the efficiency of the purification treatment at various stages.

(*iv*) To locate the cause of any sudden deterioration in quality.

(*v*) To establish the bacterial purity of final water as it leaves the purification works and to demonstrate the persistence of this high quality in the distribution system and in the premises of consumers.

Indicator Organisms

Pathogenic bacteria and other organisms are usually difficult to detect in a water supply because of their small numbers. Even in sewage effluent or polluted river water, pathogens may be present only infrequently or at irregular intervals. The indirect evidence of presence of the pathogenic bacteria in water is obtained by testing the water for *indicator organisms.* The faecal bacteria although harmless in their normal habitat, do indicate pollution if found in water and give warning that the water must be potentially dangerous. The *coliforms* and the *Eschericha coli* (E-coli) are normally used as indicator organisms, since they are present in large numbers than pathogens. They also lend themselves to numerical evaluation as well as qualitative distinction. They are also identifiable by relatively simple analytical procedures, providing the information quickly and economically.

Crude sewage contains millons of *E-coli* per 100 ml, but the number of pathogenic bacteria present will normally be very small. The absence of *E-coli* from water almost invariably justifies the assumption that water is free from pathogenic organisums. This assumption is true only for faecal pollution, and the conclusion is not applicable where the specific pollution of water is due to a urinary rather than faecal carrier. In this case the urine contains a pure culture of enteric organims and no coli-aerogenes organisms.

Coliform Index and E-coli Index

The number of coliforms in a water sample is deduced from positive results obtained in fermentation tubes inoculated with portions of varying sizes : the result so obtained is only approximate. This approximate or probable number is known as *coliform index* per millilitre or per 100 ml. For a single sample with only a few portions, the index is taken as the reciprocal of the smallest portion showing positive results. For example, suppose a test shows the following results :

10 ml	1 ml	0.1 ml	0.01 ml	0.001 ml
+	+	+	+	–

The coliform index in the above case will be the reciprocal of 0.01, *i.e.*, it will be $\frac{1}{0.01} = 100$ per ml.

If, however, a *number of tests* are carried out for each portion, the following procedure is followed for computations (Table 6.5).

1. Record the quantities tested in order of magnitudes, total number of tests and number of positive results.
2. Record ratio of positive to total successive ratios (column 4).
3. Find the difference between successive ratios (column 5)
4. Find the product of each difference and the reciprocal of the larger quantity used in computing the ratio from which difference was taken (column 6).
5. The *E-Coli index* is the sum of these products.

TABLE 6.5. E-COLI INDEX

Quantities *c.c.* *(1)*	*No. of tests* *(2)*	*Number positive* *(3)*	*Ratio* *(3) : (2)* *(4)*	*Difference between lines in col. (4)* *(5)*	*Product* *(6)*
10	30	20	0.667		
1	30	10	0.333	0.334	0.0334
0.1	30	5	0.167	0.166	0.166
0.01	30	4	0.1333	0.0337	0.337
0.001	30	1	0.05	0.0833	8.333
				Sum	8.869

∴ E-coli-Index = 8.87.

The above method of finding E-coli index has several anamolies, if the observations in column 3 do not fall on the probability curve to get consistent values in column 6. For example, if the 5 th line in colum 3 has a value 2 (instead of 1), the value in column 6 would by 3.333 instead of 8.333. Obviously, for a greater number of positive results, we should have higher index. Thus, error results from assuming a value of 2 which falls off the probability curve in column 3.

The Most Probable Number (M.P.N.)

The shortcomings of the Coliform Index or E-Coli Index have caused it to become obsolete and it has been replaced by *most probable number* (M.P.N.).

The *most probable number* (M.P.N.) of coliform or E-coli is defined as that bacterial density, which if it had been actually present in the sample under examination, would more frequently than any other, have given the observed analytical result. The M.P.N. is determined by applying laws of probabilities to the available test data. Standard tables are available to determine M.P.N. for various combination of test results. Table 6.6 gives M.P.N. of organisms per

100 ml of sample and confidence limits using 1 tube of 50 ml and 5 tubes of 10 ml. For other Tables, reference may be made to IS: 1622–1964.

TABLE 6.6

MPN OF ORGANISMS PER 100 ML OF SAMPLE USING 1 TUBE OF 50 ml AND 5 TUBES OF 10 ml

Number of positive tubes		*MPN per 100 ml*	*Limits within which MPN per 100 ml can lie*	
50 ml tubes	*10 ml tubes*		*Lower limit*	*Upper limit*
0	1	1	<0.5	4
0	2	2	<0.5	6
0	3	4	<0.5	11
0	4	5	1	13
1	0	2	<0.5	6
1	1	3	<0.5	9
1	2	6	1	15
1	3	9	2	21
1	4	16	4	40

Sampling for Bacteriological Tests

Bacteriological samples should be collected in sterile, clear glass bottles having ground glass stoppers which are protected with a piece of sterile linen, oiled silk or paper tied around the neck of the bottle. Before collecting the samples, the bottle should be sterilized in hot air oven at 160°C for one hour or in an autoclave at a pressure of at least 1 kg/cm^2 (121°C), maintained for 20 minutes. The bottle should not be opened until immediately before the sample is to be taken and the stopper, when withdrawn from the bottle should be kept in its covering and held pointing downwards. If the water to be sampled contains or is likely to contain chlorine, a minute crystal of sodium thiosulphute or 0.1 ml of sodium thiosulphate solution (0.3%) should be added to the bottle before sterilization. Bacteriological testing of water should be commenced as soon as possible after collection. Where this is not possible, the samples should be kept in ice or in a refrigerator till they are taken up for analysis. All such iced or chilled samples should be taken up for tests within 48 hours after collection.

1. TOTAL COUNT OR AGAR PLATE COUNT TEST

This test is carried out to find the number of bacteria colonies growing on a specified medium incubated for a specified period of time and at a specified temperature. The enumeration of the total bacteria is therefore known as *Colony Count*. Various methods are available, but the following three methods are in general use for water testing.

(*i*) Colony Count in *nutrient agar* after 3 days of growth at 20 – 22°C.

(*ii*) Colony Count in nutrient agar after 24 hours incubation at 37°C.

(*iii*) Colony Count after continuation of the 37°C incubation for a further 24 hours.

Out of this, the second method is more common. A suitable volume of water sample (or dilution of water sample with sterilised sample) is transferred with aseptic precautions to a sterile glass covered *Petridish* (100 mm diameter and 15 mm deep) by means of a sterile pipette. Molten nutrient agar, cooled to 45°C, is added and thoroughly mixed with the sample in the Petridish when the mixture solidifies, the Petridish is transferred to incubator (37°C), inverted and left for 24 hour for incubation. The bacterial colonies which are formed are then counted with the help of a magnifying glass. Sometimes a specially constructed box (such as Quebec colony counter) is used, which allows the examination of the plates by combined reflected and transmitted light against a dark background in a blacked out room. The results of the colony count are computed for 1 ml of the water sample. For potable water, the total count should not exceed 100 per ml.

2. THE COLIFORM TEST

As defined earlier, the coliform group includes all of the aerobic and facultaive anaerobic Gram-negative, non-spore forming, rod shaped bacteria which ferment lactose with gas formation within 48 hours at 35°C. Two standard test techniques are available for the coliform group.

(*a*) Multiple tube fermentation technique.
(*b*) Membrane filter technique.

(*a*) **Multiple tube fermentation technique.** This consists of the following tests in three stages :

(*i*) Presumptive test.
(*ii*) Confirmed test.
(*iii*) Completed test.

(*i*) **Presumptive test.** The presumptive test consists of inoculating measured amounts of water sample into tubes of Mac Conkey bile salt lactose broth and incubating the cultures at 37°C for 24 to 48 hours. If acid and gas are formed, the test is positive. If they are not formed, the test is negative. If all the tubes are negative, the conclusion is that coliform bacteria are not present in the sample. Positive results require further testing for *confirming* the presence of coliform bacteria in the sample.

The amount of water sample used in the test depends upon the anticipated bacterial purity of water. Volume of water samples in different tubes are taken in multiples of ten, such as 0.01 ml, 0.1 ml, 1 ml, 10 ml, 50 ml etc. Following are the combinations of tubes of media suggested for various types of water :

1. Raw river water–
   5 tubes of 1 ml, 5 tubes of 0.1 ml and 5 tubes of 0.01 ml.
2. Stored water, rapid filtered water or polluted well water–
   1 tube of 50 ml, 5 tubes of 10 ml and 5 tubes of 1 ml.
3. Samples from distribution system –
   1 tube of 50 ml, 5 tubes of 10 ml.

The 50 ml tubes contain 50 ml of double strength Mac Conkey broth medium, 10 ml tubes 10 ml of the double strength medium while 1 ml tubes and tubes of dilutions of 1 ml contain 5 ml of single strength medium.

All the tubes showing acid and sufficient gas to fill the concavity at the top of the Durham tube are recorded as *'presumptive positive'*. The absence of formation of acid and gas at the end of 48 hours incubation constitutes a negative test. The most probable number (MPN) of coliform organism in 100 ml of sample is then determined from the probability tables.

(*ii*) **Confirmed test.** The results obtained in the above test are only presumptive evidence of the presence of *Bact. Coli* and are in fact only presumptive evidence of the presence of any member of the coliform group or coli-aerogenes group of bacteria. Presumptive positive tubes from any sample of chlorinated water should be confirmed for the actual presence of coliform organisms. Such confirmation may not be necessary in the case of unchlorinated samples. Confirmed test is carried out either using *solid medium* or a *liquid medium.*

In the confirmed test with *liquid medium*, the presumptive culture is plated out on MacConkey agar or Eosin-methylene blue agar of Endo agar and incubated at 37°C for 24 hours in order to ensure the growth of single, discrete and pure colonies. Coliform bacteria from deeply coloured nucleated colonies with or without metallic surface luster on Eosin methylene blue agar. On Endo agar, coliform agar, coliform colonies are nucleated dark red with or with or without metallic sheen. Atypical colonies are unnucleated opaque and pink in these media. The colonies of coliform bacteria are pink in colour in MacConkey agar medium. The appearance of typical colonies on the plates within the incubation period constitutes a positive confirmed test.

In the confirmed test with liquid medium, a small portion of lactose broth showing positive presumptive test is carefully transferred to another fermentation tube containing *brilliant green bile lactose broth.* The tube is incubated at 37°C for 48 hours. The formation of gas in any amount in the inverted tube of the fermentation tube constitutes a confirmed test for the coliform organism.

(*iii*) **Completed test.** This test is applied to a typical or atypical colonies found upon the plates of solid medium or to the brilliant green lactose bile broth fermentation tubes showing gas used for the confirmed test. In the latter case, the culture from the brilliant green lactose bile broth tubes is plated out on Endo or Eosin methylene blue plates, as soon as possible after the appearance of gas. The plate is incubated at 37°C for 24 hours. From each of these plates, or from each of the plates used for confirmed test using solid medium, pick up two or more typical coliform colonies and transfer each to a MacConkey broth tube and a nutrient agar slant. If no typical colonies are present, two or more colonies considered most likely to consist of the organism of the coliform group, are picked up for the transfer. The broth tubes are incubated at 37°C for 48 hours and examined for gas production. The tubes showing gas are recorded as positive. The agar slants are incubated at 37°C for 24 to 48 hours and examined for gram-stained preparations of the growths on slants corresponding to the broth tubes that shows gas. The formation of gas in the lactose broth tube, and the demonstration of Gram-negative, non-spore forming, rod shaped bacteria in the agar culture is considered a completed positive test.

(*b*) **Membrane filter technique.** Membrane filter technique is relatively a recent development of field measurement of bacteria. It was first developed in Germany during World War II and subsequently introduced in U.S.A. by Goetz and Tsuneishi in 1951. The method permits field filtration of the water sample and shipment of filters to the laboratory on a preservative medium. The method provides a means for improved control of water quality in water treatment plant and other laboratories since it has a higher degree of precision (reproductibility) than the multiple-tube fermentation test, and it permits the examination of much larger volumes of sample than the multiple-tube-test, with increased sensitivity in coliform detection. Also, definitive results are obtained in a shorter time than with the multiple tube method.

The method consists of filtering the sample of water through a filter membrane of uniform microscopic pores small enough to retain all the coliforms and other bacteria. The membrane filter paper used is 5 cm in diameter and of 150 micron thickness. It is made of biologically inert cellulose esters containing over 500 million capillary pores of uniform dimension. The filter holding assembly consists of a seamless funnel which fastens to a receptable bearing a porous plate for support of the filter membrane. Filter membranes are sterilized prior to use. Using sterile foreceps, the sterile filter membrane is placed over the porous plate of the unit. The filtration procedure consists of passing the measured quantity of water sample through the membrane under partial vacuum. After the sample has been filtered, the funnel is rinsed three times with 20-30 ml volumes of sterile buffered dilution water. The membrane is than lifted up with sterile foreceps, gently rolled, grid side up, onto the surface of an absorbent pad containing M-endo medium. The membrane cultures are incubated at 37°C for 20 hours, in an inverted position in the incubator with 100% humidity. The colony counting exhibit a metallic-appearing surface luster. The colony countig is done either with the help of a low power microscope or with the help of any suitable colony counter. The volume of water sample to be used in the test will be governed by the expected bacterial density. When less than 20 ml of water sample is to be used, it is diluted with sterile buffered dilution water to a minimum of 30 ml just before filtration. The estimated Coliform density of the sample is recorded in terms of coliform per 100 ml sample, using relation :

$$\text{Coliform colonies per 100 ml} = \frac{\text{Coliform colonies counted} \times 100}{\text{ml sample filtered}}$$

The method is not suitable for waters with turbidity due to algae or other material.

3. Test for E-coli

As stated earlier, *E-coli* is the most frequent type present in human and animal intestine. The presence of *E-coli* in a water supply therefore indicates pollution by faecal contamination. Due to this reason, *E-coli* is also sometimes used as the *indicator organism.* The method consists of transferring a small portion of the culture from each positive presumptive tube (coliform presumptive test) as soon as gas is noticed, to one tube of brilliant green bile lactose broth, and one tube of peptone water (preheated to 44°C) and incubating these for 6 to 24 hours in a water bath maintained at 44°C. After the incubation period, the broth tubes are examined for production of gas. Half the contents of the peptone culture is mixed to it gently. The immediate appearance of the peptone culture is transferred to a test tube, and 0.2 to 0.3 ml of Kovac's regent is mixed to it gently. The immediate appearance of a red colour in the upper layer indicates a positive indole reaction. If it indicates negative reaction, the remaining original peptone culture is returned to the water bath and examined for indole production after a further period of 18 hours. The production of a gas in the broth tube and indole in the peptone water culture is a positive test for *E-coli*. If all presumptive positive tubes are tested for *E-coli,* a series of positive and negative results for *E-coli* may be obtained from which *MPN* for *E-coli* in the sample may be estimated using the *MPN* tables.

6.9. THE NUISANCE BACTERIA

There are some types of bacteria which when present in water may cause pitting and tuberculations in pipes, and render the water unsuitable for building purposes, air conditioning, paper manufacture, food and other industries. They may cause odor, tests, frothing, colour and increase in turbidity. Some of these bacteria which are of concern to a water works engineer are (*i*) iron bacteria (*ii*) sulphur bacteria (*iii*) gelatin liquefying bacteria and (*iv*) slime forming bacteria.

Iron Bacteria

Iron bacteria are those organisms which are capable of withdrawing iron present in water and depositing it in a morphologically distinct way. They oxidize ferrous to ferric iron, due to which the frictional loss in the water mains is greatly increased. The growth of these organisms in water produces slimy reddish brown deposits in tanks causing discoloration and turbidity in water. Their presence may cause pitting and tuberculations in pipe and render the water unsuitable for industrial purposes.

Sulphur Bacteria

Sulphur bacteria are autotropic bacteria which oxidize elemental sulphur or reduced sulphur compounds, obtaining their carbon requirements from carbon dioxide. They are undesirable in water used in many industrial processes and the acid produced during their metabolism may be destructive to concrete and other structures. There is also the group of anaerobic bacteria, known as *sulphate reducing bacteria* which reduce sulphate to hydrogen sulphide. They are wide spread in any type of soil, and are therefore essential agents in the bacterial corrosion of iron and steel pipes embedded in the soil. The problem is more severe in neutral, water logged clays and marshy soils where anaerobic conditions exist resulting in the need of replacement of pipe after only few years of service.

Gelatin Liquefying Bacteria

Gelatin liquefying bacteria are those micro-organisms which are capable of producing proteolytic ferments which digest and liquefy gelatin. The presence of these organisms in water is of importance in industries, such as the manufacture of photographic films, edible gelatin, glue and in food processing.

Slime Forming Bacteria

Slime forming bacteria are those non-pathogenic bacteria which may produce slime which will adhere to structures and increase either by growth or by collecting and holding insoluble debris from water supply. The presence of these slime forming organisms is especially undesirable in condenser systems, paper mills, food processing plants etc.

6.10. COMMON WATER-BORNE DISEASES

As stated earlier, water-borne diseases are caused by *pathogenic organisms* carried by water containing faecal or sewage contamination. The important water-borne diseases are : typhoid fever, para typhoid fever, the dysenteries (both amoebic and bacillary), gastro-enteritis, infectious hepatitis, schistosomiasis and Asiatic cholera. These diseases may be grouped in four heads : (*a*) bacterial diseases, (*b*) protozoal diseases (*c*) virus diseases, and (*d*) helminthic (worm) diseases.

(*a*) Bacterial Diseases

1. *Typhoid fever.* Typhoid fever is the most important endemic and epidemic water borne disease of regions where supplies are drawn from surface sources contaminated by human faeces or urine. This is caused by the bacterium *salmonella typhi* which occasionally continues to proliferate in the gall bladders of a few patients who, even after recovery from the primary infection continue to excrete the organisms in their faeces or urine for long period, even for life.

2. *Paratyphoid fever.* This fever is mostly due to ingestion of contaminated food, specially milk, dried or frozen eggs and other dairy products. This is caused by bacterium *Salmonella paratyphi A, B* or *C*. Number and infectivity of organisms released to environment are generally smaller than the typhoid fever.

3. *Cholera.* This is caused by the bacterium *vibrio cholerae*. Usually the infection is contracted by ingestion of water contaminated by infected human faecal material. Sometimes it may be contracted by contaminated food or personal contact. As compared to typhoid fever, cholera is much more violent in its onset (1 to 3 days) and often more fulminating in its clinical course.

TABLE 6.7. STANDARDS FOR QUALITY OF WATER (P = Permissible ; E = Excessive)

	I.C.M.R. Committee (P)	(E)	*W.HO. International* (P)	(E)	*U.S.P.H.S.* (P)	(E)	*A.W.W.A.* (P)	*Environmental Hygience Committee* (P)	(E)
Physical									
Turbidity (p.p.m.)	5	25	5	25	5	—	< 0.1	5	25
Colour	5	25	5	50	15	—	< 0.3	5	25
Taste and odour	Nothing disagreeable		Unobjectionable				None objectionable		
Chemical									
pH	7–8.5	6.5–9.2	7–8.5	< 6.5 or > 9.2	—	—	—	7–8.5	< 6.5 or > 9.2
Total solids (mg/l)	—	—	500	1500	500	—	< 200	500	1500
Total hardness (as $CaCO_3$) mg/l	300	600	—	—	—	—	80 to 100	300	600
Calcium (as Ca) mg/l	75	200	75	200	—	—	—	75	200
Magesium mg/l	50	150	50	150	—	—	—	50	150
Iron (as Fe) mg/l	0.3	1.0	0.3	1.0	0.3	—	< 0.05	0.3	1.0
Manganese (as Mg) mg/l	0.1	0.5	0.1	0.5	0.05	—	<0.01	0.1	0.5
Copper (as Cu) mg/l	1.0	3.0	1.0	1.5	1.0	—	< 0.2	1.0	3.0
Zinc (as Zn) mg/l	5.0	15.0	5.0	15.0	5.0	—	<1.0	5.0	15.00
Chlorides (as Cl) mg/l	250	1000	200	600	250	—	—	250	1000
Sulphate (as SO_4) mg/l	200	400	200	400	250	—	—	250	400
Phenolic substances (as Phenol) mg/l	0.001	0.002	0.001	0.002	0.001	—	—	0.001	0.002
Fluorides (as F) mg/l	1.0	2.0	0.5	1.0 to 1.5	0.8 to 1.70	—	—	1.0	2.0
Nitrates (as NO_3) mg/l	20	50	—	50 to 100	45.0	—	—	20	50

TABLE 6.7. (CONTD.)

	I.C.M.R. Committee		*W.H.O. International*		*U.S.P.H.S*		*A.W.W.A.*	*Environmental Hygiene Committee*	
	(P)	*(E)*	*(P)*	*(E)*	*(P)*	*(E)*	*(P)*	*(P)*	*(E)*
Toxic									
Arsenic (as As) mg/l		0.2	—	0.2	0.01	0.05	—	—	0.2
Barium (as Ba) mg/l	—	—	—	—	—	1.0	—	—	—
Chromium (as hexa-valent) mg/l	—	0.5	—	0.05	—	0.2	—	—	0.01
Cynides mg/l	—	0.01	—	0.01	0.01	0.2	—	—	0.01
Lead (Pb) mg/l	—	0.1	—	0.1	—	0.05	—	—	0.1
Selenium (as Se) mg/l	—	0.05	—	0.05	—	0.01	—	—	0.05
Radio-activity									
Alfa Emmitter (μc/ml)	—	10^{-9}	—	10^{-9}	—	—	—	—	10^{-9}
Beta Emmitter (μc/ml)	—	10^{-8}	—	10^{-8}	—	—	—	—	10^{-8}
Radium^{226} ($\mu\mu c$/ml)	—	—	—	10	—	3	—	—	1
Strontium^{90} ($\mu\mu c$/ml)	—	—	—	30	—	10	—	—	—
Bacteriological Five 10 ml portions	MPN<1 per 100 ml. No. sample should show an MPN of coliform bacteria in excess of 10 per 100 ml.		Not more than 10% of all portions examined shall show presence of coliform bacteria. MPN$\leq$1 per 100 ml. No two consecutive samples shall have the presence of coliform bacteria.		Not more than 10% of all samples shall show the presence of coliform. MPN not to exceed 1 per 100 ml.		Zero or less than 1.0 MPN per 100 ml. None of the samples should contain an MPN of coliform exceeding 10 per 100 ml	No coliform organisms.	

4. *Bacillary dysentery.* This is caused by bacteria of genus *shigella, Sh. dysentrial, Sh. flexneri, Sh. boydii* and *Sh. sonnei*. This disease is mostly contracted due to ingestion of food contaminated by flies or by unhygienic food handlers. It may sometimes be contracted by consuming water contaminated by human faeces.

(*b*) **Protozoal Diseases**

The two important diseases that come under this category are : *Amoebiasis and amoebic dysentery.* These are caused by the protozoon *entamoeba histolytica*. They live in the human large intestine forming cyst which are excreted in the bowel discharges of infected persons and which will live for long periods in water. Infection takes place due to ingestion of these cysts which are carried by water or flies or even human fingers.

(*c*) **Virus Diseases**

Any virus extracted and capable of causing infection when ingested could be transmitted by water. Those viruses that can multiply in the intestinal wall and that are discharged in large numbers in faeces are of more concern. The recognized viruses that are of serious concern are the following members of the *enteric virus* group : *polio viruses, coxsakie viruses, ECHO viruses, the viruses of infectious hepatitis,* the *adenoviruses* and the *reoviruses.* The poliovirus persists in the intestines of infected persons for a short time after infection which probably takes place from contaminated fingers directly or on food. *Poliomyelitis* is common where sanitation and food hygiene are poor. Delhi experienced the waterborne outbreak of *infectious hepatitis* in 1955-56.

(*d*) **Helminthic (worm) disease**

Worm infections are sporadic and occur only under grossly insanitary conditions or through gross mismanagement of sewage disposal system. The eggs and larvae of intestinal worms are discharged into faces. The common diseases that are spread are *schistosomiasis* and *swimmer's itch.* The causes of schistosomiasis is a group of trematode worms (flukes) that inhibit the viens of bladder or large intestines and discharge eggs into the urine or faeces.

6.11. STANDARDS OF PURIFIED WATER

Table 6.7 gives the physical, chemical and bacteriological standards for water quality, as suggested by the following agencies :

(*i*) Indian Council of Medical Research (I.C.M.R.) Committee.
(*ii*) World Health Organisation (W.H.O.) International.
(*iii*) United States Public Health Society (U.S.P.H.S).
(*iv*) American Water Works Association (A.W.W.A.).
(*v*) Environmental Hygiene Committee.

Recommendations of Ministy of Urban Development (MUD)

The manual on Water Supply and Treatment prepared by the Central Public Health and Environmental Engineering organisation, under the Ministry of Urban Development (MUD) India, lays down the following standards for water.

(*a*) **Physical and chemical standards**

The physical and chemical quality of water should not exceed the limits shown in Table 6.8.

TABLE 6.8
PHYSICAL AND CHEMICAL STANDARDS (MUD, INDIA)

S.N. (1)	*Characteristics* (2)		*Acceptable** (3)	*Cause for rejection*** (4)
1	Turbidity (units on J.T.U. scale)		2.5	10
2	Colour (units of platinum cobalt scale)		5.0	2.5
3	Taste and odour		unobjectionable	unobjectionable
4	pH		7.0 to 8.5	6.5 to 9.2
5	Total dissolved solids	(mg/l)	500	1500
6	Total hardness (as $CaCO_3$)	(mg/l)	200	600
7	Chlorides (as Cl)	(mg/l)	200	1000
8	Sulphates (as SO_4)	(mg/l)	200	400
9	Fluorides (as F)	(mg/l)	1.0	1.5
10	Nitrates (as NO_3)	(mg/l)	45	45
11	Calcium (as Ca)	(mg/l)	75	200
12	Magnesium (as Mg)	(mg/l)	$\ngtr 30$***	150
13	Iron (as Fe)	(mg/l)	0.1	1.0
14	Manganese (as Mn)	(mg/l)	0.05	0.5
15	Copper (as Cu)	(mg/l)	0.05	1.5
16	Zinc (as Zn)	(mg/l)	5.0	15.0
17	Phenolic compounds (as phenol)	(mg/l)	0.001	0.002
18	Anionic detergents (as MBAS)	(mg/l)	0.2	1.0
19	Mineral oil	(mg/l)	0.01	0.3

TABLE 6.8 (CONTD.)

(1)	*(2)*		*(3)**	*(4)***
	Toxic Materials			
20	Arsenic (as As)	mg/l	0.05	0.05
21	Cadmium (as Cd)	mg/l	0.01	0.01
22	Chromium (as hexavalent Cr)	mg/l	0.05	0.05
23	Cynides (as CN)	mg/l	0.05	0.05
24	Lead (as Pb)	mg/l	0.1	0.1
25	Selenium (as Se)	mg/l	0.01	0.01
26	Mercury (total Hg)	mg/l	0.001	0.001
27	Polynuclear aromatic hydrocarbons (PAH)		0.2 μ g/l	0.2 μ g/l
	Radio activity			
28	Gross Alpha activity		3 pci/l	3 pci/l
29	Gross Beta activity pCi = pico Curie		30 pci/l	30 pci/l

Notes : 1.* The figures indicated under the column *acceptable* are the limits upto which the water is generally acceptable to the consumers.

2.** Figures in excess of these mentioned under *acceptable* render the water not acceptable, but still may be tolerated in the absence of alternative and better source but upto the limits indicated under the column 'cause for rejection,' above which the supply will have to be rejected

3.*** If there are 250 mg/l of sulphates, Mg content can be increased to a maximum of 125 mg/l with the reduction of sulphates at the rate of 1 unit per every 2.5 units of sulphates.

4. It is possible that some mine and spring water may exceed these radio activity limits and in such cases it is necessary to analyse the individual radio-nuclides in order to access the acceptability or otherwise for public acceptance.

(*b*) **Bacteriological** (see Table 6.9)

(*c*) **Virological Aspects**

0.5 mg/l of free chlorine residual for one hour is sufficient to inactivate virus, even in water that was originally polluted. This free chlorine residual is to be insisted in all disinfected supplies in areas suspected of endemicity of infectious hepatitis to take care of safely of the supply from virus point of view which incidentaly takes care of safety from the bacteriological point of view as well. For farther areas, 0.2 mg/l of free chlorine residual for half an hour should be insisted

TABLE 6.9
GUIDELINE VALUES FOR BACTERIOLOGICAL QUALITY

Organism (1)	*Unit* (2)	*Guide line Value* (3)	*Remarks* (4)
(A) Piped water supplies			
1. Treated water entering the distribution system :			Turbidity < 1 NTU for disinfection with chlorine ; pH preferably 8.0 ; free chlorine residual 0.2–0.5 mg/l following 30 minutes (minimum) contact
(*i*) Faecal coliforms :	Number/100 ml	0	
(*ii*) Coliform organisms :	Number/100 ml	0	
2. Untreated water entering the distribution system :			
(*i*) Faecal coliforms	Number/100 ml	0	
(*ii*) Coliform organisms	Number/100 ml	0	In 98% of samples examined throughout the year–in the case of large supplies when sufficient samples are examined
(*iii*) Coliform organisms	Number/100 ml	3	In an occasional sample but not in consecutive sample.
3. Water in distribution system			
(*i*) Faecal coliforms	Number/100 ml	0	
(*ii*) Coliform organisms	Number/100 ml	0	In 95% of samples examined throughout the year–in the case of large supplies when sufficient samples are examined
(*iii*) Coliform organisms	Number/100 ml	3	In an occasional sample but not in consecutive sample
(B) Unpiped water supplies			
(*i*) Faecal coliform	Number/100 ml	0	
(*ii*) Coliform organisms	Number/100 ml	10	Should not occur repeatedly ; if occurence is frequent and if sanitary protection cannot be improved an alternative source should be found, if possible
(C) Emergency water supplies			
(*i*) Faecal coliforms	Number/100 ml	0	
(*ii*) Coliform organisms	Number/100 ml	0	Advise public to boil water in case of failure to meet guide line values.

PROBLEMS

1. Write a note on common impurities found in water.

2. Describe in brief various tests conducted for physical examination of water.

3. Describe in brief various important tests conducted for chemical examination of water

4. (*a*) What do you understand by Coliform index ? How do you determine it.

(*i*) Write a note on significance of MPN in water quality.

5. What objections could be there if (*i*) turbidity (*ii*) chlorides (*iii*) nitrates, and (*iv*) hardness are present in excess in water.

6. Write a note on bacteriological analysis of water.

7. Write a note on aquatic micro-organisms.

8. Classify various types of bacteria. What do you understand by the coli-aerogeneous group of bacteria ? Explain its significance.

9. Describe the Agar plate count test.

10. Describe the multiple tube fermentation technique for the determination of coliforms.

11. What do you understand by membrane filter technique. Describe.

12. What do you understands by E-coli ? How do you determine its presence in water

13. Describe various 'nuisance bacteria'

14. Write a note on various water borne diseases.

15. Give in a tabular form the standards of drinking water as described by (*i*) A.C.M.R. (*ii*) W.H.O. (*iii*) A.W.W.A. (*iv*) Manual by M.U.D.

16. Write short notes on the following.

(*i*) Hardness of water
(*ii*) pH value and its determination.
(*iii*) Coliform index.
(*iv*) M.P.N.
(*v*) Nitrogen and its compounds.
(*vi*) E-coli.

17. Differentiate between the following :

(*i*) Temporary and permanent hardness
(*ii*) B-coli and E-coli.
(*iii*) Virus and bacteria.
(*iv*) Free ammonia and albuminoid ammonia.

7

Unit Operations

7.1. INTRODUCTION

The term *'unit operations'* first became popular in chemical engineering, in which any chemical process may be resolved into a coordinated series of the so called 'unit actions'. In chemical engineering, the term 'unit operation' is applied to procedures in which changes produced are essentially *physical*. The unit operation approach has been employed in chemical engineering education for years, with considerable success.

However, in water and waste water treatment, *unit operations are those procedures in which the changes produced are physical, chemical or biological*. The unit operation approach in water and waste water treatment has the following advantages.

1. It gives better understanding of the processes inherent in the treatment, and of the capabilities of these processes in attaining the objective.
2. It helps in the development of mathematical and physical modes of treatment mechanisms and the consequent design of treatment plants.
3. In helps in the coordination of effective treatment procedure to attain desired plant performance and effluent quality.

The analysis of the physical, chemical and biological unit operations is generally achieved by (*i*) development of mathematical models and (*ii*) construction of simple physical model to produce the desired reaction.

7.2. IMPORTANT UNIT OPERATIONS

The following are important unit operations employed in water and waste water treatment.

1. ***Gas Transfer*** : Aeration.
2. ***Ion Transfer*** :
   (*a*) Chemical coagulation
   (*b*) Chemical precipitation
   (*c*) Ion exchange
   (*d*) Adsorption.
3. ***Solute Stabilization***
   (*a*) Chlorination
   (*b*) Liming.
   (*c*) Recarbonation
   (*d*) Super-chlorination
4. ***Solids Transfer***
   (*a*) Straining
   (*b*) Sedimentation
   (*c*) Floatation
   (*d*) Filtration
5. ***Nutrient or molecular Transfer***
6. ***Interfacial contact***
7. ***Miscellaneous operations*** :
   (*a*) Disinfection
   (*b*) Copper sulfating
   (*c*) Fluoridation
   (*d*) Thermal desalination.
8. ***Solid concentration and stabilization***
   (*a*) Thickening
   (*b*) Centrifuging
   (*c*) Chemical conditioning
   (*d*) Elutriation

(*e*) Biological floatation
(*f*) Vacuum filtration
(*g*) Air drying
(*h*) Heat drying
(*i*) Sludge digestion
(*j*) Incineration
(*k*) Wet combustion.

The aims of some of these operations one described in subsequent articles.

7.3. GAS TRANSFER

In this unit operation, the gases are either released/desorbed from water or are dissolved/absorbed in water. This is achieved by exposing the water through spray or bubble aeration to either air or other atmospheres, under normal, increased or reduced pressures. This is an important unit operation which is carried out for water purification to achieve the following.

(*i*) Removal of objectionable gases such as carbon dioxide, hydrogen sulfide and other volatile odorous substances, by spray or bubble aeration.

(*ii*) Deferrization and/or demanganisation of water, through the addition of oxygen by spray or bubble aeration.

(*iii*) Addition of ozone from ozone generators or chlorine gas from chlorine dispensers, either for the disinfection of waters or for the destruction of odors and tastes in waters.

(*iv*) Addition of carbon dioxide from the gas or carbon dioxide generators to *recarbonate* lime softened water.

(*v*) Removal of corrosion-promoting oxygen as well as other gases (degasification) by spraying water into vacuum chamber at ordinary temperatures or at elevated temperatures.

The first two operations of gas transfer, mentioned above, are commonly called *aeration*. In most instances, the shared engineering objective of aeration is either the *removal* of gas and other volatile substances from water or their *addition* to water, or both at the same time. Commonly used aerators fall under four categeries :

(*i*) Gravity aerators
(*ii*) Spray aerators
(*iii*) Diffusers
and (*iv*) Mechanical aerators.

7.4. ION TRANSFER

In this unit operation, ion transfer takes place by means of (*i*) chemical coagulation, (*ii*) chemical precipitation, (*iii*) ion exchange, or (*iv*) adsorption.

Chemical coagulation is aimed at the removal of colloidal particles from water. For this purpose, floc forming chemicals, known as coagulants (such as aluminium or iron salts) are added to water which results in the increase in particle size. Organic clusters of different size, shape and weight are formed, and they settle more rapidly. The coagulants themselves are soluble, but they are precipitated by transfer of their ions to substances in the water.

While chemical coagulation removes the suspended impurities from water, *chemical precipitation* throws out of solution the dissolved substances through ion transfer. The chemicals added to achieve this purpose dissolve in water, and their released ions react with the ions in water, resulting in the precipitation of dissolved impurities. The common examples of chemical precipitation are (*i*) flocculation of iron from water, by the addition of lime, (*ii*) precipitation of carbonate hardness by addition of lime (*iii*) precipitation of non-carbonate hardness by addition of soda-ash and (*iv*) precipitation of fluoride ions from water by addition of tricalcium phosphate.

Ion exchange is the reversible interchange of ions between water and a solid ion-exchange media. *Water softening* by ion exchange is an important example. To be effective, solid ion exchange must (*i*) contain ions of their own, (*ii*) be insoluble in water and (*iii*) provide enough space in their porous structure for ions to pass freely in and out of the solid. The three main groups of ion exchange processes are (*i*) water softening, (*ii*) dealkalization, and (*iii*) demineralization. Softening serves to remove the salts of calcium and magnesium. Dealkalization consists of removal of the hydroxides, carbonates and bicarbonates of calcium, sodium and magnesium. Demineralization serves the total removal of all ions. In industry, both cation and anion exchangers are employed to prepare boiler feed water and deionize or demineralize process-waters.

Adsorption is the process of ion-transfer in which *interfacial forces* remove the ions and molecules (known as adsorbates) from solution (water) and concentrate at the interface of adsorbents. The use of granular carbon or powdered activated carbon applied ahead of water treatment plant clarifiers or filters, for the adsorption of odor-and taste-producing ions and molecules, is well known. Organic contaminants are removed from water by the process of adsorption which results from the attraction and accumulation of one substance on the surface of another. The chemical nature of the carbon surface is of relatively minor significance in the adsorption of organics from water and secondary to the magnitude of the surface area of carbon available. Hence high surface area is of prime importance in adsorption. Granular activated carbons typically have surface area of 500-1400 m^2/g. The major contribution to surface area is located in the pores of molecular dimensions.

7.5. SOLUTE STABILIZATION

In this operation, water is stabilized such that objectionable solutes are converted into unobjectionable forms without removal. The common forms of solute stabilization are :

(*i*) *Chlorination* : By chlorination, hydrogen sulfide in water is oxidised into sulfate.

(*ii*) *Liming* : By passing the water through chips of marble, lime stone or dolomites, carbon dioxide in excess of equilibrium requirements is converted into soluble bicarbonate.

(*iii*) *Recarbonation*. By recarbonation of water softened by excess-time treatment, excess lime is converted into bicarbonate.

(*iv*) *Superchlorination* : By addition of chlorine dioxide to water, odour-producing substances are oxidised.

7.6. SOLIDS TRANSFER

In this unit operation, solids are removed from water. Depending upon the size of the solids, this transfer can take place by straining, sedimentation, floatation and filtration– adopted in order of the decreasing size of the solids. *Straining* removes the floating and suspended solids (such as leaves, sticks, debris etc.) that are larger in size than the openings of screens and racks. Straining out of algae is accomplished by microscreens. *Sedimentation* permits the removal of suspended solids from water, in settling tanks in which the carrying and scouring power of flowing water are reduced until suspended particles settle by gravity. *Floatation* is obtained by addition to water of flotation agents that attach themselves to suspended solids or attach suspended solids to bubbles of air and lift the particles to the surface, which are then removed from the floatation devices. *Filtration* is one of the most important operation of solids transfer which combines straining, sedimentation and interfacial contact where-in the suspended solids or flocs are transferred onto grains of the filter media containing sand, coal or other granular materials.

7.7. WATER TREATMENT PROCESSES

Raw water may contain suspended, colloidal and dissolved impurities. The purpose of water treatment is to remove all those impurities which are objectionable either from taste and odour point of view or from public health point of view. Following are the purposes of water treatment :

(*i*) to remove colour, dissolved gas and murkiness of water.

(*ii*) to remove objectionable tastes and odour.

(*iii*) to remove the disease producing micro-organisms so that water is safe for drinking purposes.

(*iv*) to remove hardness of water.

(*v*) to make it suitable for a wide variety of industrial purposes such as steam generation, brewing, dyeing etc.

For surface waters, following are the treatment processes that are generally adopted.

1. **Screening.** This is adopted to remove all the floating matter from surface waters. It is generally provided at the intake point.

2. **Aeration.** This is adopted to remove objectionable tastes and odour and also to remove the dissolved gases such as carbon dioxide, hydrogen sulphide etc. The iron and manganese present in water are also oxidised to some extent. The detailed description of this process is given in chapter 8. This process is optional and is not adopted in cases where water does not contain objectionable taste and odour.

3. **Sedimentation with or without coagulants.** The purpose of sedimentation is to remove the suspended impurities. With the help of plain sedimentation, silt, sand etc. can be removed. However, with the help of sedimentation with coagulants, very fine suspended particles and some bacteria can be removed.

4. **Filtration.** The process of filtration forms the most important stage in the purification of water. Filtration removes very fine suspended impurities and colloidal impurities that may have escaped the sedimentation tanks. In addition to this, the micro-organisms present in water are largely removed.

5. **Disinfection.** It is carried out to eliminate or reduce to a safe minimum limit, the remaining micro-organisms, and to prevent the contamination of water during its transit from the treatment plant to the place of its consumption.

6. Miscellaneous processes : These include water softening, desalination, removal of iron, manganese and other harmful constituents.

Fig. 7.1 gives the schematic layout of a water treatment plant

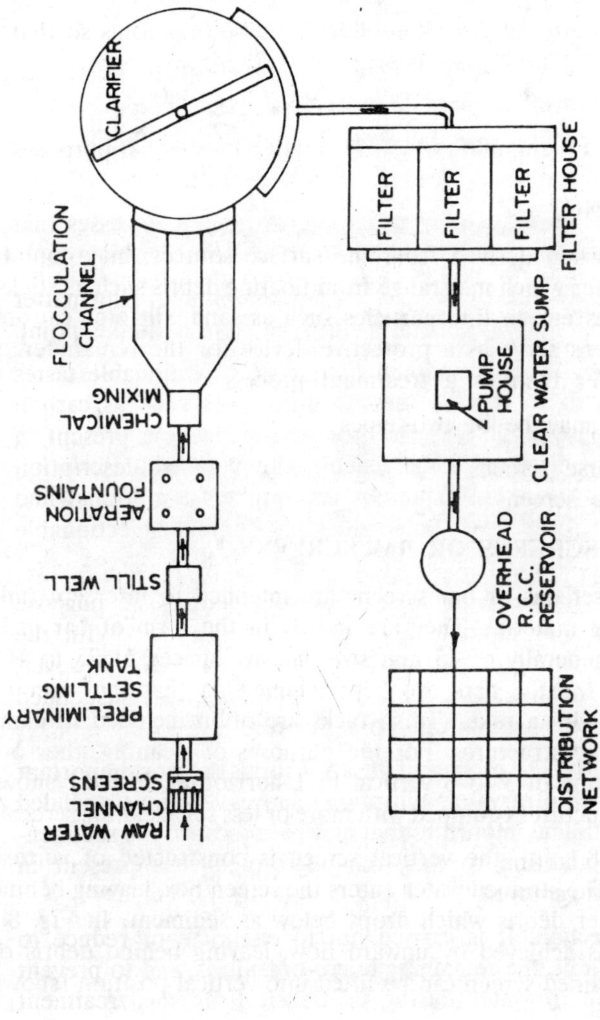

FIG. 7.1 SCHEMATIC LAYOUT OF WATER TREATMENT PLANT.

8

Screening and Aeration

8.1. SCREENING

Water, when derived from the surface sources, may contain suspended matter which may range from floating debris such as sticks, branches, leaves etc. to fine particles such as sand, silt etc., causing turbidity. Screens serve as a protective device for the remainder of the plant rather than as a treatment process.

Screens may be of two types :

(*i*) Coarse screens

and (*ii*) Fine screens.

8.2. COARSE SCREENS OR BAR SCREENS

Coarse screens or bar screens are intended to intercept only grosser floating material. They are mostly in the form of bar grill. The bars are generally of 25 mm size and are spaced at 75 to 100 mm centres. Mostly, bars are kept inclined so that they can be cleaned easily with a rake. Trash racks are often included in dams and other intake structures. For the purposes of cleaning, they are placed on a slope of 3 to 6 vertical to 1 horizontal. Fig. 8.1 shows three intake structures equipped with more or less self cleaning screens.

In Fig. 8.1 (*a*), the vertical screen is constructed of vertical wires or bars. The strained water enters the screen box, leaving behind leaves and other debris which drops below as sediment. In Fig. 8.1 (*b*), straining is achieved by upward flow, leaving behind debris or waste. The inclined screen can be lifted into vertical position (shown

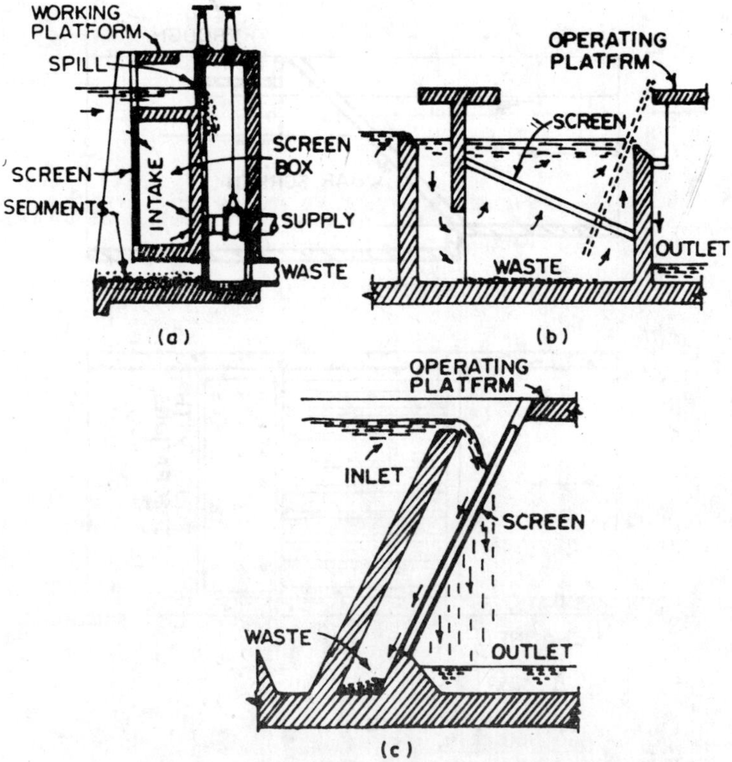

FIG. 8.1. RACKS AND SCREENS OF WATER INTAKES.

by dotted lines), for cleaning. Fig. 8.1 (*c*) shows a self flushing inclined screen. In most of the cases, the common arrangement is to slide a pair of removable screens into vertical grooves in the walls and bottom of the inlet channel.

Head loss through racks and screens

The head loss through unobstructed screens depends upon the nature of their construction (open area, blocked area, shape of screen elements etc.), as well as the approach velocity. Kirschmer (1926) has suggested the following empirical relationship and coefficients for racks with different shaped bears

$$h = \beta\,(w/b)^{4/3}\,h_v\,.\sin\theta \qquad ...(8.1)$$

where h = head loss

h_v = approach velocity head,

b = minimum width of clear opening between pairs of bars

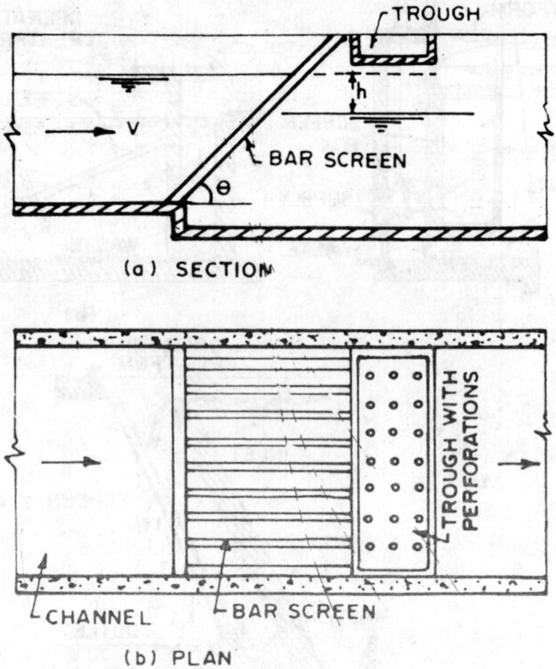

FIG. 8.2.

θ = angle of rack with horizontal

β = a shape factor

= 2.42 for sharp-edged rectangular bars

= 1.83 for rectangular bars with semi-circular upstream face

= 1.79 for circular rods.

Some manufacturers recommend dropping the channel 150 to 300 mm across a bar screen. The maximum head loss through clogged racks and screens is generally below 80 cm.

8.3. FINE SCREENS

Fine screens are used at surface water intakes, sometimes alone, some times following a bar screen. In order that fine screens do not get clogged up, some device is setup to clean it continuously. Due to this reason, fine screens are usually arranged as end less bands or drums of material perforated with holes of about 6 mm

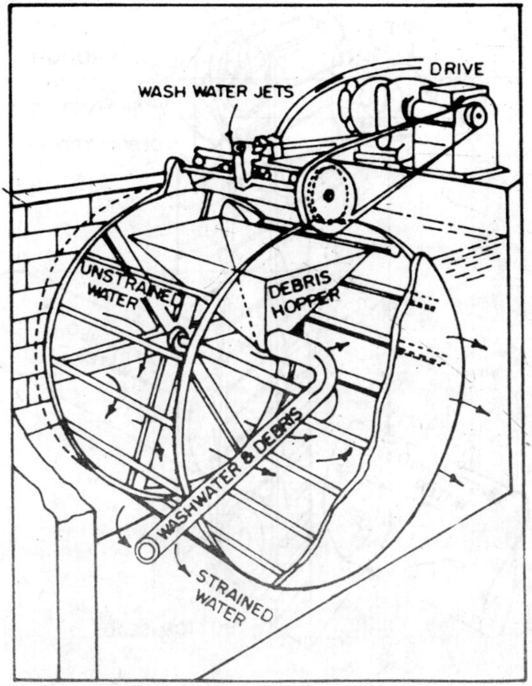

FIG. 8.3. ROTARY DRUM STRAINER.

diameter. The process is therefore known as *'automatic straining'*. Automatic strainers are self cleaning and they work continuously, eliminating solids from the liquid flowing through and disposing of these solids, also continuously. In automatic strainers, a pair of *straining fabric* is submerged in water to be strained while the remainder of the fabric remains above the water. Fig. 8.3 shows a *rotary drum strainer*, while Fig. 8.4 shows the *band screen strainer*.

8.4. MICRO-STRAINERS

A micro-strainer incorporates a specially woven stainless steel wire cloth mounted on the periphery of a revolving drum filled with continuous back washing arrangements. Two of the grades widely used have limiting apertures of 23 microns and 35 microns respectively. Micro-strainers are useful for screening stored waters which do not contain a large amount of suspended matter but which contain plankton, algae and other microscopic sized particles. They are installed to the upstream of rapid gravity or slow sand filters whose output may there by increased by as much as 50%. In operation, the revolving drum is kept submerged in the flowing water to approximately two thirds of its depth. Raw water from lake or reservoir

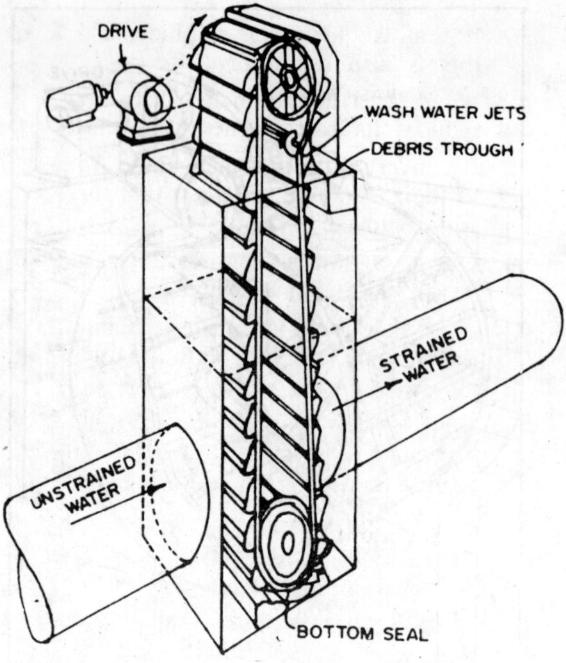

FIG. 8.4. BAND OR TRAVELLING SCREEN STRAINER.

enters through the upstream open end of the drum and flows radially outwards through the microfabric, leaving behind the suspended solid content. The solids so strained are carried upwards on the side of the fabric, with the help of a row of wash water jets which may use about 1% of total quantity of water strained. A single unit of 3 m diameter × 3 m wide drum may deal with about 50000 to 80000 litres per hour. It should be noted that microstraining cannot remove colour or finely divided matter such as clay.

8.5. AERATION

Aeration is one of the important unit operation of gas transfer. The aim of the aeration is to create extensive, new, and self-renewing interfaces between air and water, to keep interfacial films from building up in thickness

Objectives: Aeration of water is done to accomplish the following objectives.

1. It removes tastes and odours caused by gases due to organic decomposition.
2. It increases the dissolved oxygen content of the water.

3. It removes hydrogen sulphide, and hence odour due to this is also removed.
4. It decreases the carbon dioxide content of water, and thereby reduces its corrosiveness and raises its pH value.
5. It converts iron and manganese from their soluble states to their insoluble states, so that these can be precipitated and removed.
6. Due to agitation of water during aeration, bacteria may be killed to some extent.
7. It is also used for mixing chemicals with water, as in the Aeromix process and in the use of diffused compressed air.

8.6. TYPES OF AERATORS

Aeration is done by the following main types of aerators :

(*a*) Free fall aerators or gravity aerators
- (*i*) Cascade aerators
- (*ii*) Inclined apron aerators
- (*iii*) Slat tray aerators
- (*iv*) Gravel bed aerators (trickling beds)

(*b*) Spray aerators
(*c*) Air diffuser basins.

1. Cascade aerators. Cascade aerators are the simplest of the free fall aerator. Weirs and waterfalls of any kind are cascade aerators (Fig. 8.5 *a*). A simple cascade consists of a series of three or four steps – of concrete or metal. Water is allowed to fall through a height of 1 to 3 metres, and due to this it comes into close contact with air. The cascades can be either in open air, or may be in a room which has plenty of louvred air inlet. The reduction of CO_2 is usually in the range of 50 to 60%.

2. Inclined apron aerator with riffle plates : In this type of aerator, water is allowed to fall along an inclined plane/apron

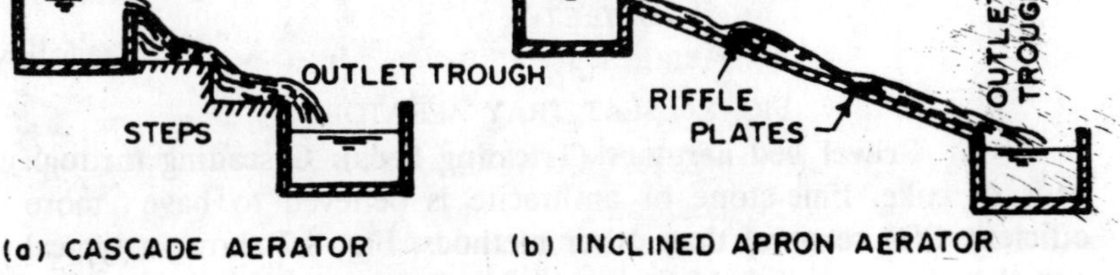

FIG. 8.5. GRAVITY AERATORS.

which is usually studded with riffle plates in herring bone fashion. The breaking-up to the sheet of water will cause agitation of water and consequent aeration (Fig. 8.5 *b*)

3. Slat tray aerators. This is most commonly used. It consists of a closed round or square structure containing a series of closely-stacked superimposed wood-slat trays (Fig. 8.6). Water enters the top of the aerator and is evenly distributed over the topmost tray. The slats in the trays are staggered so that the films of water raining over the edges of the slat in one tray fall on the centres of the slats in the tray just below. Air is supplied to the bottom of the aerator with the help of a blower, which blows it upward. A ventilator is provided at the top, which discharges air and gases to the atmosphere. Water is collected in the collector pan at the bottom, from where it flows to a catch basin or reservoir.

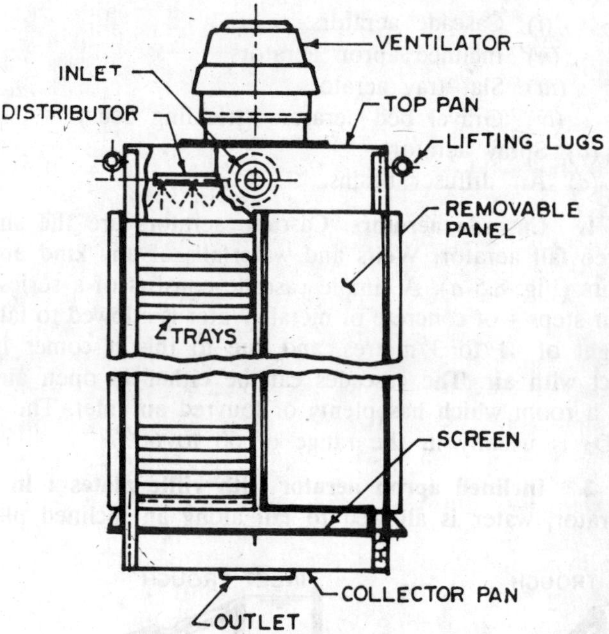

FIG. 8.6. SLAT TRAY AERATOR.

4. Gravel bed aerators (Trickling beds). Cascading through beds of coke, lime-stone or anthracite is believed to have more efficient CO_2 removal than other methods. Fig. 8.7 shows a typical gravel bed aerator in which water is applied at the top and trickles

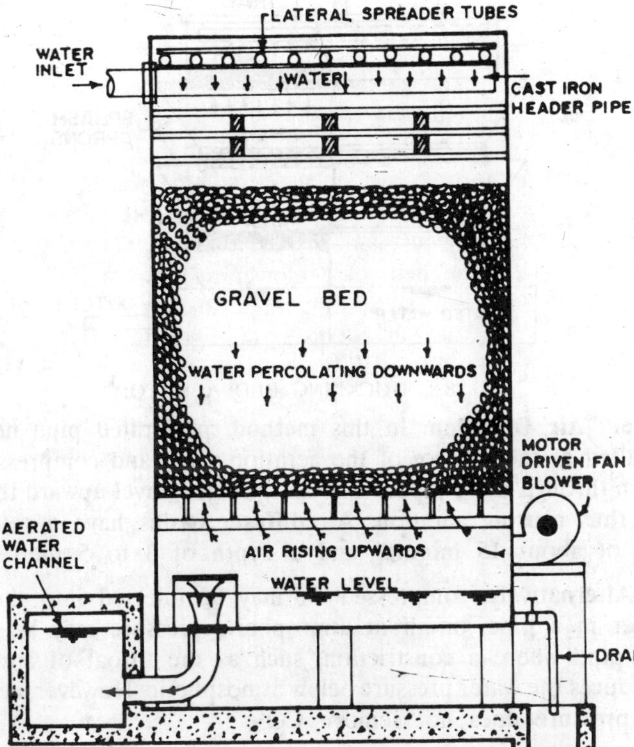

FIG. 8.7. GRAVEL BED AERATOR

down while air is blown upwards. The thickness of gravel bed may be from 1 to 1.5 m.

In another form, commonly known as *trickling beds,* three or four trays filled with coke, slag or stone are used. The thickness of bed in each tray is kept about 0.5 to 0.6 m, while the vertical distance between the bed is kept about 0.5 m. Water is applied from top through a perforated distribution pipe as shown in Fig. 8.8. During the trickling process, aeration takes place.

4. Spray Aerators. Spray aerators divide the water flow into fine streams and small droplets which come into intimate contact with the air in their trajectory. Water is sprinkled in fine jets through nozzles. It requires considerable head (0.75 to 1.5 kg/cm^2), but it reduces carbon dioxide by 70 to 90% or more.

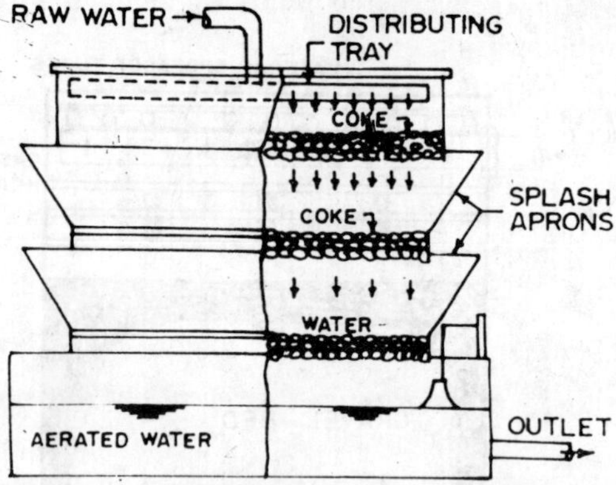

FIG. 8.8. TRICKLING BED AERATOR

5. **Air Diffusion.** In this method, perforated pipe network is installed at the bottom of the aeration tank, and compressed air is blown through these pipes. The air bubbles travel upward through water, thus causing aeration. Air diffuser basins have a retention period of about 15 minutes and a depth of 3 to 5 metres.

Alternatively, compressed air may be injected into the flow of water in a pipe, or air at atmospheric pressure may be drawn in the pipe where a constriction, such as the throat of a venturi tube reduces the water pressure below atmospheric. However, aeration under pressure does not remove CO_2.

8.7. FACTORS GOVERNING AERATION OR GAS TRANSFER

(*a*) **Rate of gas absorption**

If it is postulated that the rate of gas absorption is proporticnal to the degree of *under saturation* (or *saturation deficit*) in the absorbing liquid, we have

$$\frac{dC}{dt} = K_g (C_s - C_t) \qquad ...(8.2)$$

where $\frac{dC}{dt}$ = change in concentration, or rate of absorption, transport or transfer at time *t*.

C_s = saturation concentration of gas in water, at a given temperature (*i.e.* maximum concentration in absorption).

C_t = concentration at time *t*.

K_g = proportionality factor for existing condition of exposure.

Integrating between the limits C_0 at $t = 0$ to C_t at $t = t$, we get the basic equation.

$$C_t - C_0 = (C_s - C_0)\,[1 - \exp\,(-K_g\,t)] \qquad ...(8.3)$$

In Eq. 8.3, the proportionality factor K_g increases with temperature and the *degree of mixing* of the gas with liquid. Since the molecules of gas must pass through the gas-liquid interface, K_g is a function of A/V ratio.

$$\therefore \qquad K_g = k_g \frac{A}{V} \qquad ...(8.4)$$

where k_g = *gas transfer coefficient*, having dimension of velocity.

A/V = area of interface per unit volume of liquid

A = largest practicable area of interface between a given water volume V and air (or gas).

For absorption $C_0 < C_t < C_s$ and both $(C_t - C_0)$ and $(C_s - C_0)$ are positive

For desorption, $C_s < C_t < C_0$ and both $(C_t - C_0)$ and $(C_s - C_0)$ are negative.

The value of $K_g\ (= k_g A/V)$ must normally be determined experimentally and verified in plant-scale tests.

(*b*) **Rate of gas desorption**

Eq. 8.3 derived for rates of absorption applies also to the rate of desorption. In contrast to absorption, the rate of desorption, precipitation, release, or dissolution of a gas from a liquid becomes proportional to its degree of *oversaturation* in the liquid, or the *saturation surplus.*

(*c*) **Factors governing aeration or gas transfer.**

From Eq. 8.3, we have

$$C_t = C_0 + (C_s - C_0)\,[1 - \exp\{-k_g\,(A/V)\,t\}] \qquad ...(8.3\ a)$$

From this, we observe that gas transfer can be optimised by:

(*i*) Generating the largest practicable area A of interface between given volume (V) of water, and air or gas

(*ii*) Preventing the build-up of thick interfacial film, thereby keeping the value of k_g higher.

(*iii*) Inducing longer time (t) of exposure

(*iv*) Ventilating the aerator and its components.

Hence to ensure proper aeration it is necessary :

(*a*) to increase the area of water in contact with air *i.e.* if the water is sprayed, the smaller the droplets produced, the greater will be the area available. Similarly, if the water is being made to fall as a film over packing material in a tower, the smaller the size of the packing material, the greater will be the area available ;

(*b*) to keep the surface of the liquid constantly agitated so as to reduce the thickness of the liquid film which would govern the resistance offered to the rate of exchange of the gas ; and

(*c*) to increase the time of contact of water droplets with air or increase the time of flow which can be achieved by increasing the height of jet in spray aerator and increasing the height of tower in the case of packed madia.

Introducing factor $f = \left(\frac{A}{V}\right) t$, let us compare the efficiencies of spherical bubbles of air, rising up through water, with that of spherical droplets of water falling under gravity.

Spherical droplets of water

In the case of spray aerators, let the spherical droplets of water fall down after attaining a height h. Then, we have

$$h = \frac{1}{2} g t^2$$

$$\therefore \quad t = \sqrt{\frac{2h}{g}}$$

$$\therefore \quad \text{Factor } f_s = \left[\left(\frac{A}{V}\right) t\right]_{\text{spherical}} = \frac{6}{d}\sqrt{\frac{2h}{g}} \qquad ...(8.5)$$

(since ratio A/V for a sphere is equal to $6/d$ where d is the dia. of spherical droplets).

Spherical bubbles of air

Let the bubbles rise throgh water column of height h, with a velocity v.

$$\therefore \quad \text{Time taken, } t = \frac{h}{v}$$

$$\therefore \quad \text{Factor } f_b = \left[\left(\frac{A}{V}\right) t\right]_{\text{bubble}} = \frac{6}{d} \cdot \frac{h}{v} \qquad ...(8.6)$$

Hence, ratio $$\frac{f_b}{f_d} = \left(\frac{6}{d}\frac{h}{v}\right) \div \left(\frac{6}{d}\sqrt{\frac{2h}{g}}\right)$$

or
$$\frac{f_b}{f_d} = \frac{1}{v}\sqrt{\frac{gh}{2}} \qquad ...(8.7)$$

For illustration purpose, let the rising velocity of air bubbles be 0.3 m/sec, rising through a height $h = 3$ m.

$$\therefore \quad \frac{f_b}{f_d} = \frac{1}{0.3}\sqrt{\frac{9.18 \times 3}{2}} \approx 12.4.$$

This shows that *bubble aeration* has an advantage over falling droplets of water. However, k_g is expected to be less for bubbles than for droplets because of less turbulence at the interface.

Example 8.1. *In an experiment on the removal of carbon dioxide from water sprayed into the air in spherical droples 5 mm in dia., the initial supersaturation of the water with CO_2 was 25.8 mg/l. After 1.2 seconds of exposure, this was reduced to 11.2 mg/l. Find the coefficient of gas transfer.*

Solution.

Given $\quad C_s - C_0$ = initial super-saturation= − 25.8 mg/l.

$\quad C_t - C_s$ = 11.2 mg/l.

Adding the two, $C_t - C_0 = -25.8 + 11.2 = -14.6$ mg/l

Hence from Eq. 8.3,

$$C_t - C_0 = (C_s - C_0)\left[1 - e^{(-K_g \cdot t)}\right]$$

or $$-14.6 = -25.8\,[1 - e^{(-K_g \times 1.2)}]$$

or $$1 - e^{-1.2K_g} = 0.5659$$

or $$e^{-1.2K_g} = 1 - 0.5659 = 0.4341$$

or $$-1.2\,K_g = \log_e 0.4341 = -0.8345$$

or $$K_g = 0.6954 \text{ sec}^{-1}$$

Droplet volume per unit surface area is given by

$$\frac{V}{A} = \frac{d}{6} = \frac{0.5}{6} = 0.0833 \text{ cm}^3/\text{cm}^2$$

$\therefore$ Transfer coefficieng $k_g = K_g \cdot \frac{V}{A} = 0.6954 \times 0.0833$ cm/sec.

$\approx 5.8 \times 10^{-2}$cm/sec $\approx$ **209 cm/hour**

8.8. DESIGN OF GRAVITY AERATORS

In the case of gravity aerators, water is made to fall through the available head h, which can be put to use either in single descent or in multiple descent.

At a given instant, $\frac{dh}{dt} = v = g \cdot t$

$$\therefore \quad \int_0^h dh = g \int_0^t t\,dt$$

or
$$h = \tfrac{1}{2} g t^2 \qquad \text{...(8.8)}$$

where
h = height of fall, in metres
t = time of fall, in sec.
v = velocity in m/sec.
g = acceleration due to gravity, in m/sec^2

In a single descent through height h,

Elapsed time
$$t_1 = \sqrt{\frac{2h}{g}} \qquad \text{...(8.9 a)}$$

In n discents through the same total height h,

$$t_n = n\sqrt{\frac{2(h/n)}{g}} = \sqrt{\frac{2nh}{g}} \qquad \text{...(8.9 b)}$$

Hence we have $t_n = t_1\sqrt{n}$ ...(8.9 c)

It should be noted that the quality of exposure is poorer in multiple descents because droplets donot necessarily break away from jets of falling water as soon as they stike the air.

Example 8.2. *Find the time of exposure of water in gravity aerators falling through a distance of 4 m in (a) single descent, (b) in three descents.*

Solution.

(*a*) For single descent, we have from Eq. 8.9 (*a*),

$$t = \sqrt{\frac{2h}{g}} = \sqrt{\frac{2 \times 4}{9.81}} \approx \mathbf{0.90} \text{ sec.}$$

(*b*) For multiple descent, having $n = 3$,

$$t = \sqrt{\frac{2nh}{g}} = \sqrt{\frac{2 \times 3 \times 4}{9.81}} = \mathbf{1.56} \text{ sec.}$$

8.9. DESIGN OF FIXED SPRAY AERATORS

The following three factors affect the hydraulic performance of fixed-spray pressure aerators :

(*i*) Orifice and nozzle behavior
(*ii*) Wind effects.
and (*iii*) Pipe friction associated with multiple take-offs.

1. Orifice and nozzle behaviour

Spray aerators are normally composed of perforated or nozzle pipes which create a spray pattern. The *initial spray velocity* (v) is given by

$$v = C_v \sqrt{2gh} \quad \text{...(i) ...(8.10)}$$

where C_v = velocity coefficient (≈ 0.95)

h = orifice head or driving head.

For a pipe having multiple openings,

$$\therefore \quad Q = C\,(\Sigma a)\sqrt{2gh} \quad \text{...(ii) ...(8.11)}$$

where Q = rate of discharge.

C = discharge coefficient

= 0.8 for rounded openings

= 0.85 to 0.92 for nozzles

= 0.6 for sharp edged openings

Σa = total area of openings

If there are n openings, each of equal area a,

$$\Sigma a = na \quad \text{...(iii)}$$

The water rises either vertically or at an angle (α) and falls onto a collecting apron, after moving along a trajectory.

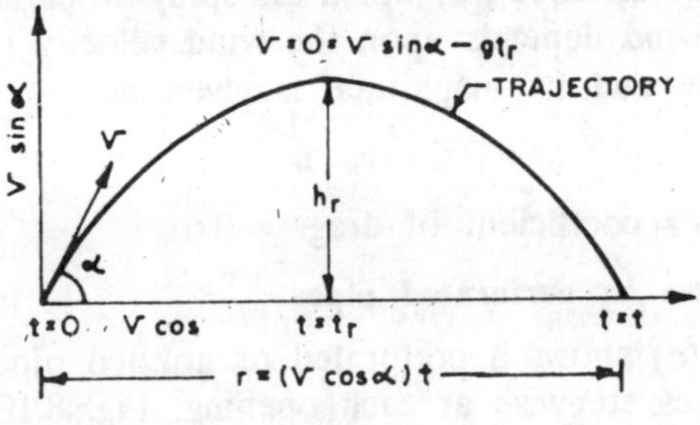

FIG. 8.9.

Let h = driving head

t_r = time of rise of spray

t = total time of exposure $= 2\,t_r$

α = inclination of jet.

At the time of rise t_r, $v = 0$

$$\therefore \quad v = 0 = v \sin\alpha - g\,t_r$$

$$\therefore \qquad t_r = \frac{1}{2}t = \frac{v \sin \alpha}{g}$$

or

$$t_r = \frac{C_v \sqrt{2gh} \sin \alpha}{g} = C_v \sqrt{\frac{2h}{g}} \,.\sin \alpha \qquad \text{...(8.12)}$$

The driving head h, carrying the spray to the top of the trajectory is

$$h = \frac{g\, t_r^2}{2\, C_v^2 \sin^2 \alpha} \qquad \text{...(8.13)}$$

The *horizontal carry* or radius of circle of spray is given by

$$r = (v \cos \alpha)\, t = 2\, v \,.\, t_r \cos \alpha$$

$$= 2 \left\{ C_v \sqrt{2gh} \right\} \left\{ C_v \sqrt{\frac{2h}{g}} \sin \alpha \right\} \cos \alpha.$$

or

$$r = 2\, C_v^2\, h \sin 2\alpha \qquad \text{...(8.14)}$$

The vertical rise h_r of the spray is given by

$$h_r = \frac{1}{2} g\, t_r^2 = C_v^2 \,.\, h \sin^2 \alpha \qquad \text{...(8.15)}$$

2. Wind effects.

The above analysis is based on the assumption that wind does not affect the trajectory. Though *air resistance* does not reduce the height or time of rise appreciably, the wind does affect performance of the spray. The distance l by which the spray droplets are carried forward by the wind depends upon the wind velocity (v_w) and the time of exposure and its magnitude is given by

$$l_w = 2\, C_D \,.\, v_w \,.\, t_r \qquad \text{...(8.16)}$$

where C_D = coefficient of drag ≈ 0.6

3. Friction in perforated pipes

Fig. 8.10 (*a*) shows a perforated or nozzled pipe in which the flow decreases stepwise at each opening. Fig. 8.10 (*b*) shows the variation of discharge Q and unit frictional-resistance (s) along the length.

For an idealized flow in a slotted pipe,

$$s = k\, Q^n \qquad \text{...(1)}$$

Also, at distance $(L - l)$ from the

$$Q = \frac{Q_e\,(L - l)}{L} \qquad \text{...(2)}$$

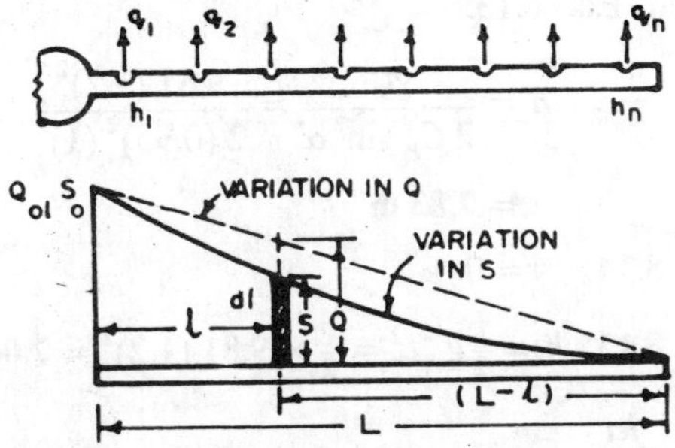

FIG. 8.10. UNIFORMLY DECREASING FLOW.

where subscript e denotes the entrant flow and resistance to it.

The total head loss h_f is given by

$$h_f = \int_0^l s\,dl = k\left(\frac{Q_e}{L}\right)^n \int_0^l (L-l)^n\,dl$$

or

$$= \frac{s_e}{L^n}\int_0^l (L-l)^n\,dl$$

or

$$h_f = \frac{s_e}{n+1}\left[L - \frac{(L-l)^{n+1}}{L^n}\right] \qquad ...(3)$$

Taking $l = L$ and $n = 2$ for the Chezy formula, we get

$$h_f = \frac{s_e}{2+1}[L] = \frac{1}{3}s_e \,.\, L \qquad ...(8.17)$$

This shows that the resistance to flow within the perforated section closely equals the resistance that would be offered to the full entrant floor by one third the length of the perforated pipe. Eq. 8.17 is also evident from the fact that area outside the parabola is one-third the area of the pertinent rectangle.

Example 8.3. *In a spray aeration process using nozzles, water is to be exposed for 2.4 seconds. Estimate the driving force required for (a) vertical jet, (b) jet inclined at 50° to the horizontal. Take velocity coefficient for nozzle as 0.95. Also, find the associated carry and rise in each case.*

Solution.

(a) *Vertical jet* :

Here $\alpha = 90°$; $\sin\alpha = 1$; $\sin 2\alpha = 0$

Also, $t_r = \frac{1}{2}t = \frac{1}{2} \times 2.4 = 1.2$ seconds.

Hence from Eq. 8.13.

$$h = \frac{gt_r^2}{2\,C_v^2 \sin^2 \alpha} = \frac{9.81\,(1.2)^2}{2\,(0.95)^2\,(1)^2}$$

$$= \mathbf{7.83\ m}$$

From Eq. 8.14 , $r = 0$

From Eq. 8.15, $h_r = \frac{1}{2} g \,.\, t_r^2 = \frac{1}{2} \times 9.81\,(1.2)^2 = \mathbf{7.06\ m}$

(b) Inclined jet

$$\alpha = 50° \,;\, \sin \alpha = 0.7660 \,;\, \sin 2\alpha = 0.9848$$

$$t_r = \frac{1}{2} t = \frac{1}{2} \times 2.4 = 1.2 \text{ sec.}$$

$$h = \frac{9.81\,(1.2)^2}{2\,(0.95)^2\,(0.7660)^2} = \mathbf{13.34\ m}$$

$$r = 2\,C_v^2\,h \sin 2\alpha = 2\,(0.95)^2 \times 13.34 \times 0.9848$$

$$= \mathbf{23.71\ m}$$

$$h_r = \frac{1}{2} g t_r^2 = \frac{1}{2} \times 9.81\,(1.2)^2 = \mathbf{7.06\ m}$$

8.10. LIMITATIONS OF AERATION

Aeration has following limitations :

1. It is not an efficient method of removal or reduction of tastes and odours caused by relatively non-volatile substances such as oils of algae etc.
2. Odour removal is 50% only when symura was causative organism.
3. Tastes and odours caused by chemicals due to industrial wastes discharged into receiving waters are not satisfactorily reduced.
4. Aeration may add more oxygen in water making it more corrosive while removing iron and manganese.
5. Iron and manganese can be precipitated by aeration only when organic matter is not present.
6. Possibility of air borne contamination in water is there.
7. Additional lime may be required to neutralise the CO_2 that would be removed by aeration.
8. Aeration is economical only in warmer months.

PROBLEMS

1. Write a note on straining. How do you determine the head loss through racks and screens ?

2. Explain the purpose of aeration in water treatment. What are its limitations.

3. Write down a mathematical expression for gas transfer and discuss the factors by which gas transfer can be optimised.

4. Explain, with the help of diagrams, various methods of aeration.

5. Show that aeration by rising bubbles is far more efficient than aeration by falling droplets.

6. Find the time of aeration of water falling through a vertical distance of 5 m in (*a*) single descent (*ii*) four descents.

Ans. [1.01 sec. ; 2.02 sec.]

7. An experiment was conducted on the removal of CO_2 from water which was sprayed into air in spherical droplets of 6 mm in diameter. In an exposure of 1.25 seconds, the supersaturation of the water with CO_2 was found to be 10 mg/l. Taking the coefficient of gas transfer as 250 cm/hour, estimate the initial super-saturation.

Ans. [23.8 mg/l]

8. Explain the principal factors that are involved in the hydraulic performance of fixed spray pressure aerators.

Estimate the magnitude of driving head that will expose water for 2.5 sec., for a jet inclined at 45°. Also, find the associated carry and rise. Take the velocity coefficient of the nozzle as 0.92

Ans [h = 18.11 m ; h_r = 7.66 m ; r = 30.66 m].

9

Sedimentation

9.1. INTRODUCTION

Surface waters which are collected as the result of runoff from rain storms over the catchment area, carry vast amounts of soil and debris. To these are added community and industrial wastes that are transported through their collecting system. Though some of the eroded particles settle at the upstream side of storage head works, they are picked up again at rising river stages. Occasionally, the silt-laden or cold waters of entrant streams dive to the bottom of impoundages and under run the main body of warmer or other wise lighter water. This results in silt appearing at outlet works faster and in greater concentration than normally expected. The canal water also carries a lot of silt along with it.

Sedimentation is the removal of suspended particles by gravitational settling. *Sedimentation tanks* are designed to reduce the velocity of flow of water so as to permit suspended solids to settle out of the water by gravity.

Plain sedimentation : When the impurities are separated from suspending fluid by action of natural forces alone, i.e. by gravitation and natural aggregation of the settling particles, the operation is called *plain sedimentation*.

Sedimentation with coagulation (Clarification) : When chemicals or other substances are added to induce or hasten aggregation and settling of finely divided suspended matter, colloidal substances, and large molecules, the operation is called sedimentation with coagulation or simply, clarification.

Chemical precipitation : When chemicals are added to throw dissolved impurities out of solution, the operation is called *chemical precipitation.*

Discrete particles : A particle that does not alter its shape, size and weight while settling or rising in water is known as discrete particle.

9.2. TYPES OF SETTLINGS

Particles may settle out of a suspension in the following four ways, depending upon the concentration of the suspension and the flocculating properties of the particles (Fig. 9.1).

1. Type I sedimentation (Discrete settling)
2. Type II sedimentation (Hindered settling)
3. Type III sedimentation (Zone settling)
4. Type IV sedimentation (Compression settling).

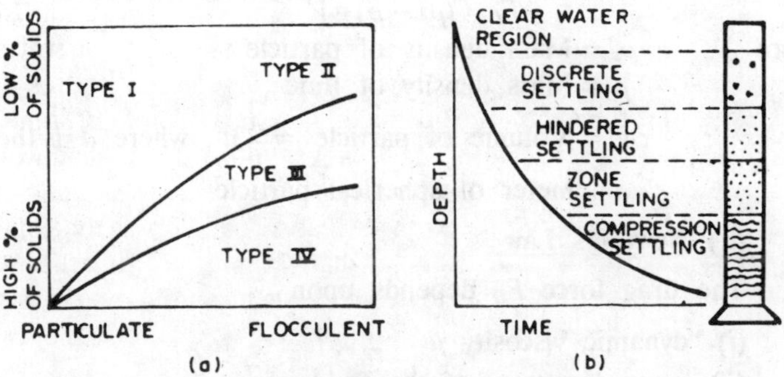

FIG. 9.1. TYPES OF SETTLINGS

Type I : Discrete settling : This corresponds to the sedimentation of discrete particles in a suspension of low solids concentration. This is also known as *free settling* since the particles have little tendency to flocculate or coalesce upon contact with each other.

Type II : Hindered settling : This type of settling refers to rather *dilute suspension* of particles that coalesce or flocculate during sedimentation process. Due to flocculation, particles increase in mass and settle at a faster rate.

Type III : Zone settling : This type of settling refers to flocculent suspension of *intermediate concentration*. Interparticle forces hold the particles together and hence the mass of the particles subside as a whole.

Type IV : Compression settling : This refers to flocculent suspension of so *high concentration* that particles actually come in contact with each other, resulting in the formation of a *structure.* Further settling can occur only by compression of the structure brought about due to weight of particles which are constantly being added to the structure.

In water treatment, only type I and type II settlings are encountered.

9.3. SETTLING OF DISCRETE PARTICLES

Type I sedimentation is concerned with the settling/removal of nonflocculating, discrete particles from water. When a discrete particle is placed in a quiescent fluid, it will accelerate until the frictional resistance (called drag force F_D) of the fluid equals the impelling force F_I (or driving force) acting on the particle. At this stage, the particle attains a uniform or *terminal velocity* and settles down with this constant velocity known as *settling velocity.*

The *impelling force* (F_I) is evidently equal to the *effective weight* of the particle :

$$F_I = (\rho_s - \rho)\, g\, V \quad ...(9.1)$$

where ρ_s = Mass density of particle

ρ = Mass density of fluid

V = Volume of particle $= \frac{\pi}{6} d^3$, where d is the diameter of spherical particle.

(*a*) **Newton's Law**

The drag force F_D depends upon

(*i*) dynamic viscosity μ

(*ii*) mass density ρ of the fluid.

and (*iii*) shape and size of the particle

The drag force is given by *Newton's law* for frictional drag in the following form :

$$F_D = C_D \cdot A \cdot \frac{\rho\, v_s^2}{2} \quad ...(9.2)$$

where F_D = Drag force

C_D = drag coefficient

= projected area of the particle$= \frac{\pi}{4} d^2$.

Equating the two, we get an equation for the settling velocity in the from

$$v_s = \sqrt{\frac{4}{3} \frac{g}{C_D} \frac{(\rho_s - \rho)\, d}{\rho}} \quad ...(9.3\ a)$$

or closely, $$v_s \approx \sqrt{\frac{4}{3}\frac{g}{C_D}(S_s - 1)\,d} \qquad ...(9.3\ b)$$

where S_s = Specific gravity of the particle.

Use of Eq. 9.3 (*a*) requires the determination of drag coefficient C_D which is related to Reynolds number R by the following observational relationships :

(*i*) For R between 0.5 to 10^4

$$C_D = \frac{24}{R} + \frac{3}{\sqrt{R}} + 0.34 \qquad ...(9.4\ a)$$

(*ii*) For high Reynolds number ($R > 10^3$ to 10^4), $C_D \approx 0.4$

(*iii*) For low, Reynolds number ($R < 0.5$)

$$C_D = \frac{24}{R}$$

where R = Reynolds number $= \dfrac{\rho v_s \,.\, d}{\mu}$.

The variation of C_D with Reynolds number R is shown in Fig. 9.2.

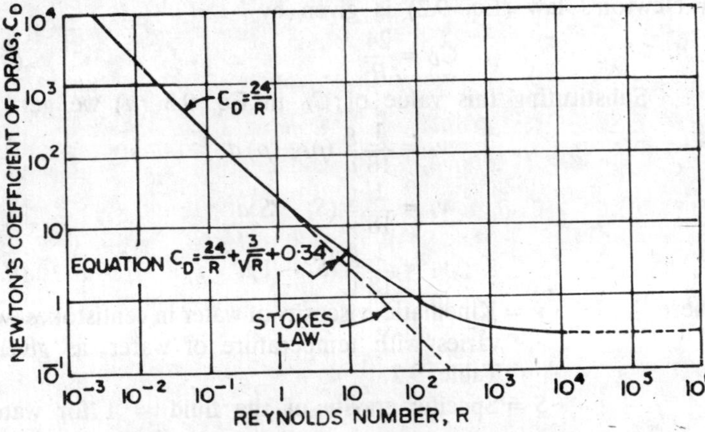

FIG. 9.2. VARIATION OF NEWTON'S C_D WITH R

Smith has given the following equation for the coefficient of drag C_D :

$$C_D = \frac{b}{R^n} \qquad ...(9.4\ b)$$

The value of b and n are given in Table 9.1 for various values of R.

TABLE 9.1. SMITH'S VALUES.

R	*b*	*n*
< 1.9	24	1
1.9 to 500	18.5	0.6
500 to 200,000	0.44	0

For Reynold's number $R > 10^3$ to 10^4, taking $C_D = 0.4$, Eq. 9.3 (*a*) takes the following form :

$$v_s = \sqrt{3.33\, g \left(\frac{\rho_s - \rho}{\rho}\right) d} \qquad \text{...(9.3)}$$

This equation is applicable for particles greater than 1 mm dia. and for Reynolds number greater than 10^3 to 10^4.

(*b*) **Stoke's Law**

Stoke's law for the drag of small settling spheres in a viscous fluid, neglecting the inertia force is given by

$$F_D = 3\,\pi\,\mu\, v_s \,.\, d \qquad \text{...(9.5)}$$

Equating this to Eq. 9.2 indicates that the drag coeficient in Newton's law (Eq. 9.2) is given by

$$C_D = \frac{24}{R} \qquad \text{...(9.4 c)}$$

Substituting this value of C_D in Eq. 9.3 (*a*) we get

$$v_s = \frac{1}{18}\frac{g}{\mu}(\rho_s - \rho)\, d^2 \qquad \text{...(9.6)}$$

or

$$v_s = \frac{1}{18}\frac{g}{\nu}(S_s - S) d^2 \qquad \text{...(9.7)}$$

or

$$v_s \approx \frac{1}{18}\frac{g}{\nu}(S_s - 1) d^2 \qquad \text{...(9.7 a)}$$

where ν = Kinematic viscosity of water in centistokes which varies with temperature of water as given in Table 9.2

S = Specific gravity of the fluid ≈ 1 for water.

TABLE 9.2 KINEMATIC VISCOSITY OF WATER

Temperature	°C °F	0 32	5 41	10 50	15 59	20 68	25 77
(centistokes)		1.79	1.52	1.31	1.15	1.01	0.90

Note. 1 centistoke = 0.01 cm²/second.

Eq. 9.7 is known as the *Stoke's law*, while Eq. 9.3 is known as *Newton's law*.

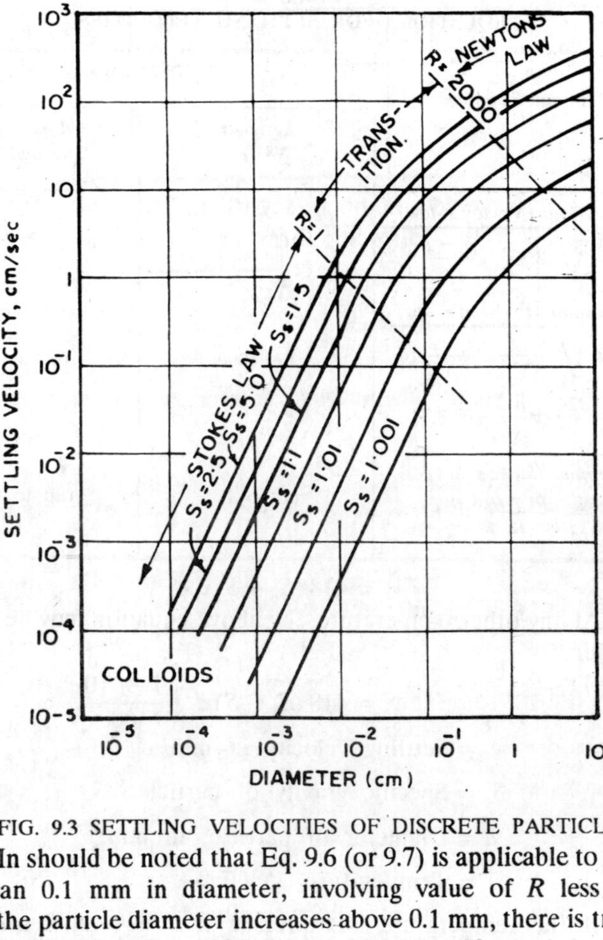

FIG. 9.3 SETTLING VELOCITIES OF DISCRETE PARTICLES

In should be noted that Eq. 9.6 (or 9.7) is applicable to particles less than 0.1 mm in diameter, involving value of R less than 1. When the particle diameter increases above 0.1 mm, there is transition range of $R = 1.0$ to $R = 1000$. Above this range, *Newton's law* applies. Fig. 9.3 gives curves for settling velocity for various specific gravities, to span the region between the Stokes range (laminar) and Newtons's turbulent range. For transition zone ($R = 1$ to 1000 and $d = 0.1$ to 1.0 mm), Eq. 9.3 (*a*) is applicable.

Table 9.3 gives the summary of applicable equations.

It should be noted that the viscosity of water depends upon temperature.

Taking $g = 981$ cm/sec^2 = 9810 mm/sec^2 and $\nu = 1.31$ centistokes at 10°C, the factor $\frac{1}{18}\frac{g}{\nu}$ comes out to be about 418. Hence Eq. 9.7 reduces to the following form, at 10°C :

TABLE 9.3.
EQUATIONS FOR SETTLING VELOCITIES.

Law and Equation	*Applicable range of*	
	Reynolds No. (R)	***d (mm) for*** $s_s = 2.65$ ***and 20°C temp.***
1. *Turbulent (Newton's law)* $v_s = \sqrt{3.33\, g \left(\frac{\rho_s - \rho}{\rho}\right) d}$	$10^3 - 10^4$	> 1
2. *Transition (Newton's law)* $v_s = \sqrt{\frac{4}{3}\frac{g}{C_D}\left(\frac{\rho_s - \rho}{\rho}\right) d}$; $C_D = \frac{24}{R} + \frac{3}{\sqrt{R}} + 0.34$	1 – 1000	0.1 to 1.0
3. *Laminar (Stoke's law)* $v_s = \frac{g}{18}\left(\frac{\rho_s - \rho}{\mu}\right) d^2$	up to 1	up to 0.1

$$v_s = 418\,(S_s - S)\,d^2 \qquad ...(9.8)$$

At any other temperature, the above equation may be modified as under :

$$v_s = 418\,(S_s - S)\,d^2 \left(\frac{T + 10}{60}\right) \qquad ...\ (9.9)$$

where v_s = Settling velocity in mm/sec

S_s = Specific gravity of particle

d = Diameter of particle, in mm

T = Temperature, in °F.

If the temperature is expressed in t degrees centigrade,

$$v_s = 418\,(S_s - S)\,d^2 . \left(\frac{3t + 70}{100}\right) \qquad ...(9.9)$$

Hazen observed that when the diameter of the particle is more than 0.1 mm, the velocity of settlement is proportional to the *first power* of the diameter and not to the second power of diameter as expressed in Stoke's law. Hence Stoke's law for particles bigger than 0.1 mm can be expressed as :

$$v_s = 418\,(S_s - S)\,d\,\frac{T + 10}{60} \qquad ...(9.10)$$

or

$$v_s = 418\,(S_s - S)\,d\,\frac{3t + 70}{100} \qquad ...(9.10\ a)$$

Effect of shape

The suspended matter is seldom spherical. They may be rod-like or disc-like spheroidal particles which possess greater surface area per unit volume than do spheres. Due to this, they settle more slowly than do spheres of equivalent volume. Also, frictional drag changes with the orientation of the particles relative to the direction of settling. At low values of R (< 10), the settling velocity of rod like and disk-like spheroidal particles are respectively 78% and 73% of the velocity of an equal volume sphere.

For particles of irregular shape,

$$\frac{A}{V} = \frac{6}{\psi d} = \frac{S}{d}$$

where ψ = *sphericity* of particle (= 1 for perfect sphere)

$$S = \frac{6}{\psi} = \textit{shape factor} \quad (= 6 \text{ for perfect-sphere})$$

Example 9.1 *Find the settling velocity of spherical silica particle of specific gravity 2.67, in water at 25°C, if the diameter of particles is 0.004 cm*

Solution :

From Table 9.2, at 25°C, ν = 0.90 centistoke= 0.009 cm²/sec. Assuming that Reynolds number will be less than 1, we have from Eq. 9.7 (*a*) :

$$v_s = \frac{1}{18}\frac{g}{\nu}(S_s - 1)d^2$$

$$= \frac{1}{18} \times \frac{981}{0.009}(2.67 - 1)(0.004)^2$$

$$= \mathbf{0.162\ cm/sec.}$$

Alternatively, from Eq. 9.9 (*a*)

$$v_s = 418\,(S_s - S)\,d^2\left(\frac{3\,t + 70}{100}\right)$$

where v_s is in mm/sec ; $S = 1$, d = dia. in mm = 0.04 mm

and t = 25°C.

$$\therefore \quad v_s = 418\,(2.67 - 1)\,(0.04)^2\left(\frac{3 \times 25 + 70}{100}\right)$$

$$= 1.62 \text{ mm/sec} = 0.162 \text{ cm/sec.}$$

Now $$R = \frac{\rho v_s d}{\mu} = \frac{v_s d}{\nu} = \frac{0.162 \times 0.004}{0.009}$$

$= 0.072 < 1$. *Hence Stoke's law is applicable.*

Example 9.2. *Solve example 9.1 if the size of the particles is 0.04 cm.*

Solution.

Since the diameter of the particle is more than 0.1 mm, Stoke's law will not be applicable. The velocity will have to be found out by Newton's law given by Eq. 9.3 (*b*), in which the drag coefficient C_D is dependent on Reynolds number R through Eq. 9.4. Since R itself is dependent on the velocity v_s, the following trial procedure is suggested by the Author.

First trial

Calculate approximate value of v_s using Eq. 9.9 (*a*),

Thus, $$v_s = 418\,(2.67 - 1)\,(0.4)^2 \left(\frac{3 \times 25 + 70}{100}\right)$$

$$= 161.9 \text{ mm/sec} = 16.19 \text{ cm/sec.}$$

Calculate approximate value of R from the relation :

$$R = \frac{\rho v_s \,.\, d}{\mu} = \frac{v_s\, d}{\nu}$$

$$= \frac{16.19 \times 0.04}{0.009} \approx 72$$

Compute C_D from Eq. 9.4 (*a*)

$$C_D = \frac{24}{R} + \frac{3}{\sqrt{R}} + 0.34 = \frac{24}{72} + \frac{3}{\sqrt{72}} + 0.34$$

$$\approx 1.027$$

Compute v_s from Eq. 9.3 (*b*)

$$v_s = \left[\frac{4}{3}\frac{g}{C_D}(S_s - 1)\, d\right]^{\frac{1}{2}}$$

$$= \left[\frac{4}{3} \times \frac{981}{1.027}(2.67 - 1)\,(0.04)\right]^{\frac{1}{2}}$$

$$= 9.22 \text{ cm/sec}$$

(against the previous value of 16.19 cm/sec.)

Second trial

Calculate new values of R and C_D

$$R = \frac{9.22 \times 0.04}{0.009} = 40.99$$

$$C_D = \frac{24}{40.99} + \frac{3}{\sqrt{40.99}} + 0.34 = 1.394$$

Calculate the value of v_s from Eq. 9.3 (*a*)

$$v_s = \left[\frac{4}{3} \times \frac{981}{1.394}(2.67 - 1)\,(0.04)\right]^{\frac{1}{2}}$$

$$= 7.92 \text{ cm/sec.}$$

Final trial

Finally, calculate the value of R, C_D and v_s :

$$R = \frac{7.92 \times 0.04}{0.009} = 35.2$$

$$C_D = \frac{24}{35.2} + \frac{3}{\sqrt{35.2}} + 0.34 = 1.527$$

$$v_s = \left[\frac{4}{3} \times \frac{981}{1.527} (2.67 - 1) (0.04)\right]^{\frac{1}{2}}$$

$$= \mathbf{7.56\ cm/sec}$$

Example 9.3. *Solve example 9.1 if the size of the particles is 0.1 cm.*

Solution.

In this case also, Newton's law will be applicable. The value of v_s will be found by trial procedure.

First trial

Calculate the approximate value of v_s using Eq. 9.9 (*a*)

$$v_s = 418 (2.67 - 1) (1.0)^2 \left(\frac{3 \times 25 + 70}{100}\right)$$

$$= 1012 \text{ mm/sec} = 101.2 \text{ cm/sec.}$$

$$R = \frac{101.2 \times 0.1}{0.009} = 1124.4$$

$$C_D = \frac{24}{1124.4} + \frac{3}{\sqrt{1124.4}} + 0.34 = 0.45$$

$$v_s = \left[\frac{4}{3} \times \frac{981}{0.45} (2.67 - 1) (0.1)\right]^{\frac{1}{2}} = 22 \text{ cm/sec.}$$

Second trial

$$R = \frac{22 \times 0.1}{0.009} = 244.6$$

$$C_D = \frac{24}{244.6} + \frac{3}{\sqrt{244.6}} + 0.34 = 0.63$$

$$v_s = \left[\frac{4}{3} \times \frac{981}{0.63} (2.67 - 1) (0.1)\right]^{\frac{1}{2}} = 18.62$$

Final trial

$$R = \frac{18.62 \times 0.1}{0.009} = 206.9$$

$$C_D = \frac{24}{206.9} + \frac{3}{\sqrt{206.9}} + 0.34 = 0.665$$

$$v_s = \left[\frac{4}{3} \times \frac{981}{0.665} (2.67 - 1) (0.1)\right]^{\frac{1}{2}}$$

$$= \mathbf{18.13\ cm/sec.}$$

9.4. TYPES OF SEDIMENTATION TANKS

Depending upon the methods of operation, sedimentation tanks are of two types :

1. The Quiescent or Fill and Draw type.
2. The Continuous flow type.

1. Fill and Draw type

In this type, the sedimentation tank is first filled with incoming water, and is allowed to rest for a certain time. During this rest period, the suspended particles settle down at the bottom of the tank. Generally, a detention time of 24 hours is allowed. At the end of the period, the clear water is drawn off through the outlet valve. The tank is then cleaned of settled particles and filled again. The cleaning may take another 6 to 12 hours. Thus, the cycle of operation takes 30 to 36 hours. Fig. 9.4 shows plan and section of such a tank. A minimum of *three* such units are required to maintain a constant supply.

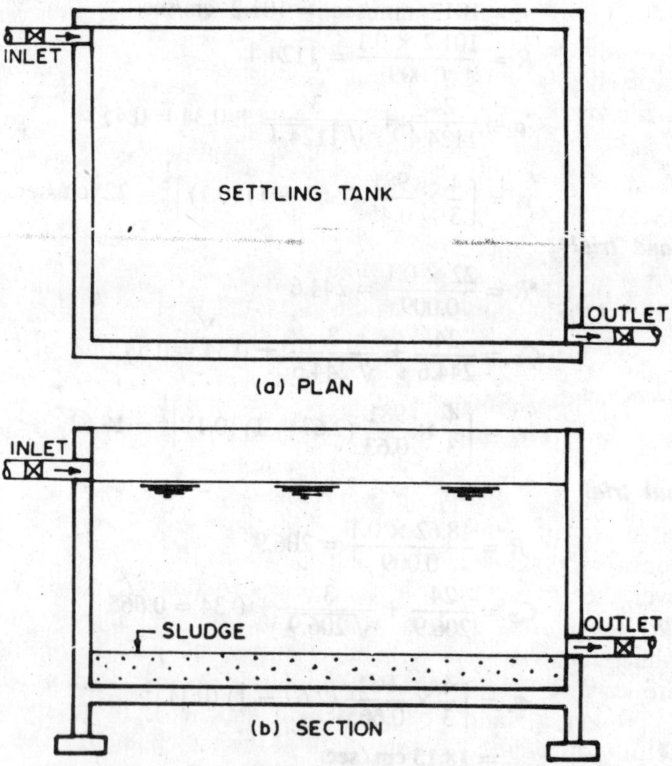

FIG. 9.4. FILL AND DRAW TYPE TANK

2. Continuous flow type tank

Out of the factors which hinder the sedimentation or settling of the particles under gravity, the velocity of flow of water is the one which can be easily controlled. In the continuous flow type, the water continuously keeps on moving in tank, though with a very small velocity during which time the suspended particles settle at the bottom before they reach the outlet. There are two types of continuous flow tanks :

(*a*) Horizontal flow tank.
(*b*) Vertical flow tank.

In the horizontal flow type, the tank is generally rectangular in plan having length equal to atleast twice the width. The water flows practically in the horizontal direction, with a maximum permissible velocity of 0.3 m/sec. Vertical flow type tanks are generally deep, circular or rectangular basins with hopper bottom.

9.5. HORIZONTAL FLOW SEDIMENTATION TANK

In the design of a horizontal flow tank, the aim is to achieve as nearly as possible the ideal conditions of equal velocity at all points lying on each vertical in the settling zone. The design of *horizontal flow tanks* is based on the following assumptions :

1. Within the setting or sedimentation zone, the particles settle exactly in the same manner as they do in a quiescent tank of equal depth.
2. The flow is horizontal and steady and the velocity is uniform in all parts of the settling zone for a time equal to the detention period.
3. The concentration of suspended particles of each size is the same at all points of the vertical cross-section at the inlet end.
4. A particle is removed when it reaches the bottom of the settling zone.

Fig. 9.5 show the vertical cross-section through an ideal horizontal flow type rectangular tank. The basin can be divided into four zones : (*i*) the inlet zone in which the influent stream disperse, over the cross-section at right angles to flow (*ii*) the settling zone (*iii*) the bottom zone or sludge zone and (*iv*) the outlet zone, in which the water and remaining suspended particles assemble and are carried to the effluent conduit.

Let L and H be the length and depth, respectively, of the settling zone. Let B be the width of the tank and Q be the discharge rate. The horizontal discharge velocity v_d is given by

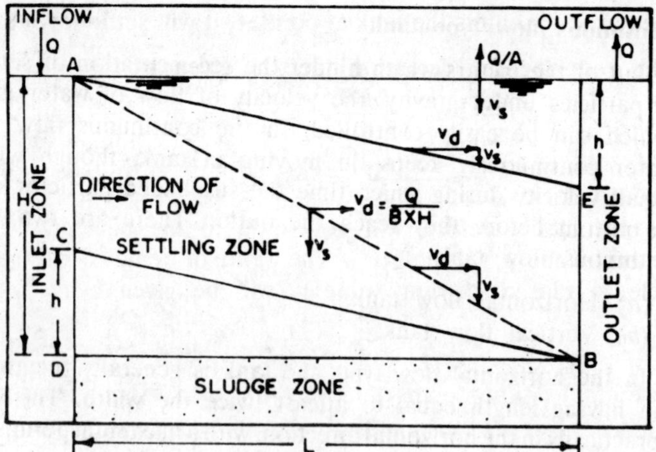

FIG. 9.5. VERTICAL CROSS-SECTION THOUGH RECTANGULAR TANK

$$v_d = \frac{Q}{H \times B} \quad \text{...}(i) \text{ ... } (9.11\ a)$$

The time of horizontal flow,

$$t_0 = \frac{L}{v_d} = \frac{LBH}{Q} \quad \text{...}(9.11\ b)$$

Consider a particle entering the tank to have a vertical falling speed v_s. The time for falling through distance H will be $\frac{H}{v_s}$. For the particle to reach the bottom before the water leaves the tank, the time of fall should be equal to time of horizontal flow. Hence

$$\frac{H}{v_s} = \frac{LBH}{Q}$$

From which $$v_s = \frac{Q}{LB} = \frac{Q}{A} \quad \text{...}(9.11)$$

Eq. 9.11 defines the *surface overflow rate* (SOR) or simply, *overflow rate*, which is numerically equal to flow divided by the plan area of the basin.

Also, from Eqs. 9.11 (*a*) and 9.11, $\frac{v_d}{v_s} = \frac{L}{H}$ ...(9.11 *c*)

Also, from Eqs. 9.11 (*b*) and 9.11, $t_0 = \frac{H}{v_s}$ ...(9.11 *d*)

Equation 9.11 suggests that all those particles, which have falling speed greater than Q/A will reach the bottom before the outlet end of tank. The time t_0 of horizontal flow, given by Eq. 9.11 *b* is known as the *detention period*. If a smaller particle, having falling speed $v_s' < \frac{Q}{A}$, enters the tank at point A, it will settle through only height h during the detention time t_0, given by

$$\frac{h}{v_s'} = t_0 = \frac{LBH}{Q}$$

$$h = \frac{LBH}{Q} v_s' = \frac{v_s'}{v_s} . H$$

This particle, therefore, will not settle to the bottom if it enters the basin above the point C. The ratio of removal of this size particle to that of settling value v_s will be given by

$$x_r = \frac{h}{H} = \frac{v_s'}{v_s} = \frac{v_s'}{Q/A} \quad \text{...(9.12)}$$

For example, suppose the tank has 400 m^2 surface area and rate of inflow is 1.6 m^3/sec., then $Q/A = 1.6/400 = 0.004$ m/sec = 0.4 cm/sec. Hence all particles having $v_s = 0.4$ cm/sec will be removed. Only 50% of the particles having falling speed $v_s' = 0.2$ cm/sec will be removed, while only 25% of the particles having falling speed of 0.1 cm/sec will be removed, and so on.

Eq. 9.12, derived by Hazen, states that for *discrete* particles and *unhindered settling,* basin efficiency is solely a function of the settling velocity of the particles and of the surface area of the basin relative to the flow rate. *The efficiency is otherwise independent of the basin depth and the detention period.*

9.6. SIZE-WEIGHT COMPOSITION AND REMOVAL

The efficiency of a sedimentation basin indicates the over all percentage removal of suspended matter at a given *overflow rate* v_s. Eq. 9.11 physically represents a settling velocity of the slowest settling particle which is 100% removed. Those particles which settle at velocities equal to or greater than SOR will be entirely removed while those which settle at lower velocities will be removed in direct proportion to the ratio of their settling velocity v_s' to v_s, assuming that they are uniformly distributed upon entering the basin.

For actual suspensions of particles with considérable variety of sizes and densities, prediction of efficiency of basin requires a settling column analysis from which the *cumulative frequency distribution curve*, similar to that of Fig 9.6 may be obtained.

According to Eq. 9.12, the weight fraction removal of particles having velocity $v_s' < v_s$ will be $\int_0^{x_s} \frac{v_s'}{v_s} dx$

Hence the overall removal from the clarified liquid will be

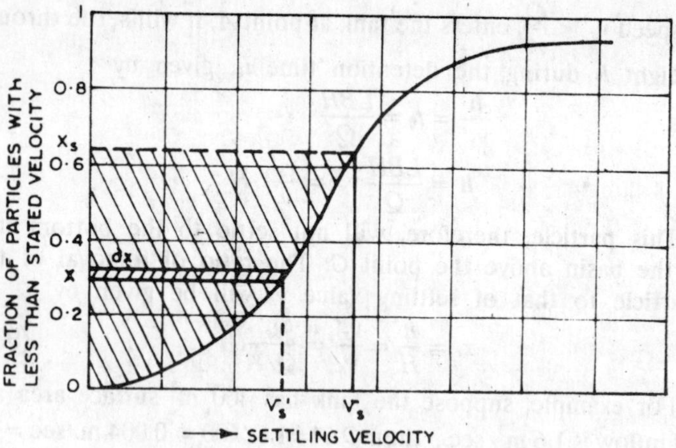

FIG. 9.6 CUMULATIVE DISTRIBUTION OF PARTICLE SETTLING VELOCITY

$$R = (1 - x_s) + \frac{1}{v_s} \int_0^{x_s} v_s' \, dx \qquad \text{...(9.13)}$$

in which $(1 - x_s)$ is the fraction with $v_s' \geq v_s$ and the integral is the fraction of particles with $v_s' < v_s$ which are removed in the basin.

The second term in the above equation can be determined by *graphical integration* of the curve, shown shaded an Fig. 9.6. The actual calculations are performed with a finite number of points on the distribution curve, so that, in practice, Eq. 9.13 is approximated by

$$R = (1 - x_s) + \frac{1}{v_s} \Sigma \, v_s' \, \Delta x \qquad \text{...(9.13 } a\text{)}$$

9.7. MAXIMUM VELOCITY TO PREVENT BED UPLIFT OR SCOUR

It is very essential that when once the particle has settled and reached the sludge zone, it should not be scoured or lifted up by the velocity of flow of water over the bed. Camp (1946) has given the following expression for the displacement velocity v_d required to start the motion of the settled particles of size d :

$$v_d = \sqrt{\frac{8 \beta g}{f} (S_s - 1) \, d} \qquad \text{...(9.14)}$$

where $\beta = 0.04$ for unigranular sand ; and 0.06 or more for non-uniform (interlocking) sticky materials.

f = Weisbach-Darcy friction factor
= 0.025 to 0.03 for setting tanks.

In terms of the setting velocity (v_s), the displacement velocity can also be expressed by the relation applicable for *fine, light and flocculent solids* :

$$v_d = v_s \left(\frac{8}{f}\right)^{\frac{1}{2}} \quad ...(9.15)$$

Taking $f = 0.025$, the above equation reduces to

$$v_d = v_s \left(\frac{8}{0.025}\right)^{\frac{1}{2}} \approx 18\, v_s \quad ...(9.15\ a)$$

A more useful or *practical relation* used is

$$v_d = 10\, v_s \quad ...(9.15\ b)$$

The ratio of length (L) to depth (H) in rectangular basin, or of surface area ($A = B \times L$) to cross-sectional area ($a = B \times H$) must be kept as follows for a basin

$$\frac{L}{H} = \frac{A}{a} = \frac{v_d \cdot t_d}{v_s \cdot t_0} = \left(\frac{t_d}{t_0}\right)\left(\frac{8}{f}\right)^{1/2} \quad ...(9.16)$$

For an ideal basin, $t_d/t_0 = 1$.

$$\therefore \quad \frac{L}{H} = \left(\frac{8}{f}\right)^{1/2} \quad ...(9.16\ a)$$

Taking a practical value of $\sqrt{8/f} = v_d/v_s = 10$; we get

$$\frac{L}{H} = \frac{A}{a} = 10 \quad ...(9.16\ b)$$

Example 9.4. *Determine the maximum displacement velocity, without the fear of settled particles to be lifted up and the ratio of length to depth of the settling unit for the following particles : (a) alum floc having $S_s = 1.1$ and diameter = 0.1 cm, and (b) anthracite coal dust having $S_s = 1.5$ and diameter = 0.01 cm. Assume a Weis-bach-Darcy friction factor $f = 0.025$ and temperature of 20°C.*

Solution :

(*a*) For *alum floc,* Eq. 9.15 is applicable.

$$\therefore \quad \frac{v_d}{v_s} = \left(\frac{8}{f}\right)^{\frac{1}{2}} = \left(\frac{8}{0.025}\right)^{\frac{1}{2}} = 17.89$$

From Stoke's law (Eq. 9.9 *a*)

$$v_s = 418\,(S_s - 1)\,d^2 \left(\frac{3\,t + 70}{100}\right), \text{ where } d \text{ is in mm}$$

$$= 418\,(1.1 - 1)\,(1)^2 \left(\frac{3 \times 20 + 70}{100}\right)$$

$$= 54.34 \text{ mm/sec} = 5.434 \text{ cm/sec.}$$

Hence $v_d = 17.89\, v_s = 17.89 \times 5.434 =$ **97.2 cm/sec.**

For an ideal basin,

$$\frac{L}{H} = \frac{v_d}{v_s} = \mathbf{17.89}$$

(*b*) For *anthracite coal,* apply Eq. 9.14, assuming $\beta = 0.04$

$$\therefore \quad v_d = \left[\frac{8 \times 0.04 \times 981}{0.025}(1.5 - 1)\, 0.01\right]^{\frac{1}{2}}$$

$$= \mathbf{7.92\ cm/sec}$$

From Stoke's law (Eq. 9.9 *a*)

$$v_s = 418\,(1.5 - 1)\,(0.1)^2\left(\frac{3 \times 20 + 70}{100}\right)$$

$$= 2.717\ \text{mm/sec} \approx 0.272\ \text{cm/sec.}$$

$$\therefore \text{ For an ideal basin, } \frac{L}{H} = \frac{v_d}{v_s} = \frac{7.92}{0.272} = \mathbf{29.15}$$

9.8. DESIGN ELEMENTS

In the continuous flow sedimentation tanks, the following are important elements of design : (1) detention period and displacement efficiency (2) overflow rate and surface loading (3) basin dimensions (4) in-let and out-let arrangement and (5) sludge removal devices.

1. Detention period and displacement efficiency.

Detention time is the theoretical time taken by a particle of water to pass between entry and exit of a settling tank. If V is the volume of the basin, Q is the discharge rate and t_0 is the detention period, we have

$$t_0 = \frac{V}{Q} = \frac{LBH}{Q} \qquad \text{...(9.17)}$$

The value of detention period depends upon the type of suspended impurities in water, and may vary from 4 to 8 hours in plain sedimentation tanks. Detention period, unless otherwise indicated, is based upon average flow.

Flowing through period (t_d) : It is the average time required for a batch of water to pass through the settling tank. It is always less than the detention period due to short circuit effects. How close this will approximate the detention period will depend upon the design of the tank. *The flowing through period* of an existing tank can be determined by placing sodium chloride in the influent and testing for the chloride content of the effluent.

Displacement Efficiency (η_d). It is defined as the ratio of 'flowing through period' to the detention period.

$$\eta_d = \frac{\text{Flowing through period}}{\text{Detention period}} = \frac{t_d}{t_0} \qquad ...(9.18)$$

It varies from 0.25 to 0.5 in plain sedimentation tanks. A well designed tank should provide a flowing through period of *at least* 30% of the detention period. *Actual detention period* used in practice should be more than twice the theoretical detention period.

2. Overflow rate and surface loading. The quantity of water passing per hour (or per day) per unit horizontal area is known as the *overflow rate* or *surface loading*. From Eq. 9.11, it is clear that velocity of settlement (v_s) is also equal to the rate of flow (Q) divided by the horizontal area (A) of the tank. Thus the tank loading or surface loading rate and the velocity of settlement are numerically equal, provided adjustment is made for units. This can be done with the help of Eq. 9.11.

$$v_s = \frac{Q}{A} = \text{surface loading.} \qquad ...(9.11)$$

Let v_s be expressed in cm/sec.

Q be expressed in litres per day ⎤ Q/A in litres per day
A be expressed in square metres ⎦ per square metre

$$\therefore \quad v_s\,(\text{cm/sec}) = \frac{(Q \times 1000)/(24 \times 60 \times 60)}{(A \times 10{,}000)} = \frac{1}{864000}\frac{Q}{A}$$

Hence surface loading $\frac{Q}{A} = 864000\, v_s$ litres/day/m^2 ...(9.11 *e*)

In other words, 1 cm per second settling velocity corresponds to a surface loading of 864000 litres per day/m^2 (or 36000 litres per hour/m^2)

Let us take an example. Sand and silt have specific gravity of about 2.65 and grain sizes of 0.001 cm or more. Their settling velocity at 26°C are as low as 0.0102 cm/sec.

Hence the associated surface loading = 864000×0.0102=8813 litres per day/m^2.

or $\quad 36000 \times 0.0102 = 367.2$ litres per hour/m^2.

Now let us consider a settling tank with a depth of 3 m. The time required to settle with the above velocity of 0.0102 cm/sec will be $\frac{300}{0.0102 \times 60} = 490$ minutes. The *actual displacement time* will be larger than this due to disturbances at the inlet and outlet and other factors. Past performance experience has shown that for 75% removal of the particles of designed category, the displacement time should be increased from 40 to 200% (average 120%) of the theoretical, and the surface loading should be reduced from 71 to 33% (average 52%) of the theoretical. (see § 9.9).

Table 9.4 gives the values of common surface loadings and detention periods, suggested by the Manual of Water Supply and Treatment (MUD) :

TABLE 9.4.
COMMON SURFACE LOADINGS AND DETENTION PERIODS

Tank Type	*Surface loading, $m^3/m^2/d$**		*Detention period, h_r**		*Particles normally removed*
	Range	*Typical design values*	*Range*	*Typical design values*	
1. Plain sedimentation	0 – 6000	15 – 30	0.01 – 15	3 – 4	Sand, silt and clay
2. Horizontal flow circular sedimentation	25 – 75	30 – 40	2 – 8	2 – 2.5	Alum and iron floc
3. Vertical flow (up flow) clarifiers	—	40 – 50	—	1 – 1.5	Flocculent

*at average design flow.

3. Basin dimensions. The surface area of the basin is determined on the basis of the overflow rate or surface loading rate.

$$\text{Thus surface area } A = \frac{\text{Volume of water in litres hour}}{\text{Surface loading rate in litres per hour/m}^2}$$

Thus a flow of 5 million litres per day and surface loading rate of 20,000 litres/day per m^2 will require a surface area of

$$\frac{5{,}000000}{20000} = 250 \text{ m}^2.$$

The length to width ratio of rectangular tanks should preferably be from 3:1 to 5:1. The depth of basin is kept between 2.5 to 5 m, with a preferred value of 3 m, inclusive of the allowance for sludge storage. The cross-sectional area is such as to provide a horizontal velocity of flow of 0.2 to 0.4 metres per minute, and normally about 0.3 metres per minute. Bottom slopes may range from 1% in rectangular tanks to about 8% in circular tanks. The slopes of sludge hoppers range from 1.2 : 1 to 2 : 1 (vertical : horizontal)

4. Inlet and outlet arrangements. The inlet and outlet arrangements should be such that minimum disturbance is caused due to influent and effluent streams. Because of greater weight due to turbidity, there is tendency for the incoming water to sink, move along the bottom and rise at the outlet. This will cause backward flow towards the inlet at the surface thereby reducing the effective detention period. An ideal inlet structure should (*i*) uniformly distribute the water as uniformly as possible across the width and depth of tank, (*ii*) mix it with the water already in the tank, to prevent density currents (*iii*) minimize large scale turbulence and (*iv*) initiate longitudinal or radial flow so as to achieve high removal efficiency. Fig. 9.7 shows various types of inlet arrangements. To achieve the above requirements, each inlet opening must face a baffle. Uniform velocity can be attained by passing the water through a training or dispersion wall perforated by holes or slots. These slots should be such that :

(*i*) velocity of flow through the slots is about 0.2 to 0.3 m/s

(*ii*) head loss is 1.7 times the velocity head

(*iii*) diameter of the hole not to be larger than the thickness of the diffuser wall.

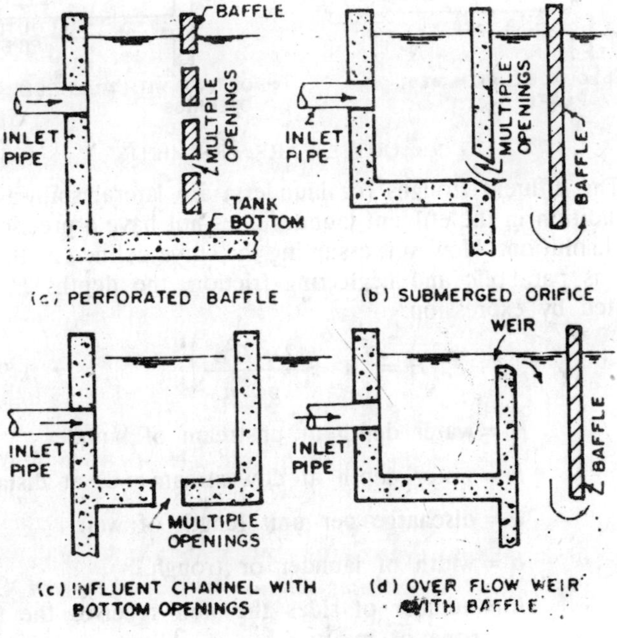

FIG. 9.7. INLET ARRANGEMENTS

Outlet arrangement consists of (*i*) weirs, notches or orifices, (*ii*) effluent trough or launder and (*iii*) outlet pipe. Weir loading rates are limited to prevent high approach velocities near the outlet. Weirs frequently consist of V-notches approximately 50 mm in depth, placed 150–300 mm on centres (Fig. 9.8 *a*), with a baffle provided in front of the weir to stop the floating matter from escaping into the effluent. The other forms of outlet arrangements are shown in Fig. 9.8 (*b*) and (*c*).

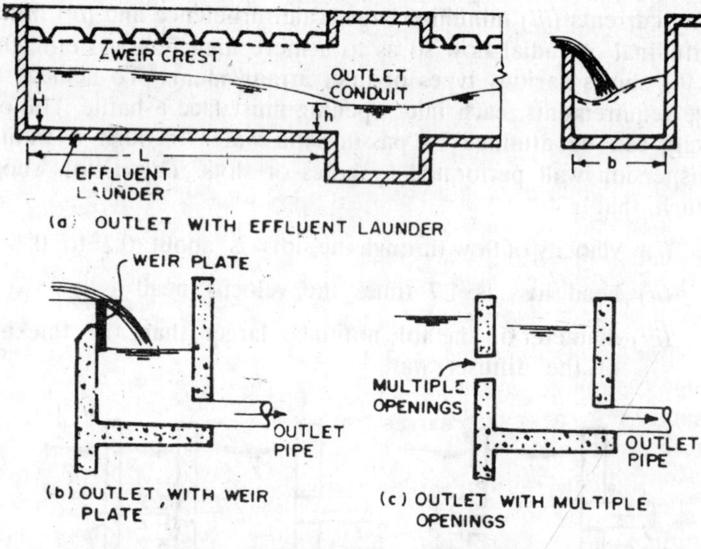

FIG. 9.8. OUTLET ARRANGEMENTS

The effluent troughs (or launders) are lateral spillways. The flow condition in the effluent launder does not have a precise mathematical solution. However, assuming the curve defined by the water surface as parabolic and neglecting friction, the depths *H*, *h* can be related by expression.

$$H = \left[h^2 + \frac{2 q^2 L^2 n^2}{g b^2 h} \right]^{1/2} \qquad ...(9.18\ a)$$

where H = water depth at upstream of launder

h = water depth at downstream end, at distance L

q = discharge per unit length of weir

b = width of launder or trough.

h = number of sides the weir receives the flow (one or two)

In the absence of any control device built into the system, it is reasonable to assume that flow at the lower end of the launder will be at its critical depth. Then

$$h = \left[\frac{Q^2}{b^2 g}\right]^{1/3} \qquad \text{...(9.18 } b\text{)}$$

where Q is the total discharge in the launder$= qL$

Sometimes, multiple troughs or launders are preferred to provide uniform distribution at low flows. There is a growing trend towards the use of effluent lauuders or troughs covering a good part of the surface of the settling basins. These are spaced at a distance of one tank depth between the troughs. The use of maximum feasible weir length in the tank from the outlet towards the inlets assists greatly in controlling density currents. Normal weir loadings are up to 300 m^3/day/m length.

5. Sludge removal. The particles settled in the basin constitute the sludge which can be removed either manually or mechanically. In the former case, the basin has to be put out of service, drained and the sludge dug out from the bottom. This method may be satisfactory where the quantity of suspended matter is small, requiring sludge clearance once in 2 or 4 months. However, where the quantity is large, the frequency of clearance will be very often, and it will not be economical to remove the sludge through manual labour. In addition to this, if the impurities settled are organic in nature, likely to decompose if allowed to accumulate for longer time, the resulting oxygen demand may cause septic conditions and produce objectionable odours. Hence mechanical or hydraulic methods are used for sludge removal. The sludge is collected and removed from the tanks as a fluid and caused to flow to its final point of disposal before being de-watered. Generally, travelling or rotating scrapers are installed to maintain the fluidity, and to transport the sludge to one or more outlet points from where it is sucked out through sludge pipe connected to sludge pump, as shown in Figs. 9.10 and 9.11.

9.9. SETTLING TANK EFFICIENCY

The hydraulic efficiency of the settling tank is the *flow through period* (t_d) which can be expected to approach the *critical detention period* (t_0) only in ideal tanks. Settling tank efficiency is reduced by the following currents :

(*i*) *Eddy currents*, set up by the inertia of the in-coming fluid

(*ii*) *Surface currents*, induced due to wind in open tanks (*iii*) *Vertical convection currents*, due to thermal gradient along the depth of the tank.

and (*iv*) *Density currents* causing cold or heavy water to under-run the basin and warm or light water to flow across its surface.

The efficiency of settling tank is dependent on the above mentioned currents, concentration of particles, shape of the basin and inlet and outlet arrangement. A well designed tank should have volumetric efficiency of atleast 70%.

Settling tanks should be capable of giving settled water having turbidity not exceeding 20 JTU, and preferably 5 JTU.

The efficiency of real basin affected by current induced short circuiting may be mathematically expressed by the following expression:

$$\frac{y}{y_0} = \frac{h}{H} = 1 - (1 + nkt)^{-1/n} \qquad \text{....(9.19)}$$

where $\frac{y}{y_0}$ = efficiency of removal of suspended particles, having settling velocity v_s

n = coefficient identifying basin performance (n = 0 to 1)

$k = \frac{1}{t_0} = \frac{v_s}{H}$

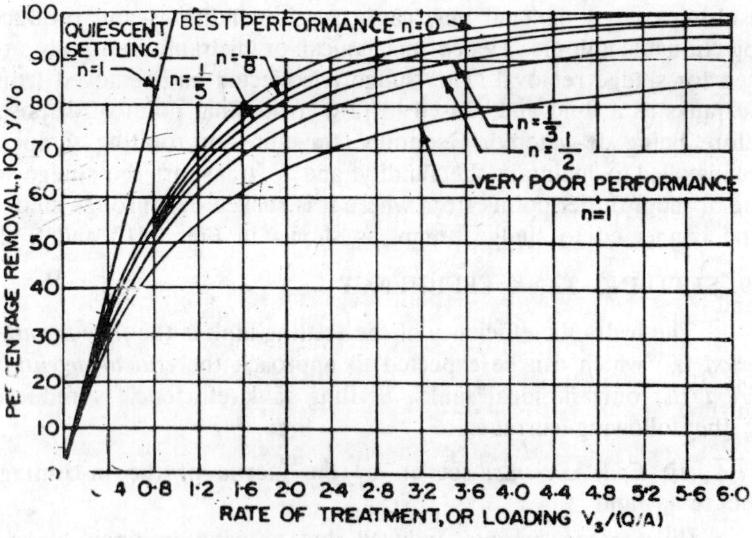

9.9. PERFORMANCE CURVES FOR SETTLING BASINS

Hence $$k.t = \frac{t}{t_0}$$

But $$\frac{t}{t_0} = \frac{v_s}{Q/A}$$

$$\therefore \quad \frac{y}{y_0} = 1 - \left[1 + \frac{n\,v_s}{Q/A}\right]^{-1/n} \quad \text{...(9.20)}$$

where Q/A = required *surface overflow rate* for real basin to achieve an efficiency of y/y_0 for a given basin performance.

Fig. 9.9 shows the performance curves for settling basins of varying effectiveness.

The values of n are assumed as 0 for best performance, $\frac{1}{8}$ for very good performance, $\frac{1}{4}$ for good performance, $\frac{1}{2}$ for poor performance and 1 for very poor performance.

From the above, we make the following observations.

(*i*) For $n = 0$ (best performance),

$$\frac{y}{y_0} = 1 - \exp(-t/t_0)$$

When $$\frac{v_s}{Q/A} = 1, \quad \frac{t}{t_0} = 1$$

$$\therefore \quad \frac{y}{y_0} = 1 - e^{-1} = 1 - \frac{1}{e} = 63\%$$

Thus maximum removal associated with $\frac{t}{t_0}\left(\text{or } \frac{v_s}{Q/A}\right)$ = unity is seen to be only 63%, instead of the 100% expected for perfect quiescence ($n = -1$)

(*ii*) For 75% removal of particles in the given settling velocity (v_s), we get from Fig. 9.9, the loading $\frac{v_s}{Q/A}$ is equal to

	1.4	for $n = 0$ (best performance)
	1.5	for $n = 1/8$
	2	for $n = 1/2$
and	3	for $n = 1$ (poor performance)

This means that for 75% removal, the *displacement time* must be from 40% to 200% *more* than with perfect quiescence, or the surface loading (or overflow velocity) must be *reduced* between $\frac{100}{1.4} = 71\%$ to $\frac{100}{3.0} = 33\%$ of the settling velocity of the particles to be removed.

Example 9.5. *A settling tank is designed for an overflow rate of 4000 litres per m^2 per hour. What percentage of particles of diameter (a) 0.05 mm (b) 0.02 mm, will be removed in this tank at 10°C ?*

Solution:

Settling velocity $v_s = \dfrac{Q}{A}$

Here $Q = 4000$ litres/hour

$$= \frac{4000 \times 10^{-3}}{60 \times 60} \text{ m}^3/\text{sec.}$$

$$A = 1 \text{ m}^2$$

$$\therefore \quad v_s = \frac{4000 \times 10^{-3}}{1 \times 3600} \text{ m/sec}$$

$$= \frac{4000 \times 10^{-3}}{1 \times 3600} \times 100 \text{ cm/sec.}$$

$$= 0.111 \text{ cm/sec.}$$

(*a*) *For 0.05 mm particles* :

Assume $S_s = 2.65$

$$\therefore \quad v_s' = 418\,(2.65 - 1)\,(0.05)^2 = 1.72 \text{ mm/sec.}$$

$$= 0.172 \text{ cm/sec}$$

$$\therefore \quad \% \text{ settled} = \frac{0.172}{0.111} \times 100 > 100\%$$

Hence all the particles of 0.05 mm dia. will settle.

(*b*) *For 0.02 mm particles* :

$$v_s' = 418\,(2.65 - 1)\,(0.02)^2 = 0.276 \text{ mm/sec}$$

$$= 0.0276 \text{ cm/sec}$$

$$\therefore \quad \% \text{ settled} = \frac{0.0276}{0.111} \times 100 = \mathbf{24.86\ \%}$$

Example 9.6. *Find the diameter of the particles with specific gravity of 1.2 removed in a tank having a surface area of 250 m^2 and treating 8 million litres of water per day. Assume temp. = 26°C.*

Solution :

$$\text{Overflow rate} = \frac{(8 \times 10^6) \times 10^{-3}}{250 \times 24} = 1.333 \text{ m}^3/\text{m}^2/\text{hour.}$$

$$\therefore \text{ Settling velocity } v_s = \frac{1.333}{1 \times 60 \times 60} \text{ m/sec}$$

$$= \frac{1.333 \times 100}{1 \times 60 \times 60} \text{ cm/sec} = 0.037 \text{ cm/sec.}$$

Applying Stoke's law,

$$v_s = 418\,(2.65 - 1)\,(d^2)\left(\frac{3 \times 26 + 70}{100}\right)$$

where $v_s = 0.37$ mm/sec

$$\therefore \qquad d^2 = \frac{0.37 \times 100}{418 \times 1.65 \times 148} = 0.00036$$

$$d = \mathbf{0.019\ mm}$$

9.10. DETAILS OF PLAIN SEDIMENTATION TANKS

Plain sedimentation tanks are usually of the following three shapes :

(*a*) Rectangular tank with horizontal flow
(*b*) Circular tank with radial or spiral flow
(*c*) Hopper bottom tank with vertical flow.

(*a*) **Rectangular tanks.** The design considerations of rectangular tanks have already been discussed. In many of these tanks, baffles are provided with the object of preventing short circuiting. Fig. 9.10 (*a*) shows the section of a *up and down* baffle tank, while Fig.9.10 (*b*) shows the plan of a *around the end baffle* tank.

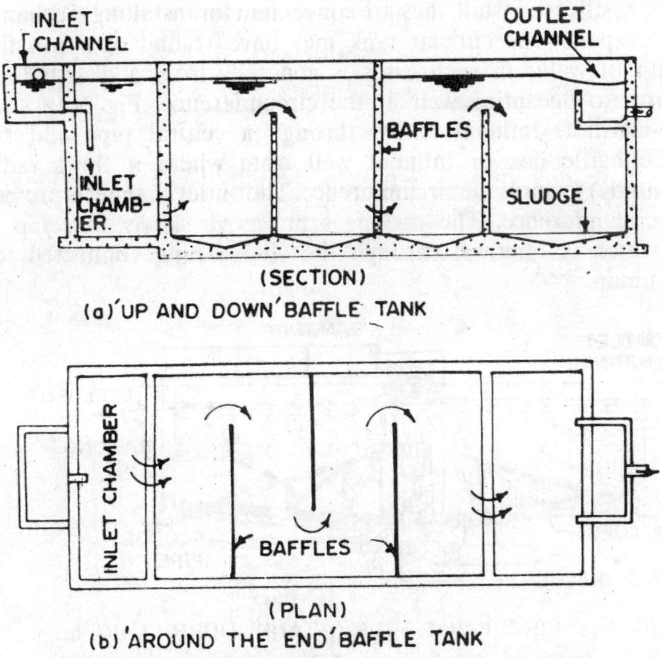

(b) 'AROUND THE END' BAFFLE TANK

FIG. 9.10. RECTANGULAR TANK WITH BAFFLES

Fig. 9.11 shows a rectangular tank, without baffles, but with a sludge hopper, and a sloping floor. Such a tank has high settling efficiency. It is provided with a mechanical sludge-removal equipment.

The sludge, scrapped by the sludge scrapers is collected in the hopper from where it is removed through the sludge withdrawal pipe connected to a sludge pump.

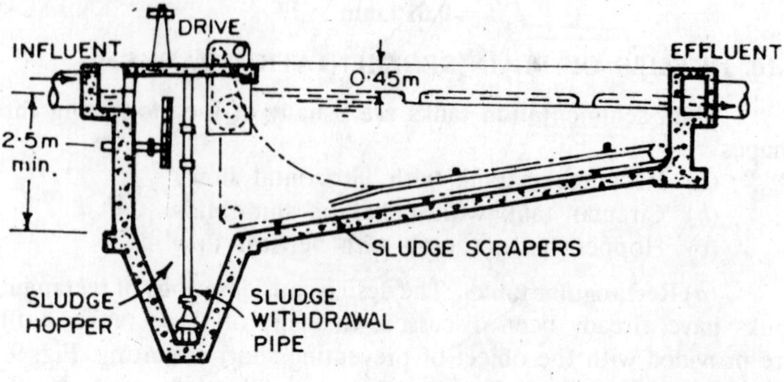

FIG. 9.11. RECTANGULAR TANK WITH SLOPING BOTTOM

(*b*) **Circular tank with radial or spiral flow.** Though circular tanks are costlier to install, they are convenient for installing mechanical sludge scrapers. The circular tank may have radial or spiral flow. The path of water in such cases is generally from a distributor at the centre to decanting weir at the circumference. Fig. 9.12 shows a Dorr-clarifier. Influent enters through a central pipe and rises upto the baffle box or influent well from where it flows radialy (horizontally) towards the circumference. The outlet channel is provided at the circumference. The racking arms move slowly to scrap the sludge which is removed through the sludge pipe connected to a sludge pump.

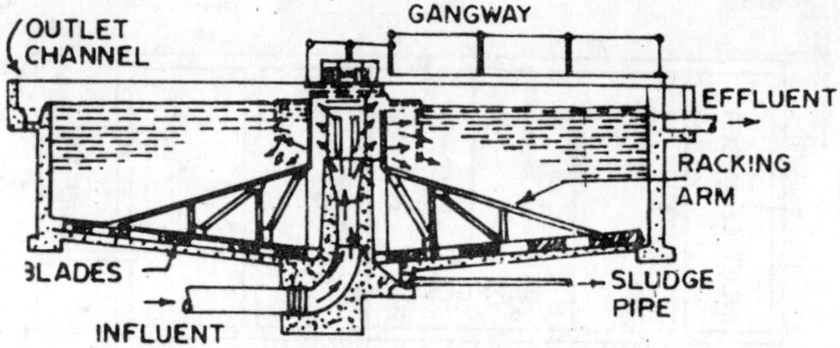

FIG. 9.12. CIRCULAR SETTLING TANK (DORR-CLARIFIER)

In the case of circular, *spiral-flow tank*, the inlet is provided at the circumference and directed at an angle between a radius and a tangent. The outlet, provided in the form of a submerged weir, is also provided at the outer circumference, but is of relatively short length. Fig 9.13 shows the plan of such a tank. The effect of the

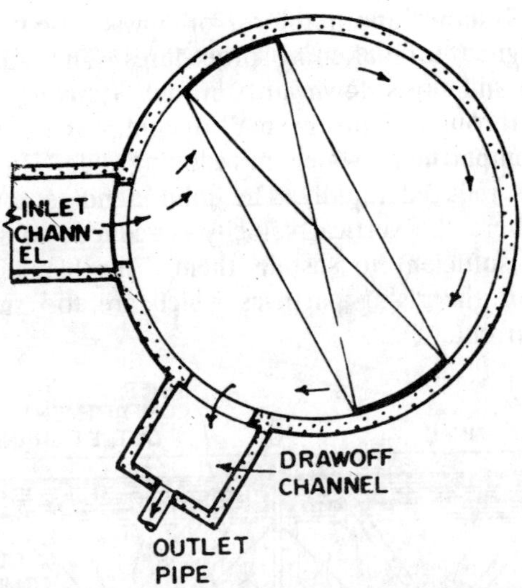

FIG. 9.13. CIRCULAR TANK WITH SPIRAL FLOW

inlet velocity is to rotate the water in the tank and to induce a spiral path for the water from inlet to outlet.

(*c*) **Hopper bottom tank with vertical flow.** Fig. 9.14 shows a hopper bottom tank. Water enters through the centrally placed inlet pipe and is deflected downwards by the action of a deflector box. Water travels vertically downwards. The sludge settles at the bottom of the hopper, from where it is removed with the help of a sludge pipe connected to a sludge pump.

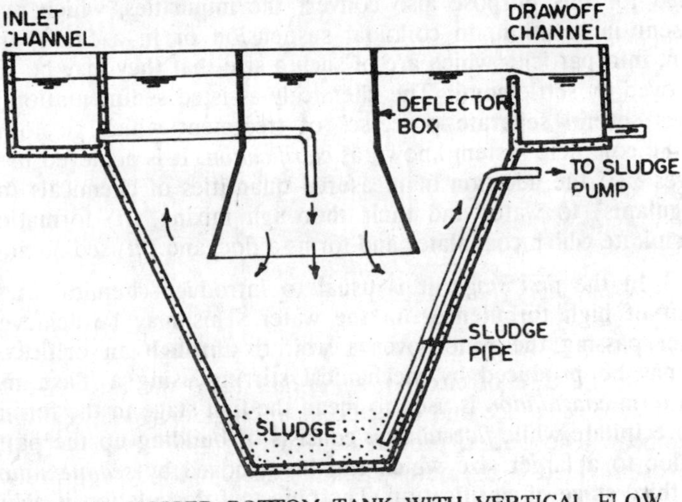

FIG. 9.14. HOPPER BOTTOM TANK WITH VERTICAL FLOW

Fig. 9.15 shows another form of upward flow settling tank, originally designed for softening procedures. The water enters at the top centre and flows downward through a mixing compartment. It then passes through relatively small openings and upward through the outside compartment which has sloping side. Due to this the vertical flow is reduced rapidly. The solids tend to form a blanket at the zone where the vertical velocity is reduced to such a value that it is just sufficient to sustain them. This blanket is effective in straining out the rising particles which are too small to settle against the current.

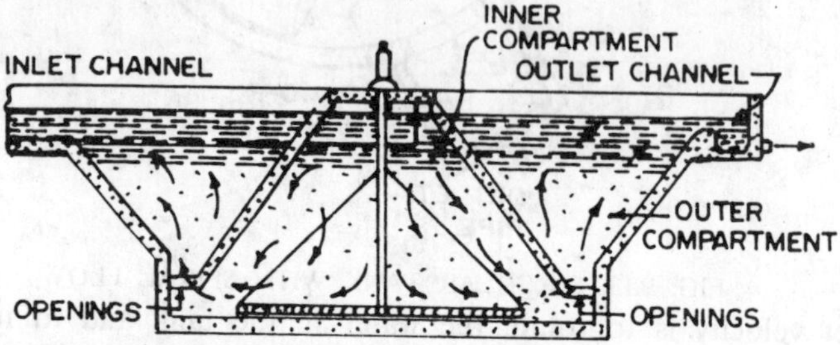

FIG. 9.15. UPWARD FLOW SETTLING TANK

9.11. SEDIMENTATION WITH COAGULATION : CLARIFICATION

The efficiency of plain sedimentation is generally very low, specially when water contains very fine suspended matter and colloidal matter. The coagulants neutralize the negative protective charge on the colloidal particles and allow them to coagulate. The chemicals added for this purpose also convert the impurities, which may be present in solution, in colloidal suspension or in a finely divided form, into particles which are of such a size that they may be readily removed by settlement. The chemicaly assisted sedimentation comprises several separate processes of treatment which go to make up the complete system known as *clarification*. It is achieved in three stages : (*i*) the addition of measured quantities of chemicals (called coagulants) to water and their thorough mixing, (*ii*) formation of precipitate which coagulates and forms a floc, and (*iii*) sedimentation.

In the *first stage,* it is usual to introduce chemical at some point of high turbulence in the water. This may be achieved by either passing the water over a weir or through an orifice plate or may be produced by mechanical stirring using a 'flash mixer'. The term *coagulation* is used to mean the first stage in the formation of precipitate while *flocculation* consists of building up the particles of floc to a larger size which can be removed by *sedimentation* in the third stage of clarification. The object of flocculation is achieved by keeping the water in a state of *controlled agitation* in *flocculators* consisting of tanks having power-operated paddles on horizontal or vertical shafts.

Coagulation aided sedimentation is required to remove the following types of impurities :

1. Miscellaneous fragments of animal and vegetable matter.
2. Plankton, mainly phytoplankton.
3. Finely divided mineral matter including clay and other colloidal 'sols'.
4. Organic colouring matter, partly in true solution and partly in colloidal form.
5. A complex mixture of colloidal and dissolved organic compounds derived from sewage and industrial effluents.
6. Bacteria and viruses.

Factors affecting coagulation

Following factors affect the coagulation :

1. Types of coagulant
2. Quantity or dose of coagulant.
3. Characteristics of water such as
   (*a*) type and quantity of suspended matter
   (*b*) temperature of water
   (*c*) pH of water.
4. Time, violence and method of mixing.

The rate of formation of floc is influenced more on the temperature than on the amount of coagulant. Thorough mixing of the coagulating chemical with the water is accomplished by a *short* period of *violent agitation*. This is then followed by a *longer* period of *conditioning*, involving gentle stirring during which the desired floc is developed.

9.12. COMMON COAGULANTS

When coagulant is dissolved in water and thoroughly mixed in it, a thick gelatinous precipitate, known as *floc,* is formed. The aluminum and ferric ions of the floc contain positive charge. Hence they attract the negatively charged colloidal particles of clay, turbidity and colour, thus helping in the removal of these impurities from water.

The following are common coagulants used :

1. Aluminum sulphate or alum,
2. Chloriniated copperas,

3. Ferrous sulphate and lime,
4. Magnesium carbonate,
5. Polyelectrolytes, and
6. Sodium aluminate.

1. Aluminium sulphate or alum

This is the most common and universal coagulant used in water works. Its chemical composition is $Al_2(SO_4)_3 \cdot 18H_2O$. It requires the presence of alkalinity in water to form the floc. Many waters have bicarbonate alkalinity naturally in them. When dissolved in water, aluminum sulphate tends to hydrolyze into aluminium hydroxide, as is evident from the following reaction :

$$Al_2(SO_4)_3\, 18H_2O + 3Ca(HCO_3)_2 = 2Al(OH)_3 + 3CaSO_4 + 18H_2O + 6CO_2 \quad ...(9.21\ a)$$

The aluminium hydroxide formed above is insoluble in water and is therefore a *floc*. There are two undesirable outcomes of the above reaction : (*i*) formation of permanent hardness by calcium sulphate, and (*ii*) formation of carbon dioxide which is corrosive to metals.

If natural alkalinity in water is insufficient to react with alum, lime is also added to water. The lime CaO unites with water to form calcium hydroxide $Ca(OH)_2$ or hydrated lime which reacts with alum as follows :

$$Al_2(SO_4)_3\, 18H_2O + 3Ca(OH)_2 = 2Al(OH)_3 + 3CaSO_4 + 18H_2O \quad ...(9.21\ b)$$

Sometimes, sodium carbonate, known as *soda ash,* is added to form alkalinity. The resulting reaction is .

$$Al_2(SO_4)_3\, 18H_2O + 3Na_2CO_2 = 2Al(OH)_2 + 3Na_2SO_4 + 3CO_2 + 18H_2O \quad ...(9.21\ c)$$

Though soda ash does not cause hardness, it is expensive than lime and is therefore less used.

Alum is found to be effective when pH of water ranges between 6.50 to 8.50. Its dose depends upon turbidity, colour, taste, pH value and temperature of water. The dose of alum varies between 10 to 30 milligrams per litre of water. Alum is preferred over other coagulants because it reduces taste and odour in addition to turbidity in water. It is cheap and the floc formed is quite tough to be broken easily. It produces crystal clear water. The main disadvantage of alum is that it is difficult to dewater the sludge formed.

2. Chlorinated copperas

Hydrated ferrous sulphate ($FeSO_4 . 7H_2O$) is traditionally referred to as *copperas.* It has too high a solubility to act as a satisfactory coagulant at the usual pH range. It is, therefore, first oxidised to ferric sulphate [$Fe_2 (SO_4)_3$] and ferric chloride ($FeCl_3$) by mixing its solution with a feed from a chlorinator before administering to the bulk of water. The resulting reactions are :

$$6FeSO_4 \, 7H_2O + 3Cl_2 = 2Fe_2 (SO_4)_3 + 2FeCl_3 + 7H_2O \quad ...(9.22)$$

The ferric sulphate and ferric chloride, produced instantaneously, are called *chlorinated copperas.* Both of these are immediately available for the formation of ferric hydroxide floc; the resulting reactions are :

$$Fe_2 (SO_4)_3 + 3Ca (OH)_2 = 3CaSO_4 + 2Fe (OH)_3 \quad ...(9.23)$$

$$2FeCl_3 + 3Ca(OH)_2 = 3CaCl_2 + 2Fe(OH)_3 \quad ...(9.24)$$

Chlorinated copperas is effective in removing colour also. The theoretical ratio of chlorine to copperas is 1 to 7.8. When used independently, ferric chloride (ferrichlor), is effective over a pH range of 3.5 to 6.5, and above 8.5. At higher pH, it is successful in removing manganese. Ferric sulphate (ferrisul), when used independently, is effective over a pH range of 4 to 7 and above 9. However, it requires special solution arrangements because of its low solubility in cold water.

3. Ferrous sulphate and lime

Ferrous sulphate or copperas can react with natural calcium bicarbonate alkalinity in water, but this is very much a delayed one. Hence lime is used with ferrous sulphate. The following reaction takes place :

$$FeSO_4 . 7H_2O + Ca(OH)_2 = Fe(OH)_2 + CaSO_4 + 7H_2O \quad ...(9.25)$$

The ferrous hydroxide $Fe(OH)_2$ thus formed, though an efficient floc, is soon oxidized by dissolved oxygen in water, and ferric hydroxide is formed :

$$4Fe(OH)_2 + O_2 + 2H_2O = 4Fe (OH)_3 \quad ...(9.26)$$

The ferric hydroxide is a satisfactory gelatinous floc, which is heavier than the one formed by alum. The effective pH range for this coagulant is 8.5 and above. Ferrous sulphate is unsuitable for the treatment of soft coloured waters because they are best coagulated at a pH range below 7.0.

4. Magnesium Carbonate and Lime

When magnesium carbonate and lime are dissolved in water, magnesium hydroxide and calcium carbonate are formed as under:

$$MgCO_3 + Ca(OH)_2 = Mg(OH)_2 + CaCO_3 \quad ...(9.27)$$

Both $Mg(OH)_2$ and $CaCO_3$ are soluble in water, resulting in the formation of sludge which is in the slurry form. Due to this, it is not commonly used. However, it is useful in removing organic colour, iron and manganese.

5. Polyelectrolytes

Polyelectrolytes are high molecular weight water-soluble polymers. They are classified as anionic, cationic and non-ionic, depending upon the charge they carry. Cationic polyelectrolytes can be used independently as effective coagulants while others are used as *coagulant-aids* along with alum etc. The cationic polyelectrolytes are available under certain trade names such as Floccal N., Magnifloc 972, Mogul 980 etc. With polyacrylamides, it is important that the unpolymerised material should be absent. The amount of a polyelectrolyte used is very small in relation to the amount of the primary coagulant. The use of a Polyelectrolytes not only broadens the pH range over which satisfactory flocculation could occur, but may also reduce the quantity of primary coagulant required. The usual dosage is 1 ppm.

6. Sodium aluminate

Sodium aluminate $Na_2Al_2O_4$ is alkaline in reaction and is used very much less often than alum because of its cost. It reacts with the salts of calcium and magnesium, as under :

$$Na_2Al_2O_4 + Ca(HCO_3)_2 = CaAl_2O_4 + Na_2CO_3 + CO_2 + H_2O \quad ...(9.28\ a)$$

$$Na_2Al_2O_4 + CaCl_2 = CaAl_2O_4 + 2NaCl \quad ...(9.28\ b)$$

$$Na_2Al_2O_4 + CaSO_4 = CaAl_2O_4 + Na_2SO_4 \quad ...(9.28\ c)$$

The coagulant remove both temporary and permanent hardness, and is effective for a pH range of 6 to 8.50 naturally available in water.

Comparison of alum and iron salts as coagulants

1. Iron salts produce heavy flocs due to which more suspended matter in removed than with alum.
2. The time required for floc formation, conditioning and settling, using iron salts is much less. Hence detention time is reduced.

3. Coagulation with iron salts is effective over a wider range of pH.
4. Use of iron salts removes hydrogen sulphide, taste and odour.
5. There is decreased tendency toward the formation of mud balls with iron salts, as compared with alum.
6. Under some conditions, iron coagulants are more economical than alum.
7. Iron salts cause staining and promote the growth of iron bacteria in the distribution system.
8. Iron salts impart corrosiveness to water, more than that imparted by alum.
9. Handling of iron salts require more skill where no skill is required in handling alum.
10. Some calcium sulphate remains in solution when natural alkalinity or addition of lime is depended on to react with the alum, resulting in the increase in the permanent hardness of water.
11. Increase in free carbon dioxide, consequent to alum coagulation may be objectionable because of its corrosive nature.
12. Unsatisfactory coagulation may result from the use of alum if sodium or potassium is present in the water.

Because of points No. 7, 8 and 9 mentioned above, alum is more frequently used as coagulant for treating raw water while iron salts are used as coagulants more frequently for treating sewage.

Example 9.7. *At a water treatment plant, 12 million litres of water is treated daily, using alum dosage of 16 mg per litres. Find (a) total quantity of alum used daily, (b) amount of carbon dioxide released.*

Solution :

(*a*) Weight of alum required $= \frac{16 \times 12 \times 10^6}{10^6} = \mathbf{192\ kg.}$

(*b*) The chemical reaction is indicated in Eq. 9.19. Let us calculate molecular weights of alum and carbon dioxide.

Molecular weight of alum $[Al_2(SO_4)_3\, 18H_2O)]$

$$= (2 \times 26.97) + (3 \times 32.066) + 16(4 \times 3 + 18) + (36 \times 1.008)$$
$$= 666.426.$$

Molecular weight of CO_2

$$= (1 \times 12.01) + (2 \times 16) = 44.01.$$

$$\therefore \quad 6CO_2 = 6 \times 44.01 = 264.06.$$

Thus, from Eq. 9.19

666.426 mg of alum releases 264.06 mg of CO_2.

$\therefore$ 192 kg will release$=\dfrac{264.06}{666.426}\times 192=$ **76.08 kg of CO_2.**

Example 9.8. *At a water treatment plant 12 million litres of water is treated daily using ferrous sulphate and lime. If the dosage of ferrous sulphate is 10 mg per litre, determine the total quantities of ferrous sulphate and lime required daily.*

Solution :

Quantity of ferrous sulphate

$$=\frac{10\times 12\times 10^6}{10^6}=\mathbf{120\ kg.}$$

The chemical reaction is indicated in Eq. 9.25, where

$$Ca(OH)_2 = CaO + H_2O$$

Molecular weight of ferrous sulphate (copperas) $FeSO_4 . 7H_2O$ is

$$= (1\times 55.85)+(1\times 32.066)+(11\times 16)+(14\times 1.008)$$
$$= 278.028.$$

Molecular weight of lime CaO

$$= (1\times 40.08)+(1\times 16)$$
$$= 56.08.$$

$\therefore$ 278.028 kg of ferrous sulphate will react with 56.08 kg of lime. Hence quantity of lime corresponding to 120 kg of ferrous sulphate$=\dfrac{56.08}{278.028}\times 120=$ **24.2 kg.**

9.13. METHODS OF FEEDING COAGULANTS

There are two methods of feeding coagulants to water : (*a*) dry feeding (*b*) wet feeding. Dry feeding is simple in operation and requires relatively less space for its working. The feeding machines are also cheaper. However, control of dose is difficult. The dosage in the wet feeding equipment can be adjusted more readily and can be easily controlled by means of automatic devices. However the chemicals which are corrosive in nature create problems in wet feeding.

Dry Feeding

Fig. 9.16 shows two common devices used for dry feeding. In each case, the coagulant, in a powder form, is kept in the tank with hopper bottom. In order to prevent the arching of the chemical, agitating plates are placed inside the tank.

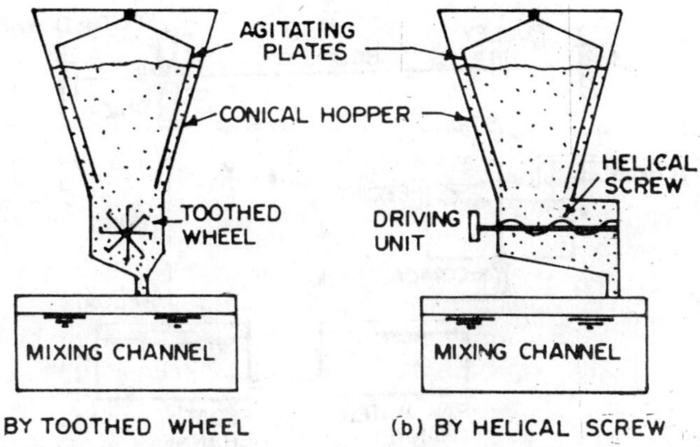

FIG. 9.16. METHODS OF DRY FEEDING.

In Fig. 9.16 (*a*), the feeding is regulated by the speed of the toothed wheel which is connected to the venturi device in the raw water pipe.

In Fig. 9.16 (*b*), the feeding is regulated by the helical screw. The drive unit of the helical screw is governed by the venturi device in the raw water pipe.

Wet Feeding

In the case of wet feeding, the solution of the coagulant, of the required strength is prepared and stored in a tank, from where it is allowed to fall in the mixing channel. The quantity of the mixture falling in the raw water channel is regulated in proportion to the quantity of flow. The regulating devices are shown in Fig. 9.17.

Fig. 9.17 (*a*) shows the wet feeding by conical plug and float. The solution of the coagulant is stored in a constant head feeding tank. The feeding of the coagulant to raw water channel is regulated through the conical plug which is governed by the movement of the pulley attached to the rod which, inturn, is rotated by the rack and pinion arrangement. The rack and pinion arrangement gets its motion through the movement of the float in the float chamber. The raw water channel and the float chamber are interconnected, and hence maintain the same level. When the flow of raw water increases, the float rises, due to which the pinion and the pulley rotate in the same direction, thereby raising the conical plug. Hence more quantity of solution of coagulant is permitted to fall in the raw water channel. The reverse is true when the flow is decreased. Thus, the dose of the coagulant is automatically controlled.

Fig. 9.17 (*b*) shows the wet feeding by adjustable weir and float. The adjustable weir is in the form of a sliding cylinder having

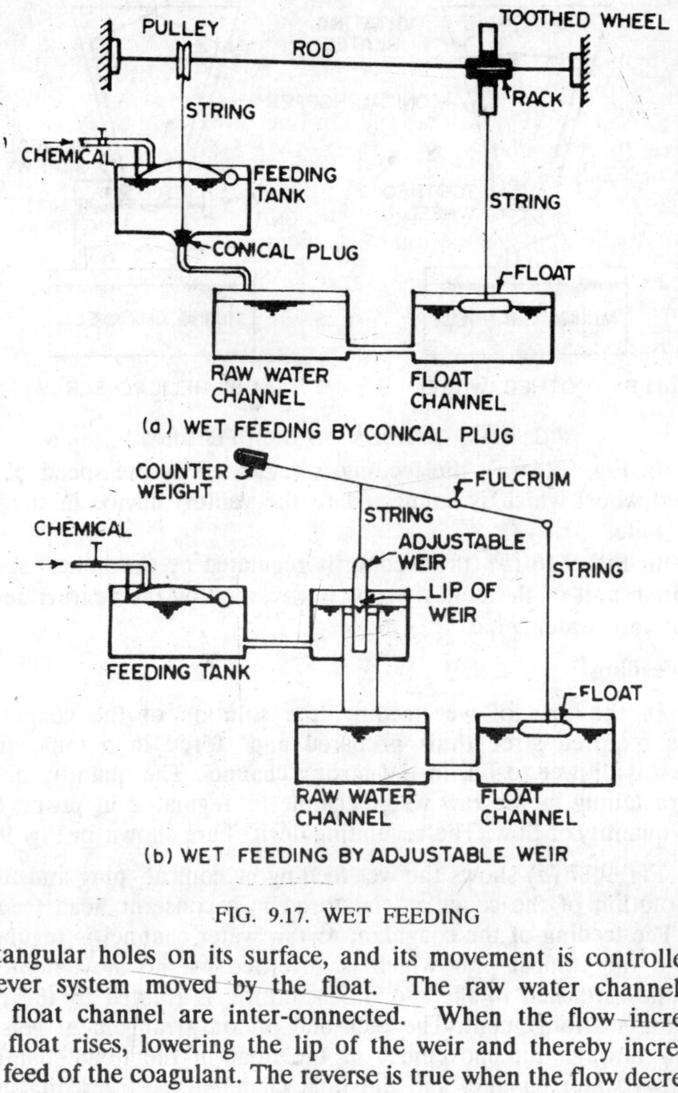

FIG. 9.17. WET FEEDING

rectangular holes on its surface, and its movement is controlled by a lever system moved by the float. The raw water channel and the float channel are inter-connected. When the flow increases, the float rises, lowering the lip of the weir and thereby increasing the feed of the coagulant. The reverse is true when the flow decreases. Thus the dose of the coagulant is automatically controlled.

9.14. MIXING DEVICES

The mixing process not only mingles the coagulants with the water but has a very great effect upon the formation of the floc. It is usual to introduce the chemical at some point of high turbulence in the water. After the dose of the chemical is added to the raw water, thorough mixing is provided by the following methods.

1. Centrifugal pump

In most of the cases, centrifugal pump is used to raise the raw water to the settling tank. The required dose of the chemical therefore can be added to the suction line of the pump. When water fed with the coagulant passes through the impeller of the pump, mixing is accomplished by the agitation. However, after the water comes out, some gentle agitation is required to get good results, and to accelerate coagulation and sedimentation.

2. Compressed air agitation

In this method the raw water fed with coagulant is agitated vigorously by diffusing compressed air from the bottom of the mixing basins.

3. Narrow mixing channel with flume

Fig. 9.18 shows a narrow mixing channel with vertical baffles. The coagulant is fed from the feeding tank. The turbulence caused by vertical baffles mixes the chemical thoroughly. A flume is some times provided to develope the hydraulic jump for causing turbulence, and also to measure the flow.

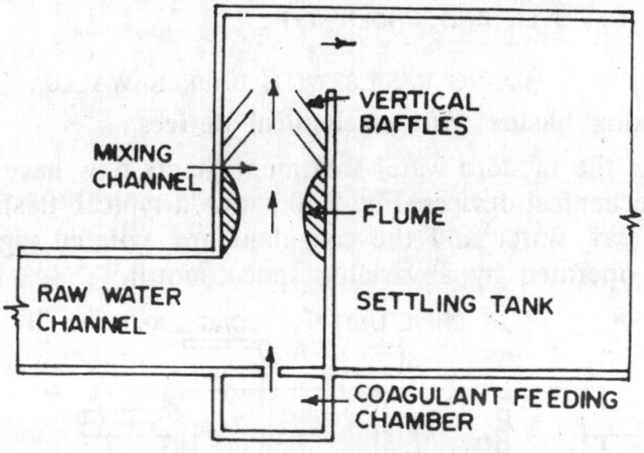

FIG. 9.18. MIXING CHANNEL

4. Mixing basins with baffle walls

Fig. 9.19 shows two types of mixing basins with baffles. In the horizontal or 'round the end type' basin [Fig. 9.19 (*a*)], the water flows horizontally for a short *distance* makes a complete turn, and continues back and forth around the ends of the baffles. This causes turbulence, and hence the mixing. Another type, known as vertical or 'over and under type', [Fig. 9.19 (*b*)], used in smaller plants, has vertically hanging baffle walls due to which water flows up and down. Mixing basins with baffle walls are not used now because of high head losses and variations in the velocities.

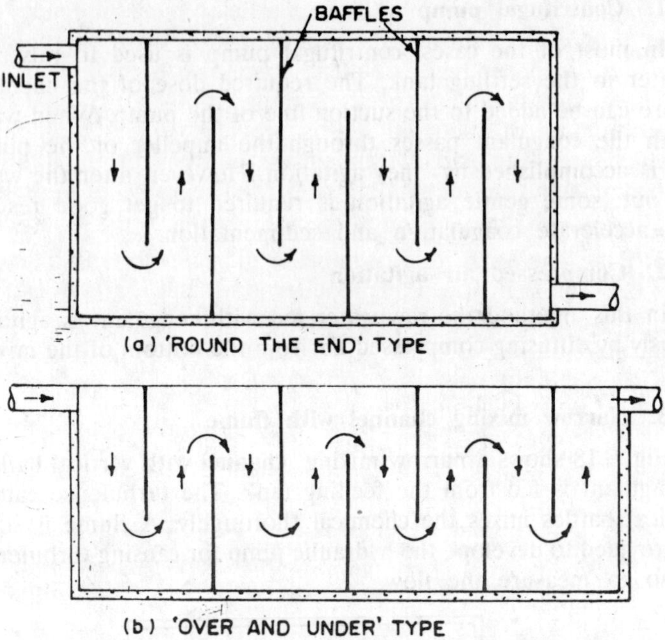

FIG. 9.19. MIXING BASINS WITH BAFFLE WALLS

5. Mixing basins with mechanical devices

Most of the modern water treatment plants now have mixing basin with mechanical devices. Fig. 9.20 shows a typical 'flash mixer' in which the raw water and the coagulant are agitated vigorously by a paddle operated by a variable speed motor.

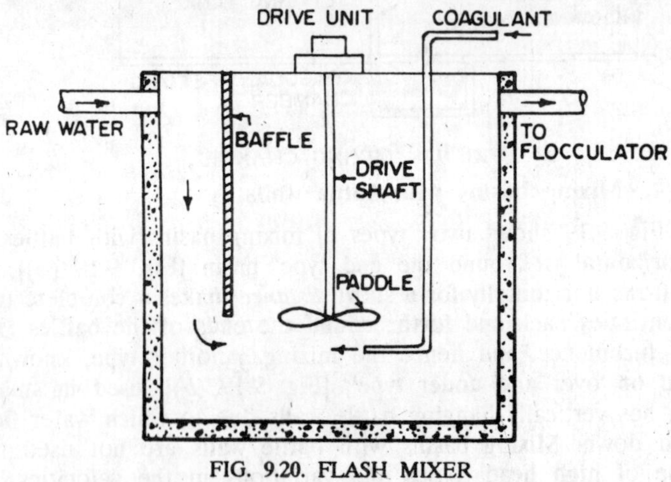

FIG. 9.20. FLASH MIXER

The intensity of mixing is dependent upon the *temporal mean velocity gradient* (G), defined in the next article. The turbulence and resultant intensity of mixing is based on the rate of power input to the water. Propeller type impellers, commonly employed in flash mixers, have high revolving speeds ranging from 400 to 1400 rpm. A detention time of 30 to 60 sec. is provided in the flash mixer units which are deep, circular or square tanks, with the ratio of hight to diameter (or side) of 1:1 to 3:1. The value of G is kept 300 s^{-1} or more. Power requirements are from 1 to 3 watts per m^3/hr of flow. The usual ratio of impeller diameter to tank diameter is 0.2 to 0.4. The shaft speed of propeller is so kept that a tangential velocity of greater than 3 m/s is imparted at the tip of the blades. Equations 9.29, 9.32, 9.33 and 9.34 derived for slow mixing in the next article are also applicable for rapid mixing by flash mixer.

9.15. FLOCCULATION

Flocculation is essentially an operation designed to force *agitation* in the fluid and induce coagulation. Basically, flocculation is a *slow mixing* or *agitating process* in which destabilised colloidal particles are brought into intimate contact in order to promote their agglomeration. Slow mixing is the hydrodynamic process which results in the formation of large and readily settleable flocs (orthokinetic flocculation) by bringing the finely divided matter into contact with the micro-flocs formed during rapid mixing. The operation of slow mixing is achieved in a basin commonly known as the *flocculator*. Flocculation is merely the provision of the type of *controlled motion* or agitation of water which will assist in the formation of settleable floc formation. The relative movement of the nuclei of solid material, caused by the agitation of the water, increases the probability of collusion between these nuclei and accelerates the rate of growth into large particles which can settle easily.

The rate of agglomeration or flocculation is dependent upon several factors such as type and concentration of turbidity, type of coagulant and its dose, and the *temporal mean velocity gradient* (G) in the basin.

The *mean velocity gradient* (G) is defined as the rate of change of velocity per unit distance normal to a section (metres per second/m) and hence a dimension of $1/T$ (generally expressed as sec^{-1}). The value of G can be computed in terms of power input by the following expression :

$$G = \sqrt{\frac{P}{\mu V}} \qquad ...(9.29)$$

where P = power dissipated (watts)
μ = absolute viscosity (N-s/m^2)
V = the volume to which P is applied. (m^3)
G = temporal mean velocity gradient. (s^{-1})

If the velocity gradient G is combined with the displacement time $t_d = V/Q$, the following non-dimensional measure of *conjunction opportunity* is obtained :

$$G \cdot t_d = \frac{V}{Q}\sqrt{\frac{P}{\mu V}} = \frac{\sqrt{PV/\mu}}{Q} \quad ...(9.30)$$

In the above expression, the term $\sqrt{PV/\mu}$ is termed as *power induced flow* while Q is known as the *displacement flow*. Hence $G\,t_d$ is, in a sense, the ratio of power induced rate of flow to displace ment induced rate of flow. The total number of particle collisions is proportional to $G\,t_d$, and hence $G\,t_d$ is a useful parameter for the design and operation of flocculation.

The desirable value of velocity gradient G in a flocculater vary from 20 to 75 sec^{-1} and of $G\,t_d$ from 2 to 6 × 10^4 for aluminium coagulants and 1 to 1.5.× 10^5 for ferric coagulants. Detention period of less than 10 minutes should be avoided; the usual detention period (t_d) varies from 10 to 30 minutes. It should be noted that very high value of G tends to shear the flocs and prevent them from building to a size that will settle rapidly while too low value of G may be unable to provide sufficient agitation to ensure effective flocculation

Again, for a given $G\,t_d$, the loading of the flocculation units is

$$\frac{Q}{V} = \frac{1}{t_d} = \frac{\sqrt{P/\mu V}}{G\,t_d} \quad ...(9.31)$$

The above relation implies that the hydraulic loading of flocculation units is not merely a function of théir relative capacity or detention time, but more significantly a function also of power input and viscosity. Hence *power input* becomes a manageable variable in numerous treatment plants.

Another useful parameter is the product of $G\,t_d$ and the floc volume concentration C (volume of floc per unit of volume of water), and this product $G\,t_d\,C$ (usually denoted as *GCT*) reflects to a certain extent the contact opportunity of the particles. The values of *GCT* are of the order of 100.

In order to have optimization aimed at saving in power consumption, *tapered flocculation* is some times practiced, in which G in a tank is made to vary from $100\,s^{-1}$ in the first stage to 50 to 60

s^{-1} in the second stage and then brought down to $20\,s^{-1}$ in the third stage in the direction of flow.

Similar to the rapid mixing units, flocculation or slow mixing can be achieved by the various methods under the following categories:

(*i*) Gravitational or hydraulic methods (such as horizontal flow baffled flocculator, vertical flow baffled flocculator, jet flocculator etc.)

(*ii*) Mechanical methods (such as paddle flocculators)

(*iii*) Pneumatic methods.

The flocculation technique most commonly used involves mechanical agitation with rotating paddle wheels or vertically mounted turbines.

The useful power input by the impeller is

$$P = F_D \,.\, v_r \qquad \text{...}(i)$$

where F_D = drag force $= C_D \,.\, F_I$

F_I = impeller force

C_D = coefficient of drag of paddle (= 1.8 for paddle with flat blades).

v_r = relative velocity of impeller and fluid.

$= v_p - v_w$

v_p = velocity of the tip of the paddles

v_w = velocity of water adjacent to the tip of the paddle.

But $F_I = \rho \dfrac{A_p v_r^2}{2}$, where A_p = area of paddles

$$\therefore \quad P = \frac{1}{2} C_D \,.\, \rho A_p v_r^3 \qquad \text{...(9.3?)}$$

Let k = ratio of fluid velocity to paddle velocity v_p.

Then $v_r = v_p - k\,v_p = (1 - k)\,v_p$ ...(*ii*)

The value of k may be taken as 0.25.

Also, we have, from Eq. 9.29

$$G = \sqrt{\frac{P}{\mu V}}$$

Hence from Eq. 9.29 and 9.32

$$G = \left(\frac{\frac{1}{2} C_D \,.\, A_p \rho\, (v_p - v_w)^3}{\mu V} \right)^{1/2} \qquad \text{...(9.33)}$$

The optimum value of G can be calculated from the following expression

$$G_{opt.}^{2.8} \,.\, t_f \,.\, c = 44 \times 10^5 \qquad \text{...(9.34)}$$

where G = optimum velocity gradient (s^{-1})

t_f = time of flocculation (min.)

c = alum concentration (mg/l)

The following *design criteria* are commonly adopted in a flocculator.

(*i*) Depth of tank : 3 to 4.5 m

(*ii*) Detention time : 10 to 40 min. ; normal : 30 min.

(*iii*) Velocity of flow : 0.2 to 0.8 m/s ; normal 0.4 m/s

(*iv*) Total area of paddles : 10 to 25% of the cross-sectional area of the tank

(*v*) Peripheral velocity of blades : 0.2–0.6 m/s ; normal 0,3 to 0.4 m/s

(*vi*) Velocity gradient (G) : 10 to 75 s^{-1}

(*vii*) Factor $G.t$: 10^4 to 10^5

(*viii*) Power consumption : 10.0 to 36.0 kW/mld

(*ix*) Outlet flow velocity : 0.15 to 0.25 m/s.

Fig. 9.21 shows a flocculating tank with power operated paddles on horizontal or vertical shafts. Fig. 9.22 shows another arrangement of the paddles in the flocculator compartment, which is normally

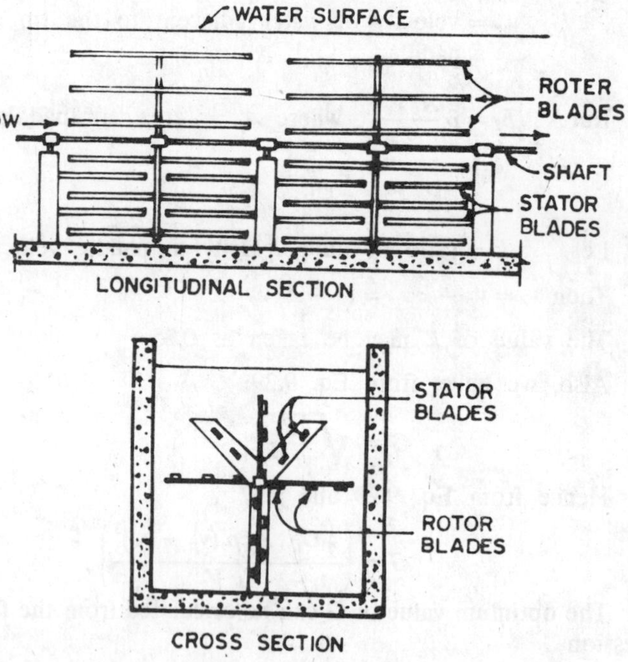

FIG. 9.21. FLOCCULATOR WITH MECHANICAL AGITATORS

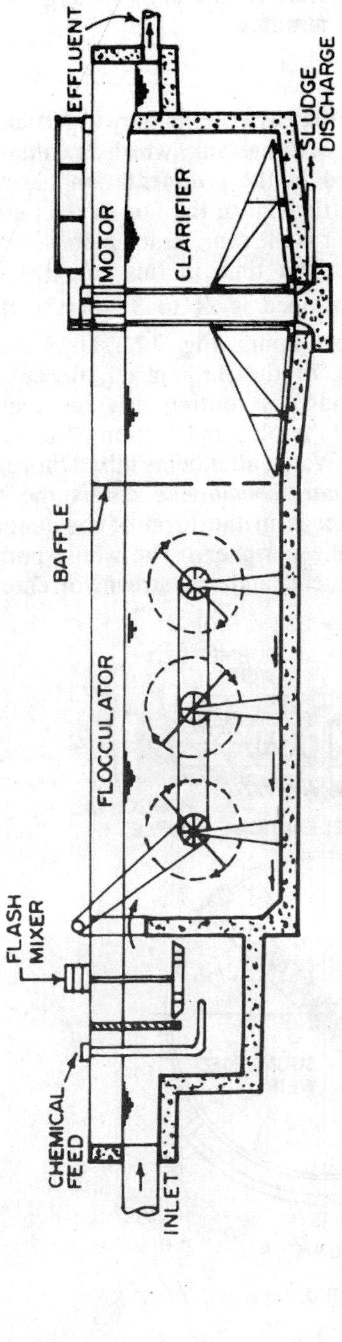

FIG. 9.22. MIXING BASIN, FLOCCULATER AND SETTLING BASIN

provided in between the flash mixer and the settling tank commonly known as clarifier. The detention time in the slow mixing unit of flocculator is normally about 40 minutes.

9.16. CLARIFICATION :

Flocculation is normally regarded as a preliminary to settlement. After flcculation, water enters the settling tank, which is commonly called a *clarifier.* Water is retained in the sedimentation basin for a sufficient period to permit the settlement of the floc to the bottom. The principles of design of a clarifier is the same as for plain sedimentation basin, except that the detention time in this case is lower. The detention period commonly adoped is $2\frac{1}{2}$ to 3 hours with an overflow rate of 1 to 1.2 metres per hour. Fig. 9.22 shows section of a sedimentation unit consisting of the flash mixer, flocculation basin and the settling basin. Sludge is continuously scrapped by mechanical means. Fig. 9.23 shows the plan and section of a circular clarifier designed by Walton & Kay. Water, after being mixed thoroughly with the coagulant in the *primary conditioner,* enters the tank tangentially at the base. The outlet is in the form of a submerged weir extending over one-third of the periphery. The whole body of water has a slow rotational movement, and any stream of entering

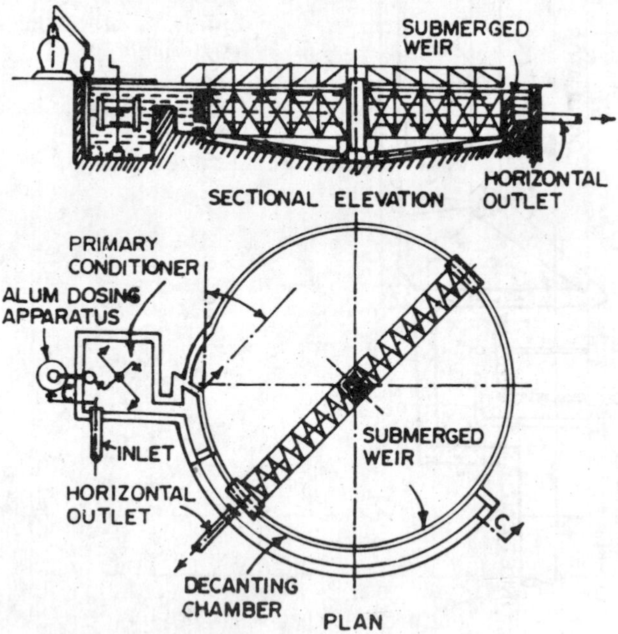

FIG. 9.23. CIRCULAR CLARIFIER (WALTON & KAY)

water is forced to spend some considerable time in smooth circular motion before it rises to, into the take off channel. In modern practice, flocculator and clarifier are provided in one single unit called *clarifloc-culater,* in which the flocculating chamber is provided in the centre and the clarifier compartment is formed by the peripheral space.

Type 2 settling : Flocculent settling

Flocculent particles resulting from coagulation will agglomerate while settling, with a resultant increase in particle size. The density of the composite particle will decrease due to inclusion of water. However, the settling velocity will increase. A suggested equation for the variation of floc density with particle size is :

$$\rho_s - \rho = k\,d^{-0.7} \qquad \text{...(9.35)}$$

where ρ_s = floc density

ρ = density of water

d = diameter of particle (or the dia. of sphere with equivalent surface area), and

k = a coefficient dependent upon the characteristics of the water and chemicals involved.

The clarification of dilute suspensions of flocculating particles is a function not only of the settling properties of the particles but also of the flocculating characteristics of the suspension. The rate of flocculation has been shown to be proportional to the mean velocity gradient in the system, the concentration of particles and, particle size. However, no mathematical relationship exists that can readily be used to analyse flocculant settling. To develop data, such as those plotted in Fig. 9.24, it is necessary to perform a settling

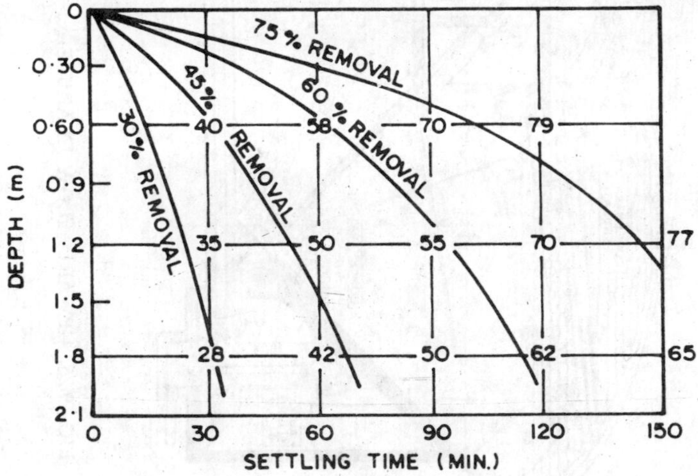

FIG. 9.24 RESULTS OF BATCH SETTING TEST

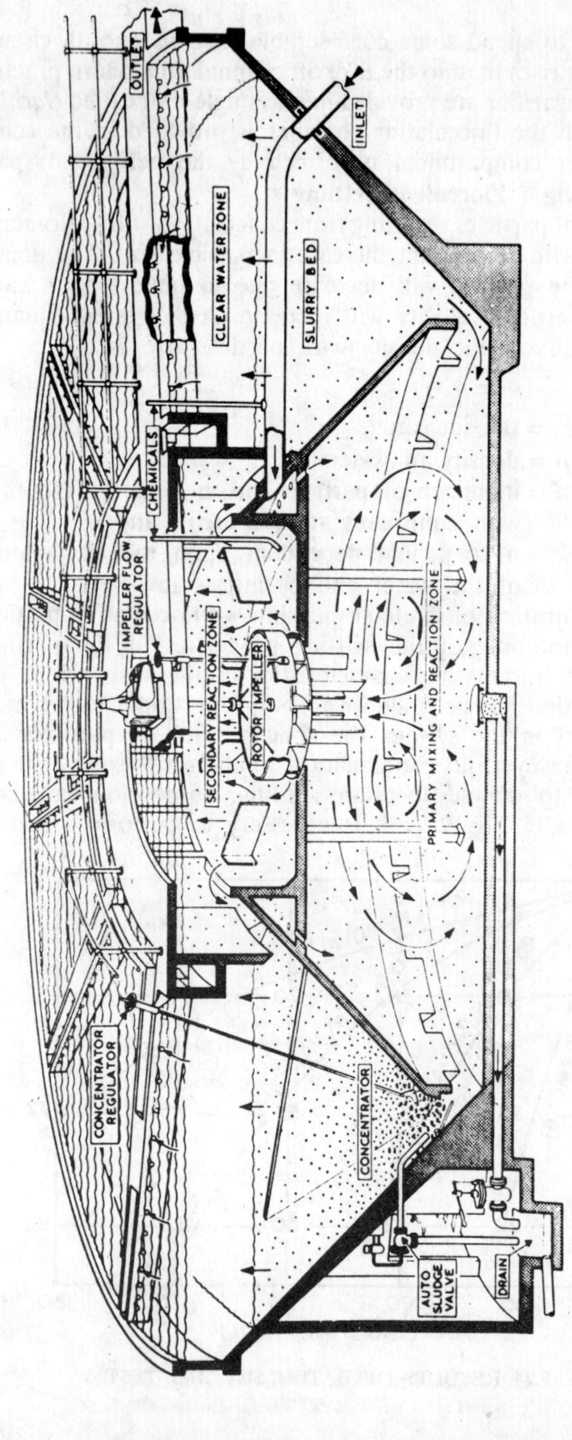

FIG. 9.25. UPWARD FLOW CLARIFIER : 'THE ACCENTRIFLOC' (PATERSON CANDY INTERNATIONAL LTD.)

column test in which a suspension is placed in a long column (glass cylinder) and is allowed to settle under quiescent conditions. Samples are withdrawn from this cylinder, at several depths at several time interval and the % removal is determined. The numerical value of % removal is located at the coordinates identifying the time and depth of the sample and iso-concentration lines connecting points of equal removal are drawn, as shown in Fig. 9.24.

9.17. SLUDGE BLANKET TANKS OR SOLID CONTACT CLARIFIERS

Sludge blanket tank is a recent development in the *vertical flow type tanks.* One such tank designed by Paterson Candy International Ltd. is shown in Fig. 9.25. The primary mixing is followed by a secondary reaction zone in which water is maintained in contact with floc produced by the chemical reaction. The water finally flows upwards, at a lower speed, in the outer zone consisting of inverted pyramid type tank (hopper bottomed tank) in which the solids gradually settle downwards by gravity. As the water rises up in the tank of increasing area, velocity gradually decreases, until at a level, the force of upward flow in the particles is counter-balanced by the downward weight of the particles, which therefore hang suspended in water, forming what is commonly known as a *sludge blanket.* In this blanket, about 1 m thick, some chemical and physico-chemical reactions are completed. In this blanket, a straining action to remove some of the finer particles also takes place. The upward flow tanks may have a detention time of 1 to 3 hours.

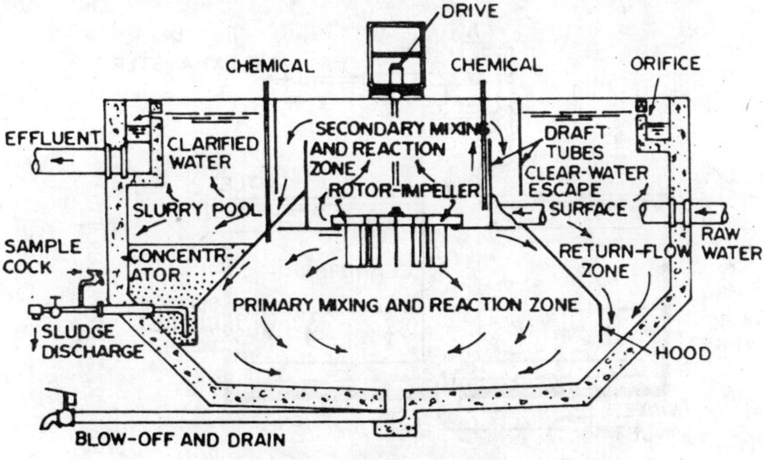

FIG. 9.26. VERTICAL-FLOW SEDIMENTATION BASIN : THE ACCELERATOR (INFILCO DEGREMONT INC.)

Fig. 9.26 shows another such type of vertical flow sedimentation tank designed by Infilco Degremont Inc. The 'accelerator' unit includes (*i*) a basin having a raw-water inlet and distribution duct, (*ii*) a primary mixing and reaction zone (*iii*) two concentric draft tubes which form the secondary mixing and reaction zone, (*iv*) a rotor-impeller for mixing and pumping driven by a motorised reducer, (*v*) an effluent channel system, and (*vi*) concentrators to accumulate and remove excess slurry.

9.18. THE PULSATOR CLARIFIER

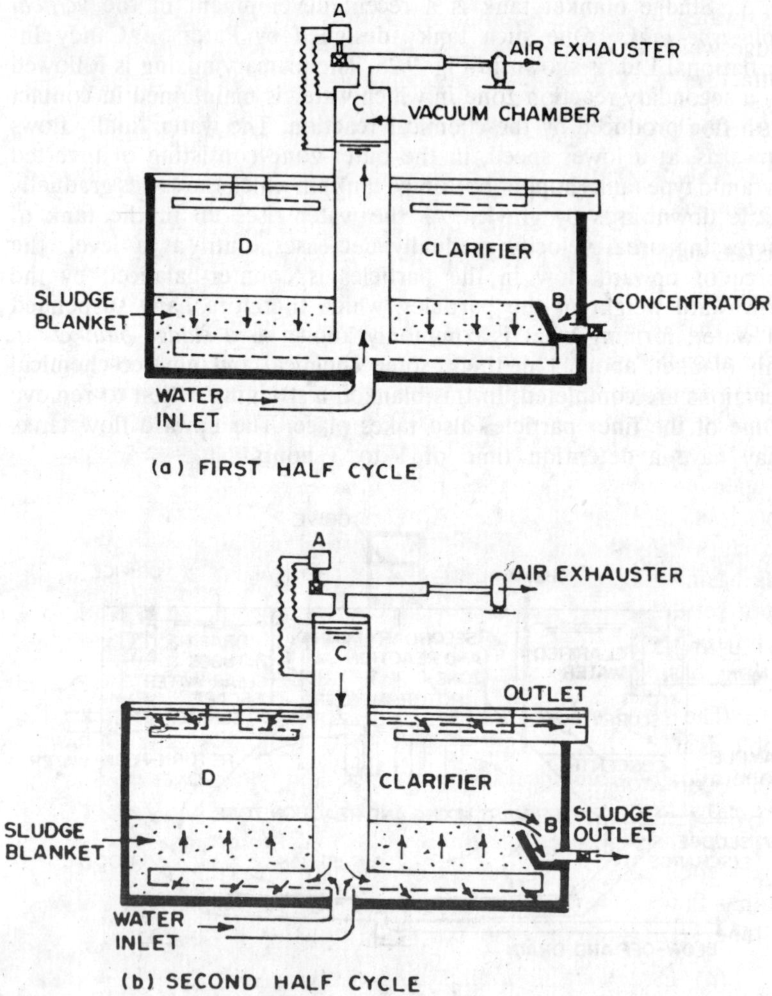

FIG. 9.27. THE PULSATOR CLARIFIER

Another form of vertical flow type sludge tank is *pulsator clarifier*, shown in Fig. 9.27, in which the pulse is generated at interval of 30 seconds to give rapid flow for 5 to 10 seconds. Due to this the sludge blanket is alternatively raised up and allowed to subside.

Water enters through the inlet at the bottom of the sludge blanket. In the first half of the cycle (Fig. 9.27 *a*), the air valve *A* is *closed* and water rises in the vacuum chamber *C*. At this stage, the water in the clarifier *D* is at rest and the sludge settles. In the next half cycle (Fig. 9.27 *b*), water in the vacuum chamber *C* rises upto upper contact and air value *A* is opened. Due to this, water in *C* falls rapidly and enters the clarifier, raising the sludge which enters concentrator *B*. When water falls to the lower contact in *C*, air value *A* is again closed.

9.19. SHALLOW DEPTH SEDIMENTATION : TUBE SETTLER

The tube settler is relatively new innovation in the development of sedimentation. In § 9.5 (Fig. 9.5), we have seen that if a smaller particle having falling speed $v_s' < \frac{Q}{A}$ enters the tank at point *A*, it will settle through only height *h* during the detention time, and hence that particle will not settle to the bottom if it enters the basin above the point *C*. However, these particles can be removed if false bottoms or trays were inserted at intervals *h*. This theory proposed by Hazen (1904) suggests that the removal of settleable solids is primarily a function of the surface area of basin and is independent of the detention time. He pointed out that by doubling the surface area by inserting one horizontal tray would double the capacity of the tank. Camp (1946) presented a design for a settling tank basin with horizontal trays spaced at 15 cm. The use of shallow depth settling basins enables the detention time of settling process to be reduced to only a few minutes in contrast to conventional settling basin designs which use 1 to 4 hour detention.

The introduction of wide shallow trays at some interval gave problems arising from the unstable hydraulic conditions, and difficulties in operation of mechanical sludge removal equipment. These problems have now been overcome by introduction of *tube settlers* or *tube clarifiers*. These settlers use tubes of 25 to 50 mm diameter, having large wetted perimeter relative to the wetted area, thereby providing laminar flow conditions for sedimentation. Two tube configurations of tube settlers are in use :

(*a*) *Essentially horizontal tube settler,* and
(*b*) *Steeply inclined tube settler.*

Fig. 9.28 (*a*) shows the horizontal tube settler in which the tubes are inclined only slightly (5°) in the direction of normal flow. The sludge settled in the tube is drained by the last portion of the filter backwash. The tube settlers of this type are used in small plants and in package plant systems.

Fig. 9.28 (*b*) shows steeply inclined tube configuration. The tubes are square, about 50 × 50 mm , made of plastic baffles inclined at 60° to the horizontal located between thin sheets of plastics. The solids settle in the tube bottom and leave the tube by sliding downward. A flow pattern is established in which the settleable solids are trapped in a downward flowing stream of concentrated solids.

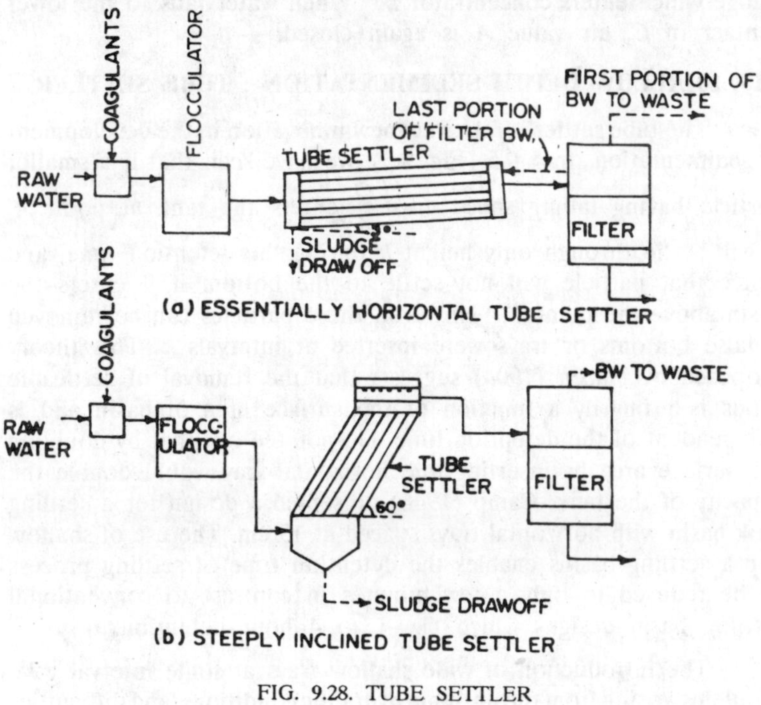

FIG. 9.28. TUBE SETTLER

Alternative approaches have been developed by various manufacturers, by incorporating steeply inclined tubes into a *modular form* which can easily be supported and installed in a sedimentation basin of rectangular or circular shape. In these innovations, the tube passages are inclined rather than inclining the whole module. The inclination of the baffles in alternate direction between one sheet and the next gives strength to the structure so that nests of such tubes may be fitted into horizontal or upward flow tanks. Fig. 9.29 shows such a tube module construction.

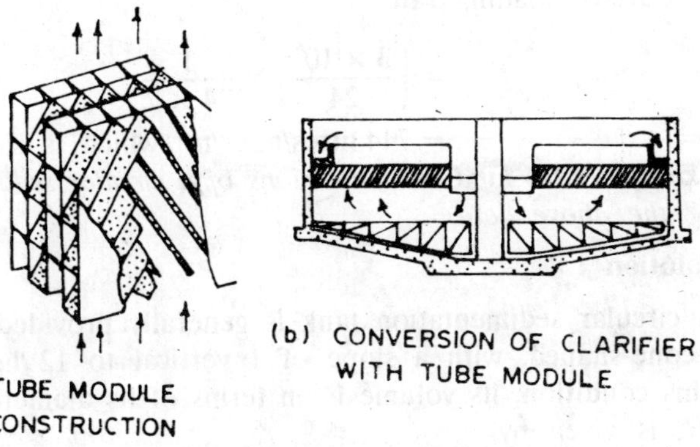

FIG. 9.29. THE TUBE SETTLER

9.20. ILLUSTRATIVE EXAMPLES

Example 9.9. *Find the dimensions of a rectangular sedimentation basin for the following data :*

(*i*) *Volume of water to be treated = 3 million litres per day.*

(*ii*) *Detention period = 4 hours.*

(*iii*) *Velocity of flow = 10 cm/min.*

Solution :

Detention time = 4 hours = 240 min.

Velocity of flow = 10 cm/min.

$\therefore$ Length of tank = 0.10 × 240 = 24 m.

Volume of water in 4 hours

$$= \frac{3 \times 10^6}{10^3} \times \frac{4}{24} = 500 \text{ m}^3$$

$\therefore$ Cross-section area

$$A = \frac{V}{L} = \frac{500}{24} = 20.8 \text{ m}^2$$

Assume a working depth of 3 m.

$$\therefore \quad \text{Width of tank} = \frac{20.8}{3} \approx 7 \text{ m.}$$

Provide an extra depth of 1 m for sludge storage and 0.5 m for free board making a total depth = 3 + 1.5 = 4.5 m.

Hence provide a settling tank of size 24 m × 7 m × 4.5 m.

Check :

Volume of water per hour

$$= \frac{3 \times 10^6}{24}$$

∴ Surface loading rate

$$= \left(\frac{3 \times 10^6}{24}\right) \times \frac{1}{24 \times 7}$$

$= 744$ litres/hour/m^2 which is satisfactory.

Example 9.10. *Find the dimensions of a circular sedimentation basin for the above data.*

Solution :

A circular sedimentation tank is generally provided with its bottom cone-shaped, with a slope of 1 vertical to 12 horizontal. Under this condition, its volume V, in terms of its diameter D and height H, is given by

$$V = D^2 (0.011\, D + 0.785\, H) \qquad \text{...(9.36)}$$

Now volume of water during detention period

$$= \frac{3 \times 10^6}{10^3} \times \frac{4}{24} = 500 \text{ m}^3$$

Let the effective depth $= 3$ m.

Substituting the values in the above expression, we get

$$500 = D^2 (0.011\, D + 0.785 \times 3)$$

Solving this by trial, we get

$$D \approx 14.2 \text{ m.}$$

Providing an extra depth of 1.5 m for sludge storage and free board, the size of the tank is 14.2 m dia × 4.5 m depth.

Example 9.11. *Design a plain sedimentation basin to treat 3 million of water litres per day so as to settle at least 75% of the particles of grain size 0.002 cm or more. Assume any suitable data required.*

Solution:

Velocity of settlement of 0.002 cm (or 0.02 mm) diameter at 26°C can be determined from Eq. 9.9 (*a*) by assuming an average specific gravity of 2.65.

$$v_s = 418\,(2.65 - 1)\,(0.02)^2 \left(\frac{3 \times 26 + 70}{100}\right)$$

$= 0.408$ mm/sec $= 0.0408$ cm/sec.

From Eq. 9.11 (*e*), 1 cm per sec. settling velocity corresponds to a surface loading of 8,64,000 litres per day/m^2.

∴ Associated surface loading

$= 86400 \times 0.0408 = 35278$ litres/day/m^2.

Assume the depth of tank as 3 m. The time required to settle with the above velocity of 0.0408 cm/sec will be

$$= \frac{300}{0.0408 \times 60} = 122.5 \text{ min.}$$

As per past experience, for 75% removal of the particles of the designed category (*i.e.* 0.002 cm), the surface loading should be reduced by an average of 52%, and the detention time increased by 120%.

∴ Design surface loading

$$= 35278 \times 0.52 = 18345 \text{ litres/day/m}^2$$

Design detention period

$$= 122.5\left(1 + \frac{120}{100}\right) \approx 270 \text{ min.}$$

$$= 4.5 \text{ hours}$$

$$\text{Surface area} = \frac{3 \times 10^6}{18345} = 163.5 \text{ m}^2$$

Volume of water in 4.5 hours

$$= \frac{3 \times 10^6}{10^3} \times \frac{4.5}{24} = 562.5 \text{ m}^3$$

Let $L = 10 \times 3 = 30$ m.

∴ $B = \dfrac{562.5}{30 \times 3} = 6.25$ m

Displacement velocity of flow

$$= \frac{30}{270} = 0.111 \text{ m/min.}, = 11.1 \text{ cm/min.}$$

which is satisfactory.

Provide an extra depth of 1 m for sludge storage and 0.5 m for free board, making a total depth

$$= 3 + 1.5 = 4.5 \text{ m}$$

Hence provide a tank of 30 m × 6.25 m × 4.5 m size.

Example 9.12. *Design a coagulation sedimentation tank to treat 10 million litres of water per day. Assume suitable data where necessary.*

Solution :

Alum and iron flocs adsorb and entrain water and reduce the specific gravity to as little as 1.002. However, these flocs may be as large as 0.1 cm in diameter, and their settling velocity at 26°C may be assumed to be about 0.12 cm per sec.

Since 1 cm per sec. settling velocity corresponds to a surface loading of 8,64,000 litres per day/ m², the associated surface loading

$$= 864000 \times 0.12$$

$$= 103680 \text{ litres/day/m}^2.$$

Reducing this by about 52%,
Design surface loading

$$= 103680 \times 0.52 \approx 54000 \text{ litres/day/m}^2.$$

$$\therefore \text{ Surface area } = \frac{10 \times 10^6}{54000} = 185.2 \text{ m}^2.$$

Provide a tank of size 20 m× 9.3 m.

Provide an effective depth of tank= 3 m, and an overall depth = 4.5 m, making an allowance of 1 m for sludge and 0.5 m for free board.

Theoretical detention time.

$$= \frac{300}{0.12 \times 60} \approx 42 \text{ min.}$$

Total actual detention time may be

$$= 42\left(1 + \frac{120}{100}\right) \approx 92 \text{ min.}$$

Hence provide the settling basin of size 20 m × 9.3 m × 4.5 m, and give a slope of 1 in 50 to the floor of the tank.

Average velocity of flow in the tank

$$= \frac{\text{Length of tank}}{\text{Detention period}}$$

$$= \frac{20 \times 100}{92} \approx 22 \text{ cm/min.}$$

which is satisfactory.

Design of floc chamber.

The section of tank is shown in Fig. 9.30.

Effective depth of tank near floc chamber

$$= 3 + 1 - \frac{20}{50} = 3.6 \text{ m.}$$

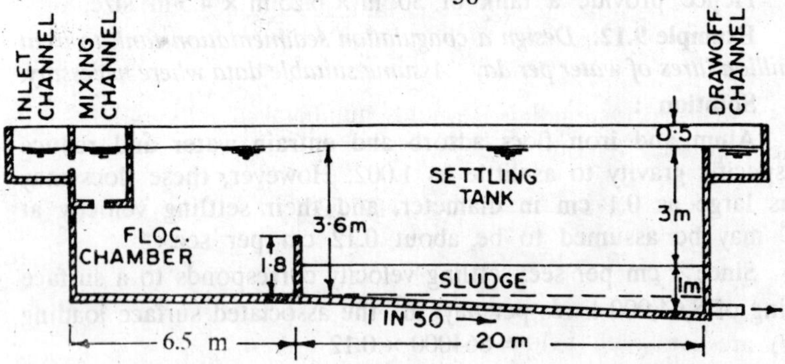

FIG. 9.30

Assume that effective depth of floc chamber is equal to half the effective depth of the settling tank near floc chamber.

∴ Effective depth of floc chamber

$$= \frac{1}{2} \times 3.6 = 1.8 \text{ m.}$$

Assume a detention period of 15 minutes for the floc chamber

∴ Total capacity required

$$= \frac{10 \times 10^6}{10^3} \times \frac{15}{24 \times 60} = 104.2 \text{ m}^3.$$

Width of floc chamber = Width of settling tank

$$= 9.3 \text{ m.}$$

∴ Length of floc chamber

$$= \frac{104.2}{9.3 \times 1.8} = 6.22 \text{ m.}$$

Keep length of floc chamber = 6.5 m.

Example 9.13. *A settling basin is designed to have a surface overflow rate of 32.6 m/day. Determine the overall removal obtained for a suspension with size distribution given below. The specific gravity of the particles is 1.2 and the water temperature is 20°C at which dynamic visocity is 1.027 centipoise and the density is 0.997 g/cm³.*

Particle size (mm)	*0.1*	*0.08*	*0.07*	*0.06*	*0.04*	*0.02*	*0.01*
Weight fraction greater than size (per cent)	*10*	*15*	*40*	*70*	*93*	*99*	*100*

(U.P.S.C. Engineering Services 1992)

Solution :

The settling velocities may be calculated from Stoke's law (Eq. 9.6) :

$$v_s = \frac{g}{18\,\mu}(\rho_s - \rho)\,d^2 = \frac{9810}{18\,(1.027)}[1.2 - 0.997]\,d^2$$

$$= 107.73\,d^2 \qquad ...(i)$$

The corresponding Reynolds number can be computed from the relation

$$R = \frac{\rho\,v_s\,d}{\mu} = \frac{0.997}{1.027}(107.73\,d^2)\,d$$

or $$R = 104.56\,d^3 \qquad ...(ii)$$

The computed values of v_s and R for various particle sizes (d) are tabulated below :

Weight fraction greater than stated (%)	10	15	40	70	93	99	100
Weight fraction finer than stated %	90	85	60	30	7	1	0
v_s (mm/s)	1.077	0.689	0.528	0.387	0.172	0.043	0.011
R	0.10	0.05	0.04	0.02	0.007	0.0008	0.0001

Since R is less than 0.5, Stoke's law is applicable and hence calculation for v_s are valid. From the above table, the *cumulative distribution curve* is drawn as shown in Fig. 9.31.

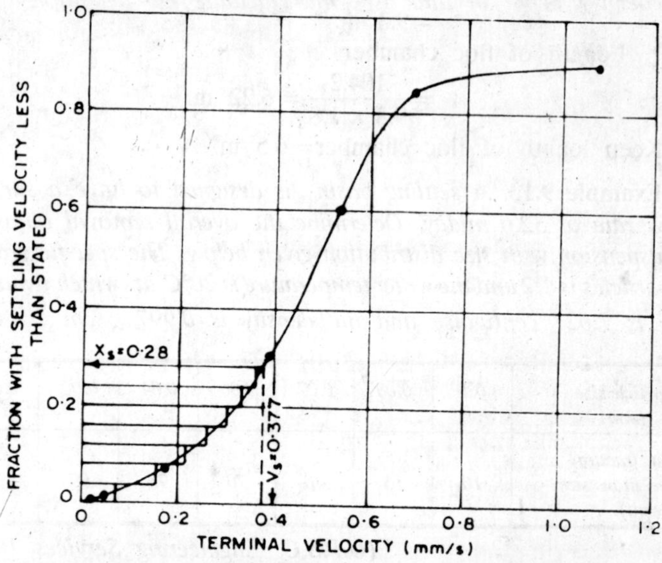

FIG. 9.31. CUMULATIVE DISTRIBUTION CURVE

$$\text{Now SOR} = v_s = 32.6 \text{ m/day} = \frac{32.6 \times 1000}{24 \times 60 \times 60} = 0.377 \text{ mm/s}$$

Hence all particles with setting velocities greater than 0.377 mm/s will be removed. Thus, from the graph (Fig. 9.31), the fraction $(1-x_s)$ is equal to 0.72. The graphical determination of $\Sigma v_s' \Delta x$ is tabulated below :

Δx	0.04	0.04	0.04	0.04	0.04	0.04	0.04
v_s'	0.06	0.16	0.22	0.26	0.30	0.33	0.36
$v_s' \Delta x$	0.0024	0.0064	0.0088	0.0104	0.0120	0.0132	0.0144

The *overall removal* is given by Eq. 9.13 (*a*)

$$R = (1 - x_s) + \frac{1}{v_s} \Sigma v_s' \Delta x$$

$$= 0.72 + \frac{1}{0.377} (0.0676) = 0.899 \text{ or } \mathbf{89.9\ \%}$$

Example 9.14. *A rectangular sedimentation basin is to handle 12 million litres/day of raw water. A sedimentation basin of width to length ratio of $\frac{1}{3}$ is proposed to trap all particles larger than 0.05 mm in size. Assume a relative density of 2.62 for the particles and 20°C as the average temperature, determine the basin dimensions. If the effective depth of tank is 3 m, calculate the detention time.*

Solution.

$$v_s = 418 (S_s - 1) d^2 \left(\frac{3t + 70}{100} \right)$$

$$= 418 (2.62-1) (0.05)^2 \left(\frac{3 \times 20+70}{100} \right) = 2.2 \text{ mm/sec}$$

$$= 0.22 \text{ cm/sec} = 7.92 \text{ m/hour}$$

From Eq. 9.11, $v_s = \dfrac{Q}{LB}$

Also, $$t_0 = \frac{L}{v_d} = \frac{LBH}{Q} = \frac{H}{v_s} \quad \text{...(Eq. 9.11 } d)$$

$$= \frac{H}{7.92} \quad \text{...}(i)$$

(where t_0 is detention time in hours)

$$Q = 12 \text{ million litres per day } = \frac{12 \times 10^6}{10^3} \times \frac{1}{24} \text{ m}^3\text{/hour}$$

$$= 500 \text{ m}^3\text{/hour}$$

$\therefore$ Capacity of tank, $V = 500 \times t_0$ ...(*ii*)

$\therefore$ $$LBH = 500 \times \frac{H}{7.92}$$

But $L = 3B$

$\therefore$ $$3B^2 = \frac{500}{7.92}$$

From which $B = 4.59 \approx \mathbf{4.60\ m}$

$\therefore$ $L = 4.6 \times 3 = \mathbf{13.8\ m.}$

If $H = 3$ m,

$$t_0 = \frac{H}{7.92} = \frac{3}{7.92} = 0.379 \text{ hours } = \mathbf{22.7\ min.}$$

Example 9.15. *In a continuous flow settling tank, 3.5 m deep and 50 m long, what flow velocity of water would you recommend for effective removal of 0.03 mm particles at 20°C. The specific gravity of particles is 2.65 and the kinematic viscosity of water may be taken as 0.01 cm²/sec.*

Solution :

$$v_s = \frac{g}{18}(S_s - 1)\frac{d^2}{\nu} \quad ...(9.7\ a)$$

$$= \frac{981}{18}(2.65 - 1)\frac{(0.003)^2}{0.01} = 0.081 \text{ cm/sec.}$$

From Eq. 9.11 (*c*)

$$v_d = \frac{L}{H} v_s$$

Taking 0.5 m F.B., $H = 3.5 - 0.5 = 3$ m ; $L = 50$ m

$$\therefore \quad v_d = \frac{50}{3} \times 0.081 = \mathbf{1.35\ cm/sec.}$$

Hence the flow velocity or displacement velocity should not be more than 1.35 cm/sec.

Example 9.16. *Two primary settling basings are 25 m in diameter with 2.5 m side water depth. Single effluent weirs are located on the peripheries of the tanks.*

For a water flow of 30000 m³/day, calculate

(*i*) *surface area and volume*
(*ii*) *overflow rate in m³/m²/d*
(*iii*) *detention time, in hours, and*
(*iv*) *weir loading in m³/m/day*

Solution :

For a circular sedimentation tank with cone-shaped bottom having a slope of 1 vertical to 12 horizontal,

$$V = D^2 (0.011 D + 0.785 H)$$

$$= 25^2 (0.011 \times 25 + 0.785 \times 2.5) = \mathbf{1398.4\ m^3}$$

Now Q per tank $= \frac{1}{2} \times 30000 = 15000$ m³/day

$$= \frac{15000}{24} \text{ m}^3\text{/hr} = 625 \text{ m}^3\text{/hr.}$$

$$\therefore \text{ Detention time } t_0 = \frac{V}{Q}$$

$$= \frac{1398.4}{625} = \mathbf{2.2375\ hours}$$

Surface area of tank $= 0.785\, D^2 = 0.785\,(25)^2$

$= \mathbf{490.63\ m^2}$

Overflow rate $= \dfrac{Q}{\text{surface area}} = \dfrac{15000}{490.63}$

$= \mathbf{30.57\ m^3/m^2/day}$

Length of weir along periphery of the tank

$= \pi D = \pi \times 25 = 78.54$ m

Weir loading per day $= \dfrac{Q}{\text{Length of weir}}$

$= \dfrac{15000}{78.54} \approx \mathbf{191\ m^3/m/day.}$

Example 9.17. *A coagulation-sedimentation plant clarifies 50 million litre of water per day. The raw water has an alkalinity equivalent of 4 mg/l of $CaCO_3$. The filter alum required at the plant is 20 mg/l. Determine the filter alum and the quick lime (containing 88% to CaO) required per year by the plant. Use the following molecular weights :*

$$(Al = 27 ; S = 32 ; O = 16 ; H = 1 ; Ca = 40 ; C = 12)$$

Solution :

(*i*) Alum required per day $= 50 \times 10^6 \times 20 = 1000 \times 10^6$ mg/day

$= 1000$ kg/day $=$ **365 tonnes/year**

(*ii*) The chemical reaction is indicated by Eq. 9.19, when water has sufficient natural alkalinity. If the natural alkalinity is insufficient, lime is added to water, due to which the reactions are :

$$CaCO_3 + H_2O + CO_2 = Ca\,(HCO_3)_2$$

Mol. wt. of $Al_2(SO_4)_3.18H_2O = (2 \times 27) + 3(32 + 4 \times 16) + 18\,(2 + 16) = 666$

Mol. wt. of $Ca\,(HCO_3)_2 = 40 + 2\,(1 + 12 + 48) = 162$

Mol. wt. of $CaCO_3 = 40 + 12 + 48 = 100$

Mol. wt. of $CaO = 40 + 16 = 56$

From Eq. 9.19, we find that alum requires 3 × 162 (= 486) parts of natural alkalinity, $Ca\,(HCO_3)_2$ for every 666 parts of alum. Also, 162 parts of natural alkalinity is equivalent to 100 parts of alkalinity as $CaCO_3$.

Hence required alkalinity (as $CaCO_3$), for water containing alum of 20 mg/l

$$= \frac{3 \times 100}{666} \times 20 = 9.01 \text{ mg/l} \qquad ...(i)$$

Natural alkalinity available as $CaCO_3 = 4$ mg/l

∴ Additional alkalinity required to be added in the form of lime

$$= 9.01 - 4 = 5.01 \text{ mg/l as } CaCO_3$$

Since 100 part of $CaCO_3$ produces 56 parts of CaO, quantity of CaO required

$$= \frac{5.01 \times 56}{100} = 2.806 \text{ mg/l.}$$

But market available quick lime contains only 88 per cent of CaO.

$$\therefore \quad \text{Quick lime required} = \frac{2.806 \times 100}{88} = 3.188 \text{ mg/l}$$

$$\therefore \text{ Quantity of quick lime per day} = 50 \times 10^6 \times 3.188 \text{ mg}$$

$$= 159.4 \text{ kg/day} \approx \mathbf{58.2 \text{ tonnes/year}}$$

9.21. DESIGN EXAMPLES

Example 9.18. ***Design of rectangular plain sedimentation tank:***

Design a rectangular sedimentation tank for the following data:

(i) Required outflow from sedimentation tank = 300 m^3/hr

(ii) Water used in desludging = 2 %

(iii) Min. size of particles to be removed = 0.018 mm

(iv) Expected removal efficiency of min. size particle = 75 %

(v) Nature of particles : Discrete and non-flocculating

(vi) Specific gravity of particles = 2.65

(vii) Assumed performance of the settling tank : Good (n = 1/4)

(viii) Kinematic viscosity of water at 20 °C = 1.01×10^{-6} m^2/s

Solution

1. Design elements

We have to design the following elements of the tank

(*i*) Size (or dimensions) of the tank

(*ii*) Design of influent structure

(*iii*) Design of effluent structure.

The tank so designed has to checked against the re-suspension of the deposited particles.

$$\text{Required outflow rate} = 300 \text{ m}^3/\text{hr}$$

$$\text{Water loss} = 2\%$$

$$\therefore \text{ Design average flow} = \left(\frac{100}{100-2}\right) \times 300 \approx 306 \text{ m}^3/\text{hr.}$$

2. *Computation of settling velocity of minimum size particles*

Assuming Stoke's law to be valid,

$$v_s = \frac{g}{18}(S_s - 1)\frac{d^2}{\nu} \qquad \text{...(Eq. 9.7 } a)$$

$$= \frac{9.81}{18}(2.65 - 1)\frac{(0.018 \times 10^{-3})^2}{1.01 \times 10^{-6}} = 2.88 \times 10^{-4} \text{ m/s}$$

$$\text{Reynold's No.} = \frac{v_s . d}{\nu} = \frac{2.88 \times 10^{-4}(0.018 \times 10^{-3})}{1.01 \times 10^{-6}}$$

$$= 0.00513 < 1$$

Hence stoke's law is applicable and the computations for velocity, done above are valid.

$$\therefore \qquad v_s = 2.88 \times 10^{-4} \text{ m/s.}$$

3. *Determination of surface overflow rate*

For an ideal tank, for 100% removal of min. size particles, the surface overflow rate will be equal to the settling velocity of min. size particles.

$$\therefore \qquad \text{SOR} = v_s = 2.88 \times 10^{-4} \text{ m/s}$$

$$= 2.88 \times 10^{-4} \times 3600 \times 24 = 24.88 \text{ m/day}$$

However, due to short circuiting, there will be decrease in the efficiency, due to which SOR will have to be *reduced.* For Eq. 9.20 we have

$$\frac{y}{y_0} = 1 - \left[1 + n\frac{v_s}{Q/A}\right]^{-1/n}$$

Here, $\frac{y}{y_0} = 0.75$ and $n = \frac{1}{4}$ (for good performance).

$$\therefore \qquad \frac{v_s}{Q/A} = \frac{1}{n}\left[\left\{1 - \frac{y}{y_0}\right\}^{-n} - 1\right]$$

$$= 4\left[\{1 - 0.75\}^{-1/4} - 1\right] = 1.657$$

$$\therefore \qquad \text{Design SOR} = \frac{Q}{A} = \frac{v_s}{1.657} = \frac{24.88}{1.657} \approx 15.02 \text{ m/d}$$

The normal values of SOR range between 15 to 30 $m^3/m^2/d$ for plain sedimentation tanks.

4. *Determination of dimensions of tank*

$$\text{Surface area of tank,} \qquad A = \frac{Q}{Q/A}$$

$$= \frac{306 \times 24}{15.02} = 488.9 \text{ m}^2$$

Let us assume L/B as 4.

Now $L \times B = 488.9$

$$\therefore \quad B = \left(\frac{488.9}{4}\right)^{1/2} \approx 11.1 \text{ m}$$

$$\therefore \quad L = 11.1 \times 4 = 44.4 \text{ m.}$$

Let us assume detention time $t_0 = 4$ hours,

$$\text{Water depth of settling zone} = \frac{Q \times t_0}{A} = \frac{306 \times 4}{11.1 \times 44.4} = 2.48 \text{ m.}$$

5. *Check against scour.*

$$v_d = \sqrt{\frac{8\beta g}{f}(S_s - 1)\, d} \qquad \text{...(9.14)}$$

Taking $\beta = 0.04$ for unigranular sand

$f = 0.03$ for settling tanks.

$$v_d = \sqrt{\frac{8 \times 0.04}{0.03} \times 9.81\,(2.65 - 1)\,(0.018 \times 10^{-3})}$$

$$= 5.57 \times 10^{-2} \text{ m/s}$$

Also, horizontal displacement velocity is given by

$$v_h = \frac{Q}{B \times H} = \frac{306/3600}{11.1 \times 2.48} = 3.09 \times 10^{-3} \text{ m/s}$$

Since $v_h < v_d$, design is OK.

6. *Design of influent structure.*

The influent structure may consist of (*i*) an influent channel (*ii*) submerged orifices, and (*iii*) baffles in front of orifices (Fig. 9.7 *b*).

Let us provide 0.75 m wide and 0.6 m deep influent channel, running across the width of the tank. Let us assume a depth in of flow of 0.4 m.

$$\therefore \text{ Velocity of flow in channel} = \frac{306/3600}{0.75 \times 0.4}$$

$$= 0.283 \text{ m/s} < 0.3 \text{ m/s.}$$

Let us provide 4 submerged orifices of size 0.20 m × 0.20 m in the inside wall of the influent channel to distribute the flow uniformly in to the settling tank.

Also, provide a baffle 1 m deep, at a distance of 0.8 m from the orifices, to reduce the disturbance.

$$\text{Head loss through the orifices} = \left(\frac{Q}{C_d\, a \sqrt{2g}} \right)^2$$

$$= \left(\frac{306/3600}{0.6 \times 4\,(0.20 \times 0.20)\sqrt{2 \times 9.81}} \right)^2$$

$$= 0.04 \text{ m}$$

7. *Design of effluent structure*

Various components to be provided in an effluent structure are : effluent weir, effluent launder (or trough), outlet box and outlet pipe or channel.

Let us assume a weir loading of 250 m^3/d/m length

$$\therefore \quad \text{Weir length} = \frac{300 \times 24}{250} = 28.8 \text{ m.}$$

Provide effluent weir of 30 m length. Let us provide 90° V-notches @ 200 mm c/c.

$$\text{No. of V-notches} = \frac{30}{0.2} = 150$$

Hence Provide 30 m length of effluent launder with 90° V-notches fixed only to one side of the launder. Let us keep width of launder = 0.3 m.

Hence critical depth h at the other end of the launder is (Eq. 9.18 *b*) :

$$h = \left(\frac{Q^2}{b^2 g} \right)^{1/3}$$

where Q = discharge from each side $= \frac{300}{2}$ m^3/hour

$$= \frac{300}{2 \times 3600} \text{ m}^3\text{/s} = 0.04167 \text{ m}^3\text{/hour}$$

b = width of launder = 0.3 m.

$$\therefore \quad h = \left[\frac{(0.04167)^2}{(0.3)^2 \times 9.81} \right]^{1/3} = 0.125 \text{ m}$$

Depth of water at the other end, neglecting the friction losses, is given by Eq. 9.18 (*a*), taking $n = 1$

$$\therefore \quad H = \left[h^2 + \frac{2 q^2 L^2}{g b^2 h} \right]^{1/2},$$

where $q.L = \frac{300}{2 \times 3600} = 0.04167 \text{ m}^2\text{/s}$

$$\therefore \quad H = \left[(0.125)^2 + \frac{2\,(0.04167)^2}{9.81\,(0.3)^2 \times 0.125} \right]^{1/2}$$

$$= 0.217 \text{ m.}$$

Let us *assume* a head loss of 0.1 m. Providing free board of 0.1 m, depth of the launder below the bottom of weir discharging into it is = 0.217 + 0.1 + 0.1 = 0.417 m. If the weir has 50 mm deep V-notches, the resulting depth of channel = 0.417+0.05 = 0.467 m. Hence provide the effluent launder of width equal to 300 mm and depth equal to 500 mm.

Example 9.19. ***Design of radial circular plain sedimentation tank:***

Design a secondary circular sedimentation tank to remove alum floc for the following data :

(i) Required output from the tank = *300 m^3/hr.*

(ii) Water used in desludging = *2%*

(iii) Min. size of alum floc to be removed = *0.8 mm*

(iv) Specific gravity of alum floc = *1.002*

(v) Expected removal efficiency of alum floc = *75%*

(vi) Assumed performance of settling tank : very good $\left(n = \frac{1}{8}\right)$

(viii) Kinematic viscosity of water at $20°C = 1.01 \times 10^{-6}\ m^2/s.$

Solution.

1. ***Design elements***

This is a typical radial flow circular tank, with central feed, wherein the water enters at the centre of the tank and emanates from the multiple ports of circular well in the centre of tank to flow radially outwards in all directions equally (Fig. 9:12). The effluent launder is provided at the outer peripheri of the tank.

Required outflow rate= 300 m^3/hr.
Water loss for desludging= 2%

$$\therefore \text{ Average design flow} = \frac{100}{100-2} \times 300 \approx 306 \text{ m}^3\text{/hr.}$$

2. ***Design of influent pipe***

Let us assume a velocity of flow of 1 m/s in the pipe .

$$\therefore \text{ Influent pipe dia.} = \sqrt{\frac{306}{3600} \times 1 \times \frac{\pi}{4}} = 0.26 \text{ m.}$$

Provide an influent pipe of 300 mm dia.

3. *Computation of settling velocity of min. size floc*

Assuming Stoke's law to be valid,

$$v_s = \frac{g}{18}(S_s - 1)\frac{d^2}{\nu}$$

$$= \frac{9.81}{18}(1.002 - 1)\frac{(0.8 \times 10^{-3})^2}{1.01 \times 10^{-6}} = 6.907 \times 10^{-4} \text{ m/s}$$

Reynolds No. $R = \frac{v_s d}{\nu} = \frac{6.907 \times 10^{-4} \times 0.8 \times 10^{-3}}{1.01 \times 10^{-6}}$

$$= 0.55 < 1$$

Hence Stoke's law is applicable and the computations for velocity done above are correct.

$$\therefore \quad v_s = 6.907 \times 10^{-4} \text{ m/s.}$$

4. *Determination of surface overflow rate*

For an ideal tank, for 100% removal of min. size flocs, the SOR will be equal to the settling velocity of min. size flocs.

$$\therefore \quad \text{SOR} = v_s = 6.907 \times 10^{-4} \text{ m/s.}$$

$$= 6.907 \times 10^{-4} \times 24 \times 3600 \approx 59.7 \text{ m/d}$$

However, due to short circuiting, there will be decrease in the efficiency, due to which SOR will have to be reduced.

From Eq. 9.20, we have

$$\frac{y}{y_0} = 1 - \left[1 + n\frac{v_s}{Q/A}\right]^{-1/n}$$

Here $\frac{y}{y_0} = 0.75$ and $n = \frac{1}{8}$

$$\therefore \quad \frac{v_s}{Q/A} = \frac{1}{n}\left[\left\{1 - \frac{y}{y_0}\right\}^{-n} - 1\right]$$

$$= 8\left[\{1 - 0.75\}^{-1/8} - 1\right] = 1.514$$

$$\therefore \quad \frac{Q}{A} = \frac{v_s}{1.514} = \frac{59.7}{1.514} = 39.43 \text{ m/d.}$$

This is within the permissible range of 30 – 40 m³/m²/d, and is therefore acceptable.

5. *Determination of dimensions of the tank*

Surface area of tank $= \frac{Q}{Q/A} = \frac{306 \times 24}{39.43} = 186.24 \text{ m}^2$

$$\therefore \quad \frac{\pi}{4}(D^2 - D_p^2) = 186.24$$

or $$\frac{\pi}{4}[D^2 - (0.3)^2] = 186.24$$

From which $D = 15.4$ m.
Providing a detention time $t_0 = 2.5$ hours,

$$\text{Depth of tank} = \frac{Q \times t_0}{A} = \frac{306 \times 2.5}{186.24} = 4.1 \text{ m.}$$

5. *Design of peripheral effluent launder*

For a peripheral effluent launder, half the flow is assumed to go about each side of the basin.

$$\text{Total length of launder} = \pi D = \pi \times 15.4$$
$$= 43.38 \text{ m.}$$

$$Q = \text{discharge from each side} = \frac{300}{2} = 150 \text{ m}^3/\text{hour}$$

$$= \frac{150}{3600} = 0.04167 \text{ m}^3/\text{s}$$

Let us provide 90° V-notches fixed to the inner side of the launder, all along its perimeter. Let us keep width of launder = 0.3 m. Hence critical depth h at the other end of the launder (Fig. 9.18 b) is.

$$h = \left(\frac{Q^2}{b^2 g}\right)^{1/3} = \left[\frac{(0.04167)^2}{(0.3)^2 \times 9.81}\right]^{1/3}$$
$$= 0.125 \text{ m.}$$

The depth of water at the other end, neglecting the friction losses, is given by. Eq. 9.18 (a) taking $n = 1$.

$$\therefore \quad H = \left[h^2 + \frac{2(qL)^2}{gb^2 h}\right]^{1/2}$$
$$= \left[(0.125)^2 + \frac{2(0.04167)^2}{9.81(0.3)^2 \times 0.125}\right]^{1/2}$$
$$= 0.217 \text{ m.}$$

Let us assume a head loss of 0.1 m. Providing a F.B. of 0.1 m, depth of launder below the bottom of weir discharging into it is $= 0.217 + 0.1 + 0.1 = 0.417$ m. If the weir has 50 mm deep V-notches, the resulting depth of launder$= 0.417 + 0.05 = 0.467$ m. Hence provide the effluent launder of width equal to 300 mm and depth equal to 500 mm.

Check for weir loading

$$\text{Weir loading} = \frac{300 \times 24}{43.38}$$

$$\approx 166 \text{ m}^2/\text{d/m} < 300 \text{ m}^3/\text{d/m}.$$

Hence ok.

Example 9.20. *Design of mechanical rapid mix unit*

Design a mechanical rapid mix unit for a design flow to be treated equal to 300 m³/h. Assume suitable permissible values for various parameters of design. Assume a temperature of 20°C.

Solution

1. ***Design parameters.***

Let us adopt the following values of various design parameters.

(*i*) *Detention time* : Normal range is 20 to 60 s. Let us provide a detention time of 30 sec.

(*ii*) *Velocity gradient* : G should be greater than $300\, s^{-1}$. Let us adopt $G = 600\, s^{-1}$.

(*iii*) *Rotational speed of impeller* : Speed should be greater than 100 rpm. Let us provide a speed of 125 rpm.

(*iv*) *Ratio of tank height to diameter* The ratio varies from 1:1 to 3:1. Let use provide a ratio of 1.5 : 1.

(*v*) *Ratio of impeller diameter to tank diameter* : Range of the ratio is from 0.2 :1 to 0.4 : 1. Let us provide a ratio of 0.4:1.

2. ***Determination of dimensions of the tank***

$$\text{Volume of tank} = \text{Flow} \times \text{detention time}$$

$$= \frac{300}{3600} \times 30 = 2.5 \text{ m}^3.$$

$$\text{But volume} = \text{area} \times \text{depth} = \left(\frac{\pi}{4} D^2\right)(1.5\, D)$$

$$\therefore \quad \frac{\pi}{4} \times 1.5\, D^3 = 2.5$$

From which $D = 1.285$ m.

Provide tank dia. = 1.3 m.

∴ Depth of tank= 1.3 × 1.5 = 1.95 m. Providing a free board of 0.2 m, total depth of tank = 1.95 + 0.2 = 2.15 m.

3. ***Computation of power requirements***

From Eq. 9.29, we get

$$P = G^2 . \mu V$$

$$\mu = 1.0087 \times 10^{-3} \text{ at } 20°\text{C}.$$

$\therefore$ $P = (600)^2 \times 1.0087 \times 10^{-3} \times 2.5$
$= 908$ watts.

$\therefore$ Power per unit volume $= \dfrac{908}{2.5} = 363$ watts/m^3

Power per unit flow of water $= \dfrac{908}{300}$
$= 3.03$ watts/m^3/hr of flow.

4. ***Determination of dimensions of flat blades and impeller***

Dia. of impeller $= 0.4\times$ tank dia.
$= 0.4 \times 1.3 = 0.52$ m

Velocity at the tip of impeller $= \dfrac{2\pi r n}{60}$ m/s

or $v_p = 2\pi \times \dfrac{0.52}{2} \times \dfrac{125}{60} = 3.40$ m/s

Again, from Eq. 9.32,

$$P = \tfrac{1}{2} C_D . \rho . A_b . v_r^3$$

where A_b = area of blades
v_r = relative velocity of blades $= (1-k)\, v_p$
v_p = velocity at the tip of impeller or blade
P_c = power spent $= 90.8$
ρ = density of water $= 998$ kg/m^3 at 20°C

Taking $k = 0.25$,
$v_r = (1 - 0.25)v_p = 0.75\, v_p = 0.75 \times 3.40$
$= 2.55$ m/s

$\therefore$ Taking $C_D = 1.8$ for blades, we get
$908 = \tfrac{1}{2} \times 1.8 \times 998 \times A_b\, (2.55)^3$

From which $A_b = 0.06097$ m^2

Providing 6 blades, area of each blade $= \tfrac{1}{6} \times 0.06097 \approx 0.01$ m^2.

Provide blades of size 0.09×0.12 m (*i.e.* 9 cm $\times$ 12 cm).

Example 9.21. ***Design of flocculator***

Design a flocculator for a design flow to be treated equal to 300 m^3/h. Assume suitable permissible values of various parameters of design. Assume a temperature of 20°C.

Solution.

1. ***Design parameters.***

The following design parameters are assumed :

(*i*) Detention period = 20 minutes.

(*ii*) Av. value of $G = 40\, s^{-1}$.

(*iii*) Speed of paddles = 4.5 rpm

(*iv*) Area of paddles = 15% of the cross-sectional area of basin.

(*v*) Velocity ratio $k = 0.25$

(*vi*) $\mu = 1.0087 \times 10^{-3}$ N-s/m^2 at 20°C

(*vii*) $\rho = 998$ kg/m^3 at 20°C.

(*viii*) Ratio of Length to width of tank = 2

2. *Computation of volume of flocculation tank*

Volume of tank V= design flow × detention time

$$= \frac{300}{60} \times 20 = 100 \text{ m}^3$$

Let the depth of tank= 0.4 B
Let the length of tank= $2B$

$$\therefore \quad B \times 2B \times 0.4B = 100$$

From which
$$B = 5 \text{ m.}$$
$$L = 5 \times 2 = 10 \text{ m}$$
$$H = 0.4 \times 5 = 2 \text{ m}$$

3. *Computation of power required*

From Eq. 9.29

$$P = G^2 . V . \mu$$
$$= (40)^2 \times 100 \times 1.0087 \times 10^{-3}$$
$$= 161.4 \text{ watts.}$$

4. *Computation of velocity differential between the paddle and water*

Let us provide revolving paddels attached to three horizontal shafts running parallel to the length. Let each shaft be located at mid-depth of the tank. Let us provide four paddles to each shaft, each running parallel to the shaft.

Max. available *outer radial* distance of each paddle

$$= \frac{B}{2 \times 3} = \frac{5}{6} = 0.833 \text{ m.}$$

Assuming 25 cm width of blade, max. value of r = 0.833 $- \frac{0.25}{2} = 0.708$ m.

Hence keep distance r from the centre line of paddles to the centre of shaft as 0.7 m.

$$\therefore \quad v_p = \frac{2\pi r n}{60} = \frac{2\pi}{60} \times 0.7 \times 4.5 = 0.3299 \text{ m/s}$$

$$\therefore \quad v_r = (1 - k)\, v_p = (1 - 0.25) \times 0.3299$$
$$= 0.2474 \text{ m/s.}$$

5. *Computation of paddle size.*

From Eq. 9.32

$$P = \frac{1}{2} C_D \rho . A_p . v_r^3$$

Taking $C_D = 1.8$ for flat blades, we have

$$161.4 = \frac{1}{2} \times 1.8 \times 998 A_p (0.2474)^3$$

From which $A_p = 11.87 \text{ m}^2$

There area total of $3 \times 4 = 12$ paddles

$\therefore$ Area of each paddle $= 11.87/12 = 0.989 \text{ m}^2$.

Let the length of each paddle $= 4.8$ m.

(slightly less than half the length of tank)

$\therefore$ Width of paddle $= \frac{0.989}{4.8} = 0.206$ m.

Hence provide 21 cm wide paddles, each of length 4.8 m.

PROBLEMS

1. What do your understand by treatment of water ? Why is it necessary ? Give an outline of various processes adopted for treatment of water.

2. Describe in brief various types of screens used for screening water.

3. Working from the fundamentals, derive Stoke's law for settlement of discrete particles in water. How do you modify the law, taking into account the temperature effect ?

4. What do you understand by plain sedimentation ? Describe the design principles of a settling tank.

5. Differentiate between Stoke's law and Newton's law. How is the drag co-efficient related to Reynolds number.

6. Find the settling velocity of silica particles of specific gravity 2.65 at 20°C, if the diameter of particles is (*a*) 0.005 cm, (*b*) 0.02 cm.

Ans. [(*a*) 0.224 cm/sec. ; (*b*) 2.7 cm/sec.]

7. Describe in brief various types of sedimentation tanks generally used.

8. Define and explain the following terms :

(*a*) detention period

(*b*) flow through period

(*c*) displacement efficiency

(*d*) surface loadings.

9. Write an expression for the maximum velocity to prevent bed uplift or scour of the settled particles. Show that for an ideal tank, the ratio of its length to depth should be equal to $\sqrt{8/f}$.

10. Design a plain sedimentation tank to treat 5 million litres water per day. Take a detention period of 8 hours and assume a depth of 3.5 m.

11. Determine the maximum displacement velocity, without the fear of settled particles being lifted up, and the ratio of length to the depth of the settling unit for the following particles :

(*a*) alum floc having $S_s = 1.15$ and diameter 0.08 cm, and

(*b*) anthracite coal dust having $S_s = 1.6$ and diameter $= 0.008$ cm.

Assume a Weisbach Darcy friction factor $f = 0.025$ and a temperature of 25°C.

Ans. [(*a*) 104.1 cm/sec, $L/H = 17.89$; (*b*) 7.76 cm/sec, $L/H = 33.30$]

12. A settling tank is designed for an overflow rate of 3,000 litres per m^2 per hour. What percentage of particles of diameter of 0.025 mm (specific gravity 2.65) will be removed at 20°C ?

Ans. [67.3%]

13. Find the diameter of particles with specific gravity of 1.4 removed in a tank having a surface area of 300 m^2 and treating 10 million litres of water per day. Assume a temperature of 20°C.

Ans. [0.02 mm.]

14. What do you understand by coagulation and flocculation ? why are they necessary ?

15. Describe various types of coagulants commonly used in water treatment.

16. Describe various methods of application of coagulants.

17. (*a*) Prove that theoretically the surface loading (Q/A), and not the depth, is a measure of effective removal of particles in a sedimentation tank.

(*b*) An ideal sedimentation tank with a surface area of 100 m^2 receives an inflow of 20 millions per day. Referring to tabuler data below for particles of specific gravity 1.20, find out what percentage of each size would settle down.

Dia. in mm	0.5	0.4	0.3	0.2	0.1
settling velocity (cm/s)	0.65	0.49	0.38	0.22	0.08

(A.M.I.E., May 1969)

18. (*a*) Why alum is commonly used coagulant ? Write down the reactions.

(*b*) Explain the mechanism of floc formation.

19. Explain various mixing devices commonly used. What is a flash mixer ?

20. Explain what is meant by flocculation ? Sketch any one type of flocculator with mechanical agitators.

21. Design a coagulation sedimentation tank to treat 16 million litres of water per day. Assume suitable data wherever necessary.

22. Write short notes on the following :

(*a*) Vertical flow sedimentation tank.
(*b*) Tube settlers.
(*c*) Sludge blanket tank.
(*d*) Inlet and outlet devices.
(*e*) Polyelectrolytes.

23. Write down the chemical reactions when the following coagulants are used :

(*a*) Alum.
(*b*) Chlorinated copperas.
(*c*) Sodium aluminate.

24. For a continuous flow settling tank 3 m deep and 60 m long, design the flow velocity to water for effective removal of 0.025 mm particles at 25°C. The specific gravity of particles is 2.65 and kinematic viscosity for water may be taken as 0.01 cm^2/sec.

(Civil Services 1985 ; Engg. Services, 1979)

Ans. [1.349 cm/sec.]

25. What is meant by coagulation ? What are the common coagulants used ? Describe the chemical reaction involved and indicate the formula (or equation) for chemical reaction, the flocs that form ; explain the reactions by stages, if any. (Civil Services, 1985)

26. (*a*) State the factors affecting the efficiency of sedimentation tank. With the help of a neat sketch, describe a sedimentation tank.

(*b*) The population of a town is 100,000 and the average per capita demand is 135 litres/day/capita. Design the coagulation-cum-sedimentation tank for supplying water to the town. The maximum demand may be taken as 1.5 times the average demand. Detention period is 5 hours for settling tank and 30 minutes for flocculent chamber. Flow rate is 900 lites/hour/m^2 of plan area. (Civil Services, 1988)

27. Describe the process of sedimentation with coagulation, as adopted in water treatment plants. Include in your answer the factors affecting coagulation, the list of commonly used chemicals and the description of mixing devices and flocculator. (Civil Services, 1989)

28. A rectangular sedimentation basin is to handle 10 million litres/day of raw water. A detention basin of width to length ratio of $\frac{1}{3}$ is proposed to trap all impurities larger than 0.04 mm in size. Assuming a relative density of 2.65 for the particles and 20° as the average temperature, compute the basin dimensions. If the depth of tank is 3.5 m, calculate the detention time.

(Civil Services, 1990)

Ans. [B =5.2 m ; L = 15.6 ; t_0 ≈ 41 min.]

29. The water works of a city treats 50 million litres/day of water in a coagulation sedimentation tank. The quantity of alum consumed is 15 mg/litres of water. If the alkalinity of raw water is equivalent to 1.0 mg/litre of $CaCO_3$, determine the quantity of alum and quick lime (containing 80% CaO) required per month by water works. (Civil Services, 1990)

Ans. [22.5 tonnes ; 6.045 tonnes]

10

Filtration

10.1. INTRODUCTION

'Filtration' is one of the most important operations in the water purification process. Though screening and sedimentation remove a large proportion of suspended matter, but they do not effectively remove fine floc particles, colour, dissolved minerals and micro-organisms. In filtration, water is passed through a filter medium in order to remove the particulate matter not previously removed by sedimentation. During filtration, the turbidity and colloidal matter of non-settleable type are removed. It precipitates the colour, and the chemical characteristics of water are changed. The bacterial content of water is considerably reduced due to the presence of an active zoological layer on the top of filtering material. James Simpson established the first filter for the Chelsea Water Company, London, in 1829. Earliest forms were of 'slow sand filter' type. Rapid sand filters were introduced in 1900-1910.

10.2. THEORY OF FILTRATION

When water is filtered through the bed of filter media, usually consisiting of clean sand, the following actions take place :

(*i*) Mechanical straining,
(*ii*) Sedimentation,
(*iii*) Biological action, and
(*iv*) Electrolytic action.

1. Mechanical straining. This is the simplest action taking place during filtration. The suspended particles are usually larger than the pore-space of the filter media, and hence they are arrested and removed, when water passes through the filter media. The major removal takes place only in the upper few centimetres of the filter media. These arrrested particles form a mat, and further help in straining the impurities.

2. Sedimentation. In mechanical straining, only those particles which are coarser than the void-size are arrested. Finer particles are arrested by 'sedimentation action'. The continuous voids of the filter media act as *'tube-settlers'* described in the previous chapter. All the colloids are thus removed by this action.

3. **Biological action.** When a filter is put into commission and the raw water is passed through it, during the first few days, the upper layers of sand grains become coated with reddish brown sticky deposit of partly decomposed organic matter together with iron, mangenese, aluminium and silica. After some time, there exists in the uppermost layer of sand a film of algae, bacteria and protozoa etc. This film is known as *schmutzdecke* or *'dirty skin'*, which acts as an extremely fine meshed straining mat. The organic impurities in water become food to different micro-organisms residing in the filter beds. The bacteria not only break-down the organic impurities, and convert them into harmless compounds, but also destroy each other and maintain a balance of life native to the filter.

4. Electrolytic action. Another function of the filter is to remove the particulate matter by electrostatic exchange. The charge of the filter medium neutralises the charge of the floc, thereby permiting the floc to be removed. During the process of 'back washing', the electrostatically neutral material is removed, and the charge of the filter media is replaced.

10.3. CLASSIFICATION OF FILTERS.

Filters are mainly classified into two classes :

1. Slow sand filter.
2. Rapid sand filter (*a*) gravity type
   (*b*) pressure type.

The difference between these types of filters lies mainly in: (*i*) head required for filtration, (*ii*) rate of filtration, (*iii*) composition of filter media, and (*iv*) method and frequency of cleaning the filter media.

10.4. FILTER MEDIA

The ideal filter medium should have a *size* and be of such *material* that it will provide satisfactory effluent, retain a maximum quantity of solids, and be readily cleaned with a minimum of wash water. The commonly used filter materials are :

(*i*) Sand

(*ii*) Anthracite

(*iii*) Garnet sand or ilmenite

(*iv*) Other locally available materials

(*i*) Sand : It is the cheapest filter medium, and is widely used. Sand used for filters should be free from clay, silt, loam, suspended matter and other organic matter. It should be fairly uniform and of the proper size. The uniformity characteristics of sand are expressed in terms of (*i*) effective size, and (*ii*) uniformity coefficient.

The *effective size* or *effective diameter* or the D_{10} size represents a size, in mm, such that 10% of the particles are finer than this size. Similarly, the sand particles finer than D_{60} size are 60% of the total weight of the sample.

The *uniformity-coefficient* (C_U) is a measure of particle range, and is given by the ratio of D_{60} and D_{10} sizes.

$$C_U = \frac{D_{60}}{D_{10}}$$

For *uniformly graded* sand, C_U is nearly unity.

The shape, size and quality of filter sand shall satisfy the following norms :

1. It whall be of hard and resistant quartz or quartzite and free of clay, fine particles, soft grains and dirt.
2. Ignition loss should not exceed 0.7 percent.
3. Soluble fraction in hydrochloric acid shall not exceed 5% by weight.
5. Specific gravity shall be in the range between 2.55 to 2.65.
6. Wearing loss shall not excced 3%.
7. Efffective size shall be
   (*a*) 0.2 to 0.3 mm for slow sand filters
   (*b*) 0.45 to 0.7 mm for rapid sand filters.
8. The uniformity coefficient shall be
   (*a*) 3.0 to 5.0 for slow sand filters
   (*b*) not less than 1.3 nor more than 1.7 in rapid sand filters.

(*ii*) *Anthracite* : Crushed anthracite has been successfuly used as filter medium as a substitute for sand in some filters and may be used in conjunction with sand and other materials in mixed media filters. It is more costly per tonne than sand, but its greater volume per unit weight and lower transportation cost per unit volume may make it more economical at some locations.

(*iii*) Garnet sand : It has high specific gravity (= 4.2) and is a dense material. Because of its high cost, it cannot be used as a sole filter material. However, it may be used as a constituent in mixed-media filter.

(iv) Other materials : Other locally available materials, such as shreded coconut husks, burned rice husks, crushed glass, slag, metallic ores etc have been used as filter material.

10.5. SLOW SAND FILTERS

'Slow sand filter' was the earliest type of filter introduced in 1829. 'Slow sand filters' are so called because the rate of filtration through them may be one-twentieth or less of the rate of filtration through rapid gravity or pressure filters. Because of low filtration rates, slow sand filters require large areas of land and are costly to install. They are also very expensive to operate due to the laborious method of bed cleaning by surface scrapping. Due to this, new construction of slow sand filters is largely confined to additions to existing slow sand filter plants.

Slow sand filters do not normally utilize coagulation to prepare water for filtration. The only pretreatment ever used on water to be filtered in a slow sand filter is primary settling in a sedimentation tank, without any coagulation. Due to this, slow sand filters are not suitable for dealing with waters containing any appreciable quantity of suspended matter. The effect of the slow sand filter is to remove turbidity by straining and to remove bacteria by biological action. It is not particularly effective in removing colloidal turbidity.

Essential features. A slow sand filter unit consists of the following parts : (1) Enclosure tank (2) Filter media (3) Base material (4) Under drainage system, (5) Appurtenances. Fig 10.1 shows the cross-section of a slow sand filter.

1. Enclosure tank. The traditional slow sand filter is an open basin, usually rectangular in size, built below finished ground level. The water tight basin or tank is constructed either in stone or brick masonry, with a coating of water proof material. The floor has a bed slope of 1 in 100 to 1 in 200 towards the central drains (Fig. 10.2). The surface area of the tank varies between 50 m^2 to 1000 m^2, and is found on the basis of filtration rate which varies between 100 to 200 litres of water per square metre. The depth of tank varies from 2.5 to 4 m.

2. Filter media. The filter media consists of sand layer, 90 to 110 cm thick. The effective size of sand varies from 0.20 to 0.35, with a common value of 0.3. The coefficient of uniformity varies from 2 to 3, the common value being 2.5. The finer the sand, better will be bacterial efficiency, but slower will be the filtration rate. Fine sand is suitable when pretreatment is poor, high bacteria and turbidity removal efficiency is desired and saving of washwater is not an important factor. However, if pretreatment is good and water to be treated is not highly polluted, coarse sand is suitable.

The sand should also contain not more than 2% of calcium and magnesium computed as carbonates.

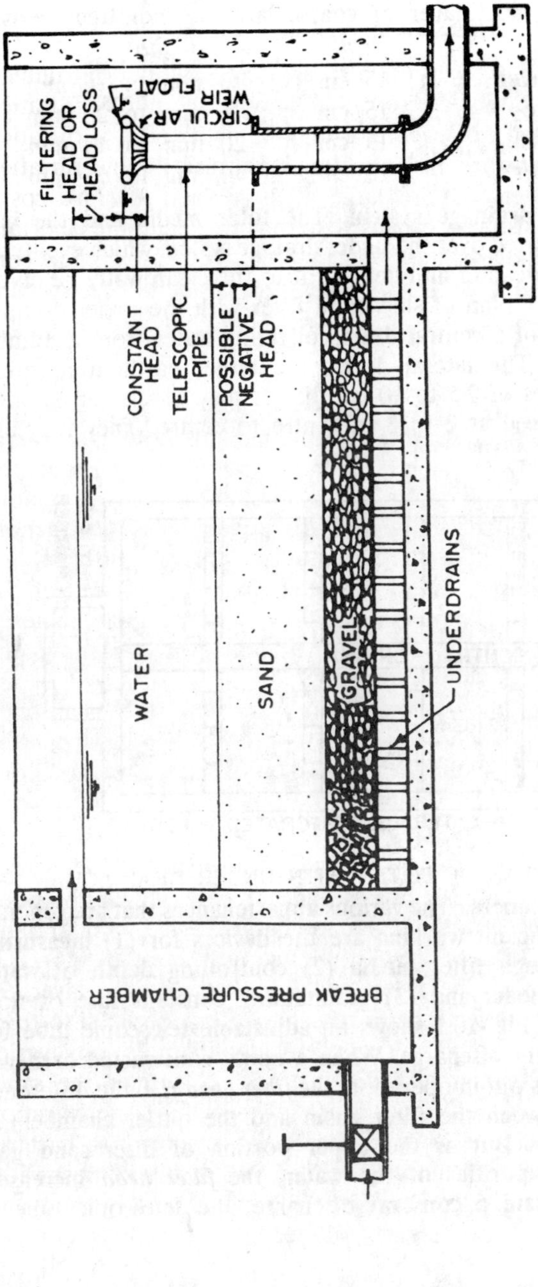

FIG. 10.1. SLOW SAND FILTER

3. Base material. The filter media is supported on base material consisting of 30 to 75 cm thick gravel bed. The gravel base is graded, and laid in layers of 15 cm with topmost layer of finer size and bottom most layer of coarse size, as indicated below :

	Depth	*Size*
Topmost layer	15 cm	3 mm to 6 mm
Intermediate	15 cm	6 mm to 20 mm
layers	15 cm	20 mm to 40 mm
Bottom layer	15 cm	40 mm to 65 mm

4. Under drainage system. The filter media and the base material are supported over the under drainge system which eventually collects the filtered water and delivers it to the clean water reservoir. Fig. 10.2 shows the plan of slow sand filter with the under drainage system consisting of a central drain collecting water from a number of lateral drains. The lateral drains are either earthenware pipes or perforated pipes of 7.5 to 10 cm dia. or may consist of patented drain devices, spaced at 2 to 3 m centre to centre. They may also be of open jointed tiles.

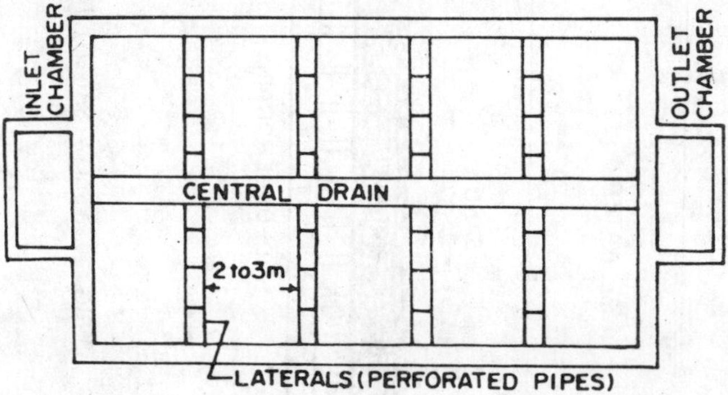

FIG. 10.2. PLAN OF FILTER WITH UNDER DRAINAGE

5. Appertenances. The various appertenances that are generally installed for efficient working are the devices for (1) measuring loss of head through filter media (2) controlling depth of water above the filter media and (3) maintaining constant rate of flow through the filter. Fig. 10.1 shows an adjustable telescopic tube for maintaining constant discharge. When a newly constructed or newly washed filter unit is put in operation, the *filter head* (*i.e.* the difference in water level between the filter basin and the outlet chamber) is only 10 to 15 cm. But as the upper portion of filter sand gets clogged due to impurities of the water, the *filter head* increases. In order to maintain a constant discharge, the telescopic tube is adjusted manually.

Working of the filter. As stated earlier, the water used for the slow rand filter should *not* be coagulated. Only plain sedimentation is used prior to filtering. Water from the plain sedimentation tanks is allowed to enter the filter basin through the inlet chamber. The depth of water over the filter media should neither be too small nor too high. Generally, it is kept equal to the thickness of the sand medium. The water passes downwards through the sand bed, through the gravel, and is then collected by the under drainage system. A slow sand filter works by a combination of both straining as well as microbioloigcal action. Vloed distinguishes three *zones of purification* in the filter media :

(*i*) the surface coating, called 'schmutzdecke',

(*ii*) the *autotrophic zone* existing a few millimetres below the schmutzdecke and

(*iii*) the *heterotrophic zone* which may extend some 30 cm in the bed.

The *schmutzdecke* or the *surface coating* is formed by reddish brown sticky deposit of partly decomposed organic matter together with iron, mangenese, aluminium and silica. This coating further absorbs organic matter in colloidal state. Thus, after 2-3 weeks of commissioning the filter, there exists in the uppermost layer of sand a film of algae, bacteria, protozoa, suspended particles and organic matter. In the *autotrophic zone*, the growing plant life breaks down organic matter, decomposes the plankton, and uses up available nitrogen, phosphates and carbon dioxide, providing oxygen in their place – thus oxidising the filtrate. In the *heterotrophic zone,* extending to a depth of 30 cm, the bacteria multiply in large number, to break down completely the organic matter, resulting in the presence of only simple inorganic substances and unobjectionable salts. The bacteria act not only to break down the organic matter, but also destroy each other, thus tending to maintain a balance of life in the filter.

The difference of water level in the filter basin and the outlet chamber is known as the *'filter head'*. Slow sand filters are worked upto a maximum filter head of either 75 cm or equal to 65 to 85% of the thickness of filter medium. In the beginning, the filtration head is only 10 to 15 cm. But as the suspended matter is arrested and the thickness of the schmutzdecke film is increased, the friction head also increases, resulting in the increase of the filtration head. When the filtration head reaches the maximum permissible value, the filter requires cleaning.

Cleaning of filter media. In order to clean the filter media, the water in the filter tank is first drained. The top layer of sand is scrapped manually and removed through a depth of 15 mm to

30 mm. The scrapped sand is either removed manually for washing, or else is placed in a portable hydraulic ejector which forces it through a hose to a washer outside the bed. The washed sand is then stored. After this, the filter is again put in operation. For that, the filter bed is refilled with filtered water from below until the sand is completely covered. This prevents entrapping of air in the sand. Thereafter, the inlet valve is opened slowly, to prevent disturbance of sand, till the desired water level is reached. The filter is then allowed to stand for atleast 12 hours. The filtrate should be run at 1/5 the normal filtration rate to the wash water drain, for 12-15 hours, to help in the formation of the biological film of sediment. Thereafter, the filtered water is diverted to the filtered-water-reservoir, but the filtration rate is kept 1/3 its normal value for another 3-4 days. After this period, the filter is run at its normal rate.

The filter requires cleaning after 20 to 40 days, depending upon impurities in the water. In each cleaning, some depth of filter sand is removed due to which the effective depth of filter medium goes on decreasing. The process of cleaning by scrapping is continued, till the thickness of the filter medium reaches a minimum value of about 60 cm. When this stage is reached, the bed is topped up with clean new sand to its original level, or the old washed sand may be replaced, if it had been adequately washed and properly stored. Raking the surface of sand will lengthen the time between scrapping. In some cases, where enough filter sand is available, a fresh layer of 15 cm thickness of sand is placed when the previous scrappings reduce the thickness of sand by 15 cm, say after 5 or 6 scrappings.

Efficiency of slow sand filters

The normal rate of flow in slow sand filters may be between 100 to 200 litres per hour per m^2 of the filter area. They are highly effective in the removal of bacterial load from water. They are expected to remove 98 to 99% of bacteria in water, under normal conditions. However, if operated at slower rates or with some preliminary treatment and chlorine, the overall bacterial efficiency may reach 99.5 to 99.9%, and the removal of coliforms is equally high. The slow sand filter is not effective in removing colloidal turbidity. The slow filters will not successfully remove turbidity over 50 ppm. For water having higher turbidity, preliminary treatment for turbidity reduction is essential. The slow sand filters do not materially reduce the true colour of water. In most cases it can remove only 20 to 25% of the colour in applied water. If the colour is higher than 25 to 30 Hazen units, slow sand filters should not be used. Slow sand filters are also not effective in removing any high concentration of manganese in solution. However, slow sand filters are generally better at removing algae from water than in coagulation followed by rapid gravity filtration. The slow sand filters are also effective in removing the 'musty' or 'earthy' odours and tastes which develop along with algae growth. The filtered water from slow sand filter contains less nutrients, and has the reduced possibility of after growth of low organisms in the distribution system.

Guide lines for design of slow sand filters

The Manual on Water Supply and Treatment prepared by Central Public Health and Environmental Engineering Organisation (MUD) gives the following guide lines (Table 10.1)

TABLE 101. GUIDE LINES FOR DESIGN OF SLOW SAND FILTERS

Description	*Recommended design value*
1. Design period	10 years
2. Number of filter beds :	
(*i*) Minimum	2
(*ii*) Areas up to 20 m^2	2
(*iii*) $A = 20$ to 249 m^2	3
(*iv*) $A = 250$ to 649 m^2	4
(*v*) $A = 650$ to 1200 m^2	5
(*vi*) $A = 1201$ to 2000 m^2	6
3. Depth of supernatant water	1.0 m
4. Filteration rate	
(*i*) Normal operation	0.1 m/hour
(*ii*) Max. over load rate	0.2 m/hour
5. Free board	0.2 m
6. Depth of filter sand	
(*i*) Initial	1.0 m
(*ii*) Final (min.)	0.4 m
7. Sand specifications	
(*i*) Effective size	0.2 to 0.3 mm
(*ii*) Uniformity coefficient	5
8. Gravel (3 – 4 layers) Depth	0.3 m
9. Under drains (made of bricks or perforated pipes)	0.2 m
10. Depth of filter box	2.7 m
11. Effluent weir level above sand bed	20–30 mm
12. Standards of performance	
(*i*) Turbidity of filtrate	1 NTU or less
(*ii*) Colour of filtrate	3 or less on the cobalt scale
(*iii*) Filter runs	When the raw water turbidity does not exceed 30 NTU, the filter runs should normally be not less than 6 to 8 weeks, with the filter head not exceeding 0.6 m.
(*iv*) Initial head loss	Should not exceed 5 cm ; a higher head loss will indicate that the entire sand bed needs over hauling

Example 10.1. *A city has a population of 20,000 with an average rate of demand as 150 litres per capita per day. Find the area of slow sand filters.*

Solution :

Max. daily demand = 1.5 × 20,000 × 150 = 450,0000 litres.

Let us assume the average rate of filtration as 150 per litres hour per m^2 of filter area.

$$\therefore \quad \text{Area of filter} = \frac{450{,}0000}{150 \times 24} = 1250 \text{ m}^2$$

Let us provide each filter unit of size 20 × 12.5 m, giving an area of 250 m^2. Total number of filter units will be 5. However, provide one unit as stand by, making a total of 6 units of each 20 × 12.5 m.

10.6. RAPID SAND FILTER : GRAVITY TYPE

The basic defect of a slow sand filter is its low filtration rate due to which large filter area is required. A rapid sand filter differs from a slow sand filter in the following aspects : (*i*) effective size and uniformity coefficient of filter medium (*ii*) rate of filtration and filtration head and (*iii*) method of cleaning the filter, and frequency of cleaning. Usually, rapid sand filtration is preceded by sedimentation with chemical coagulation. Rapid sand filters are of two types : (*a*) gravity type and (*b*) pressure type. The gravity type rapid sand filters use filter medium which is coarser than that used in slow sand filters. In the pressure type rapid filters, water is allowed to filter under pressure through the filter media. The rapid sand filters of gravity type are most commonly used in water supply projects.

Essential features

A gravity type rapid sand filter consists of the following units:

1. Enclosure tank.
2. Filter media.
3. Base material.
4. Underdrainage system, and
5. Appurtenances.

Fig. 10.3 shows the section of a typical rapid sand gravity filter, showing all the above units.

1. Enclosure Tank. The enclosure tank is usually smaller in size and is therefore placed under a roof. The tank is generally rectangular in plan, constructed either of masonry or of concrete, coated with water proof material. The depth of tank may vary between

2.5 to 3.5 m. Each unit may have a surface area of 20 to 50 m^2. They are arranged in series. The length to width ratio is normally kept between 1.25 to 1.35. In addition to the underdrainage system, the tanks also have C.I. or R.C.C. troughs spanning across the length or width of the walls for the distribution of water to be filtered during normal operation, and for collection of wash water during cleaning operation. During normal filtration operation, the troughs remain submerged.

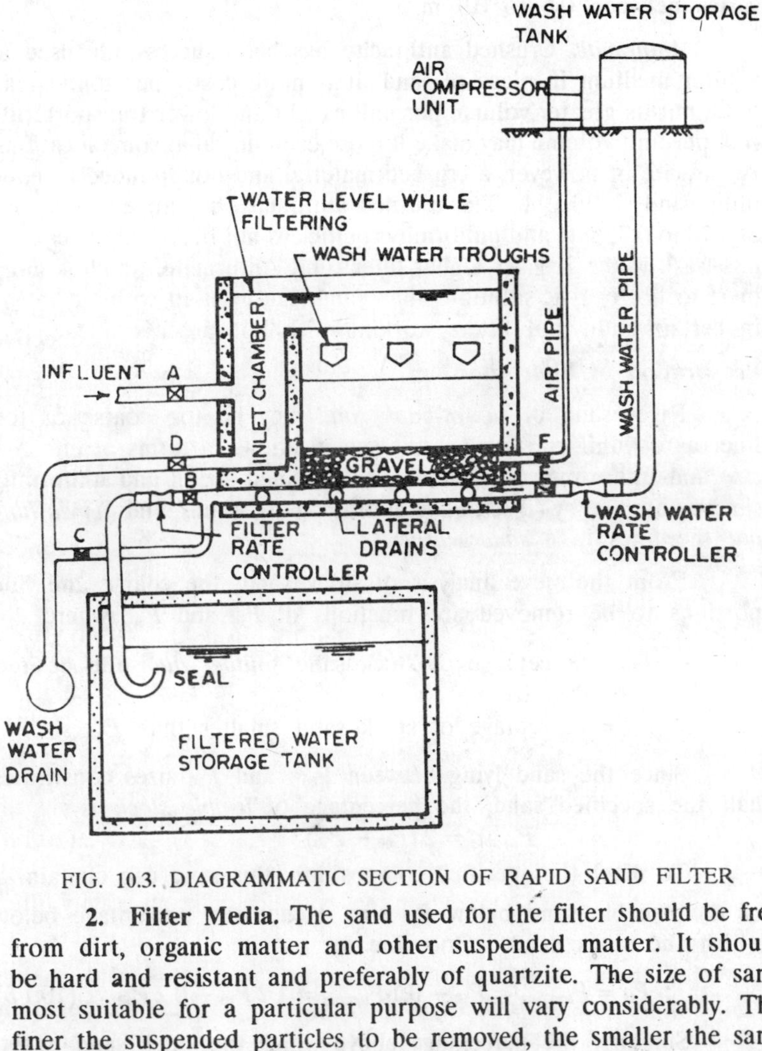

FIG. 10.3. DIAGRAMMATIC SECTION OF RAPID SAND FILTER

2. Filter Media. The sand used for the filter should be free from dirt, organic matter and other suspended matter. It should be hard and resistant and preferably of quartzite. The size of sand most suitable for a particular purpose will vary considerably. The finer the suspended particles to be removed, the smaller the sand size should be in order to increase the surface area of sand grains in the sand depth of bed. Rapid filter sand has effective size between 0.35 to 0.6 mm, the common value being 0.45 mm. The uniformity coefficient is between 1.2 to 1.7, the common vlaue being 1.5. Due to increased effective size and decreased unformity coefficient, the void space is increased, resulting in the increased rate of filtration. In back washing single media sand beds, hydraulic grading of the sand grains occur. The very fine sand accumulates at the top of the bed and the coarse particles lie below. The depth of sand media varies between 0.6 to 0.9 m.

Anthracite. Crushed anthracite has been successfully used as a filter medium in place of sand. It is more costly per tonne than sand, but its greater volume per unit weight and lower transportation cost per unit volume may make it more economical at some locations. Anthracite is, however, a crushed material and not rounded as good filter sand should be. The crushed anthracite has an effective size of 0.70 to 0.75 mm and uniformity coefficient not over 1.75. It requires less wash water. It gives higher filter runs. Anthracite, which is more inert to hot caustic solutions than sand, can be used with advantage in certain industrial filters working with hot process softners.

Preparation of filter sand

River sand or *run-of-bank sand* may be too coarse or too fine or too uniform to meet the specifications in terms of effective size and uniformity coefficient. The specified sizing and uniformity can be obtained by (*i*) *screening out* the coarse grains, and (*ii*) *washing out* the fines in a *sand washer.*

From the sieve analysis of stock sand, the coarse and fine portions to be removed are functions of P_{10} and P_{60}, where

P_{10} = percentage of stock sand *smaller than* the desired D_{10} size

P_{60} = percentage of stock sand smaller than D_{60} size.

Since the sand lying between P_{60} and P_{10} sizes constitutes half the specified sand, the percentage of *usable stock* is

$$P_{usable} = 2\,(P_{60} - P_{10}) \qquad \text{...(10.1 a)}$$

To meet the specified composition, this sand can contain $\frac{1}{10}$ of the usuable sand below P_{10} size. Hence the percentage below which the stock sand is too fine for use is

$$P_{tf} = P_{too\ fine} = P_{10} - 0.1\,P_{usable} = 1.2\,P_{10} - 0.2\,P_{60} \qquad \text{...(10.1 b)}$$

Similarly the percentage above which the stock sand is too coarse is

$$P_{tc} = P_{too\ coarse} = P_{60} + 40\ \%\ \text{of usable sand}$$
$$= P_{60} + 0.4 \times 2\,(P_{60} - P_{10})$$
$$= 1.8\,P_{60} - 0.8\,P_{10} \qquad ...(10.10)$$

From the size cumulative percentage curve, the grain size D_{tf} and D_{tc} of stock sand corresponding to $P_{too\ fine}$ and $P_{too\ coarse}$ respectively can be found. The size below D_{tf} and above D_{tc} will have to be separated out from the stock sand to bring it to the desired specification. The sizes above D_{tc} can be removed by sieving while the sizes below D_{tf} can be removed in a *sand washer* shown in Fig. 10.4.

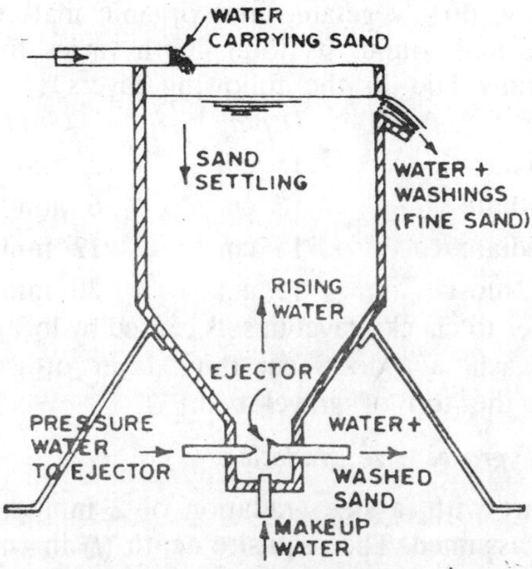

FIG. 10.4. SAND WASHER

A sand washer, shown in Fig. 10.4 is essentially an *upflow setting tank* in which an upward flow velocity is maintained to a value slightly less than the hydraulic subsidence value of the smallest particle size to be retained (*i.e.* D_{tf}) Thus particles of less than D_{tf} size are *floated out* with the flowing water. The washed sand, settling to the bottom, is ejected hydraulically or withdrawn by gravity through a shear gate.

Estimation of sand depth

The depth of sand should be such that the flocs do not break through the sand bed. The depth of sand varies from 60 to 90 cm. The depth of sand can be checked against break through of floc through sand bed by calculating the *minimum* depth required by Hudson formula :

$$\frac{Q\,d^3\,h}{l} = B_i \times 29323 \qquad ...(10.2)$$

where Q = filtration rate in $m^3/m^2/h$
d = sand size, in mm
h = terminal head loss, in metres
l = depth of sand bed in metres
B_i = break through index whose value ranges between 4×10^{-4} to 6×10^{-3} depending on response to coagulation and degree of pretreatment in filter influent.

3. **Base Material.** The filter sand media is supported on base material consisting of graded gravel layers. The gravel should be free from clay, dirt, vegetable and organic matter, and should be hard, durable and round. Its total depth varies from 45 to 60 cm, and is normally laid in the following layers :

Layer	*Depth*	*Grade size*
Top most	15 cm	2 mm to 6 mm
Intermediate	15 cm	6 mm to 12 mm
Intermediate	15 cm	12 mm to 20 mm
Bottom most	15 cm	20 mm to 50 mm

In order to check gravel upsets caused by localized high velocity during backwash, a 6 to 8 cm thick layer of garnet or ilmenite is placed at the top of gravel bed.

Estimation of gravel size gradation

To start with, a size gradation of 2 mm at top and 50 mm at bottom is assumed. The requisite depth (l) in cm of a component gravel layer of size d (mm) can be computed from the following empirical formula :

$$l = 2.54\, k\, (\log d) \qquad \text{...(10.3)}$$

where k varies from 10 to 14

4. **The under-drainage system.** The under-drainage system serves two purposes :

(*a*) It collects the filtered water uniformly over the area of gravel bed.

(*b*) It provides uniform distribution of backwash water with out disturbing or upsetting the gravel bed and filter media.

There are many under-drainage systems, such as :

(*i*) Perforated pipe system
(*ii*) Pipe and strainer system
(*iii*) Wheeler system
(*iv*) Leopald system
(*v*) Wagner system etc.

We shall discuss here only the first two systems.

(*a*) **Perforated Pipe System.** This system (Fig. 10.5), consists of a main header with several pipe laterals on both sides. In it, head differences over the bed are minimized by keeping velocities in the pipes or conduits of the system at proper value and the orifices of proper size, number and distribution. The laterals are placed at a distance of 15 to 30 cm centres. The laterals have perforations on the underside with 6 to 12 mm openings, which make an angle of 30° with the vertical as shown in Fig. 10.6. Placing openings

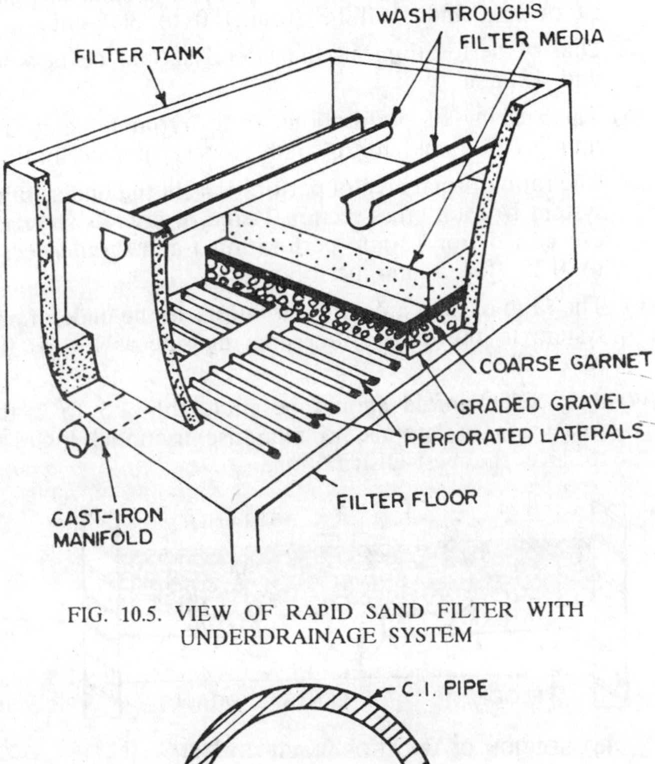

FIG. 10.5. VIEW OF RAPID SAND FILTER WITH UNDERDRAINAGE SYSTEM

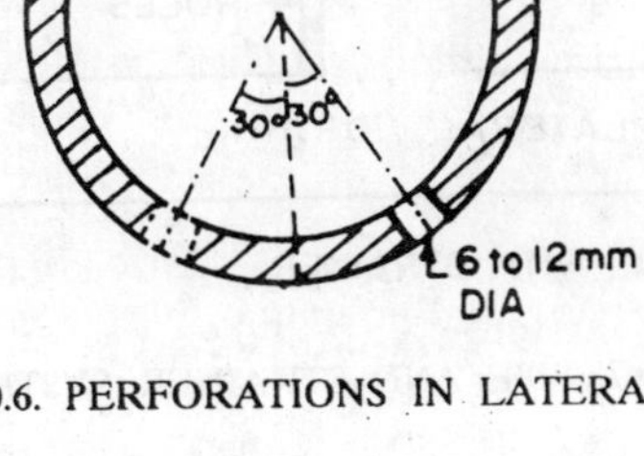

FIG. 10.6. PERFORATIONS IN LATERAL DRAIN

on the underside requires supporting the laterals on concrete blocks, 4 cm above filter bottom. The velocity of jets from the perforations during back washing is dissipated against the filter floor and in the surrounding gravel. The cast iron laterals are screwed to the central cast iron manifold or header. This system requires about 700 litres of water per minute per m^2 of filter area for back washing. This is known as *high velocity wash.* The following general rules are observed in designing the underdrainage system :

(*i*) Ratio of length to diameter of lateral should not exceed 60. The spacing of laterals closely approximates the spacing of orifices and shall be from 150 to 300 mm.

(*ii*) Dia. of perforations in the lateral should be between 5 and 12 mm.

(*iii*) The spacing of perforations varies from 80 mm (for 5 mm perforations) to 200 mm (for 12 mm perforations).

(*iv*) The ratio of total area of perforations in the underdrainage system to total cross-sectional area of laterals should not exceed 0.5 (for 12 mm perforations) and should decrease to 0.25 (for 5 mm perforations).

(*v*) The ratio of total area of perforations in the underdrainage system to the entrie filter area may be as low as 0.002 to 0.003.

(*vi*) Area of mainfold should be preferably 1.5 to 2 times total area of laterals, to minimise frictional losses and to give the best distribution.

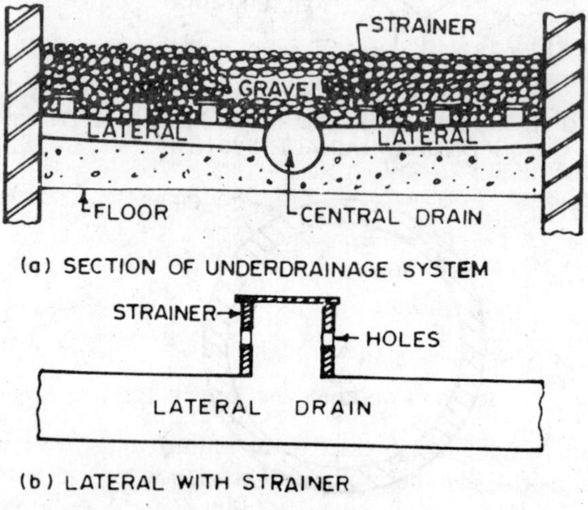

FIG. 10.7. PIPE AND STRAINER SYSTEM

(*b*) **Pipe and strainer system.** In this system, the laterals are provided with strainers, instead of drilling holes. Strainers are usually of brass or bronze and made in umbrella shape or as a cylinder having small perforations, as shown in Fig. 10.7. The strainers distribute the wash water without jet action. Some umbrella shaped strainers have a special air orifice, and are employed where an auxilliary air wash is used.

5. Appertenances

The following are important appertenances provided with a rapid sand filter unit :

(*i*) **Wash water troughs.** Wash water troughs are provided at the top of the filter to collect the backwash water after it emerges from the sand and to conduct it to the wash water gullet or drain. They are C.I. or R.C.C. troughs spanning across the width or length of the tanks. The bottom of the trough is kept above the top of the expanded sand to prevent possibility of loss of sand during back washing. At the same time, the upper edge of the trough should be placed sufficiently near to the surface of the sand so that a large quantity of dirty water is not left in the filter after completion of washing. The trough should be large enough to carry all the water delivered to it with a F.B. of 6 to 8 cm. Any submergence of the gutter will reduce the efficiency of the wash. The bottom of the trough is kept atleast 5 cm above the top level of sand. The spacing of wash mater troughs is kept between 1.5 to 2 m.

Fig. 10.8 (*a*) shows various arrangements of wash water troughs. Type (*i*) is used for only small units while type (*iv*) is used for large units. Fig. 10.8 (*b*) shows various types of troughs. They may be of cast iron, R.C.C., fibre glass, reinforced plastics or other corrosion resistant material. Normally, R.C.C. troughs are used.

The following expression is used for fixing the size of the trough with a level bottom :

$$Q = 0.8\, b\, y^{3/2} \qquad \text{...(10.4)}$$

where Q = total water received by the trough in litres per minute

b = width of the trough, in cm

y = depth of water at the upper end of the trough

Alternatively, if Q is expressed in m^3/s, and b and y are expressed in metres, Eq. 10.4 can be expressed as :

$$Q = 1.376\, b\, y^{3/2} \qquad \text{...(10.4 a)}$$

An additional 5 cm of depth should be provided to allow for free board. The V-shaped bottom, if used must be compensated by obtaining an area equivalent to the rectangular cross-section.

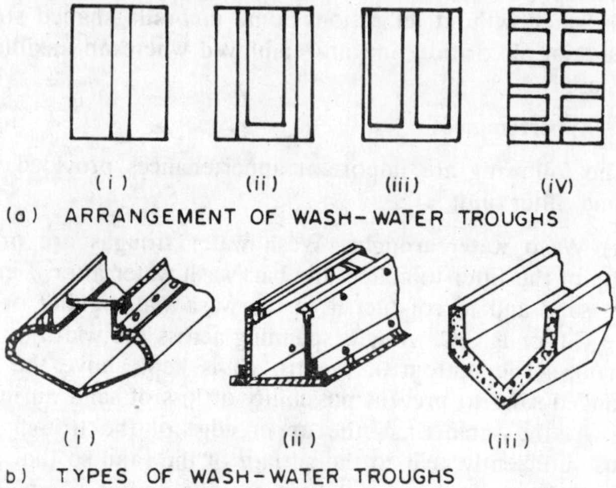

FIG. 10.8. WASH WATER TROUGHS

(*ii*) **Air compressors.** Air compressors are used to supply air for the agitation of sand grains during washing of filters. Compressor supplying the wash-air storage tank should have sufficient capacity to refill the tank between washings. The wash air storage tank should hold at least double the amount of air necessary to wash one filter. The air pressure should be sufficient to overcome frictional resistance in air pipes and depth of water above the air distribution system. The air compressor should have a capacity of supplying compressed air at the rate of 0.6 to 0.8 cubic metre per minute per m^2 of filter area for 5 minutes. In filters using air for washing, the strainer system must be adapted to the use of air, or a special air distributing system must be provided. The pipes and manifold of this system are placed immediately above the pipes of the water under drainage system.

(*iii*) **Rate control device.** For each filter unit, the flow of influent, the effluent, the wash water supply, and the wash water waste must be controlled by suitable devices. A constant rate of filtration is essential. A sudden increase in filtration rate may cause water to break through the filter material without proper treatment while a sudden reduction in the rate may relase a bubble of gas entrained in sand, causing it to make a hole through the filter bed. To ensure a uniform rate of filtration, irrespective of the head loss through the filter, rate controllers, usually of ventury type are used. Fig. 10.9 shows one such rate of flow controller.

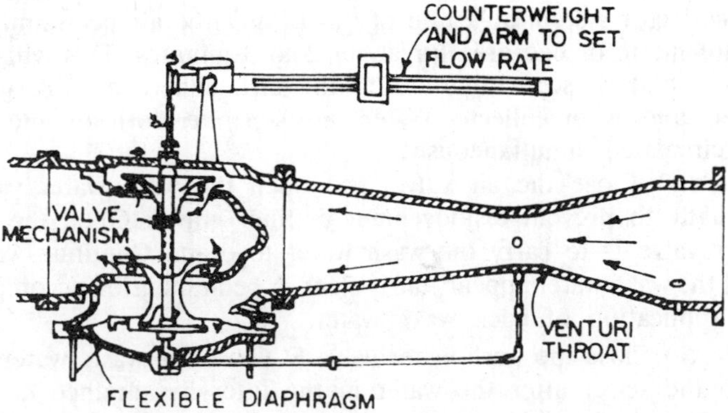

FIG. 10.9. RATE OF FLOW CONTROLLER

10.7. WORKING AND WASHING OF RAPID SAND FILTERS

Fig. 10.3 shows the diagrammatic section of a rapid sand filter, along with all accessories. The working and washing of rapid sand filter unit is controlled by six valves, marked as *A, B, C, D, E* and *F*. Valve *A* is the inlet valve or influent valve, through which water from coagulation-clarification basin enters the filter unit. Valve *B* is the filtered water storage tank valve. Valve *C* is the waste water valve to drain water from main drain. Valve *D* is the waste water valve to drain water from inlet chamber. Valve *E* is the wash-water storage tank valve, fitted after a wash water rate controller. Valve *F* is the compressed air valve.

For *normal working* of the unit, all the valves are kept *closed* except valves *A* and *B*. Valve *A* is opened to permit water from coagulation-sedimentation basin to enter the inlet chamber. Valve *B* is opened to carry the filtered water to the filtered water storage tank. The filter operates under gravity flow, there being about 2 m head (or more) of water over the bed. Filters are normally designed for a filtration rate of 3000 to 6000 litres per hour per m^2 of filter area.

Back washing. A filter is washed when the loss of head through it has reached the maximum permissible. Rapid gravity filters are washed by sending air and water upwards through the bed by reverse flow through the collector system. The following is the sequence of operations.

(*i*) Close the influent valve *A*. Allow the filter to operate till the water level reaches the edge of the troughs. Some operators permit it to fall to about 15 cm from the top of the sand.

(*ii*) Close the effuent valve *B*.

(*iii*) Open the air valve *F*, so that air blows back through the collector system at a rate of 1 to 1.5 m^3 free air per minute per square metre of bed area for about 2 to 3 minutes. This will break up the surface scum and loosen the dirt. Where air distribution system and water collector system are separate, both air and water are circulated simultaneously.

(*iv*) Close the air valve, and open the wash water valve *F* gradually to prevent dislodgement of finer gravel. Open the waste water valve *D* to carry the wash water to drain. Continue washing until the wash water appears fairly clear. A bed expansion accompanies the application of back wash water.

(*v*) Close the wash water valve *E*. Close the waste water valve *D* to the sewer after the water in the filter has drained down to the edge of the wash water trough. Allow a short period to permit material in the water to settle on the sand and form a very thin, sticky layer.

(*vi*) Open the influent valve *A* slighly. Open the valve *C* leading the filtered water to wash water drain for few minutes.

(*viii*) Close valve *C* and open valve *B*. Open the influent valve *A* fully. The filter is now back in service.

Water used for back washing should be filtered water. The total wash water used should normally not exeed 2% of the treated water. The wash water applied is of such rates that expansion is of the order of 25 to 50% of its depth. The pressure at which wash water is applied is about 5 m head of water as measured in under-drains. The rate of application of wash water may be 600 litres per sq. m of the filter surface, equivalent to a rise in the filter box of 60 cm/min for a period of 10 minutes.

Surface wash. Through repeated operations of back washing, the finer sand tends to stratify at the top. It is difficult to clean this finer sand by the conventional back washing system. Surface washing is used to clean this upper layer and also to prevent the formation of mud balls, and dead or clogged area. In the surface wash system, clean filtered water is applied from above by jets of water through nozzles of special design, at a rate of 200 to 400 litres per min. per m^2 of area, under a pressure of 0.7 to 1 kg/cm^2.

10.8. LOSS OF HEAD AND NEGATIVE HEAD

When a cleaned bed is put into operation, the loss of head through it will be small – usually of the order of 15 to 30 cm. As the water is filtered through it, impurities are arrested by the filter media, due to which the loss of head goes on increasing. The loss of head can be measured by inserting two piezometric tubes, as shown in Fig. 10.10.

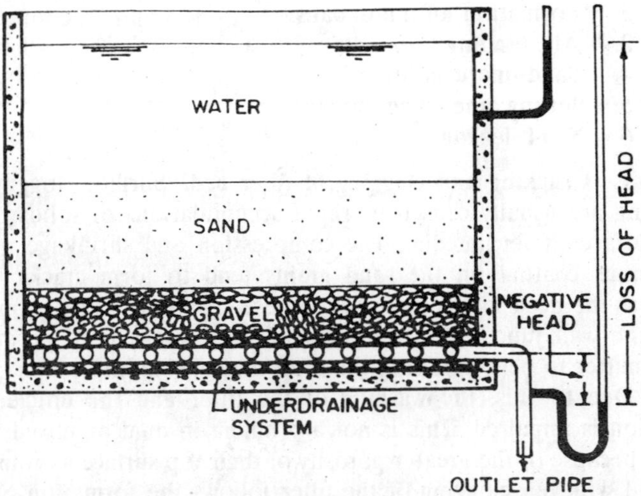

FIG. 10.10. LOSS OF HEAD

As the thickness or depth of the suspended matter on the top of sand bed increases, the head loss is increased. A stage comes when the frictional resistance exceeds the static head above the sand bed. At this, the lower portion of the media and the underdrainage system are under *partial vacuum* or *negative head.* Within the bed, the head becomes negative when the loss of head through the overlying bed depth exceed the static head. Water is then sucked through the filter media rather than being filtered through it. As shown in Fig. 10.10, the negative head is the fall of liquid level in the piezometric tube below the centre line of the underdrainage system. Due to formation of negative head, dissolved gases and air are released filling the pores of the filter and the underdrainage system. The filter then becomes *air bound,* head losses rise sharply, and filter output capacity drops rapidly.

In rapid gravity sand filters, the permissible head loss may be between 2.5 and 3.5 m, and the permissible negative head may be 0.8 to 1.2 m. A filter run is terminated when these limits are reached and the filter is then back washed. Under normal conditions, the frequency of filter wash may be between 2 and 4 days.

10.9. FILTER TROUBLES

Problems in filter operation and performance arise either from poor design or poor operation. Some of the potential filter troubles are as follows :

1. Cracking ana clogging of filter bed.
2. Formation of mud balls.
3. Air binding.
4. Sand incrustation.
5. Jetting and sand boils
6. Sand leakage.

1. Cracking and clogging of filter bed. Surface clogging and cracking are usually caused by rapid accumulations of solids on the top surface of fine media. The compression and shrinkage of soft, gelatinous coating on the sand grains tend to form cracks in the filter, as the head loss is increased. These cracks are more prominent near the wall junctions. The effect of such cracks is to permit the dirty matter to penetrate into the filter media, sometimes upto gravel layer. Due to this, the washing of the filter and the efficiency of filtration is impaired. This is not a problem in dual or mixed media filters because of the greater porosity of their top surface as compared to sand. Uneven clogging of the filter follows the formation of mud balls, cracking of the bed and the pulling away of the sand from the side walls.

2. Formation of mud balls. Mud balls are conglomerations of coagulated turbidity, floc, sand, and other binders, and are formed near the top of filter media. Mud balls are caused due to insufficient washing of sand grains. Mud may accumulate on the sand surface to form a dense mat. During washing, there may be some lateral pressure between the points where the water breaks through, and the sticky mud is pushed into lumps or balls of size of a pea upto 2 to 5 cm or more in diameter. A 50% expansion of the sand in washing is effective in minimizing the production of mud balls. Formation of mud balls can be eliminated by providing an adequate backwash flow rate and a properly designed system for auxilliary scour (surface wash).

3. Air Binding. The condition of air binding is caused by the release of dissolved air and gases from water, to form bubbles. These air bubbles occupy the void space of the filter media and the drainage system. Trouble is specially likely to occur if water is saturated or super-saturated with air in solution. The separation of this dissolved air may be due to the following reasons : (*i*) the negative head due to excessive loss of head; (*ii*) increase of temperature of water as it passes through the filter media, resulting in the reduction in its capacity to retain gases in solution; (*iii*) release of oxygen by algae. Air binding can be minimised by (*i*) avoidance of excessive negative head by providing a water depth of at least 1.5 m above the surface of the unexpanded filter bed; (*ii*) avoidance of warming of water as it passes through the plant; (*iii*) control of algae growth in the influent, and (*iv*) avoiding super-saturation of water with air.

4. Sand Incrustation. Sand incrustation may occur either due to deposition of sticky gelatinous materials from the influent water or due to an after-crystallization of calcium carbonate in case where heavy lime treatment of water is practised. Due to this, the sand grains enlarge, and the effective size of sand is changed. This can be minimied by carbonating the influent, thereby dissolving both calcium carbonate and any manganese that might have deposited on sand grains. Sodium hexa-meta-phosphate may also be used in small plants to keep calcium carbonate in dissolved state.

5. Jetting and Sand Boils. Jetting and sand boils result during backwashing when back wash water follows path of least resistance and break through to the scattered points due to small differences in porosity and permeability of sand and gravel. When jetting becomes severe, the sand boils up like quicksand and gravel as well as sand are lifted to the surface. The tendency of jetting can be minimised by surface wash or air scour in advance of bed fluidization. It can also be eliminated by use of the 8 cm layer of coarse garnet or ilmenite between fine media and gravel.

6. Sand Leakage. Sand leakage or downward migration and escape of fines result when layers of smallest gravel are displaced during backwashing. It can be minimised by properly proportioning the sand and gravel layers, and also by using coarse garnet layer between the fine media and gravel.

10.10. PERFORMANCE OF RAPID SAND FILTERS

(*i*) ***Turbidity.*** If the influent water does not have turbidity of more than 35 to 40 p.p.m., the filter can reduce the turbidity to less than 1 p.p.m. (1 NTU). Since coagulation and sedimentation always precede filtration, the turbidity of water applied to filter is always less than 35 to 40 p.p.m.

(ii) Colour. Rapid sand filters are very efficient in colour removal. The intensity of colour can be brought down below 3 on cobalt scale. Colourless water can be produced by rapid filtration after the addition of polyelectrolytes immediately in advance of filtration.

(iii) Iron and Manganese. Rapid filters remove oxidized or oxidizing iron, though it is less efficient in removing manganese.

(*iv*) Taste and Odor. Unless special treatement such as activated carbon or prechlorination is provided, rapid filters will not ordinarily remove tastes and odours.

TABLE 10.2.
COMPARISON BETWEEN SLOW AND RAPID SAND FILTERS

S. No.	*Item*	*Slow sand filter*	*Rapid sand filter*
1.	Rate of filtration	100 to 200 litres per hour per m^2	3000 to 6000 litres per hour per m^2
2.	Loss of head	15 cm initial to 100 cm final	30 cm initial to 3 m final
3.	Size of bed	Requires large area	Requires small area
4.	Coagulation	Not required	Essential
5.	Filter media of sand	Effective size : 0.2 to 0.35 mm C_u : 2 to 3 Depth : 105 cm, reduced to not less than 30 cm by scrapping	Effective size : 0.35 to 0.6 mm C_u : 1.2 to 1.7 Depth : 75 cm; not reduced by washing
6.	Base material of gravel	Size : 3 to 65 mm Depth : 30 to 75 cm	Size : 3 to 40 mm Depth : 60 to 90 cm
7.	Underdrainage system	Split tile laterals discharging into tile or concrete main drain ; or perforated pipe laterals	Perforated laterals with mains or wheeler system, or Leopald system or Wagner system for underdrainage
8.	Method of cleaning	Scrapping of top layer to 15 mm to 25 mm	Agitation and backwashing with or without compressed air
9	Amount of washwater	0.2 to 0.6% of water filtered	2 to 4% water filtered
10.	Period of cleaning	1 to 2 months	2 to 3 days
11.	Penetration of suspended matter	Superficial	Deep
12.	Supplementary treatment of water	Chlorination	Chlorination
13.	Efficiency	Very efficient in the removal of bacteria but less efficient in the removal of colour and turbidity	Less efficient in removal of bacteria, more efficient in the removal of colour and turbidity
14.	Economy	High initial cost	Cheap and economical
15.	Flexibility	Not flexible in meeting variations in demand	Quite flexible for reasonable fluctuations in demand
16.	Skilled supervision	Not essential	Essential
17.	Depreciation cost	Relatively low	Relatively high

(v) ***Bacterial Efficiency.*** Rapid filters are less efficient in removal of bacterial load, in comparison to the slow filters. The percentage of bacterial removal in a well operated plant preceded by flocculation and sedimentation is dependent on several factors, the most important of which is the concentration of organisms in the raw water. In lower ranges of bacterial loadings, removal may be from 90 to 99 percent. In all cases, filtration should be followed by adequate chlorination.

(vi) ***Filter runs*** : The filter runs should not be less than 24 hours with a head loss not exceeding 2 m.

(viii) ***Washwater consumption*** : For an efficient filters, the wash water consumption should not exceed 2 per cent of the quantity filtered in between washing.

10.11. COMPARISON OF SLOW SAND AND RAPID SAND FILTERS

The comparision of slow sand and rapid sand filter of gravity type is shown in Table 10.2.

Example 10.2 *A city has a population of 100,000 with an average rate of demand of 160 litres per head per day. Find the area of rapid sand filters.*

Solution :

Max. daily demand = 100,000 × 160 × 1.5 = 240,00,000 litres.

Let us assume an average filtration rate of 4500 litres per hour per m^2 of filter area.

$$\therefore \quad \text{Area of filters} = \frac{24000000}{4500 \times 24} = 222.2 \text{ m}^2$$

Let the size of each filter unit be 9 m × 5 m .

$$\therefore \quad \text{No. of units required} = \frac{222.5}{9 \times 5} = 5$$

Keeping one unit as standby, provide a total 6 units. These may be arranged in series with 3 units on either side.

Example 10.3. *A flat bottom trough is to receive the wash water from a section of the filter which is 2 m wide and 3 m long. The wash water rate is 700 litres per min. per sq. m. If the water is to have a depth of 25 cm at the upper end of the trough, what should be the dimensions of the trough ?*

Solution :

In Eq. 10.4 $Q = 700 \times 2 \times 3 = 4200$ litres per min.

$$y = 25 \text{ cm}$$

$\therefore$ $$Q = 0.8\, b\, y^{3/2}$$

or $$4200 = 0.8\, b\, (25)^{3/2}$$

From which $\quad b = \mathbf{42\ cm}$

Providing a free board of 5 cm, the depth of trough will be = 25 + 5 = **30 cm**

Example 10.4. *Redsign the dimensions if the wash water trough is to be V-shaped.*

Solution :

For the rectangular section, b = 42 cm and y = 25 cm as above. Let us provide a V-bottom, with a 8 cm high triangle. This triangle will have an area equivalent to a rectangle, whose depth is 4 cm. Hence y = 25 + (8 − 4) = 29 cm. Providing a free board of 5 cm, the total depth= 29 + 5 = 34 cm. The height of vertical sides will be 34 − 8 = 26 cm. The section is shown in Fig. 10.11.

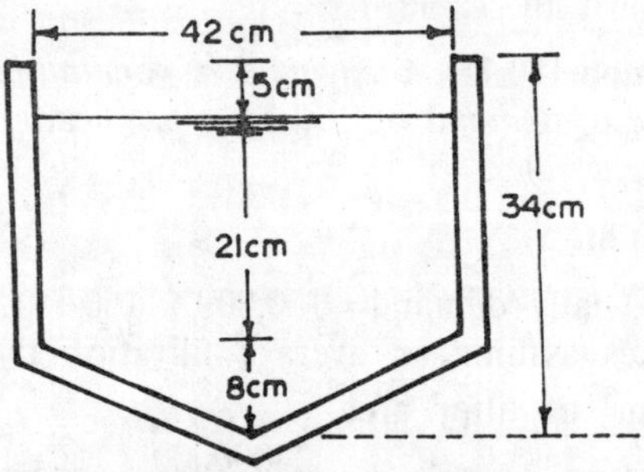

FIG. 10.11.

Example 10.5. *Design completely a rapid sand filter for a town having a total filtered water requirement of 5 million litres of water per day. Assume suitable data.*

Solution

Step 1. ***Design of filter units***

Let us assume that 3% of filtered water is used for washing of the filter every day.

$\therefore$ Filtered water requirement per day

$$= 1.03 \times 5 = 5.15 \text{ Mld}$$

Again, let us assume that 30 minutes are lost every day in washing the filter. The quantity of filtered water required per hour

$$= \frac{5.15}{(24 - 0.5)} = 0.219 \text{ Ml/hour}$$

Again, let us assume an average filtration rate of 5000 litres/h/m^2 of filter area

$$\therefore \text{ Filter area } \quad A = \frac{0.219 \times 10^6}{5000} = 43.83 \text{ m}^2$$

Let $L = \frac{4}{3}B$, and let there is 2 filter units, two being the minimum number to be provided.

$$\text{Then} \quad 2\left(B \times \frac{4}{3}B\right) = 43.83$$

$$\text{From which} \qquad B = 4.05 \text{ m}$$

Keep $B = 4$ m and $L = 5.5$ m, for each filter unit, giving total $A = 2 \times 4 \times 5.5 = 44 \text{ m}^2$.

Step 2. ***Design of under-drainage system***

The under-drainage system is shown in Figs. 10.5 and 10.7. Let us keep the total area of perforations in the under drainage system to the entire filter area as 0.003 (*i.e.* 0.3%)

$$\therefore \text{ Total area of perforations } = 0.003 \times 4 \times 5.5$$
$$= 0.066 \text{ m}^2$$
$$= 660 \text{ cm}^2$$

Let us use 12 mm dia. perforations, for which the desired total cross-sectional area of laterals is kept equal to twice the total area of perforations.

Total area of laterals $= 2 \times 0.066 = 0.132 \text{ m}^2$

Keeping area of manifold to be twice the total area of laterals,

Area of manifold $= 2 \times 0.132 = 0.264$

$$\therefore \text{ Dia. of manifold} = \sqrt{\frac{0.264 \times 4}{\pi}} = 0.58 \text{ m}$$

Let us provide 60 cm dia. manifold, laid along the length of the filter unit. Let us place the laterials at a spacing of 15 cm.

$$\text{Hence No. of laterals} = \frac{5.5 \times 100}{15} = 37.$$

Hence provide 37 laterals on either side of the central manifold, thus requiring $37 \times 2 = 74$ laterals in each filter unit.

$$\text{Length of each lateral} = \frac{(4 - 0.6)}{2} = 1.7 \text{ m}$$

Let n be the total number of perforations, each of 12 mm dia., in all the 74 laterals. Then, we have

$$n \times \frac{\pi}{4}(12)^2 = 0.066 \times (1000)^2$$

From which $n \approx 584$

$\therefore$ No. perforations in each lateral $= \frac{584}{74} = 7.9$

Hence provide 8 perforations per lateral.

$\therefore$ Area of perforations per lateral

$$= 8\left[\frac{\pi}{4}(12)^2\right] = 904.8 \text{ mm}^2$$

Area of each lateral = 2 × area of perforations per lateral

$$= 2 \times 904.8 = 1809.6 \text{ mm}^2$$

$\therefore$ Dia. of lateral $= \sqrt{\frac{1809.6 \times 4}{\pi}} = 48$ mm

Hence provide 50 mm dia. laterals, at 15 cm c/c, each having 8 perforations of 12 mm dia.

Check :

$$\frac{\text{Length of lateral}}{\text{Dia. of lateral}} = \frac{1.7 \times 1000}{50} = 34 < 60$$

Also, spacing of perforations $= \frac{1.7 \times 100}{8} = 21.25$ cm

Step 3. ***Wash water discharge and velocity***

Let the rate of washing be equivalent to a rise of 60 cm/min. This will give a wash water rate $= 0.6 \times 60 \times 1 = 36 \text{ m}^3/\text{hr}/\text{m}^2$

$\therefore$ Wash water discharge in one filter

$$= 36 \times 4 \times 5.5 = 792 \text{ m}^3/\text{h} = 0.22 \text{ m}^3/\text{s}$$

$\therefore$ Velocity of flow of wash water in laterals

$$= \frac{0.22}{74\left[\frac{\pi}{4}\left(\frac{50}{1000}\right)^2\right]} = 1.514 \text{ m/sec.}$$

Velocity of flow of wash water in the manifold

$$= \frac{0.22}{\frac{\pi}{4}(0.6)^2} = 0.78 \text{ m/sec.}$$

against a max. permissible value of 1.8 m/sec.

Step 4. ***Design of wash water trough***

Let us provide 3 troughs in a width of 4 m so that spacing of troughs = 4/3 = 1.33 m. These troughs will run parallel to the longer dimension of the filter unit.

$\therefore$ Discharge per trough $= \dfrac{0.22}{3} = 0.0733\ m^3/sec$

Let us keep width (b) of trough $= 40$ cm $= 0.4$ m

From Eq. 10.4 (a), the water depth at upper end is given by

$$Q = 1.376\, b\, y^{3/2} \qquad ...(10.4\ a)$$

$$\therefore \qquad 0.0733 = 1.376\,(0.4)\, y^{3/2}$$

From which $\quad y \approx 0.26$ m $= 26$ cm

Keeping a F.B. of 10 cm, depth of trough $26 + 10 = 36$ cm.
Hence provide wash water troughs of size 40 cm × 36 cm.

Step 5. ***Determination of depth of sand***

Let us assume 60 cm depth of sand having effective size of 0.5 mm. Let us use Eq. 10.2 for computing minimum depth so that flocs do not breack through sand bed

$$\frac{Q\,d^3\,h}{l} = B_i \times 29323$$

Here Q = rate of filtreation $= 5000 \times 2 = 10{,}000$ litres/h/m^2
$= 10\ m^3/h/m^2$

(assuming 100% overloading under emergent conditions)

$d = 0.6$ mm, assumed as mean dia.
h = terminal head loss $= 2.5$ m (assumed)
l = depth of sand, in metres.
$B_i = 4 \times 10^{-4}$ (assumed), for poor response of filtration and average degree of pretreatment.

$$\therefore \qquad \frac{10\,(0.6)^3\,2.5}{l} = 4 \times 10^{-4} \times 29323$$

From which, we get $l = 0.46$ m $= 46$ cm.

Hence the assumed depth of 60 cm will be adequate to avoid break through of floc.

Step 6. ***Size gradation of gravel***

Let the size gradation of gravel be 2 mm at top to 40 mm at bottom. The requisite depth (l) in cm of a *component gravel layer* of size d (mm) is computed from Eq. 10.3 :

$$l = 2.54\, k\, (\log d)$$

Taking $k = 12$, we have

$$l = 2.54 \times 12 \log d = 30.48 \log d.$$

When $d = 2$ mm,

$$l = 30.48\,(\log 2) \approx 9.2 \text{ cm}.$$

The depths of various gravel layers are tabulated below :

Gravel size (mm)	2	5	10	20	40
Depth (cm)	9.2	21.3	30.5	39.7	48.8
Increment (cm)	9.2	12.1	9.2	9.2	9.1

Provide a gravel depth of 50 cm.

Step 6. ***Determination of total depth of filter tank***

The total depth of filter tank

= (depth of underdrains) + (depth of gravel) + (depth of sand) + (water depth) + (free board)

Providing a water depth of 1.2 m and a free board of 0.3 m,

Total Depth = (0.6) + (0.50 + (0.6) + (1.2) + (0.3) = 3.2 m

Hence each filter tank will have a size of 4 m×5.5 m×3.2 m.

Example 10.6. *A filter unit is of size 4 m × 8 m. After filtering 8000 m^3/day in 24 hour period, the filter is back washed at the rate of 10 l/m^2/sec for 10 minutes. Compute the average filtration rate, quantity and percentage of treated water used in washing and the rate of wash water flow in each trough. The unit has 4 troughs.*

Solution

Area of filter = 4 × 8 = 32 m^2

$$\text{Rate of filtration} = \frac{8000 \times 1000}{32} = 25 \times 10^4 \text{ litres/m}^2\text{/day.}$$

Assuming that the filter is closed for 30 min. every day for back washing and cleaning,

$$\text{Av. rate of filtration} = \frac{25 \times 10^4}{23.5} = \mathbf{10638\ l/m^2/hr.}$$

Amount of water used for backwash

= 10 × 32 × 10 × 60 = **192000** litres

= 192 m^3

$$\% \text{ of filtered water} = \frac{192}{8000} \times 100 = \mathbf{2.4\%}$$

Again, total wash water discharge

= 10 × 32 = 320 litres/sec.

∴ Discharge per trough = 320/4 = **80 litres/sec.**

= 0.08 m^3/sec

Example 10.7. *The particle size distribution of the stock sand, to be used in a filter, is shown in Fig 10.12. The desired effective size of filter sand is 0.5 mm and the uniformity coefficient is 1.5. Determine the diameters of the finer and coarse particles that are required to be removed from the stock sand.*

Solution

Step 1. Desired D_{10} size= 0.5 mm

Hence, from curve of Fig. 10.12, P_{10}, corresponding to $D_{10} = 0.5$ mm is 27% .

Again, desired D_{60} size= $C_U \times D_{10} = 1.5 \times 0.5 = 0.75$ mm

Hence, from the curve, P_{60}, Corresponding to $D_{60} = 0.75$ mm is 49%.

Step 2. Thus, $P_{10} = 27\%$ and $P_{60} = 49\%$

Hence, from Eq. 10.1 (*a*), the percentage of usable sand is

$$P_{usable} = 2\,(P_{60} - P_{10}) = 2\,(49 - 27) = 44\%.$$

Step 3. From Eq. 10.1 (*b*), the % below which the stock sand is too fine is

$$P_{tf} = P_{10} - 0.1\,P_{usable} = 27 - 0.1 \times 44 = 22.6\%$$

From the curve of Fig. 10.12, the diameter D_{tf} corresponding to this is

$$D_{tf} = 0.44 \text{ mm}.$$

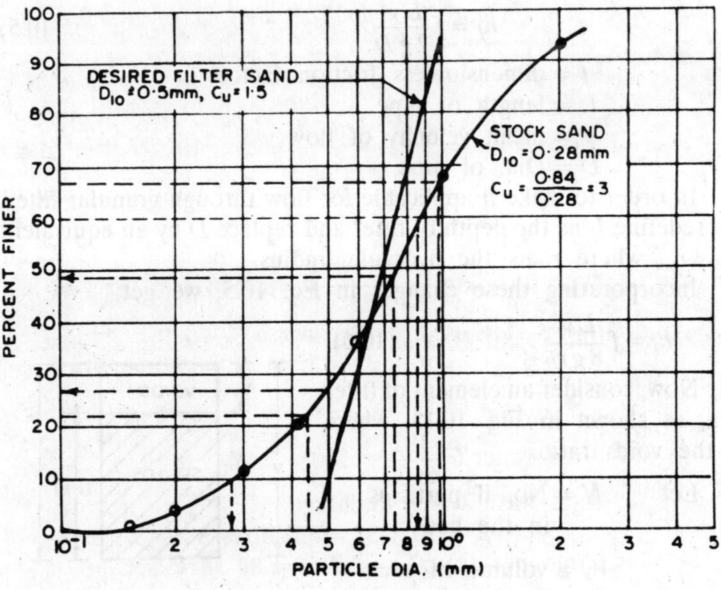

FIG. 10.12. PARTICLE SIZE DISTRIBUTION FOR STOCK SAND

Step 4 : Similarly, from Eq. 10.11 (*c*), the percentage (P_{tc}) above which the stock sand is too coarse is

$$P_{tc} = 1.8\,P_{60} - 0.8\,P_{10}$$
$$= 1.8 \times 49 - 0.8 \times 27 = 66.6\%$$

From the particle size curve, the dia. D_{tc} correponding to P_{tc} (= 66.6%) is

$$D_{tc} = 0.97 \text{ mm.}$$

Hence the stock sand finer than D_{tf} = 0.44 mm at one end of the size range and coarser than D_{tc} = 0.97 mm at the other end must be removed.

10.12. FILTRATION HYDRAULICS : CARMEN-KOZNEY EQUATION :

The filter bed can be visualised as consisting of many parallel channels formed by the inter-connecting voids between the filter media particles. However, these flow channels are irregular and do not have definite geometric configuration. The head loss through the bed of solids of the filter can be computed by two equations:

(*i*) Carmen-Kozney equation
(*ii*) Rose equation.

In order to derive Carmen-Kozney equation, let us start with the fundamental Darcy-Weisbach equation for head loss in circular pipe:

$$h_f = f\frac{L\,v^2}{2\,g\,D} \qquad ...(10.5)$$

where
f = dimensionless friction factor
L = length of pipe
v = mean velocity of flow
D = Dia. of pipe.

In order to make it applicable for flow through granular filter, let as redefine L as the depth of filter and replace D by an equivalent term $4\,r_H$, where r_H is the hydraulic radius.

Incorporating these changes in Eq. 10.5, we get

$$h_f = f\frac{L\,v^2}{8\,g.r_H} \qquad ...(10.6)$$

Now, consider an element of filter media, as shown in Fig. 10.13, where e is the voids ratio.

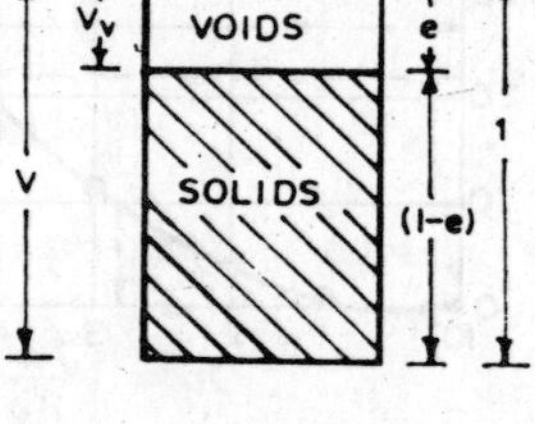

FIG. 10.13.

Let N = No. if particles in the bed

V_p = volume of each particle

$\therefore$ Volume of solids $= N V_P$

From Fig. 10.13, we observe that when the volume of solids is $(1 - e)$, the total vol. = 1. Hence when the volume of solids is $N V_p$, the total volume of the bed $= N V_p/(1 - e)$

$$\therefore \quad \text{Total channel volume} = \frac{N V_P}{1 - e} - N V_p$$

$$= \frac{e}{1 - e} N V_p$$

If A_p is the wetted surface area of each particle,
Total wetted surface area $= N A_P$

$$\therefore \quad r_H = \frac{\text{channel volume}}{\text{wetted surface area}} = \frac{e}{1 - e} \cdot \frac{V_p}{A_p} \qquad \text{..(10.7 } a\text{)}$$

For spherical particles of dia. d,

$$V_P = \frac{\pi d^3}{6} \quad \text{and } A_P = \pi d^2$$

$$\therefore \quad \frac{V_p}{A_p} = \frac{d}{6} \qquad \text{...(10.8 } a\text{)}$$

$$\text{Hence} \quad r_H = \frac{e}{1 - e} \cdot \frac{d}{6} \qquad \text{...(10.7 } b\text{)}$$

For non-spherical particles, taking ϕ_s as the *shape factor*, we have

$$\frac{V_P}{A_p} = \phi_s \cdot \frac{d}{6} \qquad \text{...(10.8)}$$

Carmen has recommended the following values of ϕ_s :

(*i*) Pulverised coal : 0.73
(*ii*) Ottawa sand : 0.95
(*iii*) Rounded sand : 0.82
(*iv*) Angular sand : 0.73

$$\text{Hence} \quad r_H = \frac{1}{1 - e} \cdot \frac{1}{\phi_s} \cdot \frac{d}{6} \qquad \text{...(10.7)}$$

Let v_s = face or approach velocity or average superficial velocity
$= Q/A$, where A is the surface area of filter.

$\therefore$ v = average velocity through filter voids $= \dfrac{v_s}{e}$

Hence from Eq. 10.6, we get

$$h_f = \left(\frac{fL}{8g}\right)\left(\frac{v_s}{e}\right)^2\left(\frac{1 - e}{e} \cdot \frac{6}{\phi_s d}\right)$$

or $$h_f = f_1 \left(\frac{L}{\phi_s d}\right)\left(\frac{1-e}{e^3}\right)\left(\frac{v_s^2}{g}\right) \quad ...(10.8)$$

which is the Carmen-Kozney equation.

The dimensionless friction factor f_1 has been related to Reynolds number (R) by the relation

$$f_1 = 150\frac{1-e}{R} + 1.75 \quad ...(10.9)$$

where $$R = \phi_s . \frac{\rho\, v_s . d}{\mu} = \phi_s \frac{v_s\, d}{\nu} \quad ...(10.10)$$

We will now consider two cases :

(*a*) Homogeneous mixed bed

(*b*) Stratified bed.

(a) Homogeneous mixed bed (unstratified bed)

For a bed of mixed size particles, such as the one found in slow sand filters, the A/V term can be evaluated on the basis of sieve analysis, using the relation

$$\left(\frac{A}{V}\right)_{av.} = \frac{6}{\phi_s}\sum_{i=1}^{n}\frac{x_i}{d_i} \quad ...(10.11)$$

where x_i = Weight fraction of particles of size d_i

n = No. of weight fractions in the mixure.

The term x_i represents the fraction by weight of particles retained between adjacent sieves, while the corresponding value of d_i is taken as the *geometric mean* size of the two size openings i.e. $d_i\sqrt{d_1 \times d_2}$ where d_1 and d_2 are adjacent sieves. Also, one value of f_1 is used to represent the entire bed.

Hence the final form of the equation is

$$h_f = \frac{f_1 L}{\phi_s}\left(\frac{1-e}{e^3}\right)\frac{v_s^2}{g}\sum_{i=1}^{n}\frac{x_i}{d_i} \quad ...10.12\ (a)$$

or $$h_f = f_1 K L \sum_{i=1}^{n}\frac{x_i}{d_i} \quad ...(10.12)$$

where $$K = \frac{1}{\phi_s}\left(\frac{1-e}{e^3}\right)\left(\frac{v_s^2}{g}\right) \quad ...(10.12\ a)$$

(*b*) Stratified bed

Here, the factor f_1 for each stratification will be different. Taking the porosity of the stratified bed as uniform and also considering the particles to be of uniform shape, Eq. 10.8 can be written as follows for a particular stratum :

$$\frac{dh_f}{dL} = K f_1 \cdot \frac{1}{d}$$

where K is given by Eq. 10.12 (b).

Hence the total head loss through the entire filter of depth L is

$$h_f = \int_0^{hf} d\,h_f = K \int_0^L \frac{f_1}{d}\,dL$$

Noting that $\quad dL = L\,x_i$

$$h_f = L\,K \sum_{i=1}^{n} \frac{f_{1i}\,x_i}{d_i} \qquad ...(10.13)$$

Eq. 10.13 is applicable to conditions encountered in a rapid sand filter.

It is to be noted that Eqs. 10.12 and 10.13 are identical except that factor f_1 has been taken inside the summation.

10.13. ROSE EQUATION

Rose gave the following equation, based on experiments :

$$h_f = f_2 \left(\frac{L}{\phi_s\,d}\right)\left(\frac{1}{e^4}\right)\left(\frac{v_s^2}{g}\right) \qquad ...(10.14)$$

where the factor f_2, which is a function of C_D, is given by

$$f_2 = 1.067\,C_D \qquad ...10.15\ (a)$$

where C_D is the coefficient of drag, connected to Reynold's number. If $R < 1.9$,

$$C_D = \frac{24}{R} \qquad ...(10.15\ b)$$

For R greater than 1.9 but less than 500,

$$C_D = \frac{b}{R^n} = \frac{18.5}{R^{0.6}} \qquad ...(10.15\ C)$$

where $\quad R = \rho\,v_s\,d/\mu = v_s\,.\,d/\nu$

Hence $\quad h_f = 1.067\,C_D \left(\frac{L}{\phi_s\,d}\right)\left(\frac{1}{e^4}\right)\left(\frac{v_s^2}{g}\right) \qquad ...(10.15)$

Rose equation is widely used for hydraulic computations for *rapid sand filters.*

Here also, we will take two cases :

(a) Homogeneous mixed bed.

(b) Stratified bed

(a) Homogeneous mixed bed

From Eq. 10.8,

$$d = \frac{6}{\phi_s}\left(\frac{V}{A}\right)_{av}$$

$$\therefore \qquad h_f = 1.067\, C_D \frac{L}{6}\left(\frac{A}{V}\right)_{av}\left(\frac{1}{e^4}\right)\left(\frac{v_s^2}{g}\right) \qquad ...(10.16)$$

where $$\left(\frac{A}{V}\right)_{av} = \frac{6}{\phi_s}\sum_{i=1}^{n}\frac{x_i}{d_i} \qquad ...(10.11)$$

$$\therefore \qquad h_f = 1.067\,\frac{C_D L}{\phi_s}\left(\frac{1}{e^4}\right)\left(\frac{v_s^2}{g}\right)\sum_{i=1}^{n}\frac{x_i}{d_i} \qquad ...(10.17)$$

(b) Stratified bed of uniform porosity

For a particular stratum,

$$\Delta h_f = \frac{1.067\, v_s^2 C_D}{g\,\phi_s\, e^4}\cdot\frac{\Delta L}{d} \qquad ...(10.18)$$

$$\therefore \qquad h_f = \sum_{i=i}^{n}\Delta h_f = \frac{1.067\, v_s^2}{g\,\phi_s\, e^4}\sum_{i=1}^{n} C_{Di}\frac{\Delta L_i}{d_i} \qquad ...(10.19)$$

where n = number of layers in stratified bed.

C_{Di} = drag coefficient for d_i

Now $\Delta L_i = x_i L$

$$\therefore \qquad h_f = \frac{1.067\, v_s^2 L}{g\,\phi_s\, e^4}\sum_{i=1}^{n} C_{Di}\frac{x_i}{d_i} \qquad ...(10.20)$$

where x_i = weight fraction of particle in i^{th} layer.

10.14. FLOW THROUGH EXPANDED BEDS

The Carmen-Kozney equations and Rose equation are applicable for beds in which voids are clear and unobstructued. However, during continuous filtration process, these voids get clogged and hence the head loss goes on increasing. The reduced flow capacity can be restored by washing out the void accumulations by fluid expansion of bed during back washing (Fig. 10.14).

Let $\bar{v}_s$ = superficial velocity

$\bar{v}_{sc}$ = critical velocity at which the bed is fluidized.

L = original thickness of bed

e = porosity of fixed bed.

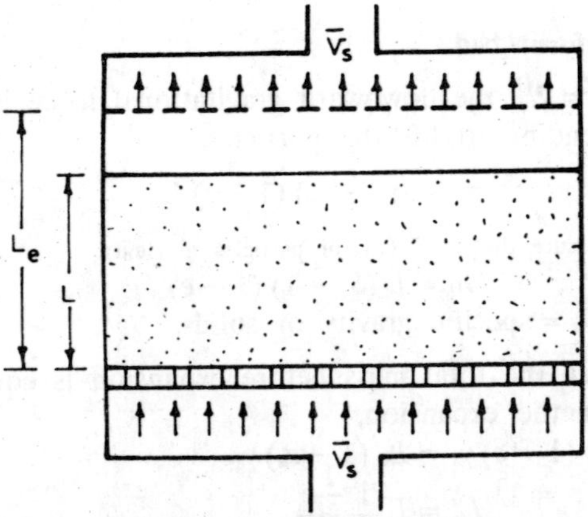

FIG. 10.14. BED EXPANSION DURING BACK WASHING

L_e = expanded thickness of bed.

e_e = porosity of expanded bed.

During back washing, the bed remains fixed at low fluid velocity. As the velocity $\bar{v}_s$ is increased, the lighter particles begin to move upwards. When $\bar{v}_s$ becomes equal to $\bar{v}_{sc}$, a given size of particle is 'suspended' or 'fluidized'. When velocity is further increased (Fig. 10.15), the particles become more widely separated. Further increase of velocity to $\bar{v}_{sc}'$, make the particles move (transport) as individuals.

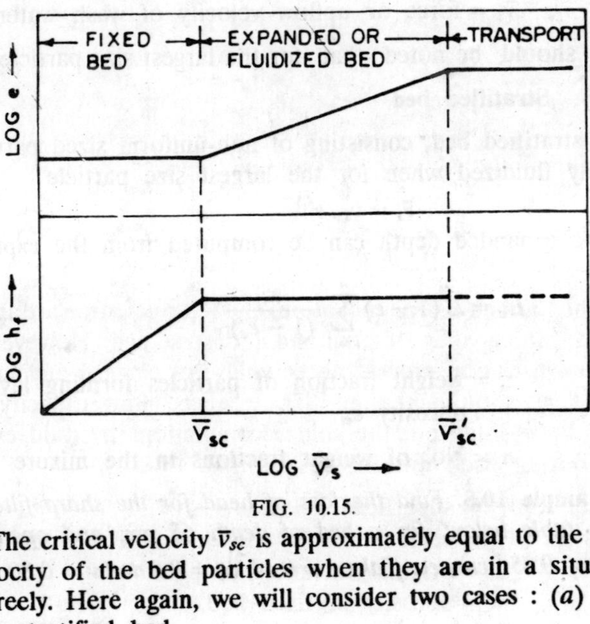

FIG. 10.15.

The critical velocity $\bar{v}_{sc}$ is approximately equal to the terminal fall velocity of the bed particles when they are in a situation to settle freely. Here again, we will consider two cases : (*a*) uniform bed (*b*) stratified bed.

(a) Uniform bed

When $\overline{v}_s = \overline{v}_{sc}$, the downward gravitational force is balanced by the force tending to lift the particles.

$$\therefore \quad \underset{\text{(frictional pressure drop)}}{h_f \cdot \gamma_w} = \underset{\text{(wt. of particles in fluid)}}{(\gamma_s - \gamma_w)(1-e)L} \quad \text{...(10.21 a)}$$

or

$$h_f = L(S_s - 1)(1-e) \quad \text{...(10.21)}$$

where S_s = specific gravity of solids.

Also, since the total mass before expansion is equal to the total mass after the expansion,

$$L(1-e)\gamma_s = L_e(1-e_e)\gamma_s$$

$$\therefore \quad L_e = L\frac{1-e}{1-e_e} \quad \text{...(10.22)}$$

Fair and Geyer have given the following expression for determining e_e :

$$e_e = \left(\frac{\overline{v}_s}{v_s}\right)^{0.22} \quad \text{...(10.23 a)}$$

or

$$\overline{v}_s = v_s \cdot e_e^{4.5} \quad \text{...(10.23)}$$

where v_s = terminal settling velocity of particles

$\overline{v}_s$ = force or upflow velocity of wash water

It should be noted that for the largest size particles $e_e = e$

(b) Stratified bed

A stratified bed, consisting of non-uniform sized particles is completely fluidized when for the largest size particle

$$\overline{v}_s = v_s \cdot e^{4.5}$$

The expanded depth can be computed from the expression:

$$L_e = L(1-e)\sum_{i=1}^{n}\frac{x_i}{(1-e_e)_i} \quad \text{...(10.24)}$$

where x_i = weight fraction of particles forming layer with porosity e_{ei}

n = No. of weight fractions in the mixure

Example 10.8. *Find the loss of head for the sharp filter sand shown in table below, for a bed of depth 65 cm, and operated at a rate of 0.45 cm/sec. Take e = 0.44, ϕ_s = 0.8 and $\nu = 0.80 \times 10^{-2}$ cm²/sec.*

Mean dia. of particles (cm × 10^2)	3.5	4.5	5.5	6.5	7.5	8.9	11.8
Wt. fraction of particles (%)	4	9	20	24	20	14	9

Solution

$$v_s = 0.45 \text{ cm/sec}$$

$$\therefore \qquad R = \frac{v_s \, . \, d}{\nu} = \left(\frac{0.45}{0.80 \times 10^{-2}} \right) d = 56.25\, d$$

For $R < 1.9$, $\quad C_D = \dfrac{24}{R}$

For $R > 1.9$, $\quad C_D = \dfrac{b}{R^n} = \dfrac{18.5}{R^{0.6}}$

The values of R, C_D and $C_D \dfrac{x}{d}$ are calculated in the tabular form shown below :

d (cm)	*x*	*R = 56.25 d*	C_D	$C_D \frac{x}{d}$
0.035	0.04	1.97	12.32	14.08
0.045	0.09	2.53	10.60	21.19
0.055	0.20	3.09	9.39	34.16
0.065	0.24	3.66	8.50	31.38
0.075	0.20	4.22	7.80	20.80
0.089	0.14	5.01	7.04	11.07
0.118	0.09	6.64	5.94	4.53
			Sum	137.21

Rose equation for head loss is

$$h_f = \frac{1.067\, v_s^2\, L}{g\, \phi_s\, e^4} \sum_{i=1}^{n} C_D \frac{x_i}{d}$$

$$= \frac{1.067\, (0.45)^2 \times 65}{981 \times 0.8\, (0.44)^4} \times 137.21 = \mathbf{65.5 \ cm}$$

Example 10.9. *Solve example 10.8 using Carmen Kozney equation.*

Solution

$$h_f = K L \sum_{i=1}^{n} f_{1i} \frac{x_i}{d_i}$$

where $\qquad KL = \dfrac{1}{\phi_s} \left(\dfrac{1-e}{e^3} \right) \left(\dfrac{v_s^2}{g} \right) L$

$$= \frac{1-0.44}{0.8\,(0.44)^3}\,(0.45)^2 \times \frac{65}{981} = 0.11$$

$$\therefore \quad h_f = 0.11\,\Sigma f_{1i}\,\frac{x_i}{d_i}$$

where $f_1 = 150\left(\frac{1-e}{R}\right) + 1.95 = \frac{150\,(1-0.44)}{R} + 1.95$

$$= \left(\frac{84}{R} + 1.95\right)$$

and $R = \phi_s \, . \, \frac{v_s \, . \, d}{\nu} = \frac{0.8 \times 0.45}{0.80 \times 10^{-2}}\, d = 45\, d$

The computations are shown in the tabular form below.

d_i (cm)	x_i	$R = 45\, d_i$	f_{1i}	$f_{1i}\frac{x_i}{d_i}$
0.035	0.04	1.575	55.28	63.18
0.045	0.09	2.025	43.43	86.86
0.055	0.20	2.475	35.89	130.51
0.065	0.24	2.928	30.67	113.24
0.075	0.20	3.375	26.84	71.57
0.089	0.14	4.005	22.92	36.06
0.118	0.09	5.310	17.77	13.55
			Sum	514.97

$$\therefore \quad h_f = 0.11\,\Sigma f_{1i}\,\frac{x_i}{d_i} = 0.11 \times 514.97$$

$$= \mathbf{56.6\ cm}$$

Example 10.10. *Determine the initial head loss in a rapid sand filter which is 65 cm deep and has a porosity of 0.42 and specific gravity of 2.66. It has a filtration rate of 160 lpm/m². Sieve analysis of filter sand, having a shape factor of 0.92 is given below.*

IS sieve No.	*1.40 – 1.00 mm*	*1.00 – 710 micron*	*710 – 600 micron*	*600 – 500 micron*	*500 – 425 micron*	*425 – 300 micron*	*300 – 250 micron*	*250 – 212 micron*	*212 – 150 micron*
% of sand retained	*1.5*	*6.5*	*15.4*	*17.8*	*20.6*	*18.2*	*12.8*	*5.0*	*2.2*

Take $\nu = 0.013\ cm^2/sec$

Solution : Filtration rate $= 160$ lpm/m^2

$= 0.16$ m/min. $= 16$ cm/min.

$= 0.267$ cm/sec.

$\therefore \quad v_s = 0.267$ cm/sec

Also, $$R = \frac{v_s \, . \, d}{\nu} = \frac{0.267 \times d}{0.013} \approx 20.5\, d$$

For $R < 1.9$, $$C_D = \frac{24}{R}$$

For $R > 1.9$, $$C_D = \frac{18.5}{R^{0.6}}$$

The computations are arranged in tabular form below.

IS sieve size	x	$d = \sqrt{d_1 \times d_2}$ *(cm)*	$R = 20.5\,d$	C_D	$C_D \, . \, \frac{x}{d}$
1.40 – 1.00 mm	0.015	0.1180	2.419	10.89	1.38
1.00 – 710 micron	0.065	0.0843	1.728	13.89	10.71
710 – 600 micron	0.154	0.0653	1.339	17.93	42.28
600 – 500 micron	0.178	0.0548	1.123	21.36	69.39
500 – 425 micron	0.206	0.0461	0.945	25.40	113.48
425 – 300 micron	0.182	0.0357	0.732	32.79	167.18
300 – 250 micron	0.128	0.0274	0.562	42.73	199.60
250 – 212 micron	0.050	0.0230	0.471	50.90	110.66
212 – 150 micron	0.022	0.0178	0.365	65.77	81.29
				Sum	795.97

Now, $$h_f = \frac{1.067}{g\,\phi_s} \frac{v_s^2 \, . \, L}{e^4} \Sigma \, C_D \frac{x}{d}$$

$$= \frac{1.067}{981\,(0.92)} \frac{(0.267)^2 \times 65}{(0.42)^4} (795.97)$$

$$= \mathbf{140.1 \ cm}$$

Example 10.11. *A bed of uniform sand, having particle size 0.65 mm dia. and specific gravity 2.66, porosity 0.42 and depth 65 cm is to be washed hydraulically. Compute (a) back wash rate so that expansion will be 50%. (b) head loss at this rate.*

Take $\nu = 1.3 \times 10^{-2}$ *cm²/sec and assume* $C_D = \dfrac{24}{R}$.

Solution :

Given : $L_e = 1.5\,L$

Now, $$\frac{L_e}{L} = \frac{1-e}{1-e_e}$$

$\therefore$ $$1.5 = \frac{1-0.42}{1-e_e}$$

From which $e_e = 0.613$

$$v_s = \frac{g}{18}\left(\frac{S_s-1}{\nu}\right)d^2 = \frac{981}{18}\left(\frac{2.66-1}{1.3\times10^{-2}}\right)(0.065)^2$$

$$= 29.4 \text{ cm/sec.}$$

Now $$e_e = \left(\frac{\overline{v}_s}{v_s}\right)^{0.22}$$

or $$\overline{v}_s = v_s \,.\, e^{4.545} = 29.4\,(0.613)^{4.545} \approx \mathbf{3.18 \text{ cm/sec.}}$$

Also, $$h_f = L\,(S_s-1)\,(1-e)$$
$$= 65\,(2.66-1)\,(1-0.42) \approx \mathbf{62.6} \text{ cm}$$

Example 10.12. *The filter of example 10.10 is to be back washed hydraulically at a rate just sufficient to fluidize the entire bed. Compute (a) the back wash rate (b) the head loss at this rate and (c) the depth of expanded bed. Take $\nu = 0.013$ cm²/sec*

Solution :

$$v_s = \left[\frac{4}{3}\frac{g}{C_D}(S_s-1)\,d\right]^{1/2}$$

For $R > 1.9$, $$C_D = \frac{b}{R^n} = \frac{18.5}{R^{0.6}}$$

$\therefore$ $$C_D = 18.5\left(\frac{\nu}{v_s d}\right)^{0.6} = \frac{1.366}{(v_s\,.\,d)^{0.6}}$$

$\therefore$ $$v_s = \left[\frac{4}{3}\frac{g\,(v_s d)^{0.6}}{1.366}(S_s-1)\,d\right]^{1/2}$$

or $$v_s^{0.7} = \left[\frac{4}{3}\frac{9.81}{1.366}(2.66-1)\,d^{1.6}\right]^{1/2} = 39.87\,d^{0.8}$$

$\therefore$ $$v_s = (39.87\,d^{0.8})^{1/0.7} \approx 193.5\,d^{1.143} \quad ...(i)$$

For the largest size particle, $v_s = 193.5\,(0.118)^{1.143} = 16.82$ cm/sec.

$\therefore$ $$\overline{v}_s = v_s\,.\,e^{4.545} = 16.82\,(0.42)^{4.545} = 0.326 \text{ cm/sec}$$

$\therefore$ $$e_e = \left(\frac{\overline{v}_s}{v_s}\right)^{0.22} = \left(\frac{0.326}{v_s}\right)^{0.22} = \frac{0.782}{v_s^{0.22}} \quad ...(ii)$$

$$\therefore \qquad h_f = L\,(S_s - 1)\,(1 - e)$$
$$= 65\,(2.66 - 1)\,(1 - 0.42) = \mathbf{62.6\ cm}$$

$$\text{Again,}\quad L_e = L\,(1 - e)\,\Sigma \frac{x}{1 - e_e}$$

Computations for $\frac{\Sigma x}{1 - e_e}$ are done in tabular form below :

Size (cm)	*x*	$d = \sqrt{d_1 \times d_2}$ *cm*	$v_s = 193.5\,d^{1.143}$ *cm/sec*	$e_e = \frac{0.782}{v_s^{0.22}}$	$\frac{x}{1 - e_e}$
0.14–0.1	0.015	0.1180	16.82	0.42	0.026
0.1–0.071	0.065	0.0843	11.45	0.46	0.120
0.071–0.06	0.154	0.0653	8.55	0.49	0.301
0.06–0.05	0.178	0.0548	7.00	0.51	0.363
0.05–0.0425	0.206	0.0461	5.74	0.53	0.441
0.0425–0.030	0.182	0.0357	4.29	0.57	0.421
0.030–0.025	0.128	0.0274	3.17	0.61	0.326
0.025–0.0212	0.050	0.0230	2.54	0.63	0.137
0.0212–0.015	0.022	0.0178	1.94	0.68	0.068
				Sum	2.203

$$\therefore \qquad L_e = L\,(1 - e)\,\Sigma \frac{x}{1 - e_e} = 65\,(1 - 0.42) \times 2.203$$
$$= \mathbf{83\ cm}$$

10.15. PRESSURE FILTERS

The pressure filter is a type of rapid sand filter which is in a closed container and through which the water passes under pressure. The pressure may vary from 3 to 7 kg/cm^2 and may be developed by pumping. It may be either horizontal type or vertical type. The diameter of vertical varies from 2 to 2.5 m and length varies from 2.5 to 8 m. The filter is operated similar to a gravity type rapid filter except that the coagulated water is usually applied directly to the filter without mixing, flocculation or conditioning. Fig. 10.16 (*a*) shows the diagrammatic sketch of horizontal type pressure filter while Fig. 10.16 (*b*) shows the vertical type pressure filter.

The uniformity co-efficient and effective size of filter sand is practically the same as that provided for rapid gravity filters, while the thickness of sand bed may vary from 50 to 60 cm. Gravel layers also follow the same practice as in gravity sand filters. The under drainage system may consist of pipe grids or false bottoms. Washing

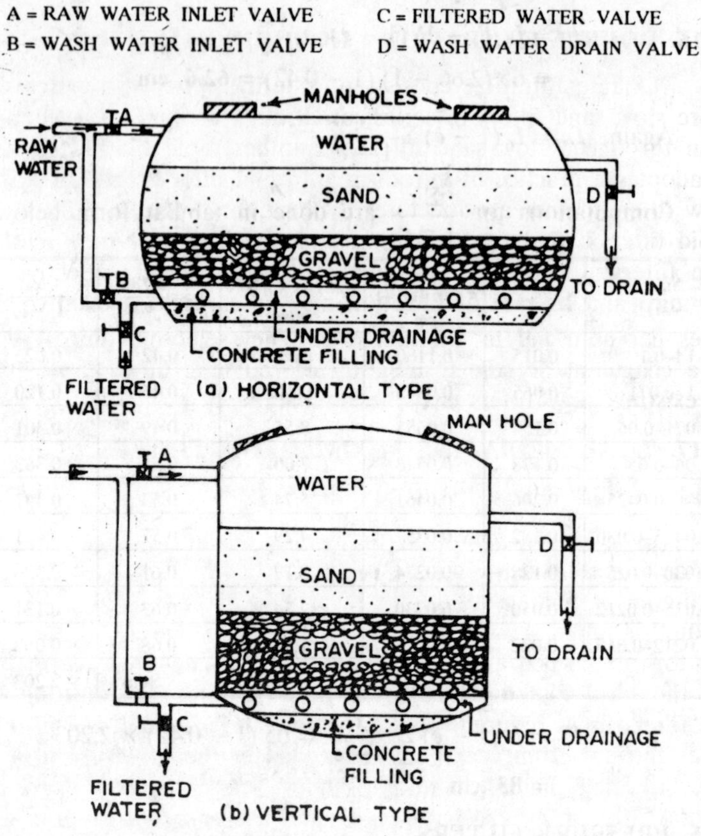

FIG. 10.16. DIAGRAMMATIC SECTIONS OF PRESSURE FILTERS

of filter media is accomplished by reversing the flow by manipulating the valves in the piping. Automatic pressure filters are also available in which backwashing is done automatically after a fixed interval of time or when the head loss has reached a given value.

Pressure filters are particularly advantageous for installations where water is received under pressure, as no pumping after filtration is required. Because the filter container is air tight, this filter may be placed on a pressure line. The only loss of head is that required to force the water through the filter. The filtration rate is much higher than the rapid gravity filter – the rate may vary from 6000 to 15000 litres per hour per m^2 of filter area. Due to this, they are considered as being unreliable in the removal of bacteria. They are, therefore, not used for treating municipal water supplies. They are mostly used in clarifying softened water at industrial plants and in treating swimming pool water that is recirculated.

10.16. DOUBLE FILTRATION : ROUGHING FILTER

Double filtration consists of filtering water through two or more slow sand filters in series, or through a rapid sand filter and then through a slow sand filter or another rapid filter. It is usual to adopt the practice of introducing a rapid filter before an existing slow filter, to increase the rate of filtration of the slow filter. The rapid filter in that case is known as a *roughing filter* or a *scrubber.* The filtering media of a roughing filter is coarser (effective size 0.8 mm) and its rate of filtration may be as high as 6000 to 8000 litres per hour per m^2 of filter area. They seldom require the use of a coagulant. By the installation of roughing filter, the capacity of existing slow filters may be doubled.

10.17. DUAL MEDIA AND MIXED MEDIA FILTERS

Dual Media Filters

In a rapid sand filter, there is a tendency of accumulation of finer sand at the top of the bed during backwashing. When water is filtered, it encounters the finer particles first, with the result that greater amounts of suspended matter and impurities are arrested near the top of the bed and the full depth of the media is not utilized. As the head loss builds up, the filter requires backwashing, and the filter run is reduced. This difficulty can be removed if filtration takes through from coarse-to-fine media, by the provision of a two layer filtering media, commonly known as *dual media.* In a dual media filter or two layer filter, a coarse layer of anthracite, about 30 to 40 cm thick, is placed above 20 to 30 cm thick layer of sand. Because of the lower specific gravity, beds of anthracite can be expanded or partially fluidised at about the same backwash rate as a sand bed when the size of the anthracite grains is about twice that of the sand grains. The size of anthracite used is usually 1.25 to 2.5 mm, placed over sand of 0.5 to 1.0 mm size. After backwashing, the two layers may have some mixing at the interface. Such a mixing is desirable since it reduces the voids size in the lower coal layer forcing it to remove floc which might otherwise pass through the coal layer. In the dual media filters, the filtrate quality can be as good as (or even better than) conventional rapid sand filtration, but filter runs can be $1\frac{1}{2}$ to 3 times as long. Alternatively, higher filtration rates can be achieved. However, a dual filter has the disadvantage of its inability to clean the deep coarse top layer with nominal rise rate. Washing can be satisfactorily done by resorting to air-water washing.

Capped Sand Filters

The performance of the existing rapid sand filters can be improved by *capping* sand filters with coal. For this, the top 15 cm layer of sand is removed and is replaced by 15 cm layer of anthracite. Such an improved filter, though inferior to the originally designed dual media filter, is better than the conventional rapid sand filter from the point of view of rate of filtation. It has been reported that "the more adverse the applied water conditions, relative to algae and floc, the more dramatic are the results obtained with anthracite capped filter run". Their performance is even much better for the worst water conditions. Savings in wash water were reported to be 100%. It has thus the benefits of longer filter runs, reduced use of washwater and better filter effluent quality.

Mixed Media filters

Mixed media filter was a result of further development of dual media filter. Ideally, the size of media particles and the pore space both should be uniformly graded from coarse to fine in the direction of flow. This is acheived in a mixed media filter using three materials : Coal at top, sand in the middle and garnet (or ilmenite) at the bottom, approximately in the proportion of 60, 30 and 10% respectively. In a total bed thickness of 75 cm the top layer of anthracite coal (specific gravity 1.4) has a thickness of 45 cm, middle layer of silica sand (specific gravity 2.65) has a thickness of 22.5 cm, and the bottom layer of garnet sand (specific gravity 4.2) of a thickness of 7.5 cm. The particle sizes range from 1.0 mm down to 0.12 mm from top to bottom of the bed.

It can be shown that for equal settling velocities, the required particle sizes for media of different density can be calculated from the relation

$$\frac{d_1}{d_2} = \left[\frac{\rho_2 - 1}{\rho_1 - 1}\right]^{1/2} \qquad ...(10.25)$$

After the backwashing, the three materials are mixed thoroughly and at each level of bed, there are particles of coal, sand and garnet. However, the top layer remains predominantly coal, the middle layer predominantly sand and the bottom layer predominantly garnet, though all the three are present at all depths. Fig. 10.17 shows the distribution of a properly designed mixed media bed, after backwashing. *Thus, a mixed media bed is not a three layered bed*. Ives has referred to mixed-media as "mixed up" media.

In a mixed media filter, the filtration is from coarse-to-fine and impurities and suspended matter are removed and stored through-

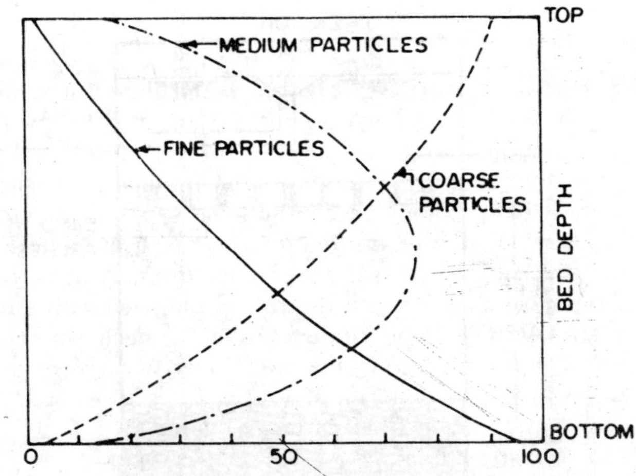

FIG. 10.17. DISTRIBUTION OF PARTICLES IN MIXED MEDIA FILTER AFTER BACKWASHING

out the full depth of bed. Due to the increased storage capacity of the mixed media bed, the length of filter run is highly increased. The total surface area of all the grains of the media in a 75 cm deep bed is usually not less that 2650 square millimetre for each square millimetre of plan area. The filteration rate in a mixed media filter is substantially increased, with greatly improved water quality.

Example 10.13. *For a mixed media filter, determine the particle sizes of anthracite and ilmenite which have settling velocities equal to that of silica sand of 0.4 mm in diameter. Take specific gravities of sand, anthracite and ilmenite as 2.65, 1.45 and 4.2 respectively.*

Solution

For the anthracite

$$d_1 = (0.4)\left[\frac{2.65-1}{1.45-1}\right]^{2/3} = 0.95 \text{ mm}$$

For the ilmenite

$$d_1 = (0.4)\left[\frac{2.65-1}{4.2-1}\right]^{2/3} = 0.26 \text{ mm}$$

Hence the anthracite smaller than 0.95 mm will remain above the silica sand of 0.5 mm, and grains of ilmenite larger than 0.26 mm would remain below the silica sand.

10.18. UPFLOW FILTERS

In the upflow filters, water flows in the upward direction. The upflow filtration has an obvious theoretical advantage, because coarse-to-fine filtration can be achieved in a single medium such

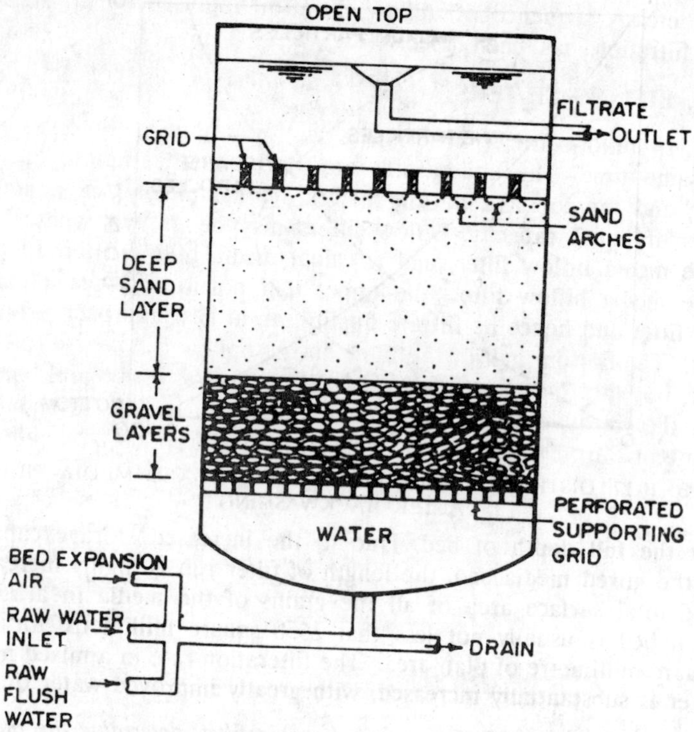

FIG. 10.18. UPFLOW FILTER WITH RESTRAINING GRID

as sand with almost perfect gradation of both pore space and grain size from coarse-to-fine in the upward direction of filtration. The filter can carry a greater load of impurity before backwashing because the large particles are held in the lower part of the media while the upper part deals with finer particles. Back-washing is also done in the *same direction* but at higher flow rates. During filtration, when the headloss above a given level exceeds the weight of the bed above that level, the bed lifts or partially fluidizes due to which the previously removed solids escape in the effluent. To minimize bed lifting, bed depth upto 2 m are used in Russia. Some plants use a metal grid, placed at the top, to prevent the sand expanding upwards during backwashing (Fig. 10.18). The media grains arch across the open space of the grid to restrain the bed against expansion. Inspite of the deep bed or provision of the grid, excessive pressure or sudden variations in pressure may break the sand bridge and cause the bed to expand. Due to this, upflow filters are not used for domestic or municipal water supply. These are used for process water, waste water treatment, deep well injection water and cooling water etc. A sequence of upflow filtration followed by downward flow filtration has been tried.

10.19. BIFLOW FILTERS

In biflow filters, filtration is in two opposite directions at the same time. The underdrains are placed in the middle of the filter, and the top and bottom halves are completely independent filters of equal capacity. Biflow filters may be of two types : (*i*) single media biflow filter, and (*ii*) dual media biflow filter. In the single media biflow filter, the upper half portion acts as a rapid sand filter and hence its filtrate quality can at best approach a rapid filter. The bottom half of the filter having upflow is a coarse-to-fine filter, but the finest materials near the outlet is some what coarser than the finest material used in rapid sand filter. Due to this the quality of filtrate is inferior to the rapid sand downflow filter above.

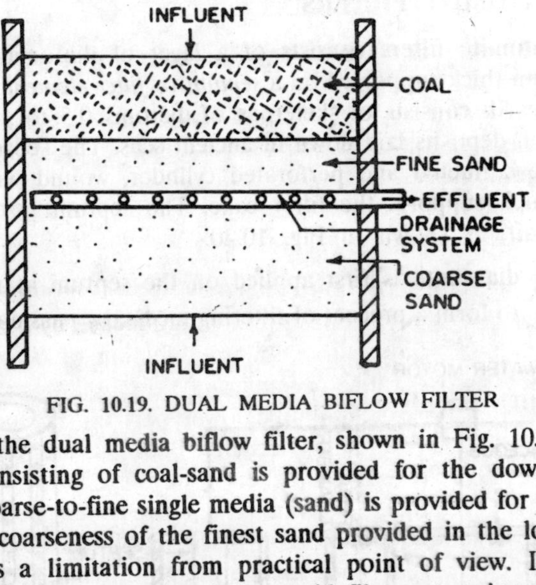

FIG. 10.19. DUAL MEDIA BIFLOW FILTER

In the dual media biflow filter, shown in Fig. 10.19, a dual media, consisting of coal-sand is provided for the downflow bed, while a coarse-to-fine single media (sand) is provided for the upflow bed. The coarseness of the finest sand provided in the lower single media has a limitation from practical point of view. If the sand is finer than that provided for a rapid filter, excessive amounts of sand will rise into coal bed during backwashing. Due to this, sand in the lower half is coarser. The quality of effluent from the either half will be inferior to that obtained from a mixed media filter.

10.20. MICRO-STRAINERS

A micro-strainer incorporates a specially woven stainless steel wire cloth mounted on the periphery of a revolving drum fitted with continous backwashing arrangements. Two of the grades widely used have limiting apertures of 23 microns and 35 microns respectively. Micro-strainers are useful for screening stored waters which do not contain plankton, algae and other microscopic sized particles. They are installed to the upstream of rapid gravity or slow sand filters whose output may thereby be increased by as much as 50%. In operation, the revolving drum is kept submerged in the flowing water to approximately two thirds of its depth. Raw water enters through the upstream open end of the drum and flows radially outwards through the microfabric leaving behind suspended solid content. The solids so strained are carried upwards on the inside of the fabric, with the help of a row of wash water jets which may use about 1% of total quantity of water strained. A single unit of 3 m diameter × 3 m wide drum may deal with about 50,000 to 80,000 litres/hour. It should be noted that micro-straining cannot remove colour or finely divided matter such as clay.

10.21. DIATOMITE FILTERS

Diatomite filters consists of a layer of diatomaceous earth about 3 mm thick supported on a *septum* or *filter element*. The diatomaceous earth consists of skeletons of diatoms 0.5 to 12 μ in size mined from deposits laid down in ancient seas. The septum consists of a hollow, flueted and perforated cylinder, wound with wire or plastic which supports the filter cake. The septum is enclosed in a filter unit, as shown in Fig. 10.20.

The diatomite is first applied on the septum at the rate of 300 gm/m^2 to form a precoat of filtering media, by passing it through

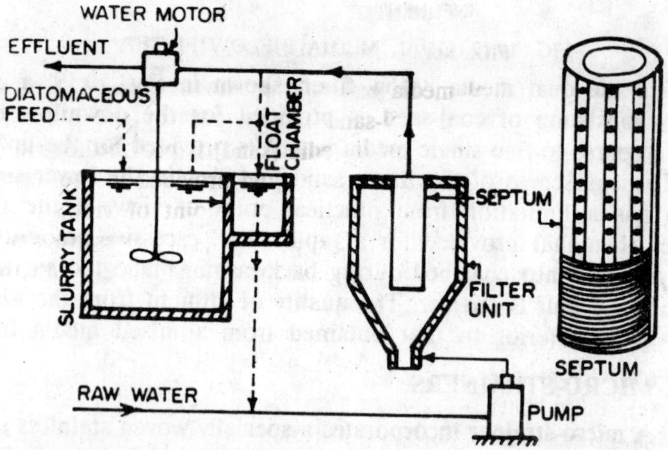

FIG. 10.20. FLOW DIAGRAM OF A SMALL DIATOMITE FILTER

water to be filtered in the form of a slurry. During this process, the filtered water is wasted. The resulting *pre-coat* is supported by the septum, which serves also as a drainage system. Water is strained through this precoat, unless the raw water contains so much turbidity that the unit may maintain itself only if diatomaceous earth, called *body feed*, is introduced continuously to the raw water, at the rate of 2 to 3 mg/litre, to preserve the open texture of the layer. The filter is backwashed by reversing the flow.

These filters are not useful for waters of high turbidity. They are not very much used for treating water for municipal supplies. They are found quite useful for temporary and shifting camps, because they require less space and are light in weight, and are therefore more mobile. They are also used as stationary units for swimming pools and general water supplies.

PROBLEMS

1. What do you understand by filtration ? What actions take place during filtration ?

2. Classify various types of filters. Differentiate between the slow sand and rapid sand gravity filters. Compare the two types of the filters in a tabular form.

3. Describe, with the help of sketches, a slow sand filter. Explain its working.

4. Describe with the help of a neat sketch, the component parts of a rapid sand gravity filter. Explain its working, including the method of washing.

5. Write note on wash water troughs. How do you design them ?

6. What do you understand by loss of head and negative head in a rapid sand filter ? What are their permissible values ? What will happen if the negative head is excessive ?

7. Explain various filter troubles. How are they avoided or solved?

8. Explain the working of a rapid sand gravity filter.

9. Write notes on the following :

(*a*) Roughing filter.
(*b*) Dual media filter.
(*c*) Mixed media filter.
(*d*) Upflow filter.
(*e*) Biflow filter.
(*f*) Microstrainers.
(*g*) Diatomite filter.

10. Explain the underdrainage system of rapid sand filter (gravity type) and mention the general rules of design for such a system.

11. Find the area of a slow sand filter for a town having population of 20000 with an average rate of demand of 100 litres per head per day.

12. A city has design population of 80,000. If the average rate of demand is 160 litres per head per day, design completely a rapid sand gravity filter including the wash water troughs and the drainage system. Assume suitable data if required.

13. The population of a city is 50,000 and the per capita consumption is 130 litres/day. Calculate the following in respect of the rapid sand filter for the above data :

(*a*) Total area of filters.

(*b*) Number and dimensions of each filter.

(*c*) Quantity of air for air wash per filter bed.

(*d*) Backwash water per filter bed after air wash.

(A.M.I.E. 1975)

14. A filter unit is 4.5 m by 9.0 m. After filtering 10000 cubic metres per day in 24 hour period, the filter is back washed at a rate of 10 l/sq. m/sec, for 15 min. Compute the average filtration rate, quantity and percentage of treated water used in washing and the rate of wash water flow in each trough. Assume 4 troughs.

(IAS 1985)

Ans. [10507 $l/hr/m^2$; 364.5 m^3; 3.645 %; 101.25 l/sec]

15. A bed of uniform sand having particle size 0.6 mm dia., specific gravity 2.65, porosity 40% and depth 70 cm is to be washed hydraulically. Compute (*a*) back wash rate so that expansion will be 50% (*b*) head loss at this rate. Take $\nu = 1.3 \times 10^{-2}\ cm^2/sec$

Assume $C_D = 24/R$.

Ans. [2.45 cm/sec ; 69.3 cm]

11

Disinfection

11.1. INTRODUCTION

When water comes out of filter plants, it may contain bacteria and other micro-organisms, some of which may be pathogenic. It is therfore necessary to disinfect water to kill bacteria and other micro-organisms, and thus prevent water-borne diseases. However, disinfection is not a substitute for filtration. Disinfection follows filtration. When the aim is to kill all the micro-organisms in water so as to make it sterile, the process is known as *sterilization*. The aim of *disinfection*, however, is to reduce the number of micro-organisms to a safe limite. Disinfection usually requires complicated mechanisms that need the attention of skilled operators to avoid breakdown and incorrect dosage.

Suitability of disinfectant

Disinfectant is a substance that is used for disinfection of water. A water disinfectant should meet the following requirments.

1. The disinfectant should be effective in killing the micro-organisms potentially present in water within the contact time available, the range of water temperatures encountered and the anticipated fluctuations in composition, concentration and condition of water being treated.
2. The disinfectant should be readily available at reasonable cost.
3. It should be safe to handle, and its method of application should be simple.

4. It should not render the water toxic, unpalatable or objectionable, aesthetically or otherwise, for its intended use.
5. It should have ability to persist in residual concentrations as a safeguard against recontamination.

Mechanism of disinfection

The *disinfectant* either destroys or inactivates the micro-organisms, by way of the following four mechanisms :

(*i*) Damage to cell wall of micro-organisms
(*ii*) Alteration of cell permeability
(*iii*) Changing the colloidal nature of the cell protoplasm, and
(*iv*) Inactivation of critical enzyme systems responsible for metabolic activities.

11.2. METHODS OF DISINFECTION

Various methods of disinfection can be broadly classified under two heads : (*a*) physical methods, and (*b*) chemical methods.

(*a*) Physical methods

Physical methods include the following :

1. Disinfection by heat : Boiling of water

2. Disinfection by light : Sunlight is a natural disinfectant. *Irradiation* by *ultraviolet rays* intensifies disinfection.

(*b*) Chemical methods

Chemical methods of disinfection include the following.

1. Oxidizing chemicals : They comprise

(*i*) The halogens – chlorine, bromine and iodine
(*ii*) Ozone, and
(*iii*) Other oxidants such as pottasium permanganate and hydrogen peroxide

2. Metal ions : Such as silver and copper ions. Silver ions are bactericidal, though they are neither viricidal nor cysticidal. Disinfection with silver ions is generally done in low concentration– as low as 15 μg per litre (parts per billion), and hence the process is very slow.

3. Alkalis and acids : It is well known that pathogenic bacteria do not last long in highly alkaline (pH >11) or highly acidic (pH 3) waters. The destruction of bacteria by caustic lime incidental to lime softening is an example.

4. Surface active chemicals : Among *surfactants*, the *cationic detergents* are only weakly destructive. The *neutral detergents* occupy an intermediate position.

Out of the various methods mentioned above for disinfection, *chlorination* is most commonly adopted. Hence in water works practice, the term chlorination is sometimes used in place of disinfection.

11.3. MINOR METHODS OF DISINFECTION

These include the following methods : (1) Boiling method (2) Excess lime treatment (3) Silver treatment (4) Ultra-violet ray treatment, and (5) Potassium permanganate tratments.

1. Boiling Method : This method is the most effective in the complete sterilization of water, since boiling of water kills all the bacteria and other micro-organisms. However, this is not possible on large scale, as for public water supplies. During epidemics, the consumers are advised to boil the water before drinking it.

2. Excess Lime Treatment : It has been found that when pH of water is greater than 9.5, E-coli present in water will die. Hence when enough lime is added to bring the pH to this figure, sterilization of E-coli and other bacteria occurs. The removal of bacteria is upto 99.93 or even 100%. However, the excess lime has to be removed by suitable method of recarbonation, before supplying it to the consumers. The necessary dose of lime is between 10 to 20 ppm of calcium oxide.

3. Silver Treatment : Water can also be disinfected with silver by the electro-catadynic action or oligodynamic action. Silver when immersed in water has been observed to exert an inhibiting action on bacterial life. In this approach, silver ions, with or without 'activators' such as palladium or gold, are deposited on particles of granular activated carbon, a material which has an exceptionally high surface area for the deposit of silver. Bacteria laden water contacting the silver impregnated carbon release minute quantities of silver – from 25 to 40 parts per billion (ppb) which serves as a disinfectant. The contact time varies from 10 to 60 minutes. However, since silver is costly, this method may be used only in small installations or for private individual houses.

4. Ultra-Violet Ray Treatment : The ultra-voilet ray offers an effective method for sterilization of water, since the light is effective in killing both the active bacteria as well as spores. Though sun is the most powerful source of ultra-voilet rays, it requires a large exposure area and long time. Hence ultra-voilet rays are generated by machines consisting of mercury-vapour lamps enclosed in a quartz globe. The destructive power of the rays begins in the blue-green region of the spectrum with a wave length of 0.490 μ and increases in effectiveness to 0.149 μ. For disinfection to be effective, water should flow in a thin clear film close to the sterilizing ray. The depth of water should not exceed 10 cm, it should be colourless and its turbidity should not be more than 15 to 20 ppm. The method is costly, and hence not commonly used except for private buildings, office buildings, institutions, and particularly swimming pools etc.

5. Potassium permanganate treatment : This is a common method of disinfection in rural areas, where mostly the water supplies are from wells which contain leser amounts of bacteria. Pottassium permanganate ($K\,Mn\,O_4$) is dissolved in a bucket of well water, and the bucket full of this water is mixed with the well water thoroughly. Potassium permanganate not only kills the bacteria but it also helps in oxidising the taste producing organic matter. The normal dose of $KMnO_4$ varies from 1 to 2 mg/l with a contact period of 4 to 6 hours. Addition of $K\,Mn\,O_4$ to water imparts it some pink colour. However well water should not be used during the first 48 hours of the addition of $K\,Mn\,O_4$ to it.

11.4. CHLORINATION

The use of chlorine has become practically universal in the disinfection of water. It is cheap, reliable, and presents no great difficulty in handling. According to AWWA, the earliest recorded use of chlorine directly for water disinfection was on an experimental basis, in 1896 at Louisville, and in 1897 at England. Its first continuous use was in Belgium, in 1902, for the dual objective of aiding coagulation and making water biologically safe.

The precise action by which chlorine kills bacteria in water is not known. Several theories have been put forward. A commonly accepted theory is the *Enzymatic Hypothesis.* According to this, chlorine compounds formed when chlorine is added to water, interfere with certain enzymes in the bacterial cells which are vital for the support of life. Enzymes invloved are created within the cell plasm, and hence action of disinfection proceeds in two stages – (*i*) penetration of the cell wall by the disinfectant, and (*ii*) its reaction with the enzymes.

When chlorine is dissolved in water at temperatures between 49°F and 212°F, it reacts to form hypochlorous and hydrochloric acids within few seconds :

$$Cl_2 + H_2 \Leftrightarrow HOCL + HCl \text{ (Hydrolysis)} \quad ...(11.1)$$

The hypochlorous acid, HOCl, ionizes or dissociates into hydrogen (H^+) ions and hypochlorite (OCl^-) ions :

$$HOCl \Leftrightarrow H^+ + OCl^- \text{ (Ionization)} \quad ...(11.2)$$

It is the hypochlorous acid and the hypochlorite ions which accomplish disinfection. Thus, when chlorine is added to water, all

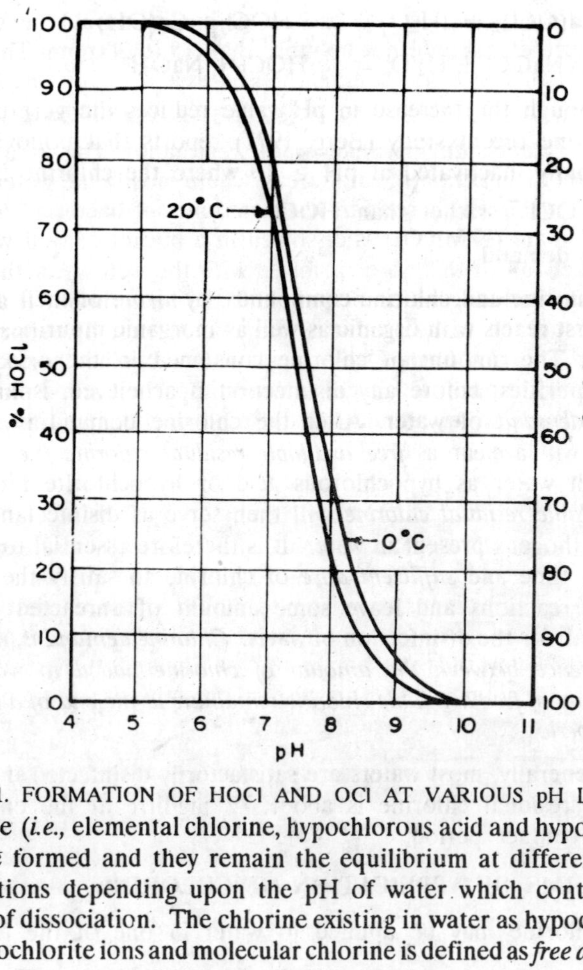

FIG. 11.1. FORMATION OF HOCl AND OCl AT VARIOUS pH LEVELS.

the three (*i.e.*, elemental chlorine, hypochlorous acid and hypochlorite ions) are formed and they remain the equilibrium at different connecentrations depending upon the pH of water which controls the amount of dissociation. The chlorine existing in water as hypochlorous acid, hypochlorite ions and molecular chlorine is defined as *free available chlorine.* The actual disinfection efficiency depends upon the ratio of hypochlorous acid to the hypochlorite ions, being greater at a high ratio. At pH < 5, chlorine exists as *elemental* or *molecular chlorine*. At pH between 5 to 10, it remains in the form of both hypochlorous acid and hypochlorite ions – the concentration of the former being reduced and that of the latter is increased as pH increases, as shown in Fig. 11.1. At pH > 10, only hypochlorite ions exist. Since the hypochlorous acid is about 80 times as effective as the hypochlorite ions, the pH of the water to be treated should be maintained at less than 7 to prevent great ionization of HOCl.

The hypochlorite salts also ionize in water and yield hypochlorite ions, which establishes equilibrium with hydrogen ions

$$Ca(OCl)_2 + 2H_2O \longrightarrow 2\,HOCl + Ca(OH)_2 \qquad ...(11.3)$$

$$NaOCl + H_2O \longrightarrow HOCl + NaOH \qquad ...(11.4)$$

Though the increase in pH value reduces the germicidal efficiency, one recent study (Berg, 1972) reports that poliovirus was more rapidly inactivated at pH $\geq$ 10 where the chlorine is in the form of OCL^- rather than HOCl.

Chlorine demand

Chlorine and chlorine compounds, by virtue of their oxidising power, first reacts with organic as well as inorganic inpurities present in water. The amount of chlorine consumed in the oxidation of these impurities, before any disinfection is acheieved, is known as *chlorine demand* of water. After the chlorine demand is fulfilled, chlorine will appear as *free available residual chlorine* (*i.e.* the one existing it water as hypochlorous acid or hypochlorite ions). The *free available residual chlorine* will then serve as disinfectant to kill to the pathogens present in water. It is therefore essential to provide sufficient time and *sufficient dose* of chlorine to satisfy the various chemical reactions and leave some amount of unreacted chlorine as residual, for the disinfection of water. *Chlorine demand is, therefore, the difference between the amount of chlorine added to water and the quantity of free available chlorine remaining at the end of a specified contact period.*

Generally, most waters are satisfactorily disinfected if the free available residual chlorine is about 0.2 mg/litre at the end of 10 minutes contact period.

11.5. FORMS OF APPLICATION OF CHLORINE

Chlorine may be applied to water in one of the following forms :

(*i*) As bleaching powder or hypochlorite
(*ii*) As chloramines
(*iii*) As free chlorine gas
(*iv*) As chlorine dioxide.

1. Bleaching powder. Bleaching powder or calcium hypochlorite, $Ca(OCl)_2$, is a chlorinated lime containing about $33\frac{1}{3}\%$ of available chlorine, when freshly made. When chlorination was first introduced, bleaching powder was exclusively used. However, bleaching powder is not stable, and it loses its strength during storage or expsoure of air. It is therefore used only on small installations or under emergency conditions. Equation 11.3 indicates the reaction with bleaching powder while Eq. 11.4 gives the reaction with sodium hypochlorite NaOCl. The process of chlorination with hopochlorites is known as *hypochlorination.* Commerical compounds such as HTH (high test hypochlorites), Pittchlor, Pittcide, Hoodchlor etc. are used instead of bleaching powder. The high-test hypochlorites, having an available chlorine contents of 65 to 70%, are stable, easily soluble and free flowing and nonhygroscopic. The oxidizing power of hypochlorite is represented in terms of *available chlorine.*

Hypochlorites are applied to water as a solution by means of hypochlorite feeding apparatus. Fig. 11.2 shows an *emergency feeder* consisting of a solution tank connected to a constant level feeding tank. The doze of the solution is adjusted by means of an adjustable pinch clamp.

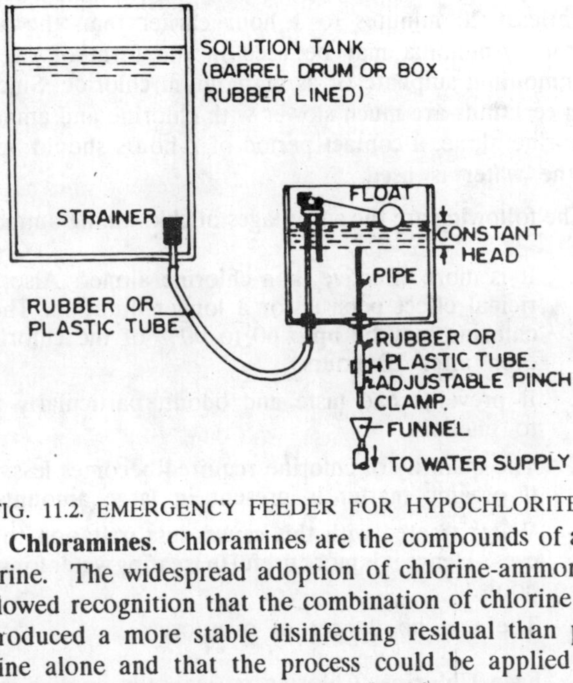

FIG. 11.2. EMERGENCY FEEDER FOR HYPOCHLORITES

2. Chloramines. Chloramines are the compounds of ammonia and chlorine. The widespread adoption of chlorine-ammonia treatment followed recognition that the combination of chlorine and ammonia produced a more stable disinfecting residual than produced by chlorine alone and that the process could be applied to limit the development of objectionable tastes. In this treatment, ammonia as NH_3 is added to the water just before the chlorine is applied. The usual proportions are 1 part of ammonia to 4.5 parts of chlorine by weight. The following reactions take place :

$$H_2O + Cl_2 = HCl + HOCl \quad ...(11.5)$$

$$NH_3 + HOCl = H_2O + NH_2Cl \text{ (Monochloramine)} \quad ...(11.6)$$

$NH_2Cl + HOCl = H_2O + NHCl_2$ (Dichloramine) ...(11.7)

$NHCl_2 + HOCl = H_2O + NCl_3$ (Trichloramine) ...(11.8)

The formation of a particular type of chloramine depends upon the pH range of water as indicated below :

pH Range	*Chloramine formed*
<4.4	Trichloramine or (Nitrogen trichloride) only.
4.4 to 5	Dichloramine only.
5 to 8	Both monochloramine and dichloramine in equal quantities at pH 7.
>8.5	Monochloramine only.

Though ammonia dissolves quickly in water, but it does not diffuse easily. Hence it should be mixed with the help of mechanical means, atleast 20 minutes to 1 hour earlier than the application of chlorine. Ammonia may be used in the form of gas or liquid or as ammonium sulphate or as ammonium chloride. Since the disinfecting reactions are much slower with chlorine and ammonia than with chlorine alone, a contact period of 2 hours should be provided before the water is used.

The following are the advantages of chloramine-ammonia treatment :

1. It is more effective than chlorine alone. Also, its bactericidal effect persists for a longer duration. The residual chlorine may be upto 60 to 90% of the chlorine added, even after 20 hours.
2. It prevents bad taste and odour, particularly those due to phenols.
3. The quantity of chlorine required becomes less, especially if organic matter is present in large amounts.
4. Water treated with this causes less irritation to nose and eyes. Hence it is more useful in treating water for swimiming pools.
5. There is no danger of overdose.

3. Free Chlorine. Chlorine is generally applied in gaseous form or in liquid form. Gaseous chlorine is a greenish-yellow poisonous substance, with a typical odour, and is about 2.48 times heavier than air. Liquid chlorine is amber coloured oily liquid and about 1.44 times as heavy as water. Unconfined liquid chlorine rapidly vapourises to gas; 1 volume of liquid yields about 462 volumes of gas. When chlorine gas is subjected to a pressure of 7 kg/cm^2, it [illegible] into liquid. Hence chlorine is stored and supplied in liquid form in metal containers under pressure of 10.5 kg/ cm^2. Since liquid chlorine is highly corrosive, the cylinders containing liquid chlorine are provided with special fittings. The cylinders are built to withstand bursting pressures of 35 kg/cm^2 corresponding to a temperature of 190° F.

Chlorine gas is respiratory irritant. It can cause varying degrees of irritation to skin, mucous membrane and respiratory system. In extreme cases, death can occur from suffocation. Hence it is desirable to have gas masks in good condition available wherever chlorine is stored or used. Liquid chlorine may cause skin and eye burns upon contact with these tissues. Dangerous compounds may be formed when chlorine comes in contact with other gases. If the room is poorly ventilated and contains carbon monoxide, escape of chlorine may result in the formation of phosgene – a highly toxic gas. Hence chlorine cylinders should be stored in a cool well-ventilated room.

The Chlorine dose depends upon the following factors :

(*i*) Organic matter present in water to be disinfected,

(*ii*) pH value of water.

(*iii*) amount of carbon di-oxide present in water,

(*iv*) temperature.

and (*v*) time of contact.

Our of these, the pH of water has the most important role to play. The State of New York recommends that for drinking water at pH 7.0, the concentration of free available chlorine after 10 minutes detention time should be 0.2 mg/litre, while at pH equal to 8.0, it should be 0.4 mg/litre. At the pH of 7 and 8 the recommended combined available chlorine residuals are 1.5 and 1.8 mg/litre, respectively after a contact period of 60 minutes.

4. Chlorine Dioxide (ClO_2). According to Ridenour and Inglos (1947), bactericidal properties of chlorine dioxide is greater than chlorine. The chlorine dioxide gas is unstable, and is therefore produced at the point of use by passing chlorine gas through sodium chlorite. The following reaction takes place :

$$2\,NaClO_2 + Cl_2 = 2NaCl + 2ClO_2 \qquad ...(11.9)$$

Until present time, the use of chlorine dioxide has been limited to special water treatment applications such as oxidation of iron, manganese, and phenolic and chlorophenolic compounds, for control of algae. Chlorine dioxide theroetically has about 2.5 times the oxiding power of chlorine. It does not react with organic materials to produce chloroform, a potential carcinogen. It also does not react with ammonia. It is entirely harmless in aqueous solution. The dosage of chlorine dioxide varies from 0.5 to 1.50 p.p m. and its action is unaffected by pH values between 6 to 10. Its efficiency is not impaired (as is that of HOCl) by a high pH environment. It is highly effective for the control of certain taste and odour problems. However, due to its high cost of production and application, it is rarely applied solely for the purpose of disinfection.

Example 11.1. *Calculate the quantity of bleaching powder required per day for disinfecting 4 million litres/day. The dose of chlorine has to be 0.5 p.p.m. and the bleaching powder contains 30% of available chlorine.*

Solution:

Chlorine required @ 0.5 p.p.m.

$$= \frac{0.5 \times 4 \times 10^6}{10^6} = 2 \text{ kg.}$$

Since the bleaching powder contains 30% of chlorine, amount of bleaching powder

$$= \frac{2 \times 100}{30} = \mathbf{6.67\ kg.}$$

Example 11.2. *Chlorine usage in the treatment of 25000 m^3/day is 9 kg/day. The residual chlorine after 10 minuts contact is 0.2 mg/l. Calculate the dosage in milligrams per litre and the chlorine demand of the water.*

Solution.

Water treated per day = 25000 m^3

$= 25 \times 10^6$ litres/day

Chlorine consumed per day = 9 kg.

$= 9 \times 10^6$ mg/day

$\therefore$ Chlorine used per litre of water $= \dfrac{9 \times 10^6}{25 \times 10^6}$

$= \mathbf{0.36\ mg/l}$

Also, residual chlorine = 0.2 mg/l

$\therefore$ *Chlorine demand* = 0.36 – 0.2= **0.16 mg/l**

11.6. APPLICATION OF CHLORINE

There are two methods of application of chlorine to water to be disinfected :

(*i*) Chlorine gas may be fed directly to the point of application to the water supply, or (*ii*) The gas may be first dissolved in a small flow of water and the solution is then fed to the point of application. The first method is less expensive but is unsatisfactory because of poor diffusions of chlorine in water. Also, it is found that at temperatures below 49.2° F (10° C) crystalline hydrates of *ice* of chlorine gas are formed. Hence when chlorine is directly fed through pipe lines and if tempreature falls down, choking of pipes leading chlorine may take place. Hence only second method is used. Chlorine is fed or applied through apparatus known as *chlorinator*. A satisfactory chlorinator should be able to feed chlorine at the desired rate irrespective of the changes in the pressure in the supply cylinder.

Fig. 11.3 shows the diagrammetic sketch of an automatic vacuum chlorinator. It essentially consists of a glass bell jar (*A*) placed in tray (*B*) in which a water seal is maintained with some constant water level. Chlorine gas is supplied to the bell jar from the chlorine cylinder through a float controlled needle valve (*C*).

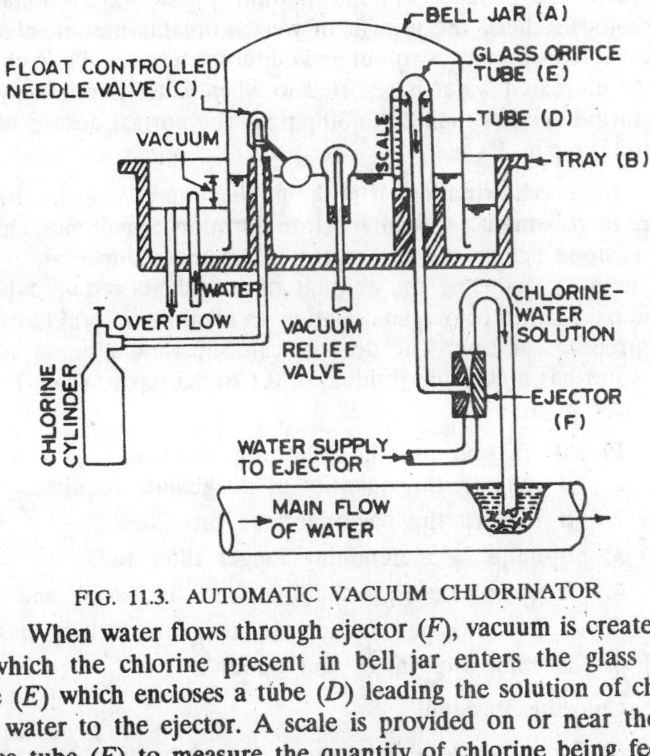

FIG. 11.3. AUTOMATIC VACUUM CHLORINATOR

When water flows through ejector (*F*), vacuum is created, due to which the chlorine present in bell jar enters the glass orifice tube (*E*) which encloses a tube (*D*) leading the solution of chlorine and water to the ejector. A scale is provided on or near the glass orfice tube (*E*) to measure the quantity of chlorine being fed. The water level in the glass orifice tube above the water level in the bell jar gives a measure of the differential head on the orifice. The chlorine-water solution from the ejector is fed directly to the main flow of water. A vacuum relief valve is also provided.

11.7. FORMS OF CHLORINATION

Depending upon the stage at which chlorine is applied to water, chlorination may be of the following forms :

(1) Plain chlorination.
(2) Pre-chlorination.
(3) Post-chlorination.
(4) Double or multiple chlorination.
(5) Break point chlorination.
(6) Super chlorination.
(7) Dechlorination.

1. Plain Chlorination. Plain chlorination is the application of chlorine to plain or raw water supply as it enters the distribution system. It also includes the chlorination of raw waters in tanks or reservoirs to check the growth of weeds, organic matter, algae and bacteria. It also removes colour and odour from water. Plain chlorination to untreated water is resorted to when water is relatively clear, with turbidities less than 20 to 30 p.p.m. The normal dose is between 0.5 to 1 p.p.m.

2. Prechlorination. It is the application of chlorine to water before its treatment – specially before filtration. Sometimes, chlorination is done before the raw water enters the sedimentation tanks. This helps in reducing the amount of coagulants required beacuse of the oxidization of oraganic matter. In alkaline water, chlorination may precede aeration. The dose of chlorine should be so adjusted that water has a chlorine residual of 0.1 to 0.5 p.p.m. when it enters the filter plant.

Prechlorination has the following advantages :

1. It reduces the quantity of coagulants required.
2. It reduces the bacterial load on filters.
3. It helps in maintaining longer filter runs.
4. It controls the algae and planktons in basins and filters.
5. It prevents putrefaction of sludge in settling basins.
6. It eliminates tastes and odour.

Chlorine Residual

(*i*) For satisfactory disinfection, prechlorination is done so as to maintain 0.3 to 0.4 mgl free available chlorine throughout the treatment.

(*ii*) At normal pH values, the free available residual at plant effluent should be 0.2 to 0.3 mg/l.

(*iii*) At higher pH of 8 to 9, the free available residual, required for complete bacterial kill, with 10 minutes contact time, should be 0.4 mg/l. However, for 30 minutes contact time, the dosage reduces to 0.2 to 0.3 mg/l.

(*iv*) For inactivation of cysts of E. histolytica, higher doses of 0.5 mg/l may be required.

(*v*) If the raw water is infected with nematodes, water should be prechlorinated for 6 hours to maintain a free available residual of 0.4 to 0.5 mg/l, to render the nematodes immobile, so that they can be removed by settling processes.

3. Post chlorination. It is the application of chlorine to water after its treatment. This is the standard form of chlorination in which chlorine is added to water as it leaves the rapid filters and before it enters the distribution system. The dose of the chlorine should be so adjusted that the residual chlorine is about 0.1 to 0.2 p.p.m. before water enters the distribution system. It is useful for protection against contamination from cross-connections.

4. Double or multiple chlorination. Double or multiple chlorination refers to the application of chlorine at two or more points in the purification process. Generally, double chlorination is resorted to, in which chlorine is applied just before water enters the sedimentation tanks, and after it leaves the filter plants. This is done specially when raw water is highly contaminated and contains large amounts of bacterial life and other organic matter. The advantages of double chlorination are similar to those of prechlorination. In addition, the maintenance of two chlorinating plants serves as a factor of safety.

5. Break Point Chlorination. When chlorine is applied to water, two actions take place one after the other :

(*i*) It kills bacteria and disinfection is effected, and

(*ii*) It oxidizes the organic matter. When chlorine is added to pure water which has no chlorine demand, a curve, such as line *A* in Fig. 11.4, is obtained between the applied chlorine and residual chlorine relationship. Generally, however, water has a chlorine demand, with the result that a curve *B* (Fig. 11.4) is obtained between residual chlorine and applied chlorine. The chlorine first performs the function of removing bacteria. During the disinfection process, the amount of residual chlorine will be less in the beginning (marked by stage I in Fig. 11.4), during which various chemicals such as ions of ferrous iron, sulphides or nitrites present in water, will be oxidised, while

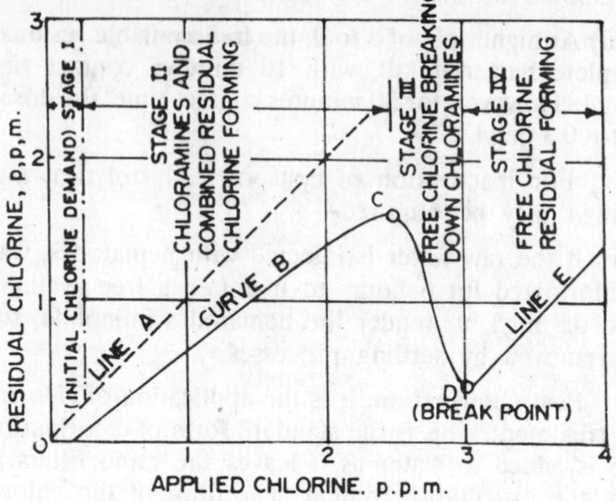

FIG. 11.4. BREAK POINT CHLORINATION

during stage II, the combined residual chlorine will gradually increase as the demand for disinfection is satisfied, till a point *C* is reached where the amount of combined residual chlorine will be maximum. This stage of point *C* is sometimes accompanied by bad taste and odour. Stage II reflects the forming of combined residuals as the ammonia or amines react with HOCl that has formed. Further increase in the applied chlorine will result in decrease in the residual chlorine, indicating the beginning of the second action, *i.e.* oxididation of organic matter present in water. During this action, the relation between applied-residual chlorine is represented by curve *CD*, wherein free chlorine breaks down chloramines changing them to nitrogen compounds while the chlorine residual actually drops. At point *D*, the bad smell and taste suddenly disappear and the oxidation of organic matter is also complete. The residual chlorine has its minimum value. Further addition of applied chlorine results in increase in the residual chlorine as represented by line *E* the slope of which will be 45°. Point *D* on the curve represents *break point* since further addition of cholrine break through water and appears as residual chlorine. Actually, upto point *C* on curve *B*, chloramines have been recorded as residual chlorine while at point *D* (break point) *true residual free chlorine* is revealed. *The break point in the chlorination of water may be defined as the point on applied-residual chlorine curve at which all, or nearly all, the residual chlorine is free chlorine.* Free chlorine residual is that part of the total residual remaining in water (after a specified contact period) that will react chemically or biologically as hypochlorous acid or hypochlorite ion.

Break point chlorination (*or free residual chlorination*) has practical significance since application of chlorine at or slightly higher than the break point concentration will have the following advantages :

(*i*) It will remove taste and odour.

(*ii*) It will have adequate chlorine residual,

(*iii*) It will leave a desired chlorine residual,

(*iv*) It will complete the oxidation of ammonia and other compounds,

(*v*) It will remove colur due to organic matter by about 30%, and

(*vi*) It will remove manganese.

The break-point stage should be determined by laboratory tests; it is represented by instantaneous yellow colour if the orthotolidine test is applied. In some cases, a distinguishable break point is not obtained, while in other cases, changes in the quality of raw water may effect rapid changes in break point. A recognizable break point may be induced by addition of ammonia to water. Generally, the break point lies between 3 to 7 p.p.m. of chlorine dose, though this is greatly influenced by the quantity of free ammonia present in water.

6. Super-chlorination. Super-chlorination is the application of chlorine beyond the stage of break point. The addition of chlorine sufficient to give a residual chlorine content of 1 to 3 p.p.m. has proved useful to destroy odours and tastes resulting from chloro-products formed between the decomposition products from vegetable matter and algae. Excess chlorine may be added at any point or points of chlorination, though it is usually applied after filtration. Super chlorination is followed by a retention or contact period of 30 to 60 minutes when the residual is discharged by means of dechlo-rination agents. Super chlorination is adopted when there is an epidemic in the locality, or when water is liable to sudden fluctuations in chlorine demand, containing a high concetration of organic impurities or when water contains cysts of *E. histolytica*, an organism causing amoebic dysentry.

7. Dechlorination. It is the process of removing excess chlorine from water before distribution to the consumers to avoid chlorine tastes. As stated above, dechlorination should be done in such a way that some residual chlorine remains in water. Dechlorination is achieved either by aeration or by the use of chemicals such as sodium thio-sulphate, sodium bio-sulphate, sodium sulphite, activated carbon, potassium permanganate or sulphur dioxide in gas or in liquid form. Sulphur dioxide and sodium disulphite are generally preferred for large supplies. Sodium dioxide gas is applied practically in the same manner as chlorine, with a contact period of dechlorination of not less than 10 to 15 minutes ; generally 0.3 to 0.6 p.p.m. of sulphur dioxide may be required. By filtering super-chlorinated water through beds of granular activated carbon, excess chlorine oxidizes the carbon to carbon dioxide, due to which odour, taste and colour are successfully removed.

Example 11.3. *Results of chlorine demand test on a raw water are given below. Determine the break-point dosage and the chlorine demand.*

Sample No.	*Chlorine dosage mg/l*	*Residual chlorine after 10 min. contact (mg/l)*
1	*0.2*	*0.18*
2	*0.4*	*0.34*
3	*0.6*	*0.48*
4	*0.8*	*0.46*
5	*0.9*	*0.27*
6	*1.0*	*0.18*
7	*1.2*	*0.38*
8	*1.4*	*0.58*
9	*1.6*	*0.78*

Solution. (Fig. 11.5) Fig. 11.5 shows the curve plotted on the basis of the above data.

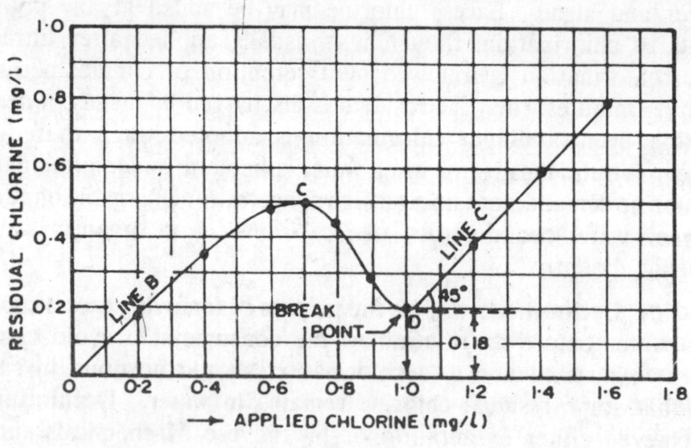

FIG. 11.5.

From the curve, we find that break point occurs at point D, at which the applied chlorine = 1.0 mg/l.

∴ Break point dosage= **1.0 mg/l**

∴ Chlorine demand at break point= 1.0 – 0.18= 0.82 mg/l.

It is observed that since the slope of curve C is 45°, the chlorine demand (= 0.81 mg/l) remains *constant* after the break point, since all additional chlorine added after point D appears as free chlorine.

Even from the data, at a dose of 1.2 mg/l, the residual chlorine = 0.38 mg/l.

Hence residual chlorine= 1.2 – 0.38 = **0.82 mg/l.**

11.8. TESTS FOR FREE AND COMBINED CHLORINE

Several tests have been developed to determine chlorine residuals in water, but the following two tests are usually employed :

1. Orthotolidine test,
2. Starch-iodide test.

The *orthotolidine method* is the common test employed to measure chlorine residuals, though the method preferred in the 'standard methods' is the cumbersome starch-iodine titration method.

1. Orthotolidine test. Orthotolidine is a colourless organic liquid that is oxidized by chlorine into a yellow coloured compound called *holoquinone* :

Orthotolidine + Chlorine → Holoquinone.

Water sample is collected after 10 minutes of chlorine dose, and 1 ml of orthotolidine solution is added to 100 ml of sample, in a glass tube. The degree of yellowness which develops will be proportional to the concentration of chlorine residual. The deeper the colour, the greater will be the amount of residual chlorine. Coloured glass standards tinted for each concentration are compared to the colour developed in the test. Rapid and accurate determination of free chlorine is usually made with the help of a *chloroscope*. The triple cell of the chloroscope is filled with water sample upto the mark and orthotolidine solution is added by means of marked pipette. The colour so formed is matched with the standard coloured disc provided with the instrument.

When the water is highly alkaline, a blue tinge is formed instead of a yellow one. In that case, the quantity of orthotolidine to be added is doubled. If chlorine has been used as the disinfecting agent, colour should be observed after 5 min. of addition of orthotolidine, while if chloramine is used as disinfecting material, the colour should be observed 15 min. after orthotolidin is added. The test should be conducted at room temperature.

Orthotolidine arsenite (OTA) method

One of the biggest limitations of the orthotolidine test is that it measures free and combined residuals indiscriminately. In order to differentiate between the free and combined residual chlorine, the *modified orthotolidine-arsenite* (OTA) *method* is used. The OTA test is based on the fact that free chlorine residuals will develop immediately (within 5 to 10 seconds) while combined residul chlorine (chloramines) will react more slowly. In the OTA test, orthotolidine is first added to the sample and is quickly followed by sodium arsenite. Within seconds, the orthotolidine will oxidize the free chlorine forming yellow holoquinone. Due to the subsequent addition of sodium arsenite, the chloramines will be quickly oxidized to form a colourless liquid. Thus, in this test, only free residual chlorine is fixed, and its concentration is determined by comparing the yellow colour with the standard one. The difference of the two tests would give combined residual chlorine.

Effect of Iron, Manganese and Nitrites

The orthotolidine test is not fool-proof, since the result is affected by the presence of interfering substances in water ; these substances include nitrite, ferric compounds, manganic compounds, organic iron compounds, lignocellulose and algae, and the effect of these substances is to increase the apparent content of chlorine. In that case, the test is performed in three steps :

(*i*) Sodium arsenite is first added to the sample, thereby dechlorinating the sample of water. The orthotolidine solution is then added and colour is noted. The colour so formed will be entirely due to the interfering substances.

(*ii*) The orthotolidine-aresenite (OTA) test is performed to arrest the colour formed by the combined residual chlorine. The colour so formed will be due to free available chlorine as well as that due to iron, manganese and other substances.

(*iii*) Finally, the standard orthotolidine test is performed, the colour developed after 5 min. will be that due to free and combined chlorine and that due to iron, manganese etc. The *free residual chlorine* is then determined by subtracting the results of (*i*) from (*ii*). The *combined residual chlorine* is determined by subtracting results of (*i*) from (*ii*).

2. Starch-iodide test. Starch-iodide test usually called the *iodometric method*, is used when the presence of nitrite, nitrogen or manganic manganese makes the orthotolidine test unsuitable. The idometric method is more precise than the orthotolidine method, particularly when the residual chlorine is greater than 1 p.p.m. The test is conducted in the following steps :

1. Collect 1 litre of water sample in a heat-proof earthenware vessel called *casserole*.
2. Add 10 ml of potassium iodide solution to the vessel and mix the contents thoroughly.
3. Add 5 ml of starch solution. Blue colour will be formed.
4. Remove the blue colour by titration with N/100 sodium thiosulphate.

The amount of chlorine in p.p.m., present in the original water sample, is determined from the following relation

Chlorine, p.p.m. = 0.3546 × (Number of ml of sodium thiosulphate required to remove blue colour)

Calorimetric methods for the iodometric determination of chlorine are also available.

11.9. FACTORS AFFECTING BACTERICIDAL EFFICIENCY OF CHLORINE

The destruction of pathogens by chlorination is dependent upon the following factors : (*i*) turbidity, (*ii*) presence of metallic compounds, (*iii*) ammonia compounds, (*iv*) pH of water, (*v*) temperature, (*vi*) time of contact, (*vii*) number and concentration of bacteria.

1. Turbidity. The effect of turbidity in water is to make it difficult to obtain free residual chlorine. Also, the penetration of chlorine and therefore the destruction of bacteria in particles of suspended matter of a turbid water may be very uncertain. Due to this reason, the application of chlorine is preferred after filtration when water will be free from turbidity.

2. Presence of metallic compounds. The presence of metallic compounds such as iron and managanese in solution in the water, utilizes large amounts of chlorine to convert these into their higher stages of oxidation which are insoluble in water. Hence it is essential to remove iron and manganese, to make chlorination useful and more meaningful.

3. Ammonia compounds. The presence of ammonia with or without organic matter may form *combined available chlorine* which is not so effective a bactericide as *free available chlorine*. Quoting from the work of Butterfield, Whitlock (1953) estimates that 25 times as much combined available chlorine must be used to achieve the same degree of kill as free chlorine in the same time. If similar doses of free and combined chlorines are used, then the combined chlorine will take *100 times as long* as the free chlorine to achieve the same degree of kill. Hoather (1949) found that in a clean well water with traditional ratio of added ammonia to chlorine of 1 to 3, the chlorine killed the bacteria more quickly than it combined with ammonia. This is illustrated in Fig. 11.6. Since ammonia is often naturally present in water, it is therefore usual to add sufficient chlorine to react with the amount of natural ammonia present and a futher dose of chlorine sufficient to create an excess of free chlorine for speedy disinfection.

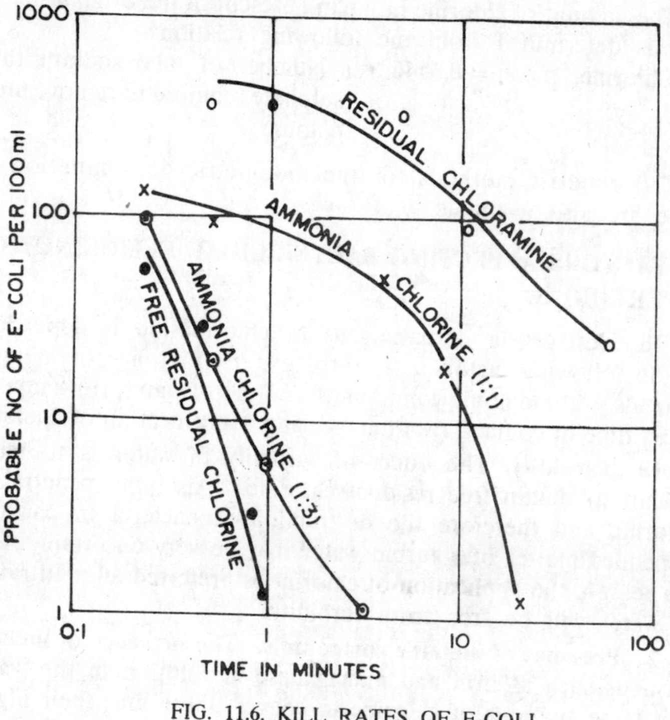

FIG. 11.6. KILL RATES OF E-COLI

4. pH of water. Increasing pH reduces effectiveness of chlorine. The effective sterlising compound, *hypochlorous acid*, is formed in greater quantities at low pH than at high pH values as is clear from table below :

pH value	*Amount of hypochlorous acid*
upto 6.7	95% of total free chlorine
At 7.0	80% of total free chlorine
At 8.0	30% of total free chlorine
At 9.0	5% of total free chlorine

5. Temperature of water. Reduction in the temperature of water results in substantial decrease in the killing power of both free and combined chlorine. In order to have 100% bactericidal activity, the requirement of residual chlorine increases with decrease in temperature and increase in pH, as is clear from Fig. 11.7.

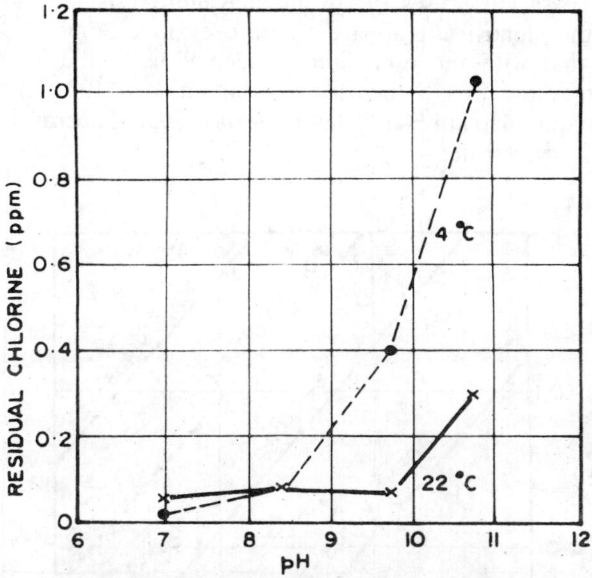

FIG. 11.7. RELATIONSHIP BETWEEN TEMPERATURE, pH AND FREE RESIDUAL CHLORINE IN PRODUCING 100% BACTERICIDAL ACTIVITY, AT 20 MINUTES

6. Time of contact. The bactericidal activity of chlorination is not instantaneous. The percentage kill of bacteria and viruses depends upon the time of contact between the chlorine and the micro-organisms, before the water is supplied to the consumers. For disinfection by free chlorine acting in clear water, a theoretical contact time of 20 minutes is required, while it is 60 min. for combined chlorine, such as chloramines etc. The contact time should be decided after taking into account all the factors such as temperature, pH value, type and concentration of micro-organisms and form of chlorination.

7. Nature and concentration of bacteria and viruses. At a given pH value and temperature, the greater the numbers of bacteria, the longer will be the time required to reduce them below a given figure. The spores of bacteria are found to be more resistant to chlorine than bacteria themselves. However, a majority of bacteria causing water borne diseases are of non-spore forming type. The typhoid bacillus and other enteric pathogenic bacteria are at least as susceptible as *E-coli* to chlorination. Some viruses are more resistant to chlorine than *E-coli*. Due to this reason, the dose of chlorine and time of contact are kept substantially higher than that required for *E-coli*. Fig. 11.8 illustrates the relative resistance of *E-coli*, *polio-virus 1*, *coxsackie virus A2* and *adenovirus 3*. While polio and coxsackie viruses are considerably more resistant than *E-coli* of HOCl, the adenovirus is apparently more sensitive. Some researchers conclude that with the exception of adenovirus 3, the amount of chlorine required to kill the viruses is about 3 to 100 times more than that required to kill *E-coli*, *Aerobacter aerogenes, Elibertella typhosa* or *Shigella dysenteriae*.

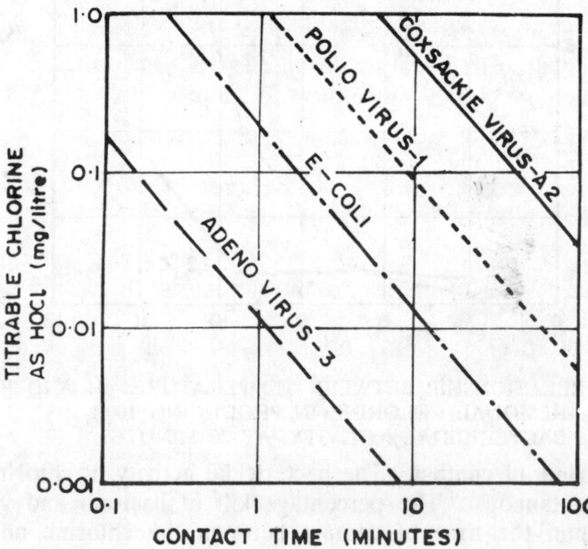

FIG. 11.8. RELATIONSHIP BETWEEN CHLORINE CONCENTRATION AND CONTACT TIME FOR 99% DESTRUCTION OF E-COLI AND THREE VIRUSES BY HOCl AT 0 – 6° C

In order to find a sterilizing agent capable of destroying cysts of *E-histolytica*, which are not satisfactorily destroyed by chlorine alone, studies have been made of cyst-penetrating power of chlorine and chloramine compounds. These compounds, arranged in order of activity are : (*i*) Chlorine Cl_2, (*ii*) Hypochlorous acid, HOCl, (*iii*) Halazone, (*iv*) Dichloramine, $NHCl_2$, (*v*) Succinchloramide, (*vi*) Monochloramine, NH_2Cl. According to Stringer and Kruse (1970), the most rapid acting cysticide is found to be hypochlorous acid (HOCl) form of free available chlorine.

The AWWA Committee report (1969) on virus in water concluded that "in the prechlorination of raw water, any enteric virus so far studied would be destroyed by a free chlorine residual of about 1.0 p.p.m., provided this concentration would be maintained for about 30 minutes and that the virus was not embedded in particulated material. In post-chlorination practices where relatively low chlorine residuals are usually maintained and in water of about 20° C and pH values not more than 8.0 – 8.5, a free chlorine residual of 0.2 – 0.3 p.p.m. would probably destroy in 30 minutes most viruses so far examined."

11.10. KINETICS OF CHEMICAL DISINFECTION

Ideal conditions :

The following ideal conditions are assumed:

(*i*) All cells of a single species of organisms are discreate units which are equally susceptical to a single species of disinfectant.

(*ii*) Both cells as well as disinfectants are uniformally dispered in water.

(*iii*) During period of contact, the discinfectant stays unchanged in chemical composition and constant in concentration.

(*iv*) Water contains no interfering substances.

Variables affecting disinfection

Under the above conditions, the process of disinfection is the function of the following variables :

(*a*) Time of contact
(*b*) Concentration of disinfectant
(*c*) Concentration of organisms
(*d*) Tempreature of disinfection

1. Time of contact

Chicks law of disinfection : Under ideal conditions, an exposed cell contains a single active centre vulnerable to a single unit of disinfectant. Chicks law states that the number of organisms (y) destroyed in unit time is proportional to the number of organisms (N) remaining

$$\frac{dy}{dt} = kN = k(N_0 - y) \qquad ...(11.10)$$

where N_0 = No. of organisms, initially present
y = No. of organisms remaining at any time t
$(= N_0 - y)$

k = Coefficient of proportionality, known as *rate constant* with dimension t^{-1}.

Integrating, $$\int_0^y \frac{dy}{N_0 - y} = \int_0^t k\,dt$$

or $$\log_e \frac{N_0 - y}{N_0} = -k\,t$$

or $$\log_e \frac{N}{N_0} = -k\,t \qquad ...(11.11)$$

or $$\frac{N}{N_0} = e^{-kt} \qquad ...(11.11)$$

When $t = 0$, $\frac{N}{N_0} = 1$ (*i.e.* 100%)

At $kt = 1$, $\frac{N}{N_0} = e^{-1} = \frac{l}{e} = 0.368 = 36.8\%$

Departures from Chick's law, even under ideal test conditions, are not uncommon. The rate of kill, rather than being constant, may increase or decrease with time

Rate of kill of organisms may be expressed as

$$\log_e \frac{N}{N_0} = -k\,t^m \qquad ...(11.12)$$

where m is constant, such that

If $m < 1$, rate of kill will decrease with time

$m > 1$, rate of kill will increase with time

Another formula that is frequently used to give the observed effects of contact time is

$$\frac{N}{N_0} = k\,t^m \qquad ...(11.13)$$

2. Concentration of disinfectant

For changing concentrations of disinfectant, the disinfecting efficiency can be expressed by

$$C^n\,t_p = \text{Constant} \qquad ...(11.14)$$

where C = time required to effect a constant percentage of kill of organism.

n = Coefficient of dilution, which is a measure of order of reaction.

When $n > 1$, efficiency of disinfectant decreases rapidly as it is diluted

$n < 1$, time of contact is more important than the dosage

$n = 1$, concentration and time are of equal weightage and a first order reaction may be in progress.

Examples of *time-concentration relationships* for chlorine as HOCl and 99% kill are :

$$C^{0.86}\, t_p = 0.24, \text{ for E.Coli}$$

$$C^{0.86}\, t_p = 6.3, \text{ for coxsacki virus } A_2$$

$$C^{0.86}\, t_p = 1.2, \text{ for poliomyelitis virus 1}$$

$$C^{0.86}\, t_p = 0.098 \text{ for adenovirus 3.}$$

3. Concentration of orgainsms (Number of organisms)

The effect of the concentration of organisms, or the number of organisms can be expressed by the following *observational equation*:

$$C^q \,.\, N_p = \text{constant} \qquad \text{...(11.15)}$$

where C = concentration of disinfectant

N_p = concentration of organisms that is reduced by a given percentage in a given time.

q = coefficient of disinfectant strength.

4. Temperature of disinfection

It is observed that increase in temperature results in a more rapid kill. The relationship between time and temperature, to effect a given percentage of kill can be expressed as :

$$\log_{10} \frac{t_1}{t_2} = \frac{E\,(T_2 - T_1)}{2.303\, R\, T_1\, T_2} \qquad \text{...(11.16)}$$

where t_1, t_2 = time for given % of kill at temperature T_1, T_2 (° K) respectively

E = activation energy, J/mol (or cal/mol).

R = gas constant,= 8.314 J/mol °K

= 1.99 Cal/mol, °K.

For $T_2 - T_1 = 10$, the useful ratio t_1/t_2 (usually called Q_{10}), is related to E as follows :

$$\log_{10} Q_{10} = \log_{10} \frac{t_1}{t_2} = \frac{E}{39000} \qquad \text{...(11.17)}$$

where E is in calories.

Relative effectiveness of the components of free available chlorine

In § 11.4, we have seen that when chlorine, in the form of Cl_2 gas, is added to water, two reactions take place : (*i*) hydrolysis and (*ii*) ionization.

(*i*) *Hydrolysis* :

$$Cl_2 + H_2O \Leftrightarrow HOCl + H^+ + Cl^- \qquad ...(11.1\,a)$$

The *stability coefficient*, or the *hydrolysis constant* (K_h) for the above reaction is

$$K_h = \frac{[HOCl]\,[H^+]\,[Cl^-]}{[Cl_2]} = 4.5 \times 10^{-4}\ (\text{mole/l})^2 \text{ at } 25°\,C \qquad ...(11.18)$$

(*ii*) *Ionization*

$$HOCl \Leftrightarrow H^+ + OCl^- \qquad ...(11.2)$$

The *ionization constant* (K_i) for this reaction is

$$K_i = \frac{[H^+]\,[OCl^-]}{[HOCl]} \qquad ...(11.19)$$

The variation in the value of K_i with temperature is given in Table below (Morris).

Temp (°C)	0	5	10	15	20	25
$K_i \times 10^8$ (moles/l)	1.5	1.7	2.0	2.2	2.5	2.7

From Eq. 11.19, we get

$$\frac{[OCl^-]}{[HOCl]} = \frac{K_i}{[H^+]} \qquad ...(11.20)$$

The quantities of HOCl and OCl that is present is water is called the *free available chlorine*. The killing efficiency of HOCl is about 40 to 80 times that of OCl^-.

The percentage distribution of HOCl at various temperatures can be computed from the relation :

$$\frac{[HOCl]}{[HOCl] + [OCl^-]} = \frac{1}{1 + \dfrac{[OCl^-]}{[HOCl]}} = \frac{1}{1 + \dfrac{K_i}{[H^+]}} \qquad ...(11.21)$$

For example, at 25°C and pH= 8,

$$\% \text{ Distribution of HOCl} = 100 \left[\frac{1}{1 + \dfrac{2.7 \times 10^{-8}}{10^{-8}}} \right] = \frac{100}{3.7}$$

$$\approx 27\%$$

Assuming the efficiencies of the two spieces to be *additive*, the *total amount* (R) of chlorine, required to produce a given percentage of kill, at a specified time, at various pH is given by

$$R = [\text{HOCl}] + [\text{OCl}^-] = [\text{HOCl}]\left[1 + \frac{(\text{OCl}^-)}{(\text{HOCl})}\right]$$

or
$$R = [\text{HOCl}]\left[1 + \frac{K_i}{(\text{H}^+)}\right] \qquad ...(11.22)$$

If C is the *killing concentration* of HOCl, we have

$$C = [\text{HOCl}] + r\,[\text{OCl}^-]$$

where r is the *proportional efficiency* of OCl^- ions relative to HOCl.

$$\therefore \quad C = \frac{R}{1 + \dfrac{K_i}{(\text{H}^+)}} + r\frac{K_i}{(\text{H}^+)} \cdot \frac{R}{1 + \dfrac{K_i}{(\text{H}^+)}} \qquad ...(11.23)$$

From which

$$R = C\,\frac{1 + \dfrac{K_i}{(\text{H}^+)}}{1 + \dfrac{r K_i}{(\text{H}^+)}} \qquad ...(11.24)$$

Example 11.4. *If 0.6 mg/litre of total chlorine is required for satisfactory disinfection of water at pH = 7.0, what dosage will be necessary at pH = 8.0. If it is given that initially, 12 min. contact time is required at pH = 7.0, find contact time at pH = 8.0 if n = 1.5 in the equation* $C^n t = k$. *Take* $K_i = 2.7 \times 10^{-8}$ *mol/lit.*

Solution

$$\frac{(\text{HOCl})}{(\text{HOCl}) + (\text{OCl})} = \frac{1}{1 + \dfrac{K_i}{\text{H}^+}} \qquad ...(11.21)$$

At pH = 7,

$$\%\ \text{HOCl} = 100\left[\frac{1}{1 + \dfrac{2.7 \times 10^{-8}}{10^{-7}}}\right] = \frac{100}{1 + 0.27} = 78.7\ \%$$

At pH = 8,

$$\%\ \text{HOCl} = 100\left[\frac{1}{1 + \dfrac{2.7 \times 10^{-8}}{10^{-8}}}\right] = \frac{100}{1 + 2.7} = 27\%$$

$\therefore$ Dose required at (pH= 8)= $0.6 \times \frac{78.7}{27} = \mathbf{1.75\ mg/l}$

For contact time= 12 min, at pH = 7.0

$$k = C^n t = (0.6 \times 0.787)^{1.5} \times 12 = 3.8938$$

Again, at pH= 8.0, we get

$$C^n t = k$$

or $$(0.6 \times 0.27)^{1.5}\, t = 3.8938$$

From which $$t = \frac{3.8938}{(0.6 \times 0.27)^{1.5}} \approx \mathbf{60\ min.}$$

Thus we find that, unless the dose is increased from 0.6 mg to 1.75 mg, the time of contact will have to be increased from 12 min. to 60 min. if the pH of water changes from 7.0 to 8.0.

11.11. IODINE TREATMENT

Low iodine residuals are more stable and persist longer in the presence of organic (or other oxidizable) material than corresponding residuals of any of the other halogens due to the following characteristics : (*i*) it possesses highest atomic weight, (*ii*) it is least soluble in water and is least hydrolyzed by water, (*iii*) it has lowest oxidation potential and reacts least readily with organic compounds. The following reactions take place, resulting in the formation of *hypoiodous acid* (HIO).

$$(I_2 + H_2O) \Leftrightarrow HIO + H^+ + I^+ \quad ...(11.25)$$

At pH values over 4.0, hypoiodous acid undergoes dissociation, resulting in the formation of *hypoiodite ion*:

$$HIO \Leftrightarrow H^+ + IO^-$$

At pH values over 8.0, HIO is unstable and will not form the hypoiodite ion but will decompose, resulting in the formation of *iodide* (I^-) and *iodate* (HIO_3) ;

$$3\,HIO + 2\,OH^- \Leftrightarrow HIO_3 + 2\,H_2O + 2I^- \quad ...(11.27)$$

As pH increases from 6 to 8, the percentage of iodine as I_2 decreases while % of iodine as HIO is increased. Hypoiodous acid formed at higher pH values is less bactericidal than free iodine. According to Scarpino (1972), 99.99% kill of most resistant of 6 bacterial strains tested could be achieved in 5 minutes with a free iodine residual of 1.0 mg/l. Rapid (5 – 10 minutes) destruction of cysts with iodine concentration of 5 – 10 mg/l have been reported for several different cysts in different neutral waters. Hypoiodous acid (HIO) destroys viruses at a rate considerably faster than that achieved by I_2 which has little viricidal activity. Though iodine's great advantage over chlorine is that it does not react with ammonia or similar nitrogenous compounds, some concern has arisen over the *possible* harmful effects of iodinated water on thyroid function and objectionable taste and odour. Iodine seems unlikely to become a municipal disinfectant because of its high cost and restricted availability when compared to chlorine. Use of iodine is at present limited to small water supplies, specially for swimming pools or troops of army etc. Certain iodine tablet preparations (*Globaline*, containing tetraglycine hydroperiodide) have been adopted by the military service, replacing Halazone chlorine releasing tablets used previously.

11.12. BROMINE TREATMENT

According to Morris (1971), bromine reacts with water similar to that of chlorine – hydrolyzing to HOBr, ionizing to OBr and reacting with ammonia to form bromanines. The bactericidal and viricidal effectiveness of HOBr has been found to be comparable to that of HOCl. However, there have been no thorough invetigations on a plant scale. Due to its greater cost and scarcity, bromine has not been used as a substitute for chlorine in municipal disinfection. Its use is restricted to small water supplies, specially for swimming pools or troops of army etc.

11.13. OZONE TREATMENT (OZONATION)

Ozone (O_3) is triatomic allotrope of oxygen. It is an unstable gas that tends to breakdown into *normal oxygen* containing two atoms of oxygen by molecule. In the process of breaking down, it releases nascent oxygen (containing single atom of oxygen) which is very effective in oxidizing the organic matter and in killing bacteria present in water. Ozone has been used for the disinfection of water supplies since the beginning of the century (1906) when it was first applied to the treatment of water for Nice, France. Nearly one thousand cities, most of them in Europe, use ozonation for drinking water treatment. However, because of its being unstable, it cannot be stored, and has to be produced at the point of use by passing dry air or oxygen between two high potential electrodes to convert oxygen (O_2) to ozone (O_3). The ozone generators convert only a small percentage (1 to 4 %) of the feed oxygen gas into ozone.

The ozone-containing air or oxygen mixture produced by the ozone generator can be introduced into the water by injecting or diffusing it into a mixing chamber. A dosage of 0.2 to 0.3 p.p.m. ozone is sufficient to kill all coliform bacteria in clean water if it is free of all oxidizable materials. For municipal water treatment applications, a dosage of 1.5 p.p.m. is often used to disinfect as well as to remove taste, odour and colour. According to Hann (1956), ozone has been found to equal or exceed chlorine in its germicidal effects. *It kills viruses even more rapidly*. Electron microscope examination indicates that viruses appear to have exploded. Ozone has been found to be many times more effective than chlorine in inactivating the virus of poliomyelitis. Ozone is also more effective than chlorine in destroying other hard to kill organisms, such as spores. The oxidation products of ozonation are usually non-poisonous and biodegradable unlike those of chlorines.

Advantages of Ozonation

1. It is a more powerful disinfectant for bacteria and viruses, than chlorine.

2. Its disinfecting action is effective over a wide temperature and pH range.

3. The bactericidal and sporicidal action is more rapid with ozone than with chlorine, and only short contact periods are required.

4. There is no danger of over treatment since ozone decomposes to oxygen.

5. It effectively reduces or eliminates colour, taste and odour. Ozonation produces a more palatable water than chlorine.

6. Ozone rapidly oxidizes organic impurities, such as alkyl-benzene-sulphonate (ABS).

7. The oxidation products of ozone are non-poisonous and biodegradable.

8. Since it is generated at the site, there are no transport and storage problems. It is not necessary to store dangerous chemicals.

Disadvantages

1. It is more expensive than chlorine.

2. It leaves no lasting residue to prevent recontamination, while chlorine provides a residual protection.

3. Because it must be produced electrically as it is needed, it is difficult to adjust treatment to variations in load, or to changes in water quality with regard to ozone demand.

4. The production of ozone is complicated when temperature and humidity are high.

The highest quality of drinking water can be obtained by ozonation followed by the introduction of a small dosage of chlorine to provide residual protection.

PROBLEMS

1. What is the difference between disinfection and sterilization ? Why is disinfection necessary ?

2. Name various disinfecting agents and explain the action of any one of these in detail.

3. What do you understand by chlorination ? Explain its action in killing bacteria.

4. What are the various forms of application of chlorine ? Write a note on hypochlorination ?

5. Explain chlorine-ammonia treatment for disinfecting drinking water. What are its advantages.

6. Explain the method of application of chlorine to water.

7. Write notes on the various forms of chlorination.

8. What do you understand by break point chlorination ? What are its advantages ?

9. Explain the orthotolidine test for residual chlorine.

10. How do you modify the orthotolidine test of water containing iron, manganese and nitrites ?

11. How do you determine the residual chlorine in water by the starch-iodide test ?

12. Explain various factors affecting bactericidal efficiency of chlorine.

13. Write a note on bactericidal and viricidal effectiveness of chlorine.

14. Explain the effectiveness of iodine and bromine treatments for disinfecting water.

15. What do you understand by ozonation ? What are its advantages and disadvantages ?

16. Write short notes on the following :

(*i*) Chlorine dioxide.
(*ii*) Excess lime treatment.
(*iii*) Ultra-violet ray treatment
(*iv*) Effect of pH on chlorination
(*v*) Disinfection of water against viruses.
(*vi*) Chloramines.
(*vii*) Prechlorination, super-chlorination and dechlorination.

12

Water Softening

12.1. INTRODUCTION

Water is said to be 'hard' when it contains relatively large amounts of bicarbonates, carbonates, sulphates and chlorides of calcium and magnesium dissolved in it. These materials react with soap, causing a precipitation which appears as a scum or curd on the water surface. No lather can be formed until enough soap has been dissolved to react with all these. Hard water causes the following troubles :

1. It causes more consumption of soap in laundry work. A wastage of 25 p.p.m. of soap takes place per one p.p.m. of hardness of water removed by it. This proves to be uneconomical for washing processes in the textile industries.

2. It leads to the modification of some of the colours, and thus affects the working of the dyeing system.

3. It causes serious difficulties in the manufacturing process, such as paper making, canning, ice manufacture, rayon industry etc.

4. It causes choking and clogging troubles of house plumbings due to precipitation of salts causing hardness.

5. It causes formation of scales on the boilers and other hot water heating system.

6. It makes food tasteless, tough or rubbery.

Water is classified with regard to its hardness, as given in Table 12.1 :

TABLE 12.1 CLASSIFICATION OF HARDNESS

Classification	*Total hardness as mg/l (or ppm) of $CaCO_3$*
1. Soft	50
2. Moderately hard	50–150
3. Hard	150–300
4. Very hard	300

The softening of water removes all the above difficulties. In addition of these, softening increases the efficiency of filtration, aids in the removal of colour, iron and manganese, makes possible the production of non-corrosive water, improves the cooking of foods and increases the efficiency of reduction of bacteria when filtration follows softening. Water must be reduced to zero hardness before soap will become effective as a detergent. In municipal water supplies, total hardness is seldom reduced below 50 p.p.m. and the carabonate or non-carbonate hardness is generally not brought below 35 p.p.m.

12.2. TYPE OF HARDNESS AND METHODS OF THEIR REMOVAL

Hardness is of two types : (*i*) temporary hardness, and (*ii*) permanent hardness. The temporary hardness is the one which is deposited when water is boiled, and is usually known as *carbonate hardness.* The permanent hardness is also known as *non-carbonate hardness,* and is mainly due to the presence of sulphates, chlorides and nitrates of calcium and magnesium. Table 12.2 shows the hardness compounds.

TABLE 12.2 HARDNESS COMPOUNDS

Causing temporary hardness	*Causing permanent hardness*
1. Calcium bicarbonate $Ca(HCO_3)_2$ 2. Magnesium bicarbonate $Mg(HCO_3)_2$	1. Calcium sulphate $CaSO_4$ 2. Magnesium sulphate $MgSO_4$ 3. Calcium chloride $CaCl_2$ 4. Magnesium chloride $MgCl_2$

(*a*) **Removal of temporary hardness :** Temporary hardness or bicarbonate hardness can be removed by the following methods : (*i*) by boiling, (*ii*) by adding lime.

The following reaction takes place during boiling :

$$Ca(HCO_3)_2 + \text{Heat} = CaCO_3 \downarrow + H_2O + CO_2 \uparrow$$

$$Mg(HCO_3)_2 + \text{Heat} = MgCO_3 \downarrow + H_2O + CO_2 \uparrow$$

However, the method has practical limitations on large scale. Hence the second method is adopted. The following reactions take place when lime is added :

$$Ca(HCO_3)_2 + Ca(OH)_2 = 2\,CaCO_3 \downarrow + 2\,H_2O$$

$$Mg(HCO_3)_2 + Ca(OH)_2 = CaCO_3 \downarrow + MgCO_3 \downarrow + 2H_2O$$

The calcium carbonate and magnesium carbonate so formed can be removed in the sedimentation tanks since these are insoluble in water.

(*b*) **Removal of permanent hardness :** Permanent hardness can be removed by one of the following methods :

1. Lime-soda process.
2. Zeolite process.
3. Demineralization or de-ionization process

12.3. LIME-SODA PROCESS

Addition of lime reduces only the bicarbonate hardness. By addition of lime and soda ash (Na_2CO_3) both the temporary and permanent (non-carbonate) hardness can be removed. The method is therefore known as *lime-soda process*, in which lime [$Ca(OH)_2$] and soda (sodium carbonate) are added to the raw water either separately or together. The process involves the thorough mixing of the chemicals with water, followed by slow agitation for 30 to 60 min. to allow completion of chemical reaction. Precipitated chemicals are removed by sedimentation or filteration or both. Lime-soda process may be operated continuously or intermittently, either cold (10° to 21° C) or hot (100° *C*). Chemical reactions and sedimentation take place rapidly at higher temperatures. The water may be *recarbonated* after sedimentation to restore carbonate balance.

The aim of the lime soda process is to make the calcium and magnesium content of the hard water take their insoluble form (such as calcium carbonate and magnesium hydroxide) so that they precipitate out. The following reactions take place. The compounds markes with (↓) are those precipitated.

$$CO_2 + Ca(OH)_2 = CaCO_3 \downarrow + H_2O \quad ...(12.1)$$

$$Ca(HCO_3)_2 + Ca(OH)_2 = 2\,CaCO_3 \downarrow + 2H_2O \quad ...(12.2)$$

$$Mg(HCO_3)_2 + Ca(OH)_2 = CaCO_3 \downarrow MgCO_3 + 2H_2O \quad ...(12.3)$$

$$MgCO_3 + Ca(OH)_2 = Mg(OH)_2 \downarrow + CaCO_3 \downarrow \quad ...(12.4)$$

$$MgSO_4 + Ca(OH)_2 = Mg(OH)_2 \downarrow + CaSO_4 \quad ...(12.5)$$

$$CaSO_4 + Na_2CO_3 = CaCO_3 \downarrow + Na_2SO_4 \quad ...(12.6)$$

$$CaCl_2 + Na_2CO_3 = CaCO_3 \downarrow + 2NaCl \quad ...(12.7)$$

$$MgCl_2 + Ca(OH)_2 = Mg(OH)_2 \downarrow + CaCl_2 \quad ...(12.8)$$

Eq. 12.1 represents the consumption of lime by carbon-dioxide present in water to be softened. Eqs. 12.2, 12.3 and 12.4 represent the removal of carbonate or bicarbonate hardness by lime. Eq. 12.4 presents excess lime treatment over Eq. 12.3 to convert magnesium carbonate into magnesium hydroxide which can be precipitated. Eq. 12.5 shows the removal of magnesium hardness by lime, though no softening takes place since magnesium and calcium ions simply interchange. Eq. 12.6 shows the removal of $CaSO_4$, either oringinally present in water or that formed by Eq. 12.5. Eq. 12.7 shows the removal of calcium chloride by soda ash.

Eq. 12.3 and 12.4 can be combined together and expressed as

$$Mg(HCO_3)_2+2Ca(OH)_2 = 2CaCO_3\downarrow + Mg(OH)_2 + 2H_2O \quad ...(12.3\ a)$$

This shows that magnesium bicarbonate requires two equivalents of lime, as against one equivalent of lime needed by others like calcium bicarbonate, magnesium chloride, magnesium sulphate, etc. Calcium sulphate does not need any lime, but if requires soda ash (Eq. 12.6).

Similarly, Eqs. 12.8 and 12.7 can be combined as follows :

$$MgCl_2 + Ca(OH)_2 + Na_2CO_3 = Mg(OH)_2\downarrow + CaCO_3\downarrow + 2NaCl \quad ...(12.8\ a)$$

This shows that for removal of magnesium chloride, both lime as well as soda ash are required.

Also, eqs. 12.5 and 12.6 can be combined as follows :

$$MgSO_4 + Ca(OH)_2 + Na_2CO_3 = Mg(OH)\downarrow + CaCO_3\downarrow + Na_2SO_4 \quad ...(12.5\ a)$$

which shows that for the removal of magnesium sulphate, both lime and soda ash are required.

Summary of chemicals required

Type of Hardness	*Chemical Required*
1. $CaCO_3$ alkalinity (as bicarbonate)	Lime
2. Calcium bicarbonate, $Ca(HCO_3)_2$	Lime
3. Magnesium sulphate, $Mg(HCO_3)_2$	Lime (two units)
4. Magnesium sulphate, $MgSO_4$	(i) Lime (ii) Soda ash.
5. Calcium sulphate $CaSO_4$	Soda ash
6. Calcium chloride $CaCl_2$	Soda ash
7. Magnesium chloride $MgCl_2$	(i) Lime (ii) Soda ash.

Amount of softening chemicals, *i.e.*, lime and soda ash depends upon the following :

(*i*) amount of free carbon dioxide present,
(*ii*) amount of half-bound carbon dioxide, which is a portion of that in the bicarbonate alkalinity,
(*iii*) non-carbonate. hardness,
(*iv*) total magnesium, and
(*v*) contents of lime and soda in the commerical chemicals.

Lime used may be either hydrated lime or quick lime. Quick lime is preferred for large plants because it is less bulky and cheaper. It can be fed by 'dry feed' equipment or added as made-up slurry. Soda ash can also be added to the water as a dry feed or a solution. As stated above, the amount of lime required for softening water is determined by the amounts of free and half-bound carbon dioxide and of magnesium that are present. The half-bound CO_2 amounts to 44% of bicarbonate alkalinity :

$$\underset{100\text{ p.p.m.}}{Ca(HCO_3)_2} = CaCO_3 + \underset{44\text{ p.p.m.}}{CO_2} + H_2O$$

The ratio of molecular weight of CO_2 and CaO (pure lime) is 44 : 56 or 1 : 1.27. For each p.p.m. of CO_2 present, 1.27 p.p.m. of CaO will be required. Since 1 p.p.m. is equal to 1 kg per million litres, 1.27 kg of CaO will be required per million litres of water, for each p.p.m. of CO_2. Since the molecular weight of CaO is 40 + 16.00 = 56 and the molecular weight of hydrated lime $Ca(OH)_2$ is 40+32.00+2 = 74, then 74 kg of hydrated lime is equivalent to 56 kg of CaO. Hence amount of hydrated lime will be equal $\frac{74}{56} \times 1.27 = 1.68$ kg per million litres of water for each p.p.m. of CO_2.

The molecular weight of CaO = 40 + 16 = 56 while molecular weight of magnesium (in magnesium salt) is 24. Hence amount of CaO required to precipitate 1 p.p.m. of magnesium is 56/24=2.33 kg per million litres of water.

Also, molecular weight of $CaCO_3$ (non-carbonate hardness) = 40+ 12+ 3× 16 = 100 while molecular weight of soda ash (Na_2CO_3) = 2 × 23 + 12 + 3 × 16 = 106. Hence amount of soda ash required to remove 1 p.p.m. of non-carbonate hardness from 1 million litres of water = 106/100 = 1.06 kg.

Example 12.1. *The analysis of a water shows the following : free CO_2 : 3 p.p.m., alkalinity : 65 p.p.m., non-carbonate hardness : 95 p.p.m., total magnesium : 10 p.p.m. Assume that it is possible to remove all but 30 p.p.m. of carbonate hardness with lime and that the finished water is to have total hardness of 80 p.p.m. Determine the amounts of chemicals required per million litres of water.*

Solution :

Permissible total hardness

$$= 80 \text{ p.p.m.}$$

Amount of carbonate hardness left

$$= 30 \text{ p.p.m.}$$

$\therefore$ Amount of non-carbonate hardness left

$$= 80 - 30 = 50 \text{ p.p.m.}$$

$\therefore$ Non-carbonate hardness to be removed

$$= 95 - 50 = 45 \text{ p.p.m.}$$

(i) Lime required for free CO_2

Molecular weight of CaO $= 40 + 16 = 56$

Molecular weight of $CO_2 = 12 + 2 \times 16 = 44$

$\therefore$ CaO required for 3 p.p.m. of CO_2

$$= 3 \times \frac{56}{44} = 3.82 \text{ mg/l} = 3.82 \text{ kg/Ml}$$

(ii) Lime required for alkalinity

Alkalinity (or carbonate alkalinity), $CaCO_3$ has molecular weight of $40+12+3\times16 = 100$ and that of pure lime CaO = 56.

$\therefore$ CaO required for 65 p.p.m. of alkalinity

$$= \frac{56}{100} \times 65 = 36.4 \text{ mg/l} = 36.4 \text{ kg/Ml}$$

(iii) Lime required to react with magnesium

Molecular weight of magnesium = 24

Molecular weight of CaO = 56

$\therefore$ CaO required for 10 p.p.m. of magnesium

$$= \frac{56}{24} \times 10 = 23.33 \text{ mg/l} = 23.33 \text{ kg/Ml}$$

$\therefore$ Total lime (CaO) required $= 3.82 + 36.4 + 23.33 =$ **63.55 kg.**

(iv) Soda ash required for non-carbonate hardness

Molecular weight of $Na_2CO_3 = 2 \times 23 + 12 + 3 \times 16 = 106$

Molecular weight of $CaCO_3 = 40 + 12 + 3 \times 16 = 100$

$\therefore$ Soda ash required for 45 p.p.m. of non-carbonate hardness

$$= \frac{106}{100} \times 45 = 47.7 \text{ mg/l} = \mathbf{47.7 \text{ kg/Ml}}$$

Example 12.2. *A sample of raw water contains, 200 mg/l alkalinity, 50 mg/l hardness as $CaCl_2$ and 75 mg/l hardness as $MgSO_4$. Compute the quantities of lime and soda required to treat 1 million litres of water. If slaked lime of 85% purity is available in place of pure lime, what will be the required quantity of slaked lime ?*

Solution : Lime is required to react with alkalinity as well as $MgSO_4$ (Eq. 12.5) while soda ash is required for both $CaCl_2$ (Eq. 12.7) as well as $MgSO_4$ (Eqs. 12.5 and 12.6).

(*i*) ***Lime required for alkalinity*** **($CaCO_3$)**

Molecular weight of CaO $= 40 + 16 = 56$

Molecular weight of $CaCO_3 = 40 + 12 + 3 \times 16 = 100$

$\therefore$ CaO required for 200 mg/l alkalinity

$$= 200 \times \frac{56}{100} = 112 \text{ mg/l} = 112 \text{ kg/Ml}$$

(*ii*) ***Lime required for*** **$MgSO_4$**

Molecular weight of $MgSO_4 = 24 + 32 + 4 \times 16 = 120$

$\therefore$ Cao required for 75 mg/l of $MgSO_4$

$$= 75 \times \frac{56}{120} = 35 \text{ mg/l} = 35 \text{ kg/Ml}$$

$\therefore$ Total lime required $= 112 + 35 =$ **147 kg/Ml**

(*iii*) ***Soda required for*** **$CaCl_2$**

Molecular weight of soda (Na_2CO_3) $= 2 \times 23 + 12 + 3 \times 16 = 106$

Molecular weight of $CaCl_2 = 40 + 2 \times 35.5 = 111$

$\therefore$ Soda required for 50 mg/l of $CaCl_2$

$$= \frac{106}{111} \times 50 = 47.75 \text{ mg/l} = 47.75 \text{ kg/Ml}$$

(*iv*) ***Soda required for*** **$MgSO_4$**

Molecular weight of soda $= 2 \times 23 + 12 + 3 \times 16 = 106$

Molecular weight of $MgSO_4 = 24 + 32 + 4 \times 16 = 120$

$\therefore$ Soda required for 75 mg/l of Mg SO_4

$$= \frac{106}{120} \times 75 = 66.25 \text{ mg/l} = 66.25 \text{ kg/Ml}$$

$\therefore$ Total soda required $= 47.75 + 66.25 =$ **114 kg/Ml**

Amount of slaked lime required

Molecular weight of slaked (or hydrated) lime $Ca(OH)_2$

$$= 40 + 2 \times 16 + 2 \times 1 = 74$$

Molecular weight of pure lime (CaO) $= 40 + 16 = 56$

$\therefore$ Quantity of slaked lime required $= \frac{74}{56} \times 147 = 194.25$ kg/Ml

But the available slaked lime has purity of 85% only.

$\therefore$ Actual quantity of market available slaked lime

$$= \frac{194.25}{0.85} = \mathbf{228.5\ kg/Ml.}$$

Note that this is more that $1\frac{1}{2}$ times the quantity of pure lime required to treat 1 million litres of water.

Example 11.3. *A raw water sample contains the following impurities :*

$CaCO_3 = 200$ ppm	$MgCl_2 = 150$ p.p.m.
$Mg(HCO_3)_2 = 120$ p.p.m.	$MgSO_4 = 100$ p.p.m.
$CaSO_4 = 100$ p.p.m.	$NaCl = 25$ p.p.m.
$Fe_2O_3 = 40$ p.p.m.	$SiO_2 = 30$ p.p.m.

Compute the annual requirements of slaked lime and soda, for treating 80,000 litres of water daily. The purity of slaked time is 85% and that of soda is 90%.

Solution : Impurities like NaCl, Fe_2O_3 and SiO_2 do not require lime and soda.

1. ***Requirements for $CaCO_3$ alkalinity :***

It will require lime only.

Molecular weight of $CaCO_3 = 40 + 12 + 3 \times 16 = 100$

Molecular weight of hydrated lime $Ca(OH) = 48 + 2(16+1) = 74$

∴ Lime required for 200 ppm of alkalinity

$$= \frac{74}{100} \times 200 = 148 \text{ mg/l} \qquad ...(i)$$

2. ***Requirements for $Mg(HCO_3)_2$***

One unit of magnesium bicarbonate requires two units of lime.

Molecular weight of $Mg(HCO_3)_2 = 24 + 1 \times 2 + 2 \times 12 + 6 \times 16 = 146$

∴ Lime required for 120 ppm of $Mg(HCO_3)_2$

$$= 2 \times \frac{74}{146} \times 120 = 121.6 \text{ mg/l} \qquad ...(ii)$$

3. ***Requirements for $CaSO_4$***

Calcium sulphate requires only soda ash.

Molecular weight of soda $(Na_2CO_3) = 2 \times 23 + 12 + 3 \times 16 = 106$

Molecular weight of $CaSO_4 = 40 + 32 + 4 \times 16 = 136$

∴ Soda required for 100 ppm of $CaSO_4$

$$= \frac{106}{136} \times 100 = 77.94 \text{ mg/l} \qquad ...(iii)$$

4. ***Requirements for*** $MgCl_2$

Magnesium chloride needs both lime as well as soda ash.
Molecular weight of $MgCl_2 = 24 + 2 \times 35.5 = 95$

$$\therefore \text{ Lime required for 150 ppm } MgCl_2 = \frac{74}{95} \times 150$$
$$= 116.84 \text{ mg/l} \qquad ...(iv)$$

and soda required for 150 ppm of $MgCl_2 = \frac{106}{95} \times 150$
$$= 167.37 \text{ mg/l} \qquad ...(v)$$

5. ***Requirements for*** $MgSO_4$

Magnesium sulphate requires both lime as well as soda ash.
Molecular weight of $MgSO_4 = 24 + 32 + 4 \times 16 = 120$

$$\therefore \text{ Lime required for 100 ppm of } MgSO_4 = \frac{74}{120} \times 100$$
$$= 61.6 \text{ mg/l} \qquad ...(vi)$$

and soda required for 100 ppm of $MgSO_4 = \frac{106}{120} \times 100$
$$= 88.33 \text{ mg/l} \qquad ...(vii)$$

$$\therefore \text{ Total quantity of lime } = 148 + 121.6 + 116.84 + 61.67$$
$$= 448.11 \text{ mg/l}$$

$$\text{Total quantity of commercial lime} = \frac{448.11}{0.85} \times 80000 \times 10^{-6}$$
$$= 42.2 \text{ kg/day}$$
$$= 42.2 \times 365 \times 10^{-3} = \mathbf{15.4 \text{ tonnes/year}}$$

$$\text{Total quantity of pure soda} = 77.94 + 167.37 + 88.33$$
$$= 333.64 \text{ mg/l}$$

$$\therefore \text{ Total quantity of commercial soda} = \frac{333.64}{0.9} \times 80000 \times 10^{-6}$$
$$= 29.66 \text{ kg/day}$$
$$= 29.66 \times 365 \times 10^{-3} = \mathbf{10.82 \text{ tonns/year}}$$

12.4. LIME-SODA SOFTENING PLANT

Lime soda softening plant includes the following units :

1. Feeding and mixing devices. These will be similar to those used for coagulation.

2. Settling basin. The settling basin usually resemble those employed for coagulation-sedimentation. However, the retention periods are somewhat longer–2 to 4 hours–to obtain greater clarification. Mechanical devices for continuous removal of the sludge are ordinarily used.

3. Recarbonation or stabilization plant. The effluent from the settling basin is supersaturated with calcium carbonate. If this is not prevented, it will precipitate on sand filters and cause enlargement of sand grains, incrustation of mains and stoppage of water meters. The carbonate balance may be partly or completely restored by recarbonation of water. This is usually accomplished by diffusing carbon dioxide gas through the water or by under-water combustion.

The following reactions take place :

$$\underset{100}{CaCO_3} + \underset{44}{CO_2} + H_2O = Ca(HCO_3)_2 \qquad ...(12.10)$$

The ratio of weight of CO_2 required to the weight of $CaCO_3$ is 44 : 100. Hence weight of CO_2 required per p.p.m. of $CaCO_3$ will be 0.44 kg per million litres of water. In practice, 25% more CO_2 is supplied to make up for the losses etc. Actual CO_2 required will be 1.25 × 0.44 = 0.55 kg per million litres of water per 1 p.p.m. of $CaCO_3$. The carbonation of water containing excess lime will form a colloidal precipitate of calcium carbonate that is less soluble at pH of 9.4 to 9.6. Hence recarbonation should not be carried below this point. Fig. 12.1 shows a small recarbonation unit.

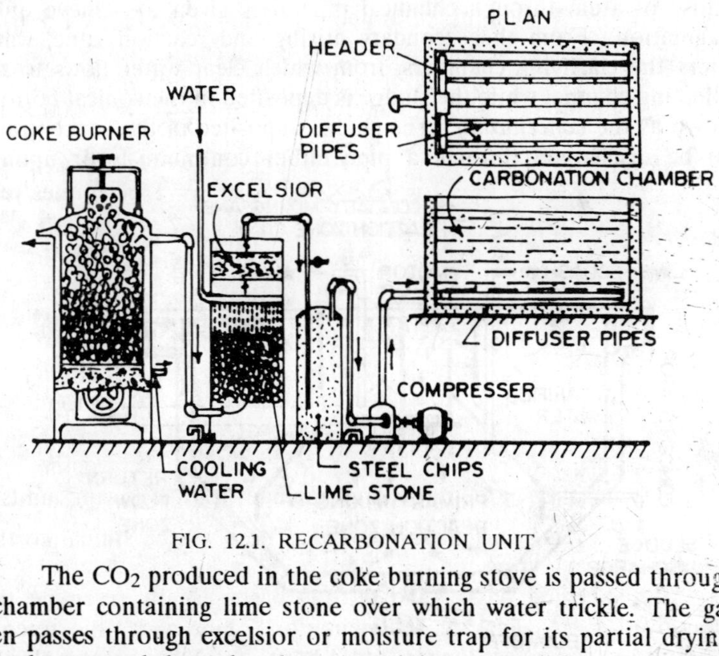

FIG. 12.1. RECARBONATION UNIT

The CO_2 produced in the coke burning stove is passed through a chamber containing lime stone over which water trickle. The gas then passes through excelsior or moisture trap for its partial drying. It is then passed through a dryer containing steel turnings, to remove remaining water and active oxygen. It is then diffused at the bottom of carbonation chamber, with the help of a compressor. The

is usually applied after water has passed through the settling basin. The reaction time allowed in the recarbonation basin should be atleast 20 minutes.

4. Filters. The recarbonated water is passed through rapid filtration to ensure complete clarification. Pressure filters may also be used.

12.5. WATER SOFTENING ACCELERATOR

An *accelerator*, or a *precipitator*, combines the mixing, flocculation and clarification in the same structure, and is widely used at lime-soda softening plant. Fig. 12.2 shows a typical accelerator. It essentially consists of (*i*) primary mixing and reaction zone, (*ii*) secondary mixing and reaction zone, (*iii*) return flow zone and concentrator, and (*iv*) clarifying chamber. A rotor impeller separates the primary and secondary mixing zones. Raw water enters the primary mixing zone through a horizontal pipe (raw water inlet) while the required dose of lime and soda ash mixture enters this zone through the chemical pipe *A*. The raw water also receives a certain quantity of settling particles from the *return flow zone*, which accelerates the precipitation. Water is drawn up in the secondary mixing and reaction zone, due to the rotation of the impeller. In the secondary zone, a dose of alum through chemical pipe *B* is given to achieve quick coagulation. From the secondary mixing and reaction zone, water enters the clarifying chambers, from which clear water flows to the collecting channel while the sludge is deposited in the conical bottom, known as the *concentrator*. The sludge deposited in the concentrator can be discharged through a pipe either continuously or upon a

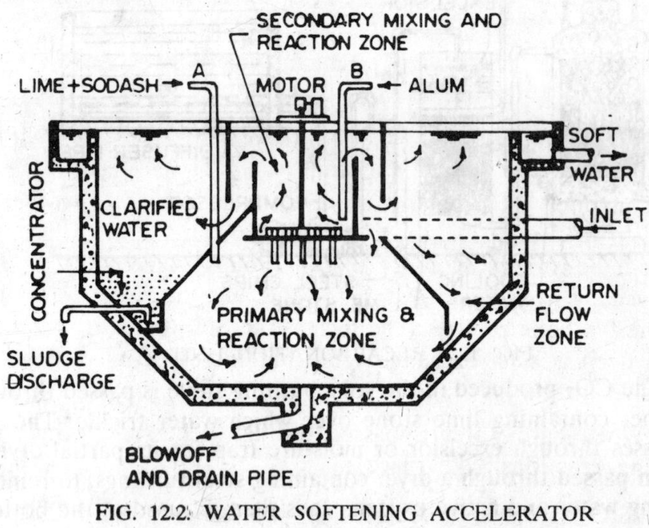

FIG. 12.2. WATER SOFTENING ACCELERATOR

schedule which may operate automatically. In the clarifying chamber, an upward velocity is kept between 7 to 9 cm per minute. Detention periods are short, varying between 1-2 hours. There are a number of units of this type sold by various manufacturers under various trade names, such as the *Accelerator* by Infilco Inc., the *Precipitator* by Permutit Company, the *Reactivator* by Graver Water Conditioning Co, the *Flocsettler* by the American Well Works and the *Hydro Treator* by the Dorr Co. In some types, the effluent passes upward through a blanket of suspended floc and other solids before it can leave the tank at the effluent weirs.

12.6. ZEOLITE PROCESS

The lime soda method is essentially a precipitation method of water softening, commonly known as Clark's system, in which some chemicals or reagents are to be added to water. In the zeolite process, also known as the *base exchange* or *ion-exchange process*, no such chemicals are added to water. Instead, hard water is passed through a bed of a special material, loosely called the *Zeolite*, which has the property of removing calcium and magnesium from the water and substituting sodium in their place by ion-exchange phenomenon. Zeolites are complex compound of aluminium, silica and soda, some forms of which are synthetic and others are naturally occurring. Natural zeolites are mainly processed from *Green Sand* (Glauconite). It is green in colour and has an exchange value of 6500 to 9000 gm of hardness per m^3 of zeolite. The common artificial zeolite is *Permutit*. It has larger grains and its colur is white. Its chemical formula is $SiO_2Al_2O_3Na_2O$.

Permutit has a high exchange value of 35000 to 40000 gm of hardness per m^3. It has the appearance of coarse sand with even, hard lustrous grains. Since it absorbs moisture from atmosphere, it should be stored in a dry place. Though glauconites or green sands usually have lower exchange value but they are more rugged than the synthetic zeolites. The increase in SiO_2 value of a synthetic zeolite decreases its exchange value but increases its resistance to aggressive attack.

When hard water is passed through a bed of Permutit, the following reactions take place :

$$2SiO_2Al_2O_3Na_2O + Ca(HCO_3)_2 = 2SiO_2Al_2O_3CaO + 2NaHCO_3 \quad ...(12.11)$$

$$2SiO_2Al_2O_3Na_2O + CaSO_4 = 2SiO_2Al_2O_3CaO + Na_2SO_4 \quad ...(12.12)$$

$$2SiO_2Al_2O_3Na_2O + CaCl_2 = 2SiO_2Al_2O_3CaO + 2NaCl \quad ...(12.13)$$

Similar reactions take place with compounds of magnesium. The effect is to reduce the hardness almost to zero. For softening municipal waters, it is usually the practice to soften only a portion of water to zero hardness and then to mix it with unsoftened water so that the resulting hardness be of about 50 to 80 p.p.m.

Regeneration. Due to continuous use of the zeolite, the sodium present in it is exhausted. At this stage, zeolite is *regenerated* by passing a solution of salt (in the form of brine solution) through it. The sodium in the brine replaces the calcium and magnesium in the exhausted zeolite, which is thus restored, and the calcium and magnesium are discharged to waste with the wash water. The zeolite is now ready for use for softening more water. The following reactions take place during *regeneration* :

$$2SiO_2Al_2O_3CaO + 2NaCl = 2SiO_2Al_2O_3Na_2O + CaCl_2 \quad ...(12.14)$$

$$2SiO_2Al_2O_3MgO + 2NaCl = 2SiO_2Al_2O_3Na_2O + MgCl_2 \quad ...(12.15)$$

Zeolite softeners : The zeolite softeners are similar to a rapid gravity or pressure filters, except that the customary layer of filter sand is replaced by 75 to 150 cm thick or more of zeolite sand. The flow of water through the zeolite may be either upward or downward. The pressure type softener is sometimes equipped with automatic regeneration controlled to regenerate either at fixed periods

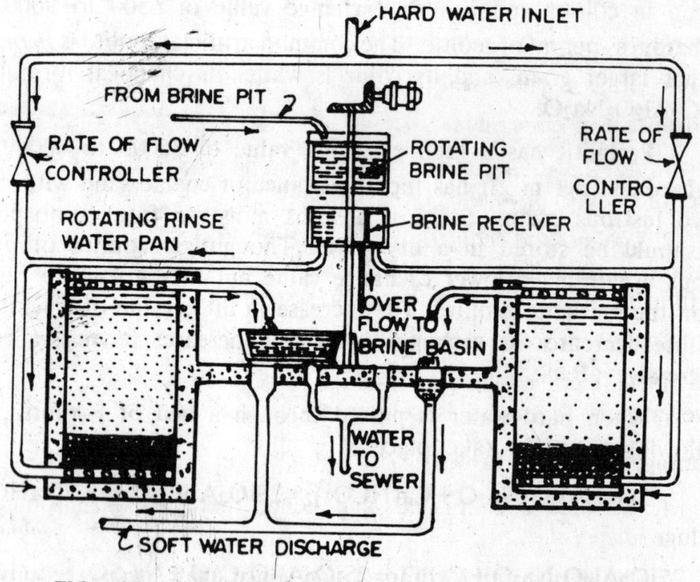

FIG. 12.3. AUTOMATIC GRAVITY ZEOLITE SOFTENER

of time or in accordance with the volume of water treated. The automatic softeners are designed for regeneration every two hours or more. Fig. 12.3 shows an automatic gravity zeolite water softener.

12.7. ADVANTAGES AND DISADVANTAGES OF LIME SODA AND ZEOLITE PROCESS

Advantages of Lime Soda Process

1. This process is suitable for turbid, chalybeate and acidic waters where zeolite process cannot be used.

2. The effluent contains lower total solids than from zeolite softening.

3. There is reduction in total mineral content of water.

4. The process has bactericidal effect.

5. This method removes iron and manganese also to some extent.

6. The method is better for excessively hard waters, particularly those high in magnesium hardness, and for water high in sodium.

7. The process can be easily added to an existing filter plant of any water supply scheme.

8. The process increases the pH value of water, which results in decrease in corrosion of the distribution system.

9. The process is more economical.

10. The process is free from patent complications.

Disadvantage of Lime Soda Process

1. A large quantity of sludge is formed in this process, and hence there are problems of its disposal.

2. The method requires recarbonation. In the absence of recarbonation, a thick layer of calcium carbonate will be deposited in the filtering media and distribution system.

3. The process requires skilled supervision.

4. It is not possible to produce zero hardness by this process. However, zero hardness is seldom required.

Advantages of Zeolite Process

1. No sludge is formed, and hence there are no problems of sluge disposal.

2. The zeolite unit is compact, and hence small space is required.

3. The unit can be easily operated. Highly skilled labour is not required.

4. In the zeolite process, any desired reduction of hardness is attainable. It is possible to get water of zero hardness. Hence this process is more useful to water softening for boilers and certain textile industries.

5. The first cost of the unit, and the operating cost are low.

6. The process is almost automatic.

7. The process is free from danger of excess chemicals in the effluent, since no chemicals are added to water.

8. The process is independent of the change in the quality of raw water.

9. There is no problem of deposition of layer of calcium carbonate or any other chemical in the distribution system.

10. The chemicals involved are cheap and easy to handle.

11. If pressure softner is used, repumping in not necessary.

Disadvantages of Zeolite Process

1. The method is unsuitable for highly turbid water, since suspended particles get deposited around the zeolite particles.

2. The process is unsuitable for water containing iron and manganese. Iron-bearing or manganese-bearing waters either deposit hydroxides on the surface of the zeolite or react with the zeolite to substitute iron or manganese for sodium in the zeolite, which is an irreversible reaction.

3. The process is unsuitable for acidic waters which irreversibly substitute hydrogen for sodium in the zeolite. The acidic water may aggressively attack the zeolite by the dissolving alumina or silica from it.

4. Zeolite softeners should be operated carefully to avoid injury to the zeolite, to the equipment, or to the quality of the water.

5. There is likelihood of growth of bacteria on the bed of zeolite. It should therefore be flushed annually with chlorinated water.

12.8. DEMINERALISATION OR DEIONISATION PROCESS

The conventional zeolites discussed in the previous article exchange the sodium cations (positive in electric charge) for the cations of calcium, manganesium and iron in the water ; the softening is essentially a cation-exchange process. In the *demineralisation* or *deionization process*, another zeolite is used which exchanges all the cations (including sodium) for hydrogen. These zeolites, available under the patented names of *Zeo-karbs*, *Catex* and *Organolite* are sometimes called *carbonaceous zeolites* since they are made from such materials as coal and lignite. Such a zeolite is also called a *hydrogen exchanger*. The chemical composition of such an exchanger is expressed as H_2Z, where H represents hydrogen ions and Z represents the organic part of the substance. *Zeokarb* is nonsiliceous and is resistant to aggressive chemical attack. Such zeolites are regenerated by the use of sulfuric acid or hydrochloric acid. The water treated with *hydrogen-zeolite* has the sulphates, chlorides and bicarbonates transformed into sulfuric, hydrochloric, and carbonic acids. The following reactions take place in the *cation exchange process* :

$$H_2Z + Ca(HCO_3)_2 = CaZ + 2CO_2 + 2H_2O \qquad ...(12.10)$$

$$H_2Z + Na_2(HCO_3)_2 = Na_2Z + 2CO_2 + 2H_2O \qquad ...(12.17)$$

$$H_2Z + CaSO_4 = CaZ + H_2SO_4 \qquad ...(12.18)$$

$$H_2Z + Na_2SO_4 = Na_2Z + H_2SO_4 \qquad ...(12.19)$$

$$H_2Z + CaCl_2 = CaZ + 2HCl \qquad ...(12.20)$$

Similar reactions take place for compounds of magnesium. The acidity in the water treated by hydrogen zeolite is un-desirable. In can be removed by (*i*) diluting treated water with raw water, (*ii*) neutralizing treated water with alkaline substance, or (*iii*) absorbing with *De-acidite* (D) or *anion exchanger* which is a zeolite-like proprietary substance regenerated with sodium carbonate or caustic soda solution. The following reactions take place in this *anion exchange process* :

$$D + (H_2SO_4 \text{ or } 2HCl) = D(H_2SO_4 \text{ or } 2HCl) \qquad ...(12.21)$$

$$D(H_2SO_4 \text{ or } 2HCl) = D + Na_2(SO_4 \text{ or } Cl_2) + H_2O + CO_2 \qquad ...(12.22)$$

The effluent from the above reaction produces only water and carbon dioxide. The water so produced has a quality equal to the best of the ordinary distilled water. It is completely free from mineral salts.

When hydrogen content of the zeolite gets exhausted it is regenerated with a solution of sulphuric acid or hydrochloric acid of suitable strength. The following reactions take place :

$$CaZ + H_2SO_4 = H_2Z + CaSO_4 \qquad ...(12.23)$$

$$CaZ + 2HCl = H_2Z + CaCl_2 \qquad ...(12.24)$$

The demineralisation process is too costly for municipal use. However, it has been found useful for manufacturing soft drinks and in industrial process where water free from mineral salts is required. However, the water to be treated should have a turbidity of less than 5 to 10 p.p.m. Since the entire process involves the use of acid for regenerating the *cation exchanger* (hydrogen exchanger) and alkali for regenerating the *anion exchanger*, the equipment used should be acid and alkali-resisting. The cylinder of the unit is generally made of transparent plastic, the pipe work is made of polythene while the valves are made of polythene or ebonite.

PROBLEMS

1. What do you understand by hard water ? Why is it necessary to remove the hardness of water. ?

2. Differentiate between permanent and temporary hardness. How do you remove temporary hardness ?

3. Explain the lime soda method of removing hardness of water. What are its advantages and disadvantages.

4. Write a note on water softening accelerator.

5. What is a zeolite ? How is it regenerated. Explain the zeolite method of water softening.

6. Compare lime-sada and zeolite process.

7. Explain the demineralisation process of water softening. Differentiate clearly between a cation exchanger and an anion exchanger.

13

Miscellaneous Treatment Methods

13.1. REMOVAL OF IRON AND MANGANESE

Iron and manganese are generally present in water supplies, either in suspension as hydrated oxides or in solution as bicarbonates. Water has iron dissolved in it as the result of carbon dioxide coming in contact with iron ore to form soluble ferrous bicarbonate. Iron in natural waters may be in ferric or ferrous condition, soluble, colloidal or insoluble. When manganese is also associated with iron, removal becomes more difficult. When they are present in amounts greater than 0.3 p.p.m., either alone or as a total, the following objectionable effects may be noted : (*i*) unpleasant taste and odour, (*ii*) cause staining of plumbing fixtures, clothing and textiles, (*iii*) accumulation of precipitated iron in water mains, (*iv*) growth of crenothrix in water mains, (*v*) cause troubles in various manufacturing processes, (*vi*) sulphate iron cause acidity and corrosive action on iron and brass. The reddish tinge in water is due to presence of iron while the brownish tinge is due to the presence of manganese.

When iron and manganese occur in water without organic matter, they can be removed by *aeration*, followed by coagulation sedimentation and filtration. By *aeration*, dissolved iron is oxidized to ferric oxide which is insoluble in water. Similarly, dissolved manganese compounds are converted to insoluble manganese compounds. The precipitated floc can be settled down in settling tanks or be further removed in gravity or pressure filters.

The following reaction takes place :

$$4Fe + O_2 + 10H_2O = 4Fe(OH)_3 \downarrow + 8H \quad \text{...(13.1 a)}$$

Hence 4 × 56 mg of Fe requires 2 × 16 mg of O_2. or 1 mg of Fe requires 0.14 mg of O_2.

A reaction period of about 5 minutes or less, at a pH of 7 to 7.5, and 0.14 mg of oxygen is required to covert 1 mg of ferric iron to ferric hydroxide.

When iron is present as ferrous bicarbonate, the following reactions take place :

$$Fe(HCO_3)_2 + 2H_2O = FeO + 2CO_2 + 3H_2O \quad \text{...(13.1 b)}$$

$$4FeO + O_2 = 2\,Fe_2\,O_3 \quad \text{...(13.1 c)}$$

$$Fe_2O_3 + 3H_2O = 2Fe(OH)_3 \downarrow \quad \text{...(13.1 d)}$$

Here also we find that 0.14 mg of O_2 is required for each mg of iron.

In case of mangenese, the following reaction takes place

$$6Mn + 3O_2 + 6H_2O = 6MnO_2 \downarrow + 12H \quad \text{...(13.2)}$$

This shows that 1 mg Mn requires 0.29 Mg of O_2.

Mangenese removal requires a pH adjustment upto 9.4 to 9.6. Prechlorination to free residual value upto 0.7 to 1.0 mg/l will affect the oxidation and precipitation of mangenese.

The above method of iron removal is most extensively used. The amount of oxygen required for the oxidation of ferrous iron is only about 0.14 part per part of iron, so that slight agitation with air is usually sufficient.

In order to accelerate the oxidation particularly, in waters with high carbon dioxide, addition of such alkaline substances as lime, soda ash or caustic soda is done. The iron removal is enhanced by catalytic action of previously deposited hydrates or oxides of iron.

Combination of iron and manganese, or iron alone, loosely bound to organic matter, make their removal difficult. The bond between them is broken by adding lime and thus raising the pH value to more than 9, so that iron and manganese can be precipitated. Alternatively, the aerated water is allowed to trickle over contact beds of coke, gravel, or crushed pyrolusite, followed by sedimentation and filtration. Filtration alone is ineffective due to biologic action dependent on a growth of organism on the filter sand. If organic acids are also present, aeration, dosage with lime, sedimentation and filtration are effective.

When water does not contain large amounts of iron or manganese, these can be removed by means of a *manganese zeolite*. As raw water passes through the bed of zeolite, the iron and manganese are oxidised to insoluble hydrated oxides that are removed by the mechanical filtering action of the zeolite bed. The bed must be washed and must be regenerated occasionally with potassium permanganate.

Oxidation of iron compounds may also be obtained by chlorination followed by sedimentation and filtration. The oxidation of iron, particularly of organic origin, can be effected by the use of potassium permanganate.

Iron and manganese can be removed in an oxidized condition, by softening water by excess lime process. A pH of 8.2 is needed for iron removal while a pH of 9.6 is needed for removal of manganese.

13.2. COLOUR, ODOUR AND TASTE REMOVAL

The water to be supplied in the public water supply scheme should be free from colour, odour and taste. The objectionable tastes and odours may be due to the following :

1. organic and vegetable matter.
2. industrial waste and domestic sewage,
3. dissolved gases,
4. dissolved mineral matter,
5. micro-organisms such as moulds, actinomycetes, iron and sulphur bacteria and algae.

Depending upon the causes, different methods can be adopted for the removal of taste and odour. Some of the methods described in earlier chapters, used for the removal of colour, odour and taste are:

(*i*) Coagulation followed by filtration.
(*ii*) Prechlorination.
(*iii*) Superchlorination followed by dechlorination.
(*iv*) Chloramine treatment.
(*v*) Use of chlorine dioxide.
(*vi*) Ozonation.
(*vii*) Removal of iron and manganese.

The other *special methods*, discussed here, are :

(*a*) Treatment by activated carbon.
(*b*) Use of copper sulphate.

Use of potassium permanganate and bleaching clay are also helpful in removing tastes and odours.

Superchlorination effectively removes tastes and odour. However, if insufficient chlorine is added, these tastes and odour may become quite severe. The excess of free chlorine breaks down the odour-producing substances of oxidation. Dechlorination is usually done to remove some of the excess chlorine, after giving sufficient contact time.

Aeration or *activated carbon* is especially applicable to the removal of tastes resulting from dissolved gases, whereas *chloramine* or *lime* is applicable in the removal of phenol tastes.

The tastes produced by micro-organisms can be objectively controlled by the use of *potassium permanganate.* It can be used alone or in combination with chlorine. Potassium permanganate has also been found useful in removing tastes and odours caused by solvent naptha wastes, rosin, soap and fermenting cereal waste.

Chlorine dioxide has been found effective against both phenolic and algae tastes. *Ozonation* improves the appearance of water, removes traces of colour, and removes odours if applied in sufficient quantity. However, it is always effective against 'earthy' and 'mouldy' tastes and odours.

13.3. ACTIVATED CARBON TREATMENT

Activated carbon is used for the control of tastes and odours in water, resulting from the presence of dissolved gases. It has also valuable colour removal properties. Activated carbon is produced by passing a carbonaceous material such as coke, charcoal, paper, char or sawdust through a heating mechansim at high temperature against a counter current of air, steam, carbon dioxide, chlorine or flue gases. Sometimes, chemical agents, such as phospheric acid or zine chloride may be added. This 'activation' and the greatly increased power of carbon to absorb gases and organic matters distinguishes activated carbon from charcoal. The activation of the carbonaceous materials removes the hydrocarbons which might interfere with the adsorption of organic matter. Activated carbon is very porous and has many carbon atoms with free valencies. It is available in granular as well as powder form, and is sold under trade names such as Darco, Minchar and Nuchar. Activated carbon must possess the properties of easy wetability, prolonged suspension and effective odour absorption capacity. Activated carbon removes organic contaminants from water by the process of adsorption. High surface area is the prime consideration in adsorption. Granular activated carbons typically have surface areas of 500-1400 m^2/g.

Activated carbon treatment has the following *functions*:

1. **It has been found effective in aiding coagulation, if adopted before filtration of water.**
2. It has valuable colour removing properties.

3. It is effective in preventing or retarding the decomposition of sludge in settling basins.
4. It is useful in removing tastes and odours due to excess chlorine, hydrogen sulphide, phenols and other elements.
5. It reduces the chlorine demand of treated water.
6. It adsorbs organic matter in water.

Activated carbon can be applied to water treatment in two ways : (*a*) as filter media (*b*) as fine powder feed.

As a filter media, granular activated carbon is used in place of the usual filter sand. Filtered water is passed through the bed of activated carbon. This will remove taste and odour, and will also adsorb organic compounds. The activated carbon filter bed, 75 to 100 cm thick is supported on about 35 to 50 cm thick graded gravel bed. When used in pressure filters, the tank interior should be coated against electrolytic corrosion. In the course of time, the carbon reaches the limit of its adsorptive powers and must be 'rejuvenated'. This is done by passing live steam into the carbon through steam pipes placed in the gravel.

The activated carbon in powder form may be applied to water at various stages of its treatment as follows :

1. It may be applied to raw water, ahead of treatment plant.
2. It may be applied in the mixing basin, either alone or fixed with other chemicals such as alum, by means of any of the types of dry chemical feed machines.
3. It may be fed before, during or after coagulation, at more than one point. This is known as *split treatment*. Usually, a portion is fed in the mixing basin and the balance just ahead of filter.
4. It may be applied just ahead of filters. Its rate of application is high when filter is washed, and becomes lower and lower as filter gets clogged.

The required dosage of powdered activated carbon is controlled by means of thresh hold odour test. For effective use, the dose must be adequate, the mixing thorough and the time of contact long enough for the material to carry out its intended work. Compared to dosing with powdered carbon, the use of carbon filter beds is more advantageous, since it is effective in removing not only the 'earthy' or 'mouldy' tastes or odours, but also in removing a wide range of complex organic substances such as pesticides and aromatic hydrocarbons.

TABLE 13.1
QUANTITY OF COPPER SULPHATE REQUIRED FOR KILLING DIFFERENT ORGANISMS

Organisms	*Odour*	*Copper sulpate (p.p.m.)*
1 *Diatomaceae :*		
Asterionella	Aromatic, fishy	0.10
Melosira		0.30
Synedra	Earthy	1.00
Navicula		0.07
2. *Chlorophyceae*		
Conferva		1.00
Scenedesmus		0.30
Spirogyra		0.20
Ulothrix		0.20
Volvox	Fishy	0.25
Xygnema		0.70
Coelastrum		0.30
3. *Cyamophyceae :*		
Anabaena	Moldy, grassy, vile	0.10
Clathrocystis	Grassy, vile	0.10
Oscillaria		0.20
Aphanizomenon	Moldy, grassy, vile	0.15
4. *Protozoa :*		
Euglena		0.50
Uroglena	Fishy, oily	0.05
Peridinium	Fishy	2.00
Chlamydomonas		0.50
Dinobryen	Aromatic, violet, fishy	0.30
Synura	Cucumber, fishy, bitter	0.10
5. *Schizomycetes :*		
Beggiafoa	Putrefactive	5.00
Crenothrix	Putrefactive	0.30

13.4. USE OF COPPER SULPHATE

Copper sulphate is used to serve two purposes :

1. Removal of colour, odour and taste from water.

2. Control of growth of algae, bacteria and aquatic weeds in the water reservoirs.

Copper sulphate is generally available in powder form or in crystal form. It may be applied either directly in the distribution pipes or in open reservoirs. In the former case, solution of copper sulphate of required strength is prepared and fed to water just ahead of the distribution system. The dose may vary from 0.3 to 0.6 p.p.m. For its application to the reservoirs, it is ground in fine powder and sprinkled on the water surface by means of a boat. Alternatively, it may be applied by means of sacks of the crystals which are dragged through water by means of boats. Table 13.1 gives the amount of copper sulphate necessary to kill various algae and other micro-organisms. The killing of algae may be troublesome because the tastes and odours will be temporarily increased. Hence smaller amounts are often successful as prophylatic doses to prevent trouble.

13.5. FLUORIDATION

It has been found that a fluoride concentration of 0.7 to 1.2 p.p.m. in water is beneficial for the prevention of dental caries in children. The allowable level of fluoride is determined by the annul average of the maximum daily air temperature, as is clear from Table 13.2.

TABLE 13.2.

TEMPERATURE EFFECT ON RECOMMENDED FLUORIDE CONCENTRATION IN WATER

Annual average maximum daily temperature (°F)	*Recommended flouride concentration (p.p.m.)*
50.0 to 53.7	1.2
53.8 to 58.3	1.1
58.4 to 63.8	1.0
61.9 to 70.6	0.9
70.7 to 79.2	0.8
79.3 to 90.5	0.7

Higher levels of fluoride (more than 3 p.p.m.) have been responsible for mottling (black to brown stains) of teeth. However, in areas where water is of low fluoride content, dental caries is high. Hence it is essential to add sufficient fluoride to their water supplies to bring the fluoride content to 1.0 to 1.2 p.p.m. The process of raising the fluoride content of water is known as *fluoridation*.

The fluoride compounds that are adopted for fluoridation are: *sodium fluoride* (NaF), *sodium silico fluoride* (Na_2SiF_6) and *hydrofluosilicic acid* (H_2SiF_6). Sodium fluoride, having 95 to 98% purity is most commonly used. For safer handling, hydrofluosilicic acid is better, since even if this is spilled on the skin, it can be removed easily and quickly by immediate washing in cold water.

The application of fluorides in water may be either in powder form or in the solution form. However, it is preferable to apply it in the solution form, since in the powder form as sodium fluoride or sodium fluosilicate it is toxic and must be contained in dust-tight hoppers and containers. It is best to apply fluoride after other treatments, but at point where thorough mixing of the fluoride with water can still take place before water leaves the treatment works. It is essential to maintain a close control of the dose.

13.6. DEFLUORIDATION.

As stated earlier, excessive concentrations (more than 3 p.p.m.) can cause dental fluorosis or mottled enamel when consumed during childhood years. When water containing 8 to 20 p.p.m. of fluoride is consumed over a long period of time, bone changes may occur, resulting in crippling fluorosis. Hence, when the fluoride concentration is more than 1 to 1.5 p.p.m., it should be removed from water. The process of reducing the fluoride concentration of water is known as *defluoridation*. Following are the principal methods of defluoridation:

1. *Calcium phosphates.* Bone has great affinity for fluorides and can be used in the filter for removal of fluorides. The bone is calcinated at 400°C to 600°C for 10 minutes followed by mineral acid treatment. It is then pulverised to pass 40 to 60 mesh (B.S.S.) and is used in the filter bed. The filter is regenerated with alkali and acid. One cubic metre of bone can treat 1 million litres of water containing 3.5 p.p.m. of fluoride.

2. *Bone charcoal.* It is essentially tri-calcium phosphate and carbon, and has been used sucessfully for the removal of fluorides.

3. *Synthetic tri-calcium phosphate.* It can be prepared from milk of lime and phospheric acid when the reaction is carefully controlled. This material has been used in contact filters for removal of fluorides. The regeneration can be done by 1% caustic soda followed by a dilute hydro-chloric acid or carbon dioxide wash to neutralise the excess alkali.

4. ***Fluorex.*** Fluorex is a special mixture of tri-calcium phosphate and hydroxyapatite. It is used as filter medium. Fluorex can be regenerated by washing with 1.5% solution of casutic soda. It can be next rinsed with twice its volume of water and the excess soda can be neutralised with the solution of carbon dioxide at 0.15% strength.

5. ***Ion-exchanger.*** There are a number of ion-exchanger materials which can be used for removal of fluorides. Benson reported that fluorides in water can be removed by successive passage through beds of cation-exchanger (Zeokarb-H) of the sulphonate coal type and an amin resin *'Nalcite B'*. Alum treated cation exchange resin from Avaram bark can be used as an effective material for removing fluorides from water.

6. ***Lime.*** Lime can also be used for reducing the fluoride content of water. Fluoride is removed alongwith the magnesium in a direct relationship to the amount of magnesium taken out in the softening process. The water must be treated to a caustic alkalinity of 30 p.p.m. and a pH value of 10.5; hence recarbonation is necessary. The process is suitable for hard waters containing less than 4 p.p.m. of fluorides.

7. ***Aluminium compounds.*** Filter alum is recognised as having a high fluoride absorption capacity, which can be further increased with coagulant aids, such as activated silica and clays. The fluorides are removed by the formation of an aluminium fluoride complex, or by absorption on the floc. Another method utilising aluminium salts involves the use of contact beds of insoluble materials impregnated with aluminium compounds. Dehydrated aluminium oxide (calcined alumina) can be used in contact beds for removal of fluorides.

8. ***Activated carbon.*** Removal of fluoride from water has also been effected by treatment with activated carbon, at pH 3.0. No removal takes place at pH 8 or above. The carbon when used as a contact bed can be regenerated with the weak acid and alkaline solution.

13.7. DESALINATION

Only about 0.5% of the earth's water is potable. The remaining 97% is ocean water and 2.5% is brakish water, both of which are impotable because of their dissolved salt content. Conversion of salt water to fresh water is not a new idea. The earliest application of desalination was the distillation of sea water on ships and the necessary apparatus was being installed on steam-ships as early as 1884. An example of the importance of desalting is the experience at Kuwait, an oil-producing Sheikhdom on the Persian Gulf on the eastern edge of the Arabian desert. The largest sea-water multistage flash distillation plant has been built for Hong Kong to give an output of 180 Ml/d, using six multistage flash evaportors. Apart from sea-water, the water from wells, containing high salt content, may also require desalination. Many wells in Rajasthan desert and Kutch area yield brakish water which needs desalination.

Following are common methods used for desalination :

1. Distillation.
2. Reverse osmosis.
3. Electrodialysis.
4. Freezing.
5. Solar evaporation.

1. DISTILLATION METHOD

This is the most commonly used method of converting saline water to potable water. In this method, saline water is boiled or evaporated in a suitable vessel to produce vapours of pure water and a liquid residue of more highly concentrated brine. The vapours are condensed to potable water. Sea-water is a complex mixture of dissolved solids, gases and marine life. The solids condense on the heat transfer surfaces and deposit a scale. Marine life can attach itself to the heat transfer areas and develop an insulating blanket, causing fouling. In order to minimize these effects, pretreatment of sea water is essential. Pretreatment of the raw brine through pH control and decarbonation will prevent the formation of magnesium hydroxide and calcium carbonate scale. Marine life is elininated by chlorinating the raw feed to kill any existing organisms and to prevent subsequent growth.

The principal factor in the design of evaporation system is the *conservation of energy*. The cost of water would be prohibitive if steam was used for evaporation on a once-through basis as in a single effect evaporator (Fig. 11.4). The following are various types of distillation plants :

(*i*) Vertical tube evaporators (VTE),
(*ii*) Multistage flash evaporators (MSF)
(*iii*) Multieffect multistage flash evaporators (MEMS), and
(*iv*) Combined processes (VTE/MSF),

(*a*) **Vertical Tube Evaporators (VTE).** Fig 13.1 (*a*) shows the single effect evaporator in which the steam from boiler heats and vaporises the salt water. The vapours are condensed to get fresh water. Fig 13.1 (*b*) shows the multiple effect evaporator in which water evaporates at the highest pressure in the first effect. This vapour is condensed in the second effect to evaporate an approximately

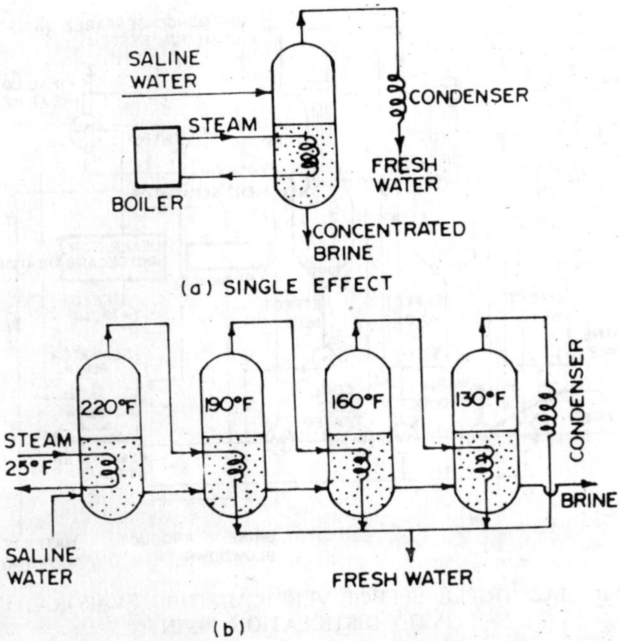

FIG. 13.1 VERTICAL TUBE EVAPORATION

equal amount of vapour. Similarly the third evaporator acts as condenser for the second and so on. The temperature in the successive evaporators will go on reducing and in order that the salt water may boil at these temperatures, the pressure in them are reduced by artifical means.

Fig. 13.2 shows a triple-effect vertical tube evaporator distillation plant, in which pretreated sea water is fed to the first of a series of stages, and flows downward through a series of tubes as a film coating the inside of the tubes. The water vapour produced in one stage serves as the steam supply for the next stage. In a recent development, fluted tubes are used in place of round tubes, thereby increasing the overall heat transfer performance 2 to 3 times.

***(b)* Multistage Flash Evaporation (MSF).** Fig 13.3 shows the principle of the multistage flash operation (MSF). After being heated to the highest temperature in the *brine heater*, the sea water passes through a number of compartments (flash chambers) wherein flashing of the brine occurs. The sea water passes while flash evaporating and cooling at successively lower pressures. The vapour released in flashing each stage condenses to heat the incoming sea water and gives fresh water. Most multistage-flash evaporators operate on a brine-recycling basis. The most successful evaporative desalting method

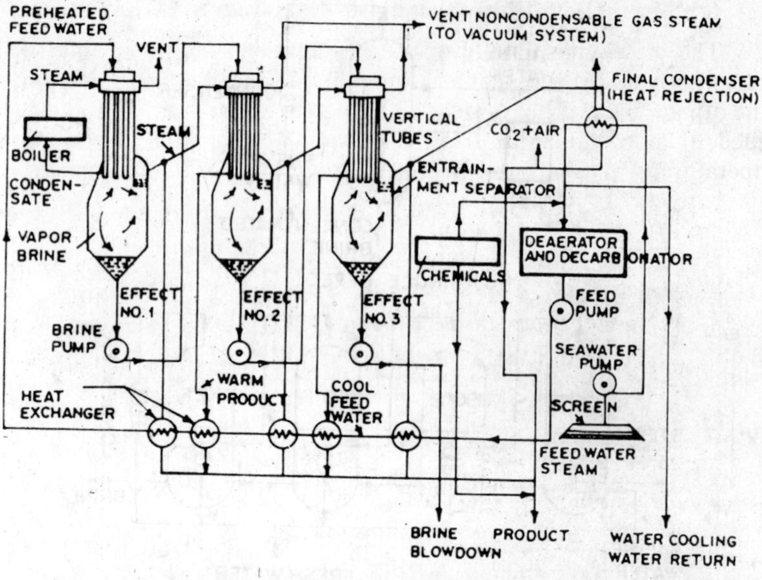

FIG. 13.2. TRIPLE EFFECT VERTICAL TUBE EVAPORATOR (VTE) DISTILLATION PLANT

is the MSF process. These units have been built with as many as 69 stages and have produced upto 20 kg of water per kg of prime steam used. The efficiency of the system can further be improved by flash enhancers and by combining the MSF system with multiple-effect evaporation process steps.

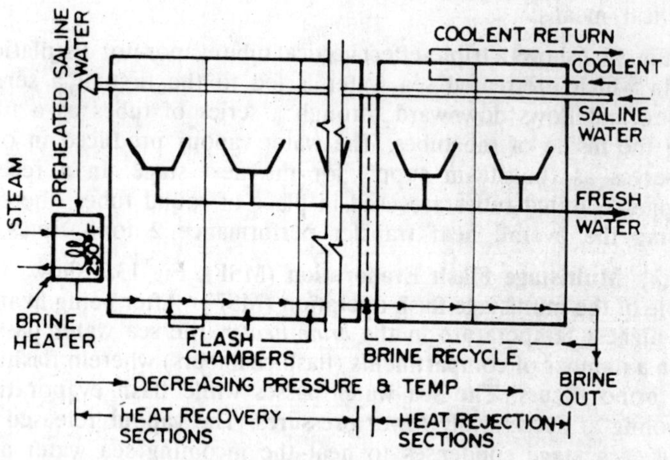

FIG. 13.3. MULTISTAGE FLASH EVAPORATION

(*c*) **Multieffect Multistage Flash Evaporation (MEMS).**

The multistage flash evaporation described above is only of *single effect* type with external heat input at one end and heat rejection at the other. The number of stages is limited by the pressure difference needed to move the brine from chamber to chamber at the lower temperature. Fig. 13.4 shows *multieffect-multistage distillation* process,

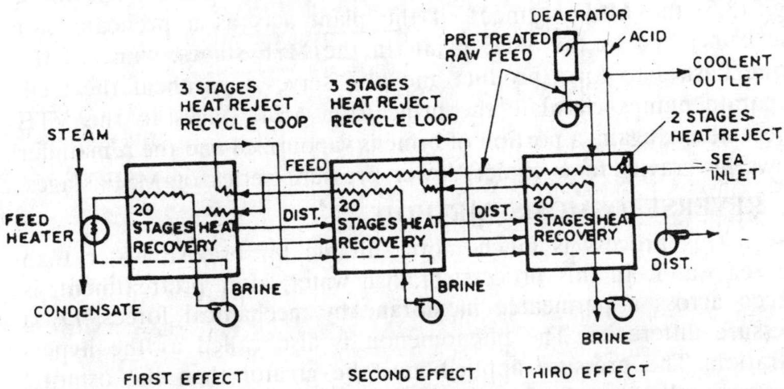

FIG. 13.4. MULTIEFFECT-MULTISTAGE DISTILLATION PROCESS

in which the overall temperature flashing range is broken into a number of temperature intervals. Within each interval, an individual single-effect multistage unit with its own heat input, heat recovery,

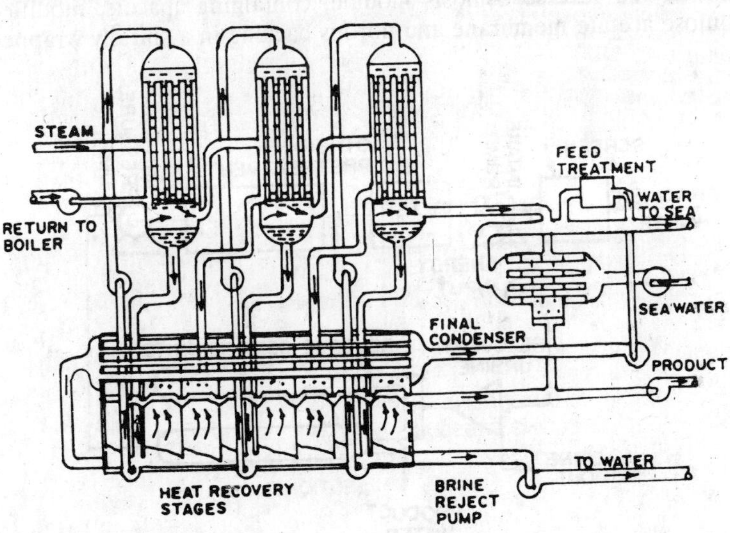

FIG. 13.5. COMBINATION OF VTE-MSF DISTILLATION PROCESS

heat rejection, and recycling stream operates. The first effect receives heat from the true external source while for the subsequent effects, the heat from the prior effect is rejected to the brine of the next effect. This gives a much better utilisation of net heat input over the single-effect process.

(*d*) Combined VTE and MSF Processes. This is the most recent innovation in desalting processes. In this process shown in Fg. 13.5, the MSF segment of the plant acts as a preheater for the brine feed to the VTE plant. In the MSF stages, some of the brine flashes to form product and also serves to preheat the feed. Separate pumps circulate the brine from MSF stages to the VTE effect where, again, a portion of brine is vapourized and the remainder flows by gravity to a cooler, lower pressure series of MSF stages.

2. REVERSE OSMOSIS METHOD

This process is much more suitable for brakish water than for sea water. In this process, brakish water, after pretreatment, is forced across a permeable membrane by mechanical forces, *i.e.* a pressure difference. The phenomenon is also called as the hyper-filtration. The pressure applied must be greater than the osmotic pressure. When salt water is fed into the reverse osmosis cells (modules), part of the water permeates the membrane and is collected as fresh water while the remainder (brine) is passed through a turbine for recovery of power before it is rejected.

Fig. 13.6 shows the out line of the unit. Fig. 13.7 shows a spiral-wound reverse osmosis module, containing spacers, modified cellulose acetate membrane and porous backing in a spirally wrapped

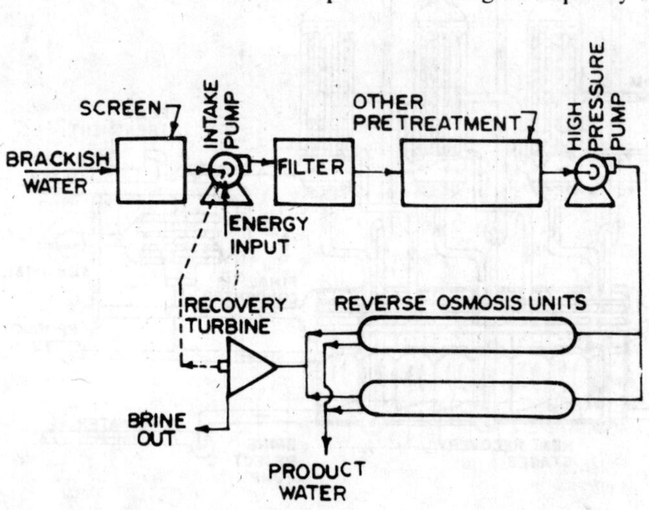

FIG. 13.6. REVERSE OSMOSIS FOR DESALINATION

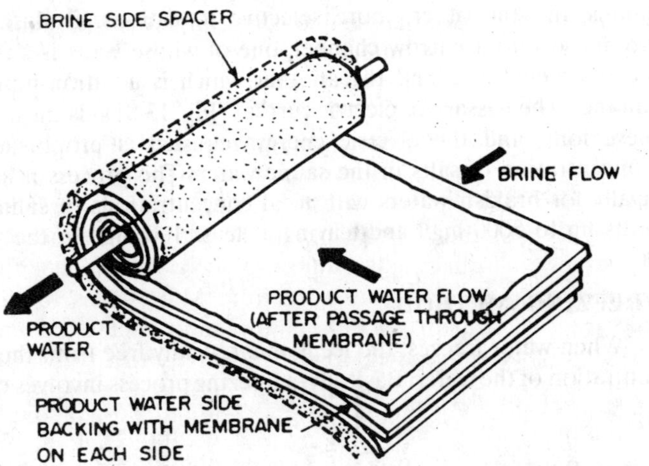

FIG. 13.7. SPIRAL WOUND CELLULOSE ACETATE MEMBRANE MODULE

double sandwich. The spiral wound design enables the utilization of far more membrane surface area than would be offered by a tubular sheet construction within the same size package.

3. ELECTRODIALYSIS METHOD

In the electrodialytic desalination process, salt ions are removed from saline water under the impetus of an electric current. The process uses membranes that are selective for the passage of ions of

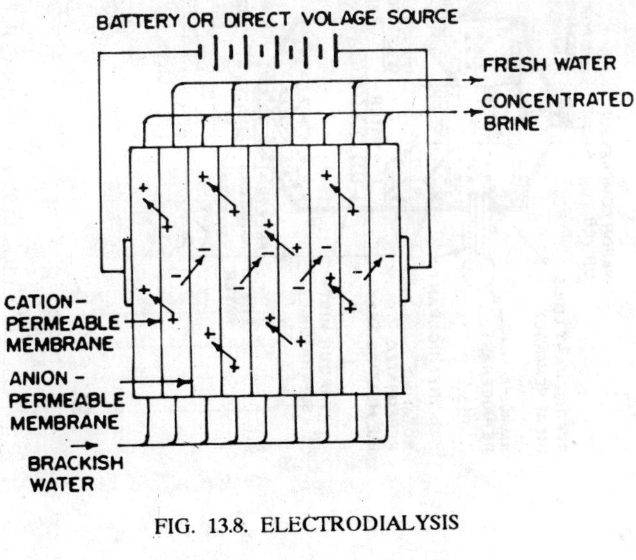

FIG. 13.8. ELECTRODIALYSIS

a given charge. Two different membranes are used – one more selective to anions and the other, more selective to cations. Brakish water is introduced into a narrow channel, one of whose walls is an anion-permeable membrane, and the another which is a cation-permeable membrane. The passage of electric current (Fig. 13.8) aids the diffusion of these ions, and the electric energy required is proportional to the concentration of salts in the saline water. The process is suitable especially for brakish waters with total concentrations of saline constituents up to 5000 mg/l and leaving at least 100 mg/l in the treated water

4. FREEZING METHOD

When water freezes, the ice is theoretically free from the saline concentration of the water. Basically, a freezing process involves cooling

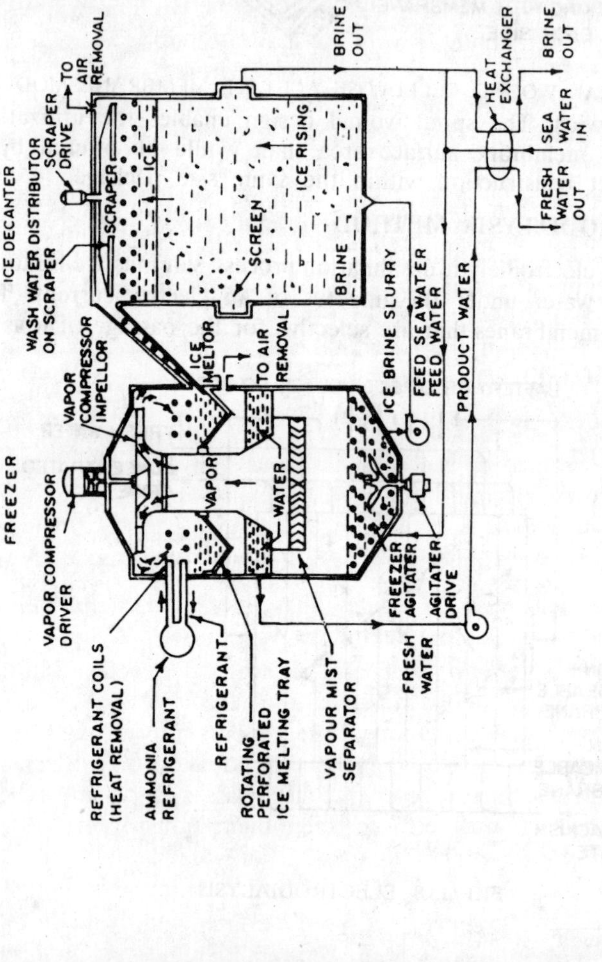

FIG. 13.9: ZARCHIN-COLT FREEZING PROCESS

the incoming sea water, freezing it to obtain fresh water ice, separating the ice and brine liquor, melting the ice to give fresh water and using the purified water and concentrated brine to chill the incoming sea water. Liquefied hydrocarbons (such as n-butane) can be used as refrigerants in direct contact with the brine. In the direct contact refrigerating system, the refrigerated butane is mixed with the sea water. In another system, known as the *pressure freezing process*, a high-boiling hydrocarbon is frozen and mixed with prechilled sea water. The ice is decanted from the mixture after the frozen hydrocarbon is melted, and is washed. It is then mixed with liquid hydrocarbon and is subjected to a pressure of about 140 kg/cm^2. The hydrocarbon freezes in consequence, while the product water from the melted ice is decanted.

Fig. 13.9 shows the Zarchin-colt freezing process which works on a *vacuum-freeze vapour compression* principle. In this process, water itself is the refrigerant. Prechilled sea water, approaching the freezing point of the brine, is introduced into a freezer. Evaporation of the water is induced by the suction of the compressor, due to which heat is absorbed from the remaining brine. The process yields ice crystals of about 0.5 mm size. The ice-brine slurry is pumped to the ice decanter where ice crystals float to the top. A rotating scrapper trims the ice layer at the top and sends the washed crystals to the freezer where they are brought in contact with the compressed water vapours. The water vapour condenses on the ice crystals suspended on a rotating perforated tray, resulting in their melting and thus yielding the desired fresh water. This cool product water is circulated through a heat exchange to cool the incoming sea water.

5. SOLAR EVAPORATION METHOD

All the desalination methods described above need a lot of thermal energy or mechanical power. In the solar evaporation, solar radiation is used as the source of heat energy. The solar enegy striking the earth is estimated to be about 250 BTU/day/m^2.

The equipment required for desalination is rather simple. Fig. 13.10 shows an experimental green-house type still, which may yield about 5 kg of fresh water/day/ m^2 of basin surface, under good sunshine conditions. Salt water is contained in a blacked bottom shallow trough which absorbs the solar energy. The vapours rise and condense on the glass or plastic surface, so inclined as to cause the collected water to flow to a common reservoir. The glass surfaces must be frequently cleaned off dust etc., to keep them transparent.

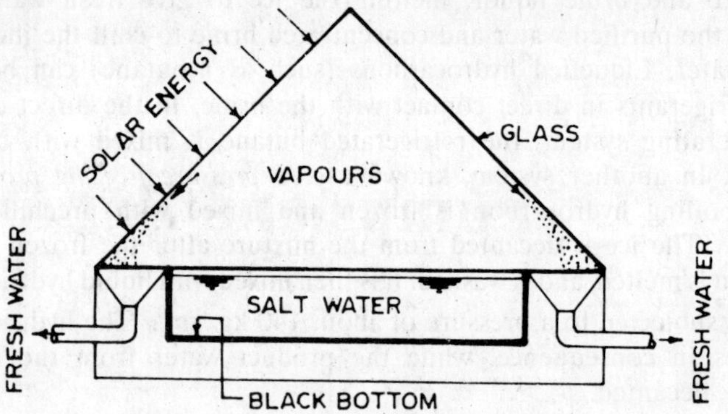

FIG. 13.10. GREEN HOUSE-TYPE STILL

PROBLEMS

1. Explain the methods of removal of iron and manganese from water. Why is their removal necessary ?

2. What are the causes of tastes and odours in water available from various sources ? Enumerate the methods of their removal.

3. Write a note on activated carbon treatment.

4. What do you understand by fluoridation ? Why is it necessary?

5. Describe various methods of removing excess fluorides from water.

6. What do you understand by desalination ? Why is it necessary ? Explain the electrodialysis method of desalination.

7. Explain the multistage flash evaporation method for desalination.

14

Pumps and Pumping

14.1. NECESSITY OF PUMPING

In very few water supply schemes, the supplies originate from mountaneous areas such that the consumers get water supply entirely by gravity. In majority of cases, pumping is required to lift the water from a river, lake or reservoir to the treatment plant. After the treatment, another lift may be required either to supply water directly into the mains or to the overhead storage tanks from where water may flow under gravity. In some cases, direct pumping may be necessary to increase pressure in the supply line. When wells are the source of supply, pumping may be necssary to lift the water from the wells.

The pumping is therefore resorted to for the following purposes:

1. To lift the raw water from the source of supply, such as lake, reservoir, river or well.
2. To lift the treated water to overhead tanks or reservoir.
3. To deliver treated water to the consumer's taps at reasonable pressure.
4. To boost the line pressure.
5. To supply water under pressure for fire hydrants.
6. For miscellaneous operations at the water treatment plants, such as (*i*) for backwashing of filters (*ii*) for pumping chemicals, and (*iii*) for dewatering of tanks, basins, sumps etc.

14.2. TYPES OF PUMPS AND THEIR CHOICE

Pumps can be classified on the basis of the following :

(*i*) mechanical principle of operation,

(*ii*) type of power required, and
(*iii*) type of service called for.

(*a*) **Classification based on mechanical principle of operation:**

Based on the principle of operation, pumps can be broadly classified into the following four types :

(*i*) Displacement pumps.
(*ii*) Centrifugal pumps.
(*iii*) Air lift pumps.
(*iv*) Miscellaneous pumps.

(*b*) **Classification based on type of power required :**

(*i*) Steam engine pumps.
(*ii*) Diesel engine pumps.
(*iii*) Electrically driven pumps

(*c*) **Classification based on the type of service called for :**

(*i*) Low lift pumps.
(*ii*) High lift pumps.
(*iii*) Deep well pumps.
(*iv*) Booster pumps.

The selection of a particular type of pump depends upon the following factors :

(*i*) Capacity of the pump.
(*ii*) Number of pump units required.
(*iii*) Suction conditions.
(*iv*) Lift (total head).
(*v*) Discharge conditions, and variations in the load.
(*vi*) Floor space requirement.
(*vii*) Flexibility of operation.
(*viii*) Starting and priming characteristics.
(*ix*) Type of drive required.
(*x*) Initial cost and running costs.

Reciprocating pumps are more suitable for lifting relatively clean water, against high and fluctuating head. Centrifugal pumps are well suited to pump under steady low head, though the water may be even turbid. They are therefore mostly used for lifting raw water from the river to the treatment plants, and then lifting treated water to the distribution reservoirs.

14.3. DISPLACEMENT PUMPS

The positive displacement pumps are those in which liquid is sucked and then it is actually pushed or displaced due to the thrust exerted on it by a moving member, which results in lifting the liquid (water) to the desired height. The pump consists of one or more chambers which are alternatively filled and emptied with the liquid to be pumped. The displacement pumps are of two types:

(*a*) Reciprocating pumps

(*b*) Rotary pumps

(*a*) **Reciprocating pumps.** This is the most common type of the displacement pumps. It essentially consists of a piston or a plunger which moves to and fro in a close fitting cylinder which is connected to the suction and delivery pipes, as shown in Fig. 14.1. The suction and delivery pipes are provided with a non-return or one-way valve which admits water in one direction only.

Fig. 14.1 (*a*) shows the *single acting reciprocating pump.* During the suction stroke, when piston or plunger moves outwards, a partial vacuum is created in the cylinder, which enables the atmospheric pressure acting on the liquid surface in the sump below to force the liquid up the suction pipe and fill the cylinder by forcing open the suction valve. During the inward motion of the plunger, the liquid in the cylinder is forced up the delivery pipe; during this operation, the suction valve *S* closes while the delivery valve *D* opens. The cycle of suction stroke and delivery stroke is then repeated. The piston or plunger is moved by a crank rotated at uniform speed by a driving engine or motor.

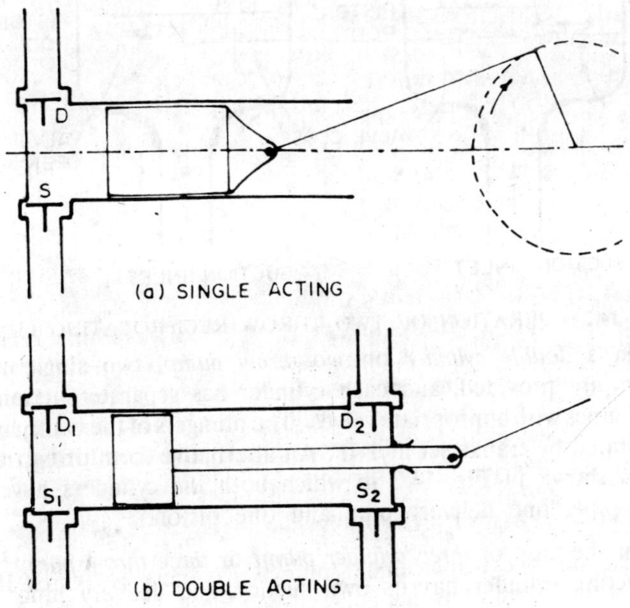

FIG. 14.1. RECIPROCATING PUMP

Fig. 14.1 (*b*) shows a *double acting reciprocating pump*, in which liquid is in contact with both the sides of the plunger. It has two suction and two delivery pipes with appropriate valves. During such stroke, suction takes place on one side of the piston while the other side delivers the liquid. Due to this, there are two suction strokes and two delivery strokes in one revolution of the crank.

Depending upon the number of cylinders provided, reciprocating pumps may be classified as follows :

(*i*) Single cylinder pump
(*ii*) Double cylinder pump
(*iii*) Triple cylinder pump
(*iv*) Duplex double action pump
(*v*) Quintuplex pump

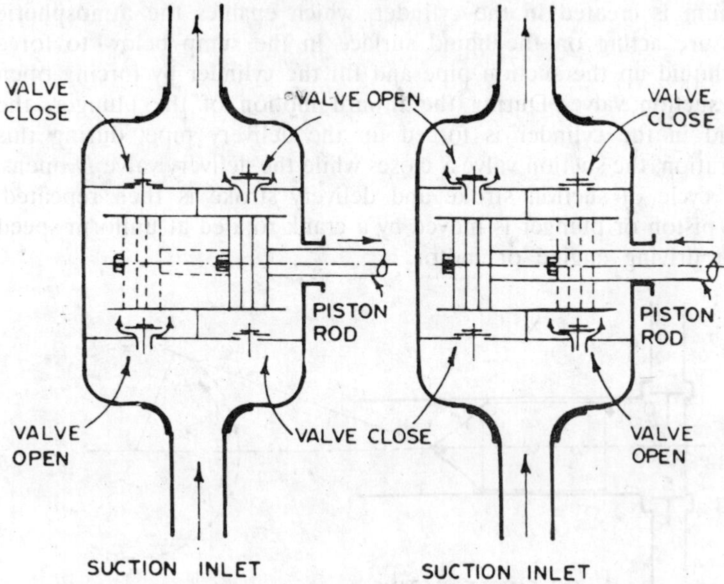

FIG. 14.2. OPERATION OF TWO THROW RECIPROCATING PUMP

In a *double cylinder* or *two throw pump*, two single acting cylinders are provided, and each cylinder has separate suction and delivery pipes with appropriate valves. The plungers of the two cylinders are operated by cranks set at 180°. An alternative form of two throw pump is shown in Fig. 14.2 in which both the cylinders have one suction pipe, one delivery pipe and one piston.

In the case of *three cylinder pump* or *three throw pump*, each single acting cylinder has its own suction and delivery pipes with appropriate valves, with their cranks set at 120°. A *duplex double action pump* is equivalent to a *four throw pump*, and is formed by combining two-double acting-double cylinder pumps. A *quintuplex* or *five throw pump* has five single acting cylinders, with separate suction and delivery, with their cranks set at 72°.

(*b*) **Rotary pumps.** In rotary pumps, two gears or cams mesh together and rotate in opposite directions. The rotating elements fit closely in the casing, so that water is drawn up the suction pipe, trapped between the teeth or cams and the casing and forced out into the discharge pipe.

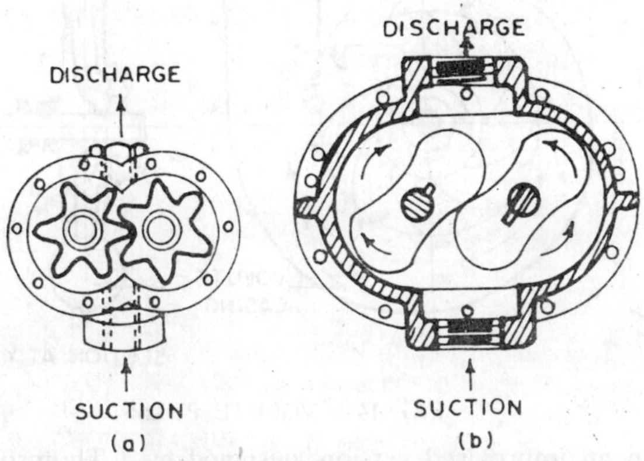

FIG. 14.3. ROTARY PUMP

In the rotary pump, no valves are required and the flow is practically continuous. The pumps which use the above principle of rotary displacement are : gear, lobe, screw, moyno and vane pumps. These pumps are not in common use in water works.

14.4. CENTRIFUGAL PUMPS

When a certain mass of liquid is made to rotate by an external force, it is thrown away from the central axis of rotation and a centrifugal head is impressed which enables it to rise to a higher level. A centrifugal pump works on this principle. The centrifugal pump employs a rotating impeller with water being drawn at the *hub* or *eye* of the pump impeller and discharged through a spiral casing by virtue of the centrifugal force developed due to the rotation of the impeller. In addition to the centrifugal action, as the liquid passes through the revolving wheel or impeller, its angular momentum changes, which also results in increasing the pressure of the liquid.

Based on the types of casing provided, centrifugal pumps are classified into two types :

(a) Volute pump
(b) Diffuser or turbine pump.

(a) **Volute pump.** In the case of volute pump, the impeller is surrounded by a spiral shaped casing, commonly known as a *volute chamber,* as shown in Fig. 14.4. The shape of the casing is such that the sectional area of flow around the periphery of the impeller gradually increases from the tongue (*T*) to the delivery pipe.

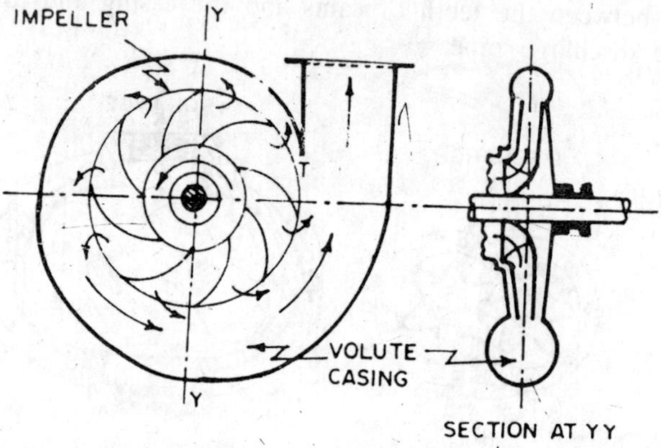

FIG. 14.4. VOLUTE PUMP

In an improvised version, designed by J. Thomson, a circular chamber, known as *vortex chamber*, is provided between the impeller and the volute chamber. The vortex chamber serves the dual purpose of reducing the velocity and increasing the efficiency of the pump by converting a large amount of kinetic energy into pressure energy.

(b) **Diffuser or Turbine pump.** In diffuser or turbine pump, a diffuser ring containing a series of guide vanes, surrounds the impeller. The diffuser ring remains stationary while the impeller rotates inside the ring. The liquid after leaving the impeller passes through the gradually enlarged passages of the guide vanes. Due to this, the velocity of flow gradually decreases while the pressure increases. After passing through the guide vanes, the liquid flows into the surrounding casing which may be either circular or volute shaped. A diffuser pump is a reverse turbine, and hence it is known as a turbine pump.

A centrifugal pump may be further classified into a *single stage* or *multi-stage* pump depending upon the number of impellers mounted on a shaft. A single stage pump has only one impeller while a multistage pump has two or more impellers connected in series. The water from the first is collected in a casing or volute and discharged into the eye of the second impeller, and so on. As many as six stages are being sometimes employed; and each stage increases the head from 30 to 45 metres. Multistage pumps are very much used in deep wells where the impeller diameter must necessarily be small.

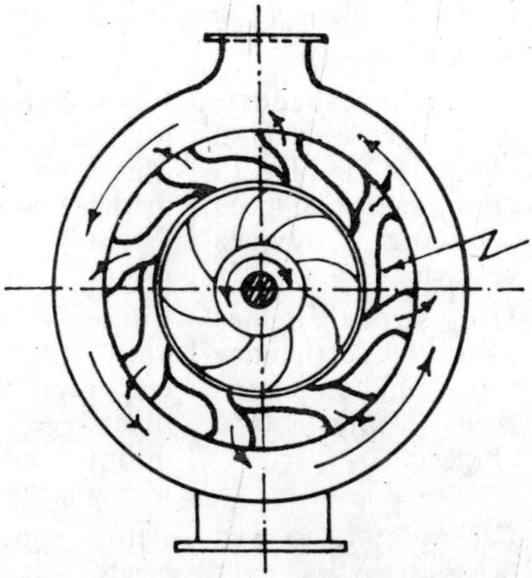

FIG. 14.5. DIFFUSER OR TURBINE PUMP

Based on the direction of flow, centrifugal pumps are classified as *radial flow pump*, *mixed flow pump* and *axial flow pump*. Normally a centrifugal pump is of radial flow type in which liquid flows through the impeller in radial direction. In axial flow type, liquid flows through the impeller in axial direction. They are used to deliver large quantities at small heads. In the mixed flow type, there is a combination of radial and axial flows.

Similarly, based on the number of entrances to the impeller, a centrifugal pump may be classified as *single suction pump* or *double suction pump*. In the double suction pump, liquid enters from both the sides of the impeller. Due to this, the thrust on the impeller is neutralised. Pumps are known as *horizontal* or *vertical* when their shafts are horizontal or vertical. The impellers of the centrifugal pump may be of two types : *open* and *closed*. Open impellers are suitable for use with liquids containing solids, such as sewage etc. However, closed impellers are more efficient, and are commonly used for water supply works.

14.5. COMPARSION OF RECIPROCATING AND CENTRIFUGAL PUMPS

The comparison between the two types is given below

	Characteristics	*Reciprocating Pumps*	*Centrifugal Pumps*
1.	*Discharge*	(*i*) Flow pulsating (*ii*) Constant discharge even if head on delivery side varies. (*iii*) Large amount of losses due to leakage through valves etc.	(i) Continous flow. (*ii*) Does not give constant discharge under variable head. (*iii*) No such trouble.
2.	*Speed*	Run at low speeds ; piston speed less than 30 m/min. If coupled to driving machines, speed reducing devices necessary.	Run at high speed (500 to 1000 r.p.m.). Direct connection with driving machines possible.
3.	*Efficiency*	Long stroke engines have 85% efficiency while smaller pumps have only 40% efficiency.	Efficiency between 40 and 85%; more efficiency if the head and discharge are maintained within narrow limits.
4.	*Head*	Can work against high heads.	Better suited to low heads.
5.	*Size*	Large size, bulky and heavy. Requires large space.	Small, compact and light.
6.	*Initial and maintenance cost*	High initial cost; maintenance cost is also high.	Low initial cost. Maintenance cost low.
7.	*Running*	Starting is difficult ; more power required ; complicated mechanism required ; operation difficult.	Starting is easy; less power required; simple mechanism; operation easier.

14.6. JET PUMP

The so-called jet pump is combination of a conventional water pump (such as a centrifugal pump) and an ejector or educer, commonly referred to as a jet. In a jet pump, a jet of water is injected at high velocity into the throat of a venturi tube, as shown in Fig. 14.6. The high velocity stream pulls the water upwards around the tip of the nozzle. This in turn creates a suction at that point. The suction draws in more water from the source which in turn is carried upward with the high velocity stream. As the water passes upward into the enlarged portion of the venturi, the velocity decreases while pressure increases. Thus velocity is converted into pressure which

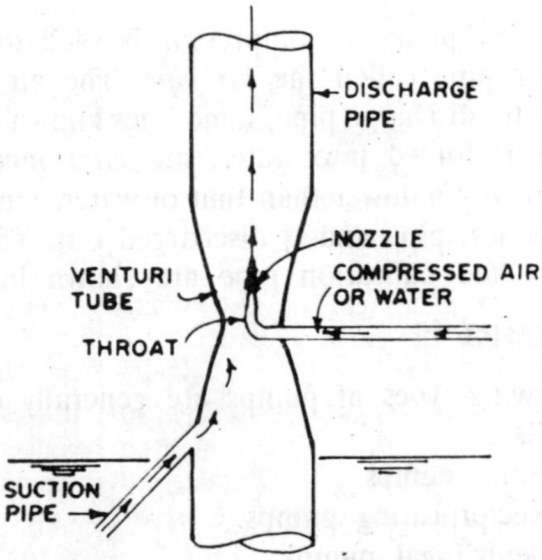

FIG. 14.6. JET PUMP

forces the water upto within suction distance of the centrifugal pump. The water then flows to the intake of the pump and is forced out the discharge side of the pump by the impeller. A byepass leads water back to the nozzle. Thus, a portion of water is constantly recirculated through the pump and the jet. A pressure regulating valve on the discharge pipe serves to maintain the correct pressure on the jet.

The main advantage of a jet pump is that the motor and the pump can be conveniently placed at the ground surface.

14.7. AIR LIFT PUMPS

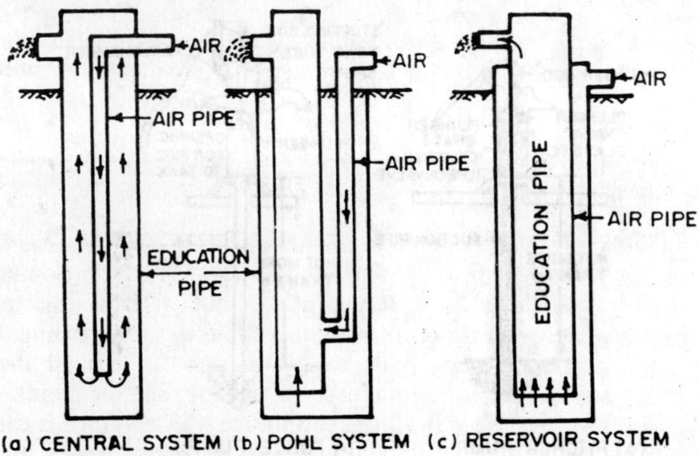

FIG. 14.7. SCHEMATIC DIAGRAM OF AIR LIFT PUMP

In an air lift pump, a compressor is used to force air into a small diameter pipe called the *air pipe.* The air pipe is placed in a bigger diameter discharge pipe, sometimes known as the *education pipe.* When air is forced into water, air-water mixture is formed whose specific gravity is lower than that of water. Due to this, water rises in the eduction pipe and is discharged out. The various ways of arranging air and education pipe are shown in Fig. 14.7.

14.8. WELL PUMPS

The following types of pumps are generally used for lifting water from wells :

(*a*) Hand pumps.
(*b*) Reciprocating pumps.
(*c*) Centrifugal pumps.
(*d*) Air lift pumps.
(*e*) Jet pumps.

(*a*) **Hand pumps**: Hand pumps are used only for shallow wells. The commonly used two types of hand pumps are (*a*) *pitcher pump* and (*b*) *force pump.* A pitcher pump essentially consists of a cylinder in which a plunger with a valve moves up and down, and a check valve at the bottom of the cylinder, as shown in Fig. 14.8 (*a*). It requires priming.

The principle of operation of a force pump is the same as that of pitcher pump, except that it is enclosed at the top and therefore can be used to force the water to higher elevations, or against pressure.

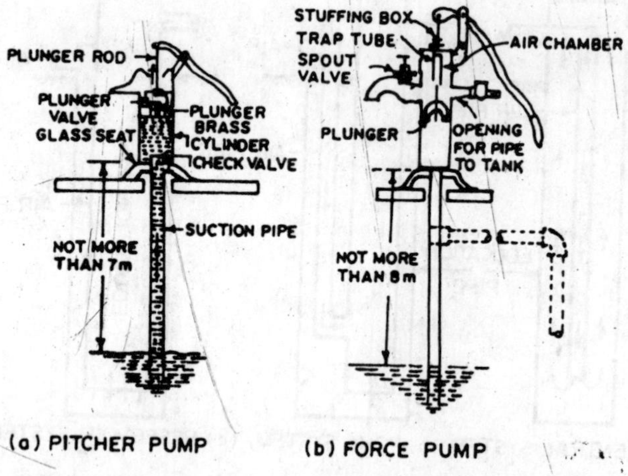

FIG. 14.8. HAND PUMPS

Such pumps also have air chamber to even out the discharge flow (Fig. 14.8 *b*).

(*b*) **Reciprocating type well pumps.** Reciprocating type pumps are used for deep wells of very small capacity and for wells of very great depth beyond the field of the deep well centrifugal pump. Wind mills may be used as source of power. The pump consists

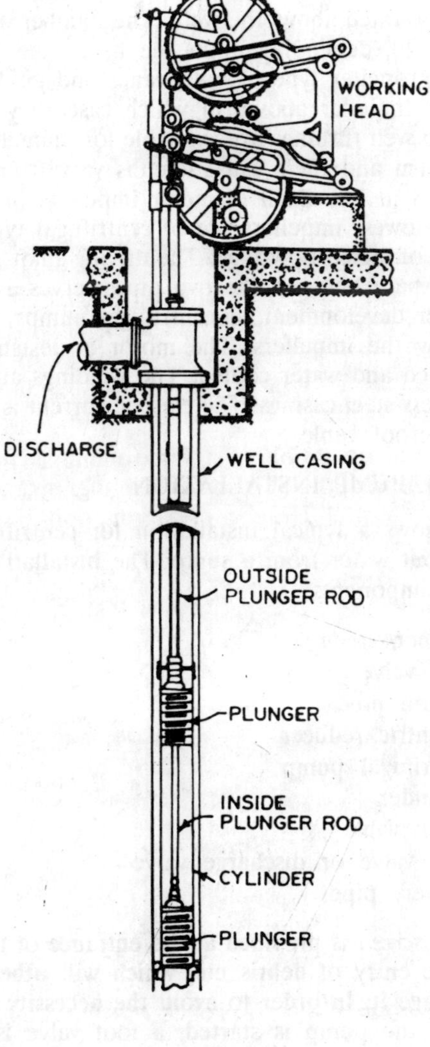

FIG. 14.9. THE TWO STROKE WELL PUMP

of three essential parts (Fig. 14.9) : (*i*) the power head, (*ii*) the pump rods, and (*iii*) working barrel or cylinder. The power head is placed on the ground surface. The pump rods run from the power head to the working barrel and connect the plungers. Fig. 14.9 shows a two stroke well pump with two plungers. Both the plungers move independently, because one of the rods is hollow and permits the solid rod to move within it.

(*c*) **Centrifugal well pumps**. Centrifugal pumps used in a deep well may have a number of impellers mounted on a shaft driven by a motor mounted above the well. The number of impellers may be as high as 20, depending upon the head, and may be of the mixed flow or propeller type or with vanes and guide passages leading to the next impeller above, in which case they are called *turbine pumps*. Deep well turbines are available for diameters varying from 10 cm to 35 cm and their usual lengths vary from 20 to 60 cm. For each 5 to 8 m lift, small diameter impellers or bowls are used in series. The lowest impellers are of centrifugal type and the upper impellers are of mixed flow type. The driving shaft is enclosed in a pipe, and the space between the two pipes serves as an outlet for water. In a later development of centrifugal pumps, the motor is placed just below the impellers. The motor is designed to run in water, is lubricated and water cooled. The windings are enclosed in water tight stainless steel casings. The electric current is conducted down by a water proof cable.

14.9. CENTRIGUAL PUMP INSTALLATION

Fig. 14.10 shows a typical installation for centrifugal pump used for pumping out water from a sump. The installation consists of the following components.

(*i*) Strainer/screen
(*ii*) Foot valve
(*iii*) Suction pipe
(*iv*) Accentric reducer
(*v*) Centrifugal pump
(*vi*) Expander
(*vii*) Check valve
(*viii*) Gate valve or discharge valve
(*ix*) Delivery pipe.

A *Strainer* or *screen* is provided at the entrance of the suction pipe to prevent the entry of debris etc, which will otherwise clog the pump and damage it. In order to avoid the necessity of outside priming every time the pump is started, a foot valve is provided at the end of the suction pipe. When the pump is stopped, the

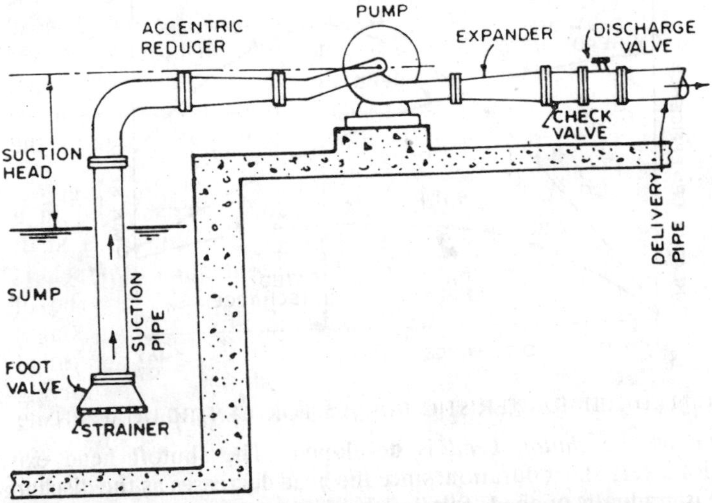

FIG. 14.10. CENTRIFUGAL PUMP INSTALLATION

foot valve will prevent water from leaving the pump. Hence foot valve serves as a *self-priming device*. The suction pipe and the delivery pipe are provided of the same diameter. However, for economy, the diameter of pump casing at the inlet end as well as the discharge end are often kept *smaller*. In order to prevent air accumulation when the pump is stopped, an *accentric reducer* is provided as shown. A *check valve* is provided at the beginning of the delivery pipe, to prevent back flow through the pump, if there is a power failure. In addition to this, a *gate valve* is also provided in the beginning of the delivery pipe. The *gate valve* is initially kept closed, and is then gradually opened as the pump picks up the speed, so that the discharge pipe is filled with water and full back pressure exists on the pump. If this is not done, and if the pump is started with the discharge valve open, with the discharge pipe empty, the motor may be overloaded, resulting in its burning out.

14.10. CHARACTERISTICS OF CENTRIFUGAL PUMP

Fig. 14.11 shows typical characteristic curves for centufugal pump. Curse I is between head and discharge, curve II is between the efficiency (η) and discharge while curve III is between BHP and the discharge.

As the discharge increases, the developed head decreases. A given pump will not work below a certain *minimum head*, hence maximum discharge of the pump will be limited by that minimum head. Initially, when the discharge valve is closed, maximum head,

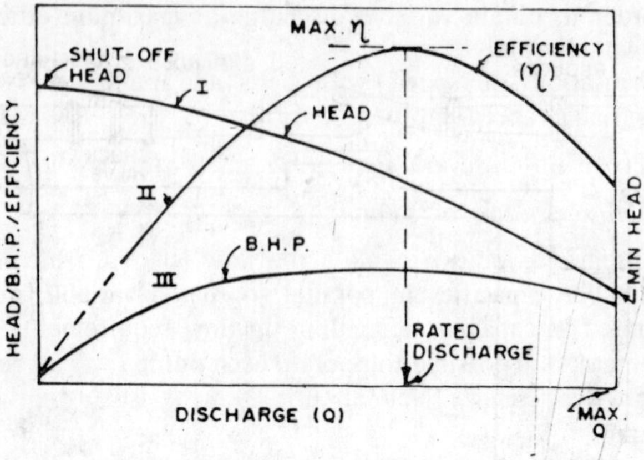

FIG. 14.11. CHARACTERISTIC CURVES FOR CENTRIFUGAL PUMP.

known as the *shutoff head* is developed. The shutoff head exists only for a very short duration, since the head decreases as the discharge valve is gradually opened. Similarly, as is clear from curve II, efficiency of the pump increases as the discharge Q increases, till it reaches its maximum value, and then decreases. The discharge at the maximum efficiency is known as the *rated discharge.*

Pumps may be more explicitly defined by the parameter called *specific speed* which is defined as the speed at which a pump will discharge a unit flow under a unit head at maximum efficiency. It is expressed by

$$N_s = \frac{N\sqrt{Q}}{H^{3/4}} \qquad ...(14.1)$$

where Q = discharge in litres per second

H = Head in meters

N = rotational speed, in rpm.

TABLE 14.1 SPECIFIC SPEED

Pump Type	*N_s Range*
Centrifugal	Upto 2600
Mixed flow	2600–5000
Axial flow	5000 to 10000

14.11. MULTIPLE PUMP SYSTEMS

When variable discharge is required from a pump working at a particular speed, it cannot function at its maximum efficiency.

Hence in order to obtain variable discharge at maximum efficiency, there are two alternatives :

(*i*) Variation of speed with belts and pulleys or with a variable speed motor

or (*ii*) Provision of several pumps of small capacity in parallel.

Parallel operation of pumps

The preferred way to regulate the flow is to provide several pumps of smaller capacities in parallel so that a variable number can be operated at capacity, depending on flow requirements. With this arrangement, it is possible to operate each pump (and all pumps) near maximum efficiency. Such an arrangement is shown in Fig. 14.12.

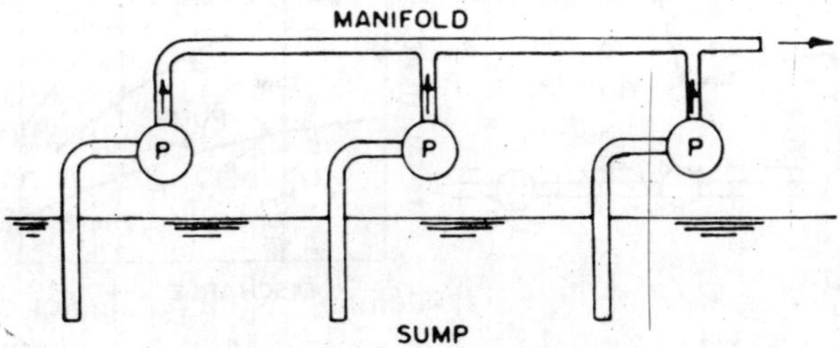

FIG. 14.12. PUMPS OPERATING IN PARALLEL

By automatic switching according to the level in the suction well, any of the pumps can be brought into operation. In predicting the head versus discharge curve for parallel operation, it is assumed that the head across each pump is the same. Thus, at any arbitrary head, the individual pump discharges are added as shown in Fig. 14.13.

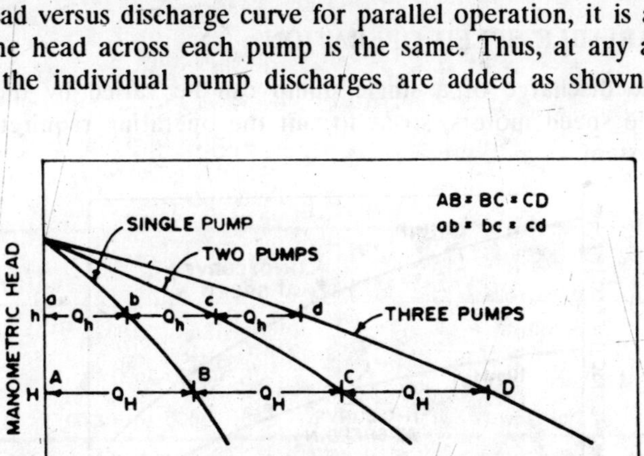

FIG. 14.13. CHARACTERISTIC CURVES FOR IDENTICAL PUMPS IN PARALLEL

Series operation of pumps

Such an operation is commonly adopted in multistage and borehole pumps, wherein the discharge from the first pump (or stage) is delivered to the_inlet of the second pump (or stage) and so on. Thus, in series operation, the same discharge passes through each pump while the pressure (or head) is boosted in doing so. Fig. 14.14 (*a*) shows the series configuration. Fig. 14.14 (*b*) shows the resulting head versus discharge characteristics which are clearly obtained by adding the individual pump manometric heads at arbitrary discharges.

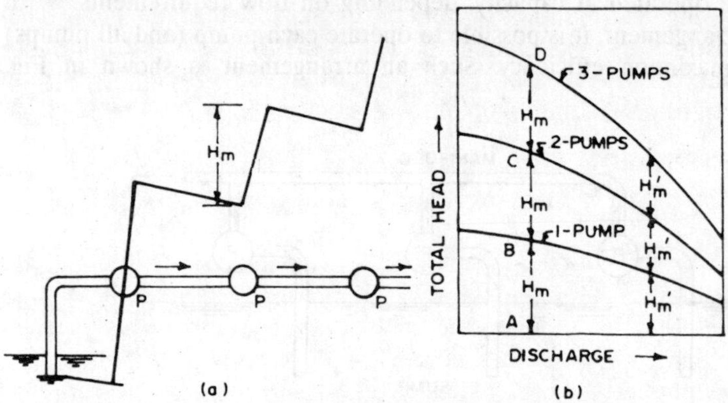

FIG. 14.14. PUMPS OPERATING IN SERIES

It is to be noted that all the pumpes in series operation are to be operated simultaneously.

14.12. VARIABLE SPEED OPERATION

The discharge of a single pump can be varied by the use of variable speed motors, so as to suit the operating requirements of the system.

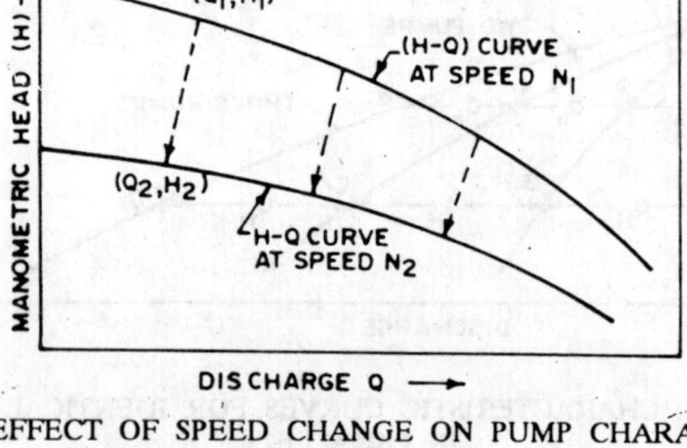

FIG. 14.15. EFFECT OF SPEED CHANGE ON PUMP CHARACTERISTICS

Using dimensional analysis and dynamic similarity criteria, it can be shown that if a pump delivers a discharge Q_1 at a total rise in head H_1 runing at a speed N_1, the corresponding values of Q_2 and H_2, when the pump is running at speed N_2, are given by the following equations :

$$Q_2 = Q_1 \left(\frac{N_2}{N_1}\right) \quad \text{....(14.2)}$$

and

$$H_2 = H_1 \left(\frac{N_2}{N_1}\right)^2 \quad \text{...(14.3)}$$

Fig. 14.15 shows the effect of speed change on pump characteristics. In constructing the characteristic curve for speed N_2, several pairs of values of Q_1, H_1 from the curve for N_1 can be obtained and transformed into homologous points Q_2, H_2 on the N_2 curve.

14.13. SUCTION LIFT LIMITATIONS : CAVITATION

Cavitation is the phenomenon of cavity formation or the formation and collapse of cavities. Cavities are considered to develop when the absolute pressure in a liquid reaches the vapour pressure of the liquid at the operating temperature. Cavitation can occur at the inlet to a pump and on the impeller blades, particularly, if the pump is mounted above the level in the suction well. Cavitation causes physical damage, reduction in discharge and increase in noise. Under severe conditions, cavitation can result in break down of pumping equipment. To avoid cavitation, the pressure head at inlet should not fall below a certain minimum which is influenced by the further reduction in pressure within the pump impeller.

Fig. 14.16 shows the head conditions in the suction pipe.

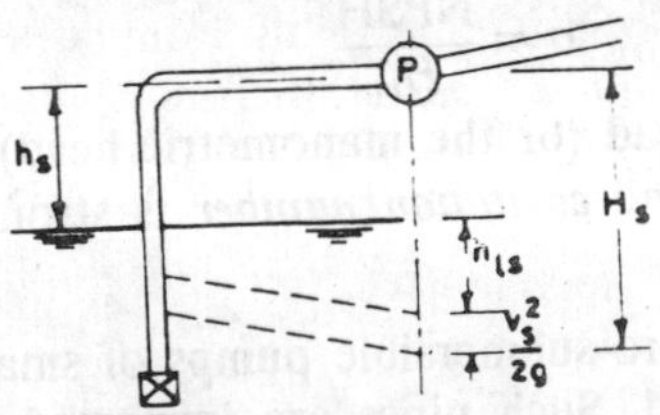

FIG. 14.16. HEAD CONDITIONS IN SUCTION PIPE.

Let h_s = static suction lfit

H_s = manometric suction head

h_{ls} = head loss in suction pipe.

v_s = velocity in suction pipe.

p_s = pressure at inlet

p_v = vapour pressure

p_a = ambient atmospheric pressure

The term $\frac{p_s - p_v}{\rho g}$, which is the absolute head at the pump inlet above vapour pressure, is known as *net positive suction head*, NPSH.

$$\therefore \qquad \text{NPSH} = \frac{p_s - p_v}{\rho g} = \left(\frac{p_a}{\rho g} - H_s\right) - \frac{p_v}{\rho g} \qquad ...(14.4)$$

where
$$H_s = h_s + h_{ls} + \frac{v_s^2}{2g} \qquad ...(14.5)$$

valves of NPSH can be obtained from pump manufacturer and are derived from full scale or model tests.

As the net positive suction head (NPSH) for a pump is reduced, a point is reached where cavitation becomes deterimental. This point is usually referred to as the minimum net positive suction head $(\text{NPSH})_{\text{min}}$.

Again, substituting the value of H_s in Eq. 14.4, we get

$$\text{NPSH} = \frac{p_a - p_v}{\rho g} - h_s - \left(h_{ls} + \frac{v_s^2}{2g}\right) \qquad ...(14.6)$$

The minimum value of static lift is then given by

$$h_{s,\,min} = \frac{p_a - p_v}{\rho g} - (\text{NPSH})_{min} - \left(h_{ls} + \frac{v_s^2}{2g}\right) \qquad ...(14.7)$$

The required NPSH value for any pump is supplied by the manufacturer. While installing the pump, it is essential to check that the available NPSH is *greater* than the manufacturers requirement.

Thoma introduced a *cavitation number* (σ) as under :

$$\sigma = \frac{\text{NPSH}}{H_m} \qquad ...(14.8)$$

where H_m is total head (or the manometric head) across the pump. Thoma found that the *cavitation number* is strongly related to the specific speed.

Recently, electro-submersible pumps of small to medium size have been introduced. Such pipes are immersed in the sump well to the specified depth, thus avoiding the use of suction pipes and thereby eliminating the problems of cavitation and cooling.

14.14. SYSTEM HEAD CURVE

In order to effectively design a pumping station, knowledge of the *total dynamic head* (TDH) against which the pump must operate, is essential. The TDH is represented by the following equation

$$\text{TDH} = H_{ST} + h_{ls} + h_{ld} \qquad ...(14.8\ a)$$

where H_{ST} = total *static* head (or lift) *i.e.* elevation difference between the pumping source and point of delivery

h_{ls} = head loss in suction pipe

h_{ld} = head loss in delivery pipe (*i.e.*, friction, valves, velocity head etc.)

TDH = total dynamic head H which is defined as the rise in total head across the pump.

Eq. 14.8 (*a*) is also sometimes expressed as follows :

$$\text{TDH} = H_{ST} + H_L + H_v \quad ...(14.8)$$

where H_L = total head loss (including valve losses) through the suction and delivery pipes.

H_v = velocity head $(v^2/2g)$ at the discharge end.

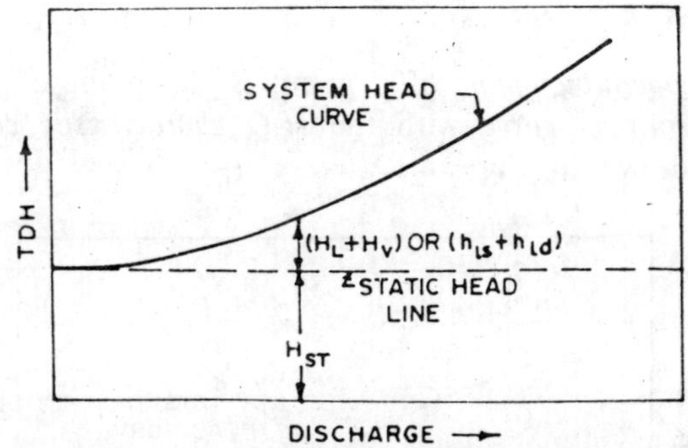

FIG. 14.17. SYSTEM HEAD CURVE.

A plot of TDH versus discharge is known as *system head curve*, shown in Fig. 14.17. Each point on the system head curve denotes the head comprising of (*i*) static head H_{ST}, (*ii*) head loss, H_L and (*iii*) velocity head, H_v. It should be noted that the system head curve will vary with discharge (Q) since H_L and H_v are both a function of discharge. In addition, the static head H_{ST} may also vary as a result of fluctuating water levels. Obviously, the static head is more at low water level (LWL) and less at high water level (HWL) of the sump well. The corresponding system curves are shown in Fig. 14.18.

The friction head (H_f) is the sum of (*i*) head losses in the entire length of the piping, from the foot valve to the final point of delivery piping (*ii*) loss in the valves (foot valve, non-return valve and isolating valves), and (*iii*) losses in bends tees, elbows, reducers etc.

It should be clearly noted that the system head curve will change by any changes made in the system, such as change in length or size of piping etc.

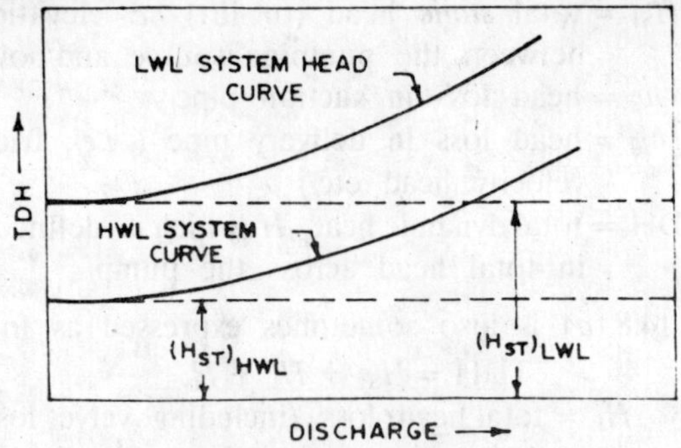

FIG. 14.18. SYSTEM HEAD CURVES

14.15. OPERATING POINT OR OPERATING RANGE OF A PUMP

The *operating point* of a pump is the point of intersection of the *system-head-curve* with the *H-Q characteristic curve* of the pump, as shown in Fig. 14.19.

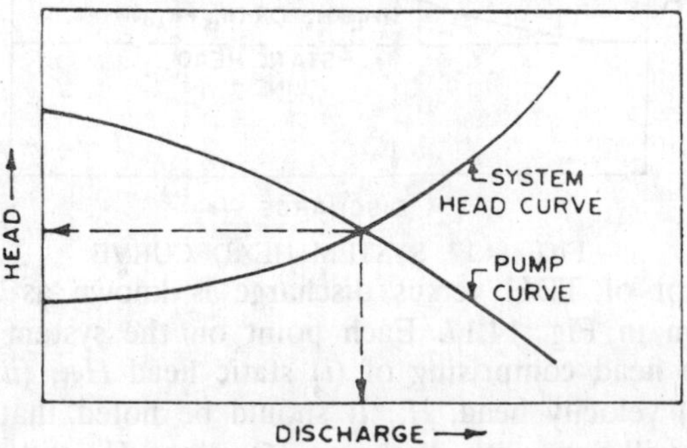

FIG. 14.19. OPERATING POINT OF PUMP

As indicated earlier, the system head curve will change due to any change made in the system. If the system head curve changes, the operating point will shift. This is more particular when the water level in the sump fluctuates. There will be two system-head-curves corresponding to LWL and HWL, as shown in Fig. 14.20. Hence if the level of water in the suction would deplete during pumping from HWL to LWL, the operating point of the pump would vary from a low-head high- dishcarge point (P_1) to high-head low-dishcarge point (P_2) as shown in Fig. 14.20. The range between points P_1 and P_2 is known as the *operating range* of the pump, due to fluctuations in water level in the sump.

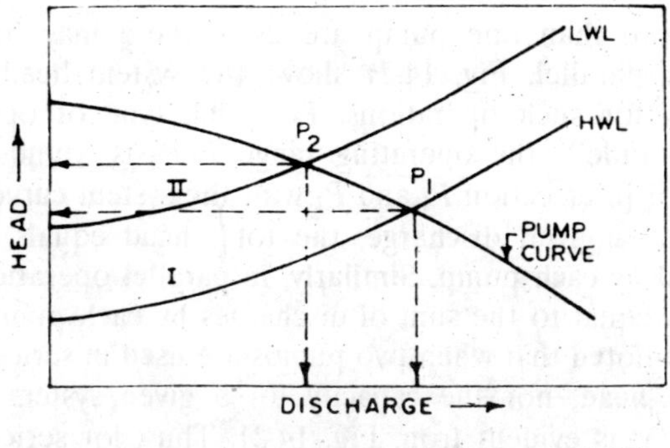

FIG. 14.20. OPERATING RANGE OF PUMP

14.16. SELECTION OF PUMPING UNITS

Generally, the system-head-curve is given or known, and it is required to find a pump (or pumps) to deliver the anticipated flows. To find this, the system head curve and the pump characteristic curve (*i.e.* *H-Q* curve) are plotted on the same sheet and the operating point (Fig. 14.19) is found. The operating point (*P*) gives the head and flow at which the pump will be operating. Hence a pump is selected so that the operating point is as close as possible to the peak efficiency.

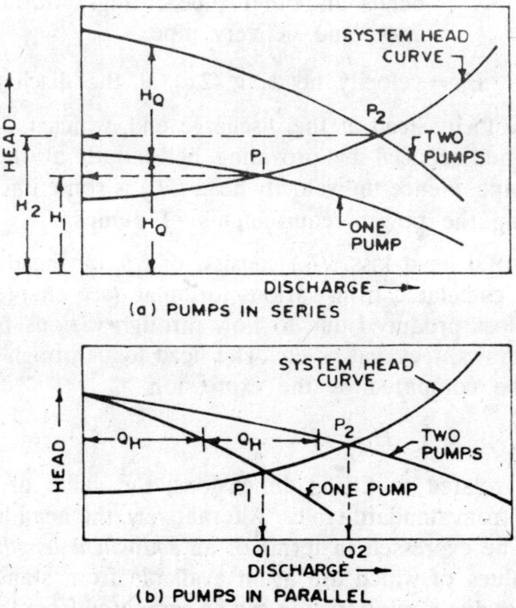

FIG. 14.21. SELECTION OF PUMPING UNITS

If more than one pump are used, these may be connected in series or parallel. Fig. 14.21 shows the system head curves and H-Q curves for such operations. For each type of operation (*i.e.* series or parallel), the operating range $P_1 P_2$ is found by locating the points of intersection P_1 and P_2 with the system curve. For series operation at a given discharge, the total head equals the sum of heads added by each pump. Similarly, in parallel operation, the total discharge is equal to the sum of discharges by each pump. However, it should be noted that when two pumps are used in series or parallel, neither the head, nor the capacity for a given system head curve is doubled, as is evident from Fig. 14.21. Thus, for series operation, $H_2 < 2H_1$. Similarly, for parallel operation, $Q_2 < 2Q_1$.

14.17. POWER REQUIREMENTS OF PUMPS

As seen earlier, the total lift, or the *total dynamic head* (TDH) against which the pump is required to operate is given by

$$\text{TDH} = H = H_{ST} + H_L + H_v \qquad \text{...(14.8)}$$

where H_{ST} = total static head (or lift), *i.e.* the elevation difference between the pumping source and point of delivery.

H_L = total head loss (including losses through valves, bends and other pipe fittings) through the suction and delivery pipe.

H_v = velocity head $(v^2/2g)$ at the discharge head.

The velocity head at the discharge end, which is quite small, can be further reduced by providing bell-mouth at the discharge end of the pipe. Hence the velocity head (H_v) is sometimes neglected in computing the power requirements of pumps.

The total head loss (H_L) consists of friction head loss in the pipe line as calculated from various formulae (see chapter 15), plus the energy loss produced due to flow through various fittings such as bends ellbows, tees, valves etc. The head loss through the fittings (H_{LF}) can be computed by the expression

$$H_{LF} = K \cdot \frac{v^2}{2g} \qquad \text{...(14.9)}$$

where K is related to fitting dimensions, the value of which may be obtained from standard tables. Alternatively, the head loss through fittings can be expressed in terms of an ***equivalent length of straight pipe***, the values of which are again available from standard tables. The head loss due to pipe friction can be computed from the following expression

$$h_f = \frac{4f'lv^2}{2gd} = \frac{flv^2}{2gd} \quad \text{...(14.10)}$$

(where $f = 4f'$)

Substituting $V = \frac{Q}{A} = \frac{4Q}{\pi d^2}$, we get

$$h_f = \frac{4f'l}{2gd}\left(\frac{4Q}{\pi d^2}\right)^2$$

or

$$h_f = \frac{f'lQ^2}{3.025\,d^5} = \frac{flQ^2}{12.1\,d^5} \quad \text{...(14.11)}$$

where l = length of the pipe (m)
d = diameter of the pipe (m)
f' = coefficient of friction
f = coefficient of friction $= 4f'$

If W kg is the weight of water lifted in one second through a height H metres, the work done by the pump

$= W \times H$ m-kg/sec

$= Q \times w \times H$ m-kg/sec

∴ Water horse power (WH.P) in metric unity

$$= \frac{QwH}{75} \quad \text{...(14.12)}$$

(Since 1 metric H.P.= 75 kg-m/sec)

Brake horse powe (B.H.P)

$$= \frac{QwH}{75\,\eta_P}$$

where η_P = Efficiency of pump.

If the pump is driven by electric motors, the electric horse power (E.H.P.) is given by :

$$\text{E.H.P.} = \frac{\text{B.H.P}}{\text{motor efficiency}} = \frac{\text{B.H.P.}}{\eta_e} = \frac{QwH}{75\,\eta_P\,\eta_e} = \frac{QwH}{75\,\eta} \quad \text{...(14.14)}$$

where η = overall efficiency

Since one metric H.P. = 736 watts = 0.736 kilowatt

No. of kilowatts= E.H.P. × 0.736 ...(14.15 a)

$$= 0.736\frac{QwH}{75\,\eta} = 9.81 \times 10^{-3}\frac{QwH}{\eta} \quad \text{...(14.15 b)}$$

Taking $w = 1000\ \text{kg/m}^3$

$$\text{Power in kilowatts} = \frac{9.81\,QH}{\eta} \quad \text{...(14.15)}$$

Alternatively electrical power required is given by

$$P_e = \frac{w \rho Q H}{\eta} \text{ watts} = \frac{1000 \times 9.81\, QH \times 10^{-3}}{\eta} \text{ kilowatts}$$

or $$P_e = 9.81 \frac{QH}{\eta} \text{ kilowatts} \qquad ...(14.15)$$

Installed capacity of pumps. The efficiency of pumping engines depends upon the amount of load. At full load, the efficiency may be 80%; at half load it may be upto 70% while the quarter load efficiency is generally upto 50%. The total installed capacity at a pumping station depends on the following three factors :

(*a*) The maximum rate of hourly demand.

(*b*) Balancing storage.

(*c*) Operation schedule.

The total capacity is essentially kept larger than the maximum rate of demand. The total capacity is divided into suitable smaller units so that different units may be put into operation or cut off depending upon vaiations in the flow requirements. If there is balancing storage, capacity required would be small, since the pumps can work at average hourly rate of demand. Generally, the complete pump installation is so designed that the total H.P. required could be divided into at least three units, to suit the minimum, average and maximum demands.

14.18. ECONOMICAL DIAMETER OF PUMPING MAINS

We have seen that the required H.P. of pump depends upon the *total dynamic head* (*H*) given by Eq. 14.8. For a fixed value of static head (H_{ST}), the TDH varies with the friction head loss (H_f) and the velocity head (H_v). Both H_f and H_v depends upon the velocity of flow of water in the pumping mains. If the diameter of the pumping mains is decreased (to reduce the cost of pipe line), thereby increasing the velocity of flow for a given fixed discharge,

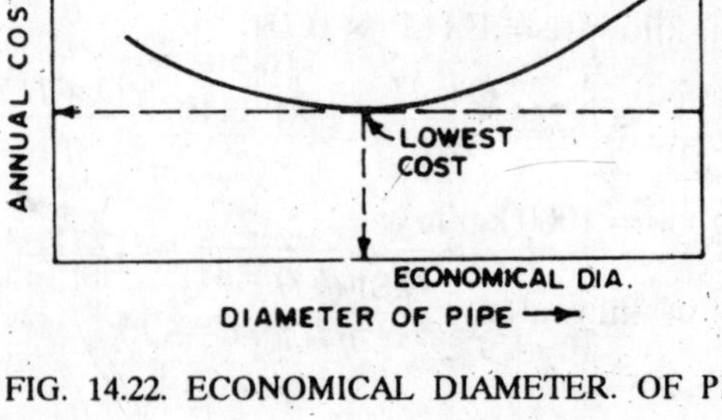

FIG. 14.22. ECONOMICAL DIAMETER. OF PIPE.

both H_f and H_v will increase and hence TDH will increase. This will result in higher cost of pumping. For optimum conditions, one must choose such a diameter of pipe, which together with the pumping cost, will make the total annual expenses the *minimum*. Such a diameter is known as the *economical diameter* of the pumping mains.

Generally, the velocity of flow in pipe may vary from 0.8 to 1.8 m/sec. For a rigorous analysis, the total cost of pipe and pumping can be determined for various values of velocity of flow, and a curve, such as the one shown in Fig. 14.22, can be plotted. It will be seen that, initially the total annual cost decreases as the pipe diameter is increased, reaches its minimum value and then again increase. The pipe diameter corresponding to the least cost is then selected. Alternatively, the followig empirical formula given by Lea may be adopted :

$$D = a\sqrt{Q} \qquad \text{...(14.12)}$$

where D = diameter of pipe (m)
Q = discharge to be pumped (m)
a = 0.97 to 1.22.

Example 14.1. *A city has a population of 1,50,000. Water is to be supplied at the rate of 160 litres per head per day. If the static lift of the pump is 40 metres, calculate the B.H.P. of motor. The rising main is 300 m long and its diameter is 50 cm. Assume that motor efficiency is 85%, pump efficiency is 60%, f = 0.04 and the peak hour demand is 1.5 times the average demand.*

Solution :

$$\text{Average demand} = 150000 \times 160$$

$$= 24 \times 10^6 \text{ litres/day} = 24000 \text{ m}^3/\text{day}$$

$$= \frac{24000}{24 \times 60 \times 60} = 0.278 \text{ m}^3/\text{sec}$$

$\therefore$ Peak hour demand $= 1.5 \times 0.278 = 0.417$

Static lift $\quad H_{ST} = 40$ m

Frictional head loss $\quad H_f = \dfrac{f l Q^2}{12.1\, d^5} = \dfrac{0.04 \times 300\,(0.417)^2}{12.1\,(0.50)^5}$

$$= 5.56 \text{ m} = H_L$$

$$\text{Velocity of flow} = \frac{4Q}{\pi d^2} = \frac{4 \times 0.417}{\pi (0.5)^2} = 2.12 \text{ m/sec}$$

$$\therefore \quad H_v = \frac{v^2}{2g} = \frac{(2.12)^2}{2 \times 9.81} = 0.23 \text{ m}$$

$$\therefore \quad H = H_{ST} + H_L + H_v = 40 + 5.56 + 0.23 = 45.79 \text{ m}$$

Overall efficiency $\eta = 0.85 \times 0.6 = 0.51$

$$\text{B.H.P. of motor} = \frac{Q\,w\,H}{75\,\eta}$$

$$= \frac{0.278 \times 1000 \times 45.79}{75 \times 0.51} = \mathbf{332.8}$$

Example 14.2. *Water is to be lifted from a tube well to an overhead tank. Find the B.H.P. of pumping units, for the following data :*

Discharge from tube well	:	*60 litres/sec*
R.L. of ground	:	*201.50 m*
R.L. of water in tube well	:	*180.00 m*
Depression head during pumping	:	*4 m*
R.L. of bottom of overhead tank	:	*220.50 m*
Depth of water in tank	:	*3.50 m*
Length of rising main	:	*100 m*
Coefficient of friction	:	*0.04*
Combined efficiency of pump and motor	:	*70 %*
Velocity of water in rising main	:	*2 m/sec.*

Solution :

$$Q = 60 \text{ litres/sec} = 0.06 \text{ m}^3\text{/sec}$$

If d = dia. of rising main, in metres, we have

$$Q = \frac{\pi}{4} d^2 \times v$$

or
$$d = \sqrt{\frac{4Q}{\pi v}} = \sqrt{\frac{4 \times 0.06}{\pi \times 2}} = 0.195 \text{ m}$$

$$\approx 20 \text{ cm.}$$

Now static head

$$H_{ST} = (201.50 - 180.00) + (220.50 - 201.50 + 3.50) + \text{depression head } (= 4 \text{ m})$$

$$= 21.50 + 22.50 + 4 = 48 \text{ m}$$

$$H_f = \frac{f\,l\,Q^2}{12.1\,d^5} = \frac{0.04\,(100)(0.06)^2}{12.1\,(0.20)^5}$$

$$= 3.72 = H_L$$

$$\text{Velocity head} = \frac{v^2}{2g} = \frac{(2)^2}{2 \times 9.81} = 0.20 \text{ m}$$

$$\therefore \quad H = 48 + 3.72 + 0.20 = 51.92 \text{ m}$$

$$\text{B.H.P.} = \frac{Q\,w\,H}{75\,\eta} = \frac{0.06 \times 1000 \times 51.92}{75 \times 0.7}$$

$$= \mathbf{59.34}$$

Hence adopt *three units* of 20 H.P. each. One unit of 20 H.P. may be provided as a stand by unit.

Example 14.3. *For water supply of a town, water is pumped from a river 2 km away into a reservoir. The maximum difference of levels of water in river and the reservoir is 25 m. the population of the town is 80000 and per capita water demand is 125 litres per day. If the pumps are to operate for a total of 8 hours and the efficiency of pumps is 80%, determine the H.P. of the pumps. Assume friction factor as 0.03 and velocity of flow as 2 m/s, and maximum daily demand as 1.5 times the average demand.*

Solution

$$\text{Average water requirement} = 80000 \times (125 \times 10^{-3})$$
$$= 10000 \text{ m}^3/\text{day} .$$

$$\text{Max. daily water requirement} = 1.5 \times 10000 = 15000 \text{ m}^3/\text{day}$$

Since pumping time is 8 hours per day,

$$Q = \frac{15000}{8 \times 60 \times 60} = 0.5208 \text{ m}^3/\text{s}$$

Now $$v = \frac{Q}{A} = \frac{4\,Q}{\pi\,d^2}$$

$$\therefore \quad d = \sqrt{\frac{4\,Q}{\pi\,v}} = \sqrt{\frac{4 \times 0.5208}{\pi \times 2}} = 0.576, \text{ say } 0.58 \text{ m.}$$

Now $$H_f = \frac{f\,l\,v^2}{2\,g\,d}$$

$$\therefore \quad H_f = \frac{0.03 \times 2000\,(2)^2}{2 \times 9.81 \times 0.58} = 21.09 \text{ m} = H_L$$

The velocity head and other minor losses can be neglected in comparison to the above value of H_f.

Again, $$H_{ST} = 25 \text{ m}$$

$$\therefore \quad H = H_{ST} + H_L = 25 + 21.09 = 46.09 \text{ m}$$

$$\therefore \quad \text{H.P. of pump} = \frac{Q\,w\,H}{75\,\eta} = \frac{0.5208 \times 1000\,(46.09)}{75 \times 0.8}$$

$$\approx \mathbf{400}$$

Example 14.4. *Water is to be raised from a tube well to an overhead storage tank. Find the B.H.P. of the electro-motor pumping set from the following data :*

Discharge from tube well	*= 40 litres per second*
Velocity in rising main	*= 2 m per second*
Length of rising main	*= 250 m*
R.L. of ground	*= 100.00 m*
R.L. of water in the well	*= 86.00 m.*
Depression head	*= 4.00 m*
R.L. of water level in over head tank	*= 112 m*
Coefficient of friction	*= 0.0485*
Combined efficiency of electro-pumping set	*= 70%*

Solution :

$$Q = 40 \text{ litres per second} = 0.04 \text{ m}^3/\text{sec}$$

Pipe dia. $d = \sqrt{\dfrac{4Q}{\pi v}} = \sqrt{\dfrac{4 \times 0.04}{\pi \times 2}} \approx 0.16 \text{ m}$

$$H_{ST} = (100-86) + (112-100) + \text{Depression head}$$
$$= 14 + 12 + 4 = 30 \text{ m}$$

$$H_f \approx \frac{f\,l\,Q^2}{12.1\,d^5} = \frac{0.0485 \times 250 \times (0.04)^2}{12.1\,(0.16)^5}$$
$$= 15.29 \text{ m}$$

$$H_v = \frac{v^2}{2g} = \frac{(2)^2}{2 \times 9.81} = 0.20 \text{ m}$$

$\therefore \quad H = H_{ST} + H_f + H_v = 30 + 15.29 + 0.2 = 45.49 \text{ m}$

Now B.H.P. $= \dfrac{Q\,w\,H}{75\,\eta} = \dfrac{0.04 \times 1000\,(45.49)}{75 \times 0.70} = \mathbf{34.66}$

Hence adopt 3 units of 12 B.H.P. each, along with one stand by unit of 12 B.H.P.

Example 14.5. *A pump is to deliver water from an under ground tank against a static head of 30 m. The suction pipe is 40 m long and is of 30 cm diameter with Darcy-Weisbach friction factor f= 0.02. The delivery pipe is of 25 cm diameter, 1500 m long and has f = 0.022. The pump characteristics can be expressed as*

$$H_p = 100 - 6000\,Q^2$$

where H_p *= pump head in metres and* Q *= discharge in* m^3/s *Calculate the head and the discharge of the pump.*

Solution

Darcy-Weisbach formula for friction loss is

$$H_f = \frac{f L v^2}{2 g d} = \frac{f L Q^2}{2 g \left(\frac{\pi}{4}\right)^2 . d^5}$$

Hence head loss in suction pipe is

$$H_{fs} = \frac{0.02\,(40)\,Q^2 \times 16}{2 \times 9.81\,(\pi^2)\,(0.30)^5} = 27.2\,Q^2 \qquad ...(i)$$

Head loss in delivery pipe is

$$H_{fd} = \frac{0.022 \times 1500\,Q^2 \times 16}{2 \times 9.81\,(\pi)^2\,(0.25)^5} = 2792.1\,Q^2 \qquad ...(ii)$$

Static head $H_L = 30$ m

$\therefore$ Total dynamic head $H = H_{ST} + H_L = H_{ST} + H_{fs} + H_{fd}$

$= 30 + 27.2\,Q^2 + 2792.1\,Q^2 = 30 + 2819.3\,Q^2$

But this should be equal to the pump head H_p.

$\therefore \quad H = H_p$

or $\quad (30 + 2819.3\,Q^2) = 100 - 6000\,Q^2$

$\therefore \quad 8819.3\,Q^2 = 70$

From which $\quad Q = \sqrt{\dfrac{70}{8819.3}} =$ **0.089 cumecs**

$= 89$ litres/second

$\therefore \quad H_p = 100 - 6000\,Q^2 = 100 - 6000\,(0.089)^2$

= **52.38 m.**

Check : $\quad H_{fs} = 27.2\,Q^2 = 27.2\,(0.089)^2 = 0.22$

$H_{fd} = 2792.1\,Q^2 = 2792.1\,(0.089)^2 = 22.16$

$H_L = 30$ m

$\therefore \quad H = 0.22 + 22.16 + 30 = 52.38$ m $= H_p$

Example 14.6. *Water is pumped from a low level reservoir to a high level reservoir through a main pipe line of 0.4 m dia. and of 1500 m length. The pump is located at the low level reservoir. At a point along the main line at a distance of 400 m from the high level reservoir, a branch line of 0.3 m dia. and 300 m length takes off to discharge 150 litres/sec in the atmosphere.*

Level of water surface in high level reservoir = + 50.00 m

Level of water in the open end of the 0.3 m dia branch line = + 45.00 m

Level of water in the low level reservoir = + 25.0 *m.*
Darcy's friction factor for each pipe = 0.035

Determine the flow rate into the high level reservoir and the theoretical H.P. of the pump, assuming the delivery valve of the pump to be at + 27.0 *m.*

Solution

Fig. 14.23 shows the positions of low reservoir, high reservoir, main pipe line and the branch pipe line. Let *A* be the position of delivery valve of pump (Elev. + 27.0), *B* be the position of junction point, *C* be the position of free discharge point (Elev. + 45.00 m) and *D* be the level of the reservoir.

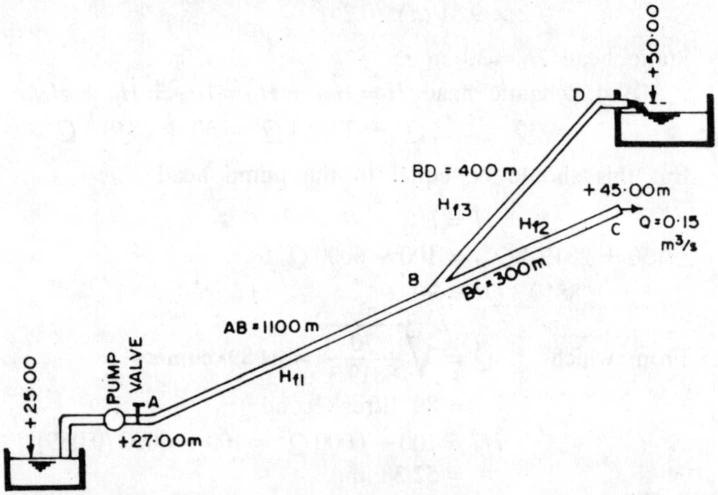

FIG. 14.23.

Let head loss in $AB = H_{f1}$
head loss in $BC = H_{f2}$
and head loss in $BD = H_{f3}$

Applying Bernouli's theorem between *B* and *D*, we get

$$Z_B + P_B = Z_D + P_D + H_{f3}$$

or
$$Z_B + P_B = +\ 50.00 + \frac{p_a}{w} + H_{f3} \quad ...(i)$$

Also, applying Bernoulli's theorem between *B* and *C*, we get

$$Z_B + P_B = Z_C + P_C + H_{f2}$$

or
$$Z_B + P_B = +\ 45.00 + \frac{p_a}{w} + H_{f2} \quad ...(ii)$$

Equating (*i*) and (*ii*), we get

$$50.00 + \frac{p_a}{w} + H_{f3} = 45.00 + \frac{p_a}{w} + H_{f2}$$

$$\therefore \quad H_{f3} = H_{f2} - 5.00 \quad ...(iii)$$

But $H_{f2} = \frac{f L v^2}{2 g d}$

where $v = \frac{4Q}{\pi d^2} = \frac{4 \times 0.15}{\pi (0.3)^2} = 2.122$ m/s

$$\therefore \quad H_{f2} = \frac{0.035 \times 300 (2.122)^2}{2 \times 9.81 (0.3)} = 8.03 \text{ m.}$$

$$\therefore \quad H_{f3} = H_{f2} - 5.00 = 8.03 - 5.00 = 3.03 \text{ m} \quad ...(a)$$

But head loss between *B* and *D* is given by

$$H_{f3} = \frac{f L v^2}{2 g d} = \frac{0.035 \times 400 v^2}{2 \times 9.81 \times 0.4} = 1.7839 v^2 \quad ...(b)$$

$$\therefore \quad 1.7839 v^2 = 3.03$$

From which $v = 1.30$ m/s.

$\therefore$ Discharge in $BD = A \times V = \frac{\pi}{4} (0.4)^2 \times 1.3 =$ **0.1634 m³/sec**

= 163.4 litres/sec.

$\therefore$ Discharge through $AB = 0.1634 + 0.15 = 0.3134$ m³/sec

and velocity through $AB = Q/A = \frac{4Q}{\pi (0.4)^2} = 2.494$ m/sec.

Now total dynamic head (TDH) to pump water to point *D* is given by

$$H_D = H_{ST} + H_{f1} + H_{f3}$$

where $H_{f1} = \frac{0.035 \times 1100 (2.494)^2}{2 \times 9.81 \times 0.4} = 30.51$ m

$$\therefore \quad H_{ST} = 50.00 - 25.00 = 25.00 \text{ m}$$

$$H_{f3} = 3.03 \text{ m}$$

$$\therefore \quad H_D = 25.00 + 30.51 + 3.03 = 58.54 \text{ m}$$

Similarly, total dynamic head required to pump water to point *C* is

$$H_C = H_{ST} + H_{f1} + H_{f2}$$

where $H_{ST} = 45.00 - 25.00 = 20.00$ m

$H_{f2} = 8.03$ m.

$$\therefore \quad H_C = 20.00 + 30.51 + 8.03 = 58.54 \text{ m}$$

Note that $H_D = H_C = H$, as expected.

H.P. required to pump $Q_C = 0.15\ m^3/s$ through head H_C is

$$(H.P.)_C = \frac{w\,Q_C\,H_C}{75} = \frac{w\,Q_C\,H}{75}$$

Similarly, H.P. required to pump $Q_D = 0.1634\ m^3/s$ through head H_D is

$$(H.P.)_D = \frac{w\,Q_D\,H_D}{75} = \frac{w\,Q_D\,H}{75}$$

$$\therefore \text{ Total H.P. of pump} = \frac{w\,H}{75}(Q_C + Q_D)$$

or $$\text{H.P.} = \frac{Q\,w\,H}{75}, \text{ where } Q = \text{total discharge}$$

$$= \frac{0.3134 \times 1000 \times 58.54}{75} = \mathbf{244.62}$$

Example 14.7. *A centrifugal pump, running at 1400 rpm has the characteristics given below.*

Discharge (lps)	*11.0*	*16.5*	*22.1*	*27.7*	*33.3*	*38.8*	*44.3*
Head (m)	*25.1*	*24.3*	*23.4*	*22.5*	*20.8*	*18.4*	*15.4*
Efficiency η (%)	*64*	*69*	*72*	*73*	*71*	*68*	*61*

Draw the operating characteristics of the pump and determine its specific speed.

Solution.

Fig. 14.24 shows the characteristic curves *i.e.* $H\text{–}Q$ and $\eta - Q$ curves.

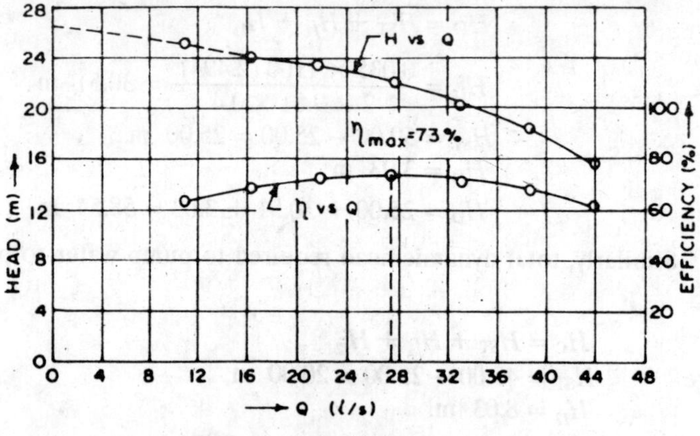

FIG. 14.24.

From the curve, we get max. efficiency $\approx 73\%$, when it is discharging 27.7 lps at a head of 22.5 m.

Now $N = 1400$ rpm.

$Q = 27.7$ lps, at max. efficiency

$H = 22.5$ m

$$\therefore \quad N_S = \frac{N\sqrt{Q}}{H^{3/4}} = \frac{1400\sqrt{27.7}}{(22.5)^{3/4}}$$

$$= \mathbf{713\ rpm}$$

Example 14.8. *From a clear water reservoir 2.5 m deep and maximum water level at 32.00 m, water is to be pumped to an elevated reservoir at 80.00 m, at a constant rate of 8,10,000 litres per hour. The distance is 1200 m. Give the economical diameter of the rising main and the water horse power of the pump. Neglect minor losses and take $f = 0.04$.*

Solution :

Discharge $Q = 810000$ litres/hour

$$= \frac{810000 \times 10^{-3}}{60 \times 60} = 0.225\ \text{m}^3/\text{s}$$

For economical dia. of rising main, use Lea's formula (Eq. 14.12) :

$$d = a\sqrt{Q}$$

Taking $a = 1.20$,

$$d = 1.20\sqrt{0.225} = \mathbf{0.57\ m}$$

Let the pump be fixed at max. water level (= 32.00 m) of the sump.

$\therefore$ Max. suction head = 2.5 m

Delivery head = 80 − 32 = 48 m

$\therefore$ Max. static head $H_{ST} = 2.5 + 48 = 50.5$ m

Again, head loss in pipe is given by

$$H_f \approx \frac{f l Q^2}{12.1\, d^5}$$

$$= \frac{0.04 \times 1200\,(0.225)^2}{12.1\,(0.57)^5} = 3.34\ \text{m}$$

Neglecting other losses, the total head is

$$H = H_{ST} + H_f = 50.5 + 3.34 = 53.84\ \text{m}.$$

$$\therefore \quad \text{Water H.P.} = \frac{Q\,w\,H}{75} = \frac{0.225 \times 1000 \times 53.84}{75}$$

$$= \mathbf{161.52}$$

Example 14.9. *A pumping station, situated at an elevation of 610 m uses pumps which require 32 kPa positive suction pressure (NPSH) when delivering water at 20°C. Determine the allowable suction lift of these pumps if the entrance and friction losses are 12 kPa. Take barometric pressure at 610 m altitude as 94 kPa and vapour pressure of water at 20°C as 2.35 kPa.*

Solution

The barometric pressure at 610 m altitude is 94 kPa. However, since the barometric pressure may further be reduced by *storm activity*, it is recommended to reduce the above value by about 3.5 kPa.

$$\therefore \quad p_a = 94 - 3.5 = 91.5 \text{ kPa}$$

$$p_v = 2.35 \text{ kPa}$$

Expressing Eq. 14.6 in terms of pressure, we have

$$(\text{NPSH})_{pressure} = p_a - p_v - p_l - p_s$$

where p_l = losses in the suction pipe = 12 kPa

p_s = suction pressure.

$(\text{NPSH})_{pressure}$ = net positive suction pressure of pump = 32 kPa (given)

$$\therefore \quad 32 = 91.5 - 2.35 - 12 - p_s$$

From which $p_s = 45.15$ kPa

$$\therefore \quad \text{Allowable suction lift} = \frac{45.15}{9.81} = \mathbf{4.6 \ m}$$

Example 14.10. *Tests on a physical model pump indicated a cavitation number of 0.12. A homologous (i.e. geometrically and dynamically similar) unit is to be installed at a place where the atmospheric pressure is 980 mb and the vapour pressure head is 0.24 m. The suction pipe is 8 m long and of 20 cm dia., fitted with a foot valve at the bottom and a 90° elbow at the top leading into the pump inlet. The total head at the operating discharge of 40 l/s is 28 m. Calculate the maximum permissible suction head and suction lift. Adopt the following values :*

(i) *Coefficient of friction : 0.06*

(ii) *Foot valve head loss : 4 $v_s^2/2g$*

(iii) *Bend head loss : 1.0 $v_s^2/2g$*

Solution :

$$p_a = 980 \text{ mb} = 0.98 \times 10.198 = 9.994 \text{ m of water}$$

$$\therefore \quad \frac{p_a - p_v}{\rho g} = 9.994 - 0.24 = 9.754 \text{ m of water}$$

From Eq. 14.8, $\sigma = \dfrac{\text{NPSH}}{H_m}$

$$\therefore \qquad \text{NPSH} = \sigma H_m = 0.12 \times 28 = 3.36 \text{ m}$$

From Eq. 14.4, we have

$$\text{NPSH} = \frac{p_a - p_v}{\rho g} - H_S$$

$$\therefore \qquad H_S = \frac{p_a - p_v}{\rho g} - \text{NPSH} = 9.754 - 3.36 = \mathbf{6.394 \text{ m}}$$

Now suction lift $h_s = H_S - h_{ls} - \dfrac{v_s^2}{2g}$ ...(*i*)

$$Q = 40 \text{ l/s} = 40 \times 10^{-3} \text{ m}^3/\text{s}$$

$$A = \frac{\pi}{4} d^2 = \frac{\pi}{4} (0.2)^2 = 0.0314 \text{ m}^2$$

$$\therefore \qquad v_s = \frac{Q}{A} = \frac{40 \times 10^{-3}}{0.0314} = 1.273 \text{ m/s.}$$

$$\therefore \qquad \frac{v_s^2}{2g} = \frac{(1.273)^2}{2 \times 9.81} = 0.083 \text{ m}$$

$$h_f = \frac{f l v_s^2}{2 g d} = \frac{0.06 \times 8 (1.273)^2}{2 \times 9.81 (0.2)} = 0.198 \text{ m.}$$

$$\text{Valve head loss} = 4 \frac{v_s^2}{2g} = 4 \times 0.083 = 0.332 \text{ m}$$

$$\text{Bend head loss} = 1.0 \frac{v_s^2}{2g} = 0.083 \text{ m}$$

$$\therefore \qquad H_{ls} = 0.198 + 0.332 + 0.083 = 0.613 \text{ m.}$$

Substituting the values in Eq. (*i*) above

$$h_s = 6.394 - 0.613 - 0.083 = \mathbf{5.698 \text{ m}}$$

Example 14.11. *A pump lifts water from a low level reservoir to a high level reservoir with a static lift of 8 m. The total length of pipe, 25 cm in diameter is 1.5 km. The pump characteristics are given in the table below.*

Discharge (l/s)	*0*	*8*	*16*	*24*	*32*	*40*	*48*
Manometric head H_m (m)	*20*	*18.6*	*16.7*	*14.0*	*10.9*	*7.4*	*3.6*
Overall efficiency (%)	–	*48*	*69*	*75*	*70*	*60*	*47*

Compute the steady discharge between the tanks and the power consumption. Take coefficient of friction $f = 0.0403$ and losses in valves. bends, velocity head etc equal to $5.4\ v^2/2g$.

Solution

The data about $H_m - Q$ curve and $\eta - Q$ curve are given, and these two curves are shown in Fig. 14.25. The system curve can be plotted by computing the system head, corresponding to various discharges, by Eq. 14.8 :

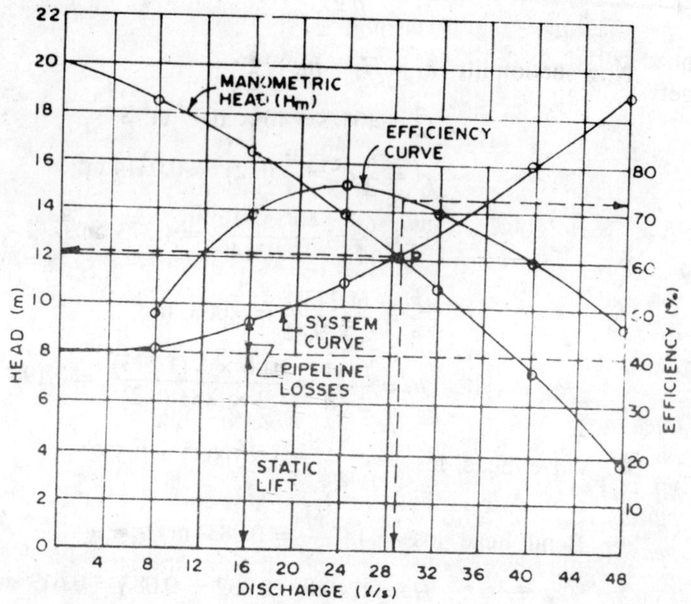

FIG. 14.25.

$$(\text{TDH}) = H = H_{ST} + H_L + H_v$$

where H_{ST} = total static head = 8 m

$$H_L + H_v = (\text{friction loss}) + (\text{Misc. losses})$$

Friction loss $\quad h_f \approx \dfrac{f\,l\,Q^2}{12.1\,d^5} = \dfrac{0.0403 \times 1500}{12.1\,(0.25)^5}\,Q^2 \approx 5120\,Q^2$

Misc. losses, $\quad h_m = 5.4\,\dfrac{v^2}{2g} = \dfrac{5.4}{2 \times 9.81} \times \left(\dfrac{4\,Q}{\pi\,d^2}\right)^2 = 0.446\,\dfrac{Q^2}{d^4}$

$$= \frac{0.446}{(0.25)^4}\,Q^2 = 114.2\,Q^2$$

where Q is in cumecs. If, however, Q is expressed in litres/s,

$$h_f = 5120\,Q^2\,(10^{-3})^2 = 5.12 \times 10^{-3}\,Q^2 \text{ (m)} \quad ...(i)$$

and $\quad h_m = 114.2\,Q^2\,(10^{-3})^2 = 1.142 \times 10^{-4}\,Q^2 \text{ (m)} \quad ...(ii)$

The values of h_f, h_m and H for various values of Q (in l/s) are tabulated below.

Q (l/s)	8	16	24	32	40	48
h_f	0.328	1.311	2.949	5.243	8.192	11.796
h_m	0.007	0.029	0.066	0.117	0.183	0.263
H_{st}	8.0	8.01	8.0	8.0	8.0	8.0
H	8.335	9.34	11.015	13.36	16.375	20.059

The intersection of the manometric head curve with the system head curve gives the *operating point P*, corresponding to which we get

$$Q = \mathbf{28.8\ l/s} = 0.0288\ \text{m}^3/\text{s}$$

$$H = 12.2\ \text{m}$$

$$\eta = 73\ \%$$

Now power concumption is given by

$$P_e = 9.81 \frac{QH}{\eta}\ \text{(kilowatts)} \qquad ...(14.15)$$

$$= \frac{9.81 \times 0.0288 \times 12.2}{0.73} = \mathbf{4.72\ kN}$$

Example 14.12. *Two identical pumps, having characteristics tabulated below, are installed at a pumping station to deliver water to an over head tank through a 25 cm pipe line 2 km long. The static lift is 12 m. Calculate the discharge and power consumption, if the pumps are to be connected in parallel. Take f= 0.0484 and minor head losses, including velocity head, equal to 8.0 $v^2/2g$.*

Discharge (l/s)	0	8	16	24	32	40
Manometric head (m)	25	23	19.6	15.6	11.6	7
Overall efficiency (%)	–	50	66	57	40	25

Solution

The manometric head curve for single pump is shown by curve I in Fig. 14.26. As discussed in § 14.11, the manometric head curve for two pumps in parallel will be such that for a given head, the discharge will be double the discharge of a single pump under that head. The data for the manometric head curve for two pumps are tabulated below.

Manometric Head (m)	23	19.6	15.6	11.6	7
Discharge (l/s)	16	32	48	64	80

The manometric head curve for two pumps in parallel is shown by curve II in Fig. 14.26.

The system head is given by

$$H = H_{ST} + h_f + h_m$$

where h_m = minor head loss $= 8.0 \dfrac{v^2}{2g} = \dfrac{8.0}{2 \times 9.81}\left(\dfrac{4Q}{\pi d^2}\right)^2$

$$= 0.661 \frac{Q^2}{d^4} \text{ when } Q \text{ is in m}^3\text{/sec.}$$

$$= 0.661 \frac{(Q \times 10^{-3})^2}{(0.25)^4} = 1.6922 \times 10^{-4} Q^2 \text{ when } Q \text{ is in l/s} \quad ...(i)$$

$$h_f \approx \frac{f\,l\,Q^2}{12.1\,d^5} = \frac{0.0484 \times 2000\,Q^2}{12.1\,(0.25)^5}$$

$$= 8192\,Q^2 \text{ when } Q \text{ is in m}^3\text{/sec}$$

$$= 8192\,(Q \times 10^{-3})^2 = 8.192 \times 10^{-3} Q^2 \text{ when } Q \text{ is in l/s} \quad ...(ii)$$

$$H_{ST} = 12 \text{ m.}$$

$$\therefore \quad H = 12 + 8.192 \times 10^{-3} Q^2 + 1.6922 \times 10^{-4} Q^2$$

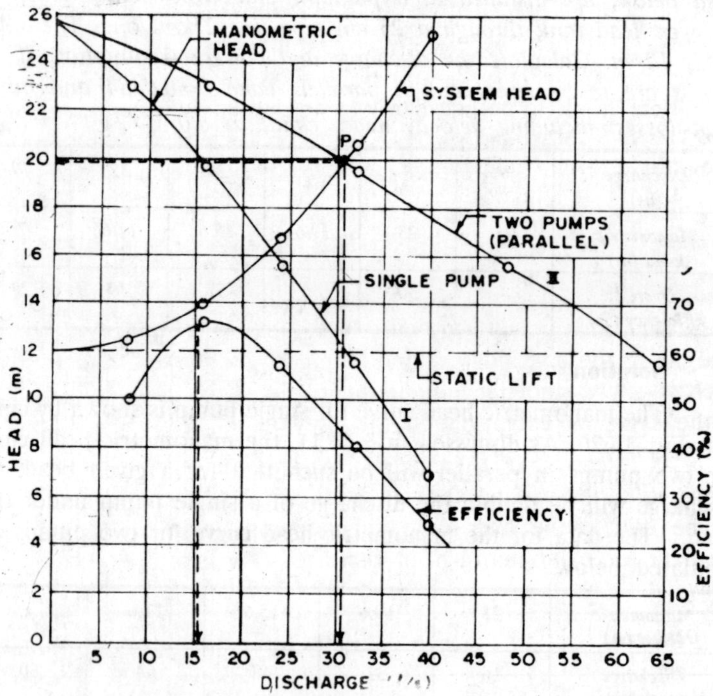

FIG. 14.26. PARALLEL OPERATION OF PUMPS

The values for the system head curve are tabulated below:

Q *(l/s)*	8	16	24	32	40
h_f *(m)*	0.524	2.097	4.719	8.389	13.107
h_m *(m)*	0.011	0.043	0.097	0.173	0.271
h_T *(m)*	12.0	12.0	12.0	12.0	12.0
H *(m)*	12.535	14.140	16.816	20.562	25.378

The operating point *P* is obtained by the intersection of system curve with the manometraic head curve II. From the co-ordinates of point *P*, we get

$$Q = \mathbf{30.6}\ \text{l/s} \ \text{and}\ H = 19.9\ \text{m}$$

Thus, each pump will discharge 30.6/2 = 15.3 l/s, corresponding to which the efficiency comes out to be 65% from the efficiency curve.

$$\text{Hence power consumption} = \frac{9.81\,Q\,H}{\eta}\ \text{kW} \qquad \text{...(14.15)}$$

$$= \frac{9.81 \times 30.6 \times 10^{-3} \times 19.9}{0.65} = \mathbf{9.19\ kW}$$

Note : It is to be noted that for single pump operation, operating Q = 23 l/s, H = 16.2 m and η = 59 %, corresponding to which power required will be = 6.2 kW.

Example 14.13. *Determine the discharge and the power consumption, if the two pumps of example 14.12 are connected in series.*

Solution

The manometric head curve for the single pump will be the same, as in the previous example. Similarly, the system head curve, which is always independent of the pump characteristics, will also be the same as computed in the previous example.

As discussed in § 14.11, the manometric head curve for two pumps in series will be such that for a given discharge, the head will be double the one for a single pump under that discharge. The data for manometric head curve for two pumps are tabulated below :

Discharge (l/s)	0	8	16	24	32	40
Manometric head (m)	50	46	39.2	31.2	23.2	14

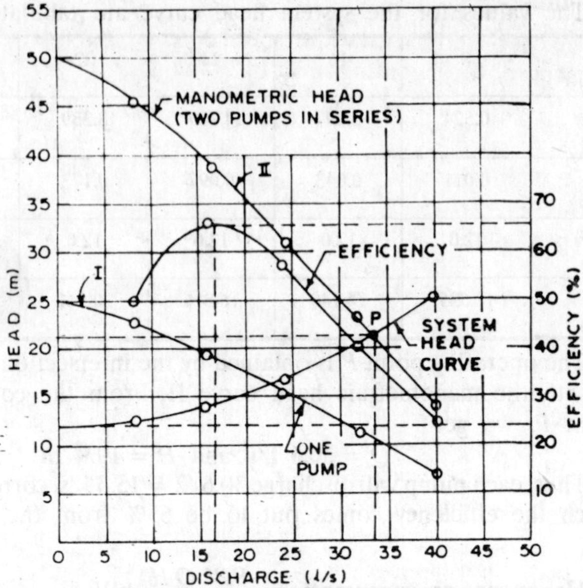

FIG. 14.27. TWO PUMPS IN SERIES.

The operating point P is obtained at the intersection of the system head curve with the manometric head curve for two pumps. The co-ordinates of point P are found to be :

$$Q = \mathbf{33.5}\ \mathbf{l/s}$$

$$H = 21.5\ \text{m}$$

The discharge of each pump will be 33.5/2= 16.75 l/s, corresponding to which η comes out to be 65.5 %.

Hence $$P_e = \frac{9.81\, Q\, H}{\eta}\ \text{kW}$$

$$= \frac{9.81(33.5 \times 10^{-3})\, 21.5}{0.655} = \mathbf{10.79\ kW}$$

Example 14.14. *A variable speed pump is used to pump water against a static lift of 10 m, through a 25 cm dia. pipe line of 1500 m length. The characteristics of the pump at 1500 rpm are as under.*

Discharge (l/s)	0	8	16	24	32	40	48	54
Manometric head (m)	40.0	39.0	37.5	35.0	31.0	25.5	17.5	5.5

Determine the total head of the pump, and the discharge at pump speeds of 1000 rpm and 750 rpm. Take friction factor f= 0.0484 and minor losses $= 9\, v^2/2\, g$.

Solution :

The characteristic curve for speed N_2 can be constructed, using that for speed N_1, by using Eq. 14.3 and 14.2 :

$$H_2 = H_1 \left(\frac{N_2}{N_1} \right)^2 \quad \text{...(a)}$$

and

$$Q_2 = Q_1 \left(\frac{N_2}{N_1} \right) \quad \text{...(b)}$$

For example, taking $N_1 = 1500$ and $N_2 = 1000$, we have

$$H_2 = H_1 \left(\frac{1000}{1500} \right)^2 = 0.4444\, H_1 \quad \text{...(i)}$$

and

$$Q_2 = Q_1 \left(\frac{1000}{1500} \right) = 0.6667\, Q_1 \quad \text{...(ii)}$$

Similarly, taking $N_1 = 1500$ and $N_3 = 750$, we have

$$H_3 = H_1 \left(\frac{750}{1500} \right)^2 = 0.25\, H_1 \quad \text{...(iii)}$$

and

$$Q_3 = Q_1 \left(\frac{750}{1500} \right) = 0.5\, Q_1 \quad \text{...(iv)}$$

With the help of the above equations, the following table can be constructed for the pump characteristics at N_1, N_2 and N_3.

Speed	***Pairs of characteristic values***								
$N_1 = 1500$ rpm	$Q_1 \rightarrow$ $H_1 \rightarrow$	0 40.0	8 39.0	16 37.5	24 35.0	32 31.0	40 25.5	48 17.5	54 5.5
$N_2 = 1000$ rpm	$Q_2 \rightarrow$ $H_2 \rightarrow$	0 17.78	5.33 17.33	10.67 16.67	16.0 15.55	21.33 13.78	26.67 11.33	32.00 7.78	36.00 2.44
$N_3 = 750$ rpm	$Q_3 \rightarrow$ $H_3 \rightarrow$	0 10	4.00 9.75	8.00 9.38	12.00 8.75	16.00 7.55	20.00 6.38	24.00 4.38	27.00 1.38

Hence the manometric head curves for $N_1 = 1500$ rpm, $N_2 = 1000$ rpm and $N_3 = 750$ rpm can be plotted, as shown in Fig. 14.27.

In order to plot the system head curve, we have

$$H = H_{ST} + h_f + h_m$$

where h_f = friction loss $\approx \dfrac{f\,l\,Q^2}{12.1\, d^5} = 1500 = \dfrac{0.0484 \times 1500\, Q^2}{12.1\,(0.25)^5}$

$= 6144\, Q^2$ when Q is in m^3/sec

$= 6144\, (Q \times 10^{-3})^2 = 6.144 \times 10^{-3}\, Q^2$ when Q is in l/s.

$$h_m = 9 \frac{v^2}{2g} = \frac{9}{2g} \left(\frac{4Q}{\pi d^2} \right)^2 = 0.7436 \frac{Q^2}{d^4} \quad \text{(when } Q \text{ is in } m^3/s\text{)}$$

$$= 0.7436 \frac{(Q \times 10^{-3})^2}{(0.25)^4} = 1.904 \times 10^{-4} Q^2 \text{ (when } Q \text{ is in l/s)}$$

The values of h_f, h_m and H are tabulated below.

Q (l/s)	8	16	24	32	40	48	54
h_f (m)	0.39	1.57	3.54	6.29	9.83	14.16	17.92
h_m (m)	0.01	0.05	0.11	0.19	0.30	0.44	0.56
H_{ST} (m)	10.0	10.0	10.0	10.0	10.0	10.0	10.0
H (m)	10.4	11.62	13.65	16.48	20.13	24.6	28.48

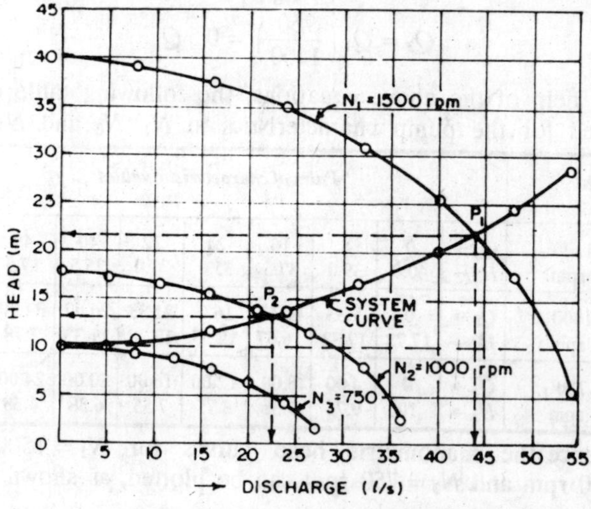

FIG. 14.28. VARIABLE SPEED PUMP

From Fig. 14.28 we find that the system curve intersects N_1 and N_2 curves at points P_1 and P_2 respectively, having the following operating values :

Point P_1 : $Q_1 = \mathbf{44}$ **l/s**
$H_1 = \mathbf{21.5}$ **m**

Point P_2 : $Q_2 = \mathbf{22.5}$ **l/s**
$H_2 = \mathbf{13.0}$ **m**

If $N_3 = 750$ rpm, there is no intersection and hence the pump will not discharge.

PROBLEMS

1. Mention different types of pumps used in water works practice and explain their merits and demerts.

2. Explain the working of a reciprocating pump. What type of reciprocating pump is used for a deep tube well.

3. Describe the principle of working of centrifugal pumps ? Classify various types of centrifugal pumps.

4. Explain how do you compute the power requirements of a pumping unit.

5. Design raw water pumping sets for water supply of a town from the following data :

Population to be served	= 10,000
Average rate of water supply	= 150 litres/capita/day
Seasonal variation factor	= 2
R.L. of the bottom of the tank	= 330.0 m
R.L. of the high flood level	= 328.0 m
R.L. of low water level	= 320.0 m
Duration of pumping in hours per day	= 16
Water to be pumped above bottom level of the tank	= 3 m
Efficiency of the motor	= 90%
Efficiency of the pump	= 70%

Neglect minor losses. *(A.M.I.E.)*

6. Water is required to be pumped from a reservoir for a colony having population of 8000. Following data is available ·

(i) Rate of water supply	:	140 litres/head/day
(ii) Design discharge	:	1.6 × average discharge
(iii) Hours of pumping	:	16
(iv) Bottom R.L. of sump	:	200.50 m
(v) Bottom R.L. of foot valve	:	200.50 m
(vi) Low water level of sump	:	202 m
(vii) Max. water level of sump	:	204 m
(viii) R.L. of centre line of pump	:	206.50 m
(ix) R.L. of ground	:	205.50 m
(x) Full supply level of tank	:	265.0 m
(xi) Length of rising main	:	400 m
(xii) Velocity of flow in pipe line	:	1.5 m/sec
(xiii) Minor losses	:	10% of major losses
(xiv) Friction facto:	:	0.04
(xv) Overall efficiency	:	70%

Find the H.P. of the pump and draw neat sketches of the installations.

7. Explain, with the help of neat sketches, the working of a jet pump.

8. Water has to be supplied to a town having one lakh population at the rate of 150 litres per capita per day from a river 2000 m away. The difference in elevation between the lowest water level in the sump and reservoir is 36 m. If the demand has to be supplied in 8 hours, determine the size of the main and the brake horse power of the pumps required. Assume maximum demand as 1.5 times the average demand. Assume $f = 0.0075$, velocity in pipe 2.4 m/sec and efficiency of pump 80 percent.

(U.P.S.C Engg. Services, 1969)

9. Water is pumped from a low level reservoir to a high level reservoir through a main pipe line of 0.45 m dia. and 1400 m length. The pump is located at low level reservoir. At a point along the main line at a distance of 450 m from high level reservoir, a branch line of 0.3 m dia. and 360 m length takes off to discharge 180 l/sec in the atmosphere.

Level of water surface in high level reservoir = + 30.00 m
Level of water in the open end of the 0.3 m dia. branch line = + 25.50 m
Level of water in low level reservoir = + 18.00 m
Darcy's frictional coefficient for both the pipes = 0.032

Determine the flow rate into the high level reservoir and the theoretical H.P. of the pump, assuming the delivery level of the pump to be at + 20.00 m.

(U.P.S.C. Engg. Services, 1988)

10. A pump is to deliver water from an under ground tank against a static head of 40 m. The suction pipe is 50 m long and is of 25 cm diameter with Darcy-Weisbach friction factor $f = 0.02$. The delivery pipe is of 20 cm diameter, 1600 m long and has $f = 0.022$. The pump characteristics can be expressed as

$$H_P = 100 - 6000\,Q^2$$

where H_P = pump head in metres and Q = discharge in m^3/s.
Calculate the head and discharge of the pump.

(U.P.S.C. Civil Services, 1991)

11. For water supply of a town, water is pumped from a river 3 km away into a reservoir. The max. difference of levels of water in river and the reservoir is 20 m. The population of the town is 50000 and per capita water demand is 120 litres per day. If the pumps are to operate for a total of 8 hours and the efficiency of pumps is 80%, determine the horse power of the pumps. Assume friction factor as 0.03, the velocity of flow as 2 m/s and max. daily demand as 1.5 times the average daily demand.

(UPSC Engg. Services, 1992)

15

Conveyance of Water

15.1. INTRODUCTION

Water is conveyed or transported from the source to the community through various types of *conduits*. These may be either open or closed types depending upon whether the necessary energy is to be provided by gravity or by pumping. The various types of conduits used are : open channels, tunnels and pipe lines. Open channels and tunnels are generally used to convey raw water from the source to the water treatment plants. However, the transported water must be safeguarded against pollution by inferior water sources. This is a special problem when open channels or conduits operating at low pressures are used. Due to this, pipe lines are invariably used for conveyance of water. Another advantage of using pipe lines is the reduction in conveyance losses, such as evaporation and seepage losses.

Open channels are designed to convey water under conditions of atmospheric pressure. They are so constructed that balanced cut and fill is obtained. Depending upon the available slopes, designed velocities and the value of water, they may be unlined or lined. Modern trend is to line the channels with brick tiles, concrete, bituminous material, butyl rubber, vinyl, soil-cement etc. to reduce the resistance to flow, minimise seepage and lower the maintenance costs. *Pressure tunnels* are constructed to cross rivers and valleys. They may be operated under pressure or act as open channels. *Pipe lines* usually follow the profile of the ground quite closely. Pipelines used in important transportation systems may require gate valves, check valves, air release valves, drains, surge control equipment, expansion joints, insulation joints, manholes and pumping stations.

15.2. PIPES

Pipe is a circular closed conduit through which the water may flow either under gravity or under pressure. When pipes do not run full, they run under gravity, such as in sewer lines. However, in supply, pipes mostly run under pressure. Pipes may be made of the following materials :

1. Cast iron.
2. Wrought iron.
3. Steel.
4. Galvanised iron.
5. Cement concrete.
6. Asbestos cement.
7. Plastic.
8. Lead.
9. Copper
10. Wood.

15.3. CAST IRON PIPES

Cast iron pipes are used in great majority of water distribution mains because of centuries of satisfactory experience with it. Cast iron pipe is resistant to corrosion and accordingly is long lived – its life may be over 100 years. Cast iron pipes are manufactured by two methods : (*i*) ordinary sand moulding process, and (*ii*) centrifugal process. In sand casting, molten metal is poured into sand lined moulds that are set in either a vertical or horizontal position. Horizontally cast pipes are referred to as *McWane pipes*, while those cast in vertical position are referred to as *pit cast*. McWane pipes are 100 per cent stronger in tension and 50 per cent stronger in rupture than pit cast pipes. In the centrifugal process, pipes may be cast either in a sand mould or in a water cooled metal mould. The former are known as *sand spun pipes* while the latter are called *Delavaud pipes*. Centrifugal process produces a finer grained and denser structure with a more uniform thickness; hence these pipes are denser and lighter. Cast iron pipes are generally cast in lengths of 3 to 6 metres.

Table 15.1 (*a*) gives the hydrostatic test pressure for vertically cast pipes, as prescribed by IS : 1537-1976. Table 15.2 gives the weights and dimensions for class A vertically cast C.I. pipes (spigot and socket type), while Table 15.3 gives those for class B.

Table 15.1 (*b*) gives the hydrostatic test pressures for centrifugally spun cast iron pressure pipes.

TABLE 15.1 (*a*).
HYDROSTATIC TEST PRESSURE FOR VERTICALLY CAST PIPES

Nominal Diameter	*Test pressure* (*kg/cm*^{2})			
	Socket and spigot pipes		*Flanged pipes*	
	Class A	*Class B*	*Class A*	*Class B*
Upto 300 mm	20	25	20	25
300 to 600 mm	20	25	15	20
600 to 1000 mm	15	20	10	15
1000 to 1500 mm	10	15	10	10

TABLE 15.1 (*b*).
HYDROSTATIC TEST PRESSURES FOR CENTRIFUGALLY CAST PIPES

Class	*Socket and spigot pipes*		*Flanged pipes*	
	Hydrostatic test pressure at works	*Max. hydrostatic test pressure after installation*	*Hydrostatic test pressure at works*	*Max. hydrostatic test pressure after installation*
A	35	12	–	–
A	35	18	35	18
B	35	24	35	24

Joints in cast iron pipes

The following types of joints are provided in C.I. pipes :

(*i*) **Spigot and socket joint.** Fig. 15.1 shows the spigot and socket (bell) joint. The enlarged end of the pipe is known as the *socket* or *bell.* The socket accommodates the normal or the spigot end. The hemp or sterilised yarn is used to maintain the alignment and molten lead heated to 400°C is then poured and caulked up using a series of chisels. The quantity of lead varies from about 5 kg for 15 cm dia. pipe to about 55 kg for a 120 cm dia. pipe. These joints are also known as the *run-lead joint.*

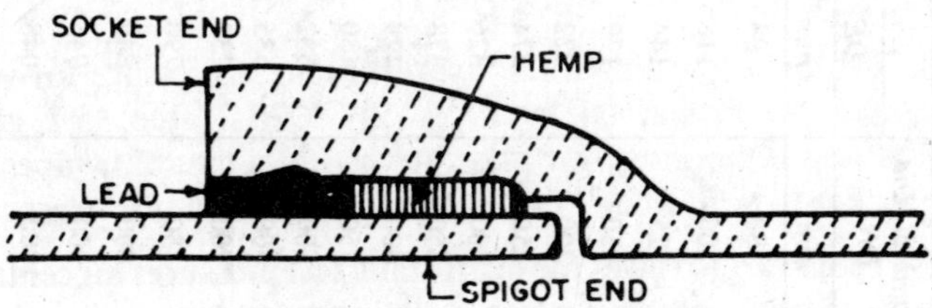

FIG. 15.1. SPIGOT AND SOCKET JOINT

TABLE 15.2. WEIGHTS AND DIMENSIONS FOR VERTICALLY CAST C.I. SOCKET AND SPIGOT PIPES – CLASS A

Nominal Dia. DN	Barrel			Wt. of socket	Total wt. for one working length *l* in metres				
	DE	Thickness, e	Wt. for 1 m		3.66	4	4.88	5	5.5
mm	mm	mm	kg	kg	kg	kg	kg	kg	kg
80	98	7.9	16.0	5.5	64	69.5	—	85.5	—
100	118	8.3	20.5	7.1	84	89	—	109	120
125	144	8.7	26.4	9.2	106	115	—	141	155
150	170	9.2	33.2	11.5	133	144	—	178	194
200	222	10.1	48.1	16.8	193	209	251	257	281
250	274	11.0	65.0	22.9	261	283	340	348	380
300	326	11.9	84.0	29.8	337	366	440	450	492
350	378	12.8	105.0	37.5	422	458	550	563	615
400	429	13.8	128.7	46.3	517	561	674	690	754
450	480	14.7	156.0	56.0	627	680	817	836	914
500	532	15.6	181.0	66.0	728	790	949	971	1061
600	635	17.4	241.4	89.3	973	1055	1267	1296	1417
700	738	19.3	311.6	116.8	1257	1363	—	1675	1830
750	790	20.2	348.9	131.7	1409	1527	—	1876	2051
800	842	21.1	389.1	147.8	1568	1704	—	2093	2288
900	945	22.9	474.3	182.6	1917	2080	—	2554	2791
1000	1048	24.8	570.0	222.3	2303	2502	—	3072	3357
1100	1152	26.6	672.4	265.6	2726	2955	—	3628	3966
1200	1256	28.4	783.1	313.3	3175	3446	—	4229	4623
1500	1567	33.9	1222.1	501.1	4974	5389	—	6612	7223

TABLE 15.3. WEIGHTS AND DIMENSIONS FOR VERTICALLY CAST C.I. SOCKET AND SPIGOT PIPE – CLASS B

Nominal Dia. DN	Barrel			Wt. of socket	Total wt. for one working length *l* in metres				
	DE	Thickness, e	Wt. for 1 m		3.66	4	4.88	5	5.5
mm	mm	mm	kg	kg	kg	kg	kg	kg	kg
80	98	8.6	17.3	5.5	69	74.5	—	92	—
100	118	9.0	22.0	7.1	88	95	—	117	128
125	144	9.5	28.7	9.2	114	124	—	153	167
150	170	10.0	35.9	11.5	143	155	—	191	209
200	222	11.0	52.1	16.8	207	225	271	278	304
250	274	12.0	70.6	22.9	281	305	368	376	411
300	326	13.0	91.4	29.8	364	395	476	487	533
350	378	14.0	114.5	37.5	457	495	596	610	667
400	429	15.0	139.5	46.3	557	604	727	744	814
450	480	16.0	169.0	56.0	675	732	881	901	986
500	532	17.0	196.7	66.0	786	853	1026	1026	1148
600	635	19.0	262.9	89.3	1052	1141	1372	1404	1535
700	738	21.0	338.2	116.8	1355	1470	—	1808	1977
750	790	22.0	380.6	131.7	1525	1644	—	2035	2225
800	842	23.0	423.1	147.8	1696	1840	—	2263	2475
900	945	25.0	516.6	182.6	2073	2249	—	2766	3024
1000	1048	27.0	619.2	222.3	2489	2699	—	2318	3628
1100	1152	29.0	739.0	265.6	2970	3192	—	3923	4326
1200	1256	31.0	851.6	313.3	3430	3725	—	4578	4993
1500	1567	37.0	1313.1	501.1	5380	5834	—	7167	7833

(*ii*) **Flanged joint.** This type of joint is used in C.I. pipes with flanges. The pipes are carefully aligned before bolts are inserted and the flanges are pulled together. A rubber ring of diameter lesser than he bolt-circle diameter is inserted between the flanges as shown in Fig. 15.2. The contact should be between the clean dry metal flanges and the clean dry rubber, and no oil or grease etc. should be inserted.

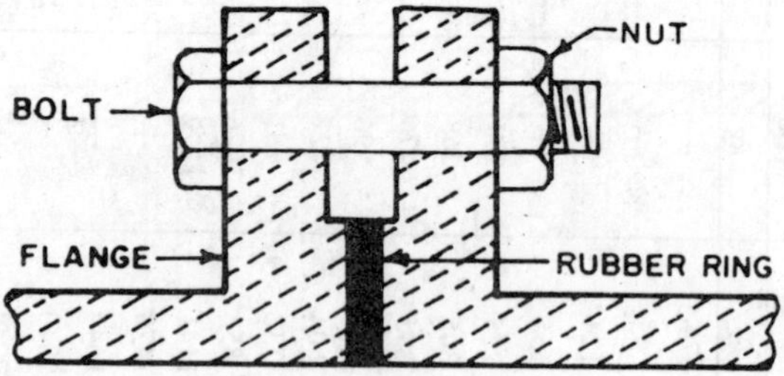

FIG. 15.2. FLANGED JOINT

(*iii*) **Tyton joint.** Fig. 15.3 shows *O-ring joint* or the Tyton joint in which a specially shaped rubber ring is fitted into the inside of the socket of one pipe and the spigot of the next pipe is then forced home. A special lubricant to the chamfered spigot is applied for forcing it inside.

(*iv*) **Coupled joints.** Two types of coupling joints are used : (*a*) Viking Johnson coupling shown in Fig. 15.3 (*a*), and (*b*) victaulic coupling shown in Fig. 15.3 (*b*). In both these, the pipes do not have sockets or spigots. Such joints are highly flexible.

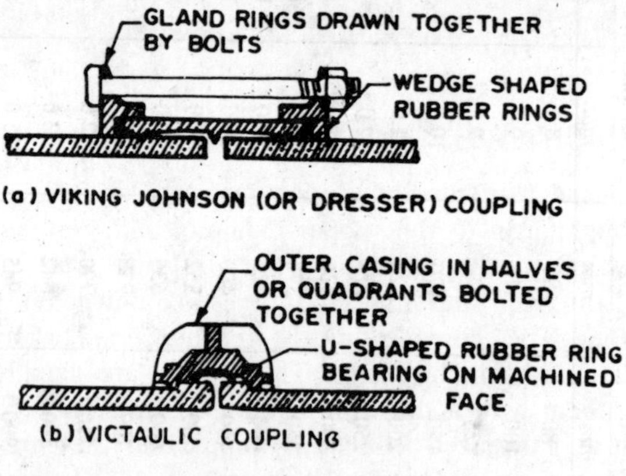

FIG. 15.3. COUPLINGS

(*v*) **Expansion joints.** Expansion joints allow the pipes to expand or contract freely under the changing temperature conditions. Fig. 15.4 shows such a joint, in which the space between the socket and spigot is filled with an elastic gasket.

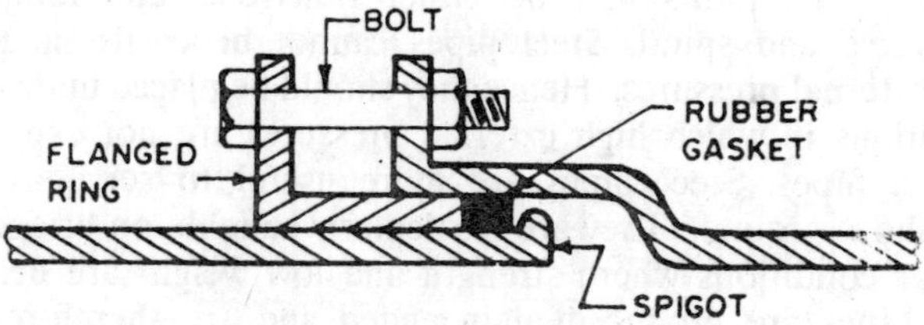

FIG. 15.4. EXPANSION JOINTS

Advantages of C.I. pipes

C.I. pipes have the following advantages ;

1. C.I. pipes are of moderate cost.
2. Their jointing is easier.
3. They are resistant to corrosion.
4. They have long life.
5. Service connections can be easily made in them.

Disadvantages

1. They are subject to tuberculation in certain waters, due to which their carrying capacity is reduced to as much as 70%.
2. They are heavier, and hence uneconomical when their diameter is more than 120 cm.
3. They can not be used for pressures greater than 7 kg/cm^2.
4. They are fragile.

15.4. WROUGHT IRON AND GALVANISED IRON PIPES

Wrought iron pipes are manufactured by rolling flat plates of the wrought iron to the proper diameter and welding the edges. Such pipes are much lighter than the C.I. pipes and can be more easily cut, threaded and worked. They look much neater, but are much costlier. They corrode quickly, and hence are used principally for installation within buildings. These pipes are usually protected by coating them with a thin film of molten zinc. Such coated pipes are known as galvanized iron (G.I.) pipes, and they are commonly jointed by screwed and socketed joints.

15.5. STEEL PIPES

Steel pipes of small diameter can be made from the solid, but the larger sizes are made by rivetting or welding together the edges of suitably-curved plates, the sockets being formed later in a press. The joints may be either transverse and longitudinal or transverse and spiral. Steel pipes cannot be easily made to resist high external pressures. Hence they should be placed under favourable conditions in which high external pressures are not expected to fall on the pipes. Steel pipes are more useful to serve as big mains laid above ground. They are particularly suitable on bridges or under similar conditions where strength and low weight are desired. Most steel pipes are produced plain ended and are therefore connected with Viking Johnson couplings (Fig. 15.5 *a*). The life of steel pipe under ordinary conditions is greatly dependent on the quality of protective coating which is provided on both the external as well as internal surface.

The plate thickness t required to resist the internal pressure p (including water hammer) is given by

$$t = \frac{p\,R}{f_t\,\eta} \qquad \text{...(15.1)}$$

where R = radius of the pipe
f_t = permissible tensile stress
η = joint efficiency.

The critical external pressure that will cause buckling in the steel pipe without stiffening rings is given approximately by the expression :

$$P_{cr} = 20\,\frac{E\,I}{D^3} = \frac{5\,E\,t^3}{3\,D} \qquad \text{...(15.2)}$$

where P_{cr} = External pressure per unit length
E = modulus of elasticity of the metal
I = moment of inertia ($t^3/12$)
t = thickness of pipe
D = Diameter of the pipe

15.6. CEMENT CONCRETE PIPES

Cement concrete pipes may be either plain or reinforced, and are best made by the spinning process. They may be either precast, or may be prepared at the site. The plain cement concrete pipes are used for heads up to 7 m while reinforced cement concrete pipes are normally used for heads upto 60 m. For heads higher than this, prestressed concrete pipes are used. Reinforced concrete pipes are used only for large water supply lines, but they are not suitable for distribution systems. Water tightness in low pressure reinforced concrete pipe is obtained by providing sufficient thickness of good concrete (1 : 1 : 2 mix) and longitudinal reinforcement equal to at least 0.25% of the cross-sectional area of the concrete. The whole section should be designed for tensile and bending stresses. The circumferential reinforcement is designed to resist the hoop tension. Water tightness of high pressure concrete pipe may be obtained by insertion of a thin steel cylinder in the pipe walls either with or without prestressed reinforcement.

Indian Standard (IS : 458-1961) classifies the concrete pipes into six classes as given in Table 15.4.

TABLE 15.4. CLASSIFICATION OF CONCRETE PIPES

Class	*Description*	*Condition where normally used*
NP 1	Unreinforced concrete, non-pressure pipes.	For drainage and irrigation use, above ground or in shallow trenches.
NP 2	Reinforced concrete, light duty, non-presure pipes	For culverts carrying light traffic
NP 3	Reinforced concrete, heavy duty, non-pressure pipes.	For culverts carrying heavy traffic, such as Railway loadings.
*P*1	Reinforced concrete pressure pipes tested to a hydrostatic pressure of 2.0 kg/cm^2 (20 m head).	For use on gravity mains, the actual working pressure not exceeding 2/3 of the test pressure.
*P*2	Reinforced concrete pressure pipes tested to a hydrostatic pressure of 4.0 kg/cm^2 (40 m head)	For use on pumping mains, the actual working pressure not exceeding 1/2 the test pressure.
*P*3	Reinforced concrete pressure pipes tested to a hydrostatic pressure of 6.0 kg/cm^2 (60 m head).	Same as above.

The ends of the concrete pipes are so manufactured that they are suitable for butt end joint for all classes of pipes. *NP* 1 class of pipe may have spigot and socket ends.

Advantages

Concrete pipes have the following advantages :

1. They are more suitable to resist the external loads and loads due to backfilling.
2. The maintenance cost is low.
3. The inside surface of pipes can be made smooth, thus reducing the frictional losses.

4. The problem of corrosion is not there.
5. Pipes can be cast at site, and hence the transportation problems are reduced.
6. Due to their heavy weight, the problem of floatation is not there when they are empty.
7. Expansion joints are not normally required.

Disadvantages

1. Unreinforced pipes are liable to tensile cracks, and they cannot withstand high pressure.
2. The tendency of leakage is not ruled out as a result of its porosity and shrinkage cracks.
3. It is very difficult to repair them.
4. Precast pipes are very heavy, and it is difficult to transport them.

15.7. ASBESTOS CEMENT PIPES

Asbestos cement pipes are manufactured from asbestos fiber and portland cement combined under pressure to form a dense homogeneous structure having strong bond between cement and the fiber. Such a pipe is claimed to be completely impervious to passage of water through its walls. Their tensile and compressive strengths used in the design are 200 and 600 kg/cm^2 respectively. They are generally manufactured under proprietary names. These pipes are finished with plain ends, the exterior of the barrel being manufactured to receive the coupling. The joints used in the asbestos cement pipes are of simplex type, consisting of a sleeve which fits over the plain ends of the lengths, water-tightness being obtained by two rubber rings compressed between the sleeve and the pipe barrels as shown in Fig 15.5. The joint thus becomes flexible and watertight. The pipe can be tapped and threaded for service connections in the same manner as iron pipe.

Advantages of asbestos cement pipes

1. They have smooth internal surface, due to which the frictional losses are reduced.
2. They are light and can be easily transported.
3. They can be easily cut, fitted or jointed.
4. Service connections can be easily taken, since they can be easily drilled and tapped.
5. They are anti-corrosive.
6. No expansion joints are required.

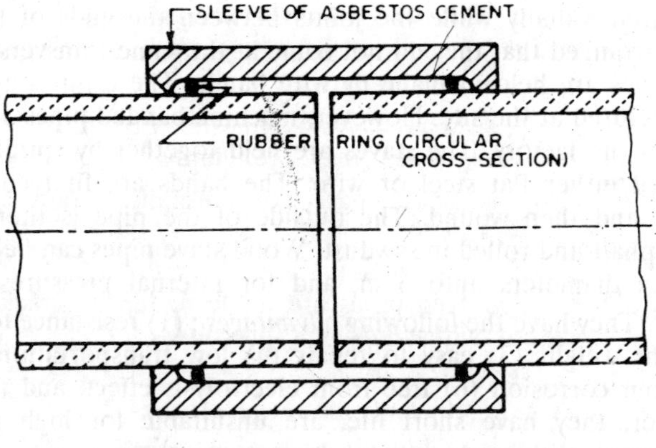

FIG. 15.5. SIMPLEX JOINT FOR ASBESTOS CEMENT PIPES

Disadvantages

1. They are soft and brittle. They are very weak under impact loading due to moving traffic.
2. They are not durable.
3. They cannot be laid in exposed places.
4. They are costly.

15.8. COPPER AND LEAD PIPES

Copper pipes are very costly, and their use is restricted for conveyance of hot water in the interior of buildings, and for making gooseneck in the service connections. These pipes do not sag or bend due to hot water. They are also resistant to corrosion by acidic or alkaline waters. Copper pipes can withstand high internal pressure.

Lead pipes are very soft and can be easily bent. Due to this property, they are extensively used in service connections and for internal plumbing. Due to the cumulative poisoning effect of lead, these are not used as water supply pipes. They are also not suitable for conveying acidic water. They are used for apparatus required for alum and chlorine doses.

15.9. WOOD-STAVE PIPES

Wood stave pipes are prepared of staves or planks of wood held together by steel bands. They have been used for many years for water supply pipes, though they were replaced due to lack of capacity. Bored wood pipes are still used in industrial plants where highly corrosive liquids are to be conveyed. Wood stave pipes are of two types : (*a*) *continuous stave*, and (*b*) *machine banded*. In the continuous stave pipe, the joints between the sides of the staves are placed radially while the joints between the ends of the staves are so arranged that they do not come at the same transverse section. The staves are held in plane by wire bands. These pipes are usually manufactured at the site. However, machine-banded pipes are usually made in the factory. The staves are held together by spirally wound bands of either flat steel or wire. The bands are first coated with asphalt and then wound. The outside of the pipe is then covered with asphalt and rolled in sawdust. Wood stave pipes can be manufactured in diameters upto 6 m, and for internal pressures upto 10 kg/cm^2. They have the following *advantages* : (1) resistance to freezing (2) light weight (3) easy to repair (4) low transportation cost (5) free from corrosion (6) free from electrolysis effect, and (7) cheap. However, they have short life, are unsuitable for high pressures, and cannot withstand external high pressure.

15.10. PLASTIC PIPES

Plastic pipes are of recent origin. They are made of synthetic resins of high molecular weight, polymerized from simple compounds by heat, pressure and catalysis. Plastic pipes and tubes are manufactured from a number of polymerisation products such as *polythene, Polypropylene, Polyvinyl chloride* and *Cellulose acetate butyrate*. In India, Polythylene and Polyvinyl chloride (P.V.C.) pipes are manufactured at present.

Advantages

1. They are free from corrosion.
2. They are free from damage due to freezing and thawing of water in closed pipe.
3. They are cheaper.
4. They are highly resistant to acidic waters.
5. They are light in weight, and easy to handle and transport.
6. They are smooth and possess low hydraulic resistance.
7. They are good electric insulators.
8. Their jointing, bending and installation is easy.
9. They have adequate strength, and are resistant to shock.
10. They are durable, and are unaffected by age, sunlight or weather.

Disadvantages

1. They have low resistance to heat.
2. The coefficient of expansion is high.
3. Some types of plastics impart taste to water.

15.11. STRESSES IN PIPES

Pipes may be subjected to following types of pressures :

(*i*) Internal water pressure

(*ii*) External pressures due to overburden etc.

(*iii*) Temperature stresses

(*iv*) Water hammer

(*v*) Longitudinal stresses due to flow around bends.

(*vi*) Stresses due to foundation reaction.

1. INTERNAL WATER PRESSURE

The internal water pressure causes hoop tension in the pipe shell, its magnitude being given by

$$f = \frac{p\,d}{2\,t} \qquad \text{...(15.3)}$$

where p = internal pressure of water,
d = diameter of the pipe
t = thickness of pipe shell.

The above expression is true only for pipes of small diameters in which the pressure p may be assumed to be uniform. In pipes of large diameter, due to the difference in levels, the pressure is maximum at bottom and minimum at the top. If p is the pressure at top the problem can be looked upon as if the pipe is acted upon by a uniform pressure p throughout causing only hoop tension, and in addition, there are stresses due to the weight of water assumed to just fill the pipe without any pressure at top.

2. STRESSES DUE TO EXTERNAL LOADS

When water supply pipes are buried underground, they are subjected to weight of the backfill and the external loads if the pipe is buried under a road. According to Marston, the load W on buried pipe, due to earth filling, is given by

$$W = C\,.\,\gamma\,.\,B^2 \qquad \text{...(15.4)}$$

where γ = unit weight of the backfill material (kg/m^3)
W = load per metre length of pipe
B = width of the trench
C = coefficient, which depends upon the type of soil and the ratio of the depth to the trench width. (Table 15.5)

It is assumed that the load W caused by the overlying material is distributed uniformly over the arch of the pipe. The extreme fibre stresses at the end of the horizontal diameter are given approximately by the following expression suggested by Williams :

$$S = \frac{3\,W D}{8\,t^2} \pm \frac{W}{2\,t} \qquad \text{...(15.5)}$$

where D = diameter of the pipe.

In the above formula, plus sign is used to obtain compression while minus sign is used for tension. To obtain maximum fibre stresses on the vertical points, the quantity $\frac{W}{2t}$ is not used in Eq. 15.5.

TABLE 15.5. VALUES OF C

Ratio of depth to trench width	*Sand and damp top soil*	*Saturated top soil*	*Wet clay*	*Saturated clay*
0.5	0.46	0.46	0.47	0.47
1.0	0.85	0.86	0.88	0.90
1.5	1.18	1.21	1.24	1.28
2.0	1.46	1.50	1.56	1.62
2.5	1.70	1.76	1.84	1.92
3.0	1.90	1.98	2.08	2.20
3.5	2.08	2.17	2.30	2.44
4.0	2.22	2.33	2.49	2.66
5.0	2.45	2.59	2.80	3.03
6.0	2.61	2.78	3.04	3.33
7.0	2.73	2.93	3.22	3.57
8.0	2.81	3.03	3.37	3.76
9.0	2.90	3.14	3.52	3.98
10.0	2.92	3.17	3.56	4.04
12.0	2.97	3.24	3.68	4.22
15.0	3.01	3.30	3.77	4.38
Very great	3.03	3.33	3.85	4.55

Marston proposed the following formula for computing additional load on the conduits due to loads applied to the surface of the fill

$$W_t = \frac{1}{L_e} \cdot I_e \cdot C_t \cdot P \qquad \text{...(15.6)}$$

where W_t = Average load on pipe due to wheel load (kg/m)

L_e = Effective length of the pipe. It is taken equal to the actual length of a precast segment section of pipe if less than 1 m. If the precast length is more than 1 m, the maximum value of L_e is recommended as 1 m.

I_e = Impact factor

C_t = Load coefficient

P = Concentrated wheel load, on the embankment.

The load coefficient C_t is dependent on the length and width of the conduit cross-section and the depth of cover over it and may be evaluated in accordance with the Boussinesq formula of stress distribution.

3. TEMPERATURE STRESSES

Temperature stresses are caused in metal pipes when they are laid above ground. When they are subjected to temperature variations, change in the length of the pipe will take place. If this change in length is restricted due to fixidity or any other reason, the pipe will be subjected to the following stress :

$$f_t = E_p \cdot \alpha \cdot \Delta T \quad \text{...(15.7 a)}$$

where f_t = temperature stress
E_p = modulus of elasticity of pipe material (Table 15.6)
α = coefficient of thermal expansion
ΔT = temperature variation

The value of coefficient of expansion (α) for conduits per °C may be taken as 11.9×10^{-6} for steel, 8.5×10^{-6} for cast iron and 10×10^{-6} for concrete.

The resulting longitudinal force is given by

$$P_t = \pi (d + t)\, t \,.\, f_t \quad \text{...(5.7 b)}$$

where d is the internal dia. and t is the thickness of pipe.

4. STRESSES AT BENDS

When water flows through a pipe bend, an outward thrust is exerted on the pipe due to the combined action of (*i*) unbalanced internal pressures and (*ii*) centrifugal force exerted by the moving water. Their combined action cause a tendency to pull the pipe apart longitudinally. Hence external resistance in the form of buttress or anchorage should be provided or joints must be made sufficiently strong to overcome the tension.

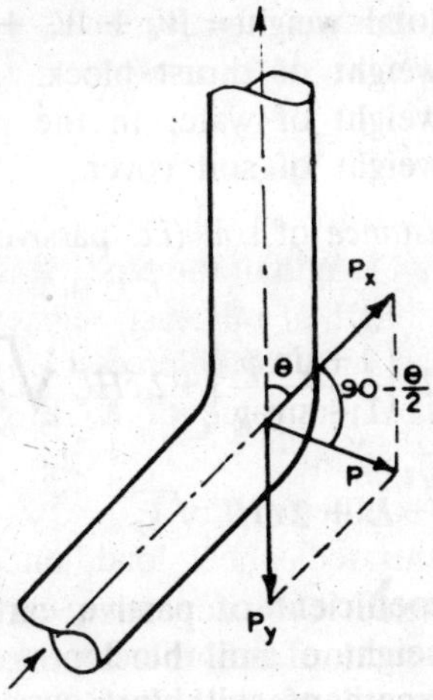

FIG. 15.6. THRUST AT BEND

The total tension (T) in the pipe at the bend is given by

$$T = pA + \frac{wAV^2}{g} \quad ...(15.8)$$

where
p = internal water pressure
A = area of section of pipe
w = unit weight of water
V = velocity of flow of water

If θ is the angle of bend, or the change in the direction of the pipe, the horizontal thrust P is given by

$$P = 2A\left(\frac{wV^2}{g} + p\right)\sin\frac{\theta}{2} \quad ...(15.9)$$

Anchorages are required to resist the above horizontal thrust. Anchorages are designed as concrete buttress or 'kick blocks', to resist the unbalanced pressure by their weight. The resistance offered by the pipe joints themselves, by the friction of the pipe exterior and by the bearing value of the soil in which the block is buried may be taken into consideration, if the cost of the block is to be minimised.

The *total resistance* offered by the block consist of the following three components :

1. *Frictional resistance* on the bottom of the thrust-block and the circumference of the pipe line :

$$P_f = \mu W \quad ...(15.10)$$

where
μ = coefficient of friction, the value of which is usually taken as 0.3
W = total weight $= W_b + W_w + W_s$
W_b = weight of thrust block.
W_w = weight of water in the pipe.
W_s = weight of soil cover.

2. *Lateral resistance* of soil (*i.e.* passive pressure) against the block :

$$P_p = \gamma\frac{H^2}{2}L\left(\frac{1+\sin\phi}{1-\sin\phi}\right) + 2cHL\sqrt{\frac{1+\sin\phi}{1-\sin\phi}} \quad ...(15.11\ a)$$

or

$$P_p = k_p\frac{\gamma H^2}{2}L + 2cHL\sqrt{k_p} \quad ...(15.11)$$

where
k_p = coefficient of passive earth pressure.
H = height of soil block
L = length of soil block

ϕ = angle of internal friction for the soil
c = unit cohesion.
γ = unit weight of soil mass

3. *Lateral resistance* of the soil when the thrust block is free to yield away from the soil mass, *i.e.*, active earth pressure on the portion of the projected pipe.

$$P_a = \gamma . h \left[\frac{1 - \sin \phi}{1 + \sin \phi} \right] - 2c \sqrt{\frac{1 - \sin \phi}{1 + \sin \phi}} \quad ...(15.12\ a)$$

or

$$P_a = k_a . \gamma . h - 2c\sqrt{k_a} \quad ...(15.12)$$

where k_a = coefficient of active earth pressure
h = height of earth mass = dia. of pipe

Hence the *total resistance* is given by

$$P_r = P_f + P_p + P_a \quad ...(15.13)$$

$$\text{Factor of safety} = \frac{P_r}{P} \quad ...(15.14)$$

which should be greater than 2.

A thrust block should be designéd so that the line of action of the resultant of the resisting forces coincides with the line of thrust of the pipe, so as to prevent overturning.

Thrust blocks are reinforced along its external surfaces and around the pipe embedment zone. The minimum surface reinforcement should be 5 kg/m^2 of the surface area. The spacing of the bars should not exceed 500 mm.

Thrust blocks are needed not only at the changes in horizontal or vertical alignment of the pipe line, but also at fittings which may not be able to transmit longitudinal forces such as flexible couplings.

5. WATER HAMMER

Water hammer effect is produced when the velocity of water in the pipe is checked suddenly by sudden closure of a valve. When the valve is closed, there will be sudden rise in pressure due to the momentum of the moving water being destroyed. The rise in pressure in some cases may be so large that the pipe may even burst. It is, therefore, essential to take into account this pressure rise in the design of the pipes. The maximum water hammer pressure $P_{h,max}$ is given by the expression :

$$p_{h,max} = \frac{w}{g} U_p . V_0 \quad ...(15.15)$$

where U_p = velocity of pressure wave generated

$$= \frac{V_s}{\sqrt{1 + \frac{K_w}{E_p} \cdot \frac{d}{t}}} \quad ...(15.16)$$

$$V_s = \sqrt{\frac{K_w}{\rho}} = \text{velocity of sound wave in water} \quad ...(15.17)$$

≈ 1433 m/s

$\rho = w/g$

K_w = bulk modulus of water (2.07×10^8 kg/m^2)

E_p = modulus of elasticity of pipe material (kg/m^2)

d = diameter of pipe (m)

t = thickness of pipe (m)

V_0 = normal velocity of flow in the pipe line (m/s), before sudden closure

From Eqs. 15.15, 15.16 and 15.17, we get

$$p_{h,max} = \frac{w}{g} \cdot \frac{V_s}{\sqrt{1 + \frac{K_w}{E_p} \cdot \frac{d}{t}}} \cdot V_0 \quad ...(15.18)$$

or $$P_{h,max} = \frac{1000}{9.81} \times \frac{1433}{\sqrt{1 + \frac{K_w}{E_p} \cdot \frac{d}{t}}} \cdot V_0$$

$$\approx 14.6 \times 10^4 \frac{V_0}{\sqrt{1 + \frac{K_w}{E_p} \cdot \frac{d}{t}}} \text{ kg/m}^2 \quad ...(15.19\ a)$$

or $$p_{h,max} \approx \frac{14.6\, V_0}{\sqrt{1 + \frac{K_w}{E_p} \cdot \frac{d}{t}}} \text{ kg/cm}^2 \quad ...(15.19)$$

The above expression has been derived, considering the elasticity of pipe. If, however, *elasticity of pipe is neglected*, then

$$U_p = \sqrt{\frac{K_w}{\rho}} = V_s = 1433 \text{ m/s}$$

$$\therefore \quad p_{h,max} = \frac{w}{g} \cdot \sqrt{\frac{K_w}{\rho}} \cdot V_o \quad ...(15.18\ a)$$

or $$p_{h,max} = \frac{1000}{9.81} \times 1433\, V_0 = 14.6 \times 10^4\, V_0 \text{ kg/m}^2 \quad ...(15.18\ b)$$

If the actual time of closure (T) is *greater than the critical time* T_c, *the water hammer pressure* (p_h) *is reduced approximately in the proportion* T_c/T.

If L is the distance of the valve from the reservoir, the *critical time* T_c (*i.e.* the time taken by the wave in going to the reservoir and coming back to the valve) is given by

$$T_c = \frac{2L}{U_p} \quad ...(15.20)$$

$$\therefore \quad p_h = p_{h,max} \cdot \frac{T_c}{T} \quad ...(15.21)$$

The values of E_p for different materials of pipe are given in Table 15.6.

TABLES 15.6. VALUES OF E FOR PIPE MATERIAL

Material	*E* (kg/m^2)
1. Poly ethylene (soft)	1.2×10^7
2. Poly ethylene (hard)	9×10^7
3. PVC	3×10^8
4. Concrete	2.8×10^9
5. Asbestos cement	3×10^9
6. Reinforced cement concrete	3.1×10^9
7. Prestressed concrete	3.5×10^9
8. Cast iron	7.5×10^9
9. Ductile iron	1.7×10^{10}
10. Wrought iron	1.8×10^{10}
11. Mild steel	2.1×10^{10}

In small pipes, a water hammer pressure of a magnitude 7 kg/cm^2 may be assumed.

15.12. CORROSION IN PIPES

When a metal, immersed in water or in an electrolyte, goes into solution, the action is known as corrosion. The pipe corrosion may be internal (due to flowing water) or external (due to surrounding media) and may lead to disintegration of the metal. Corrosion is an electrochemical process. Iron ions tend to go into the water and replace hydrogen ions, the iron acting as an anode and the hydrogen forming at the cathodic points. The factors that contribute to pipe corrosion are : (*i*) *pH* of flowing water (*ii*) presence of water in surround soil and its *pH* (*iii*) composition of pipe material (*iv*) temperature (*v*) oxygen, and (*vi*) soil bacteria. In water supply pipes, the corrosion may result in tuberculation with accompanying reduction in its carrying capacity, leaks in pipe, disintegration of pipe etc. The following hypothesis have been advanced to explain corrosion:

1. Galvanic or bimetallic corrosion. A metal goes into solution when it is immersed in corrosive water or in the solution of an electrolyte. This is because the cations of the metallic element tend to migrate through the electrolytically neutral solution and deposit on the surface of an electro-negative metal. Also, there is migration of anions towards anode. The reaction is represented by the following equations by Camp :

$$\text{Metallic element} \underset{\text{reduction}}{\overset{\text{oxidation}}{\Leftrightarrow}} \text{cations} + \text{electrons}$$

$$\underset{\text{(metal)}}{Fe} \Leftrightarrow Fe^{++} + 2 \text{ electrons}$$

$$Fe + 2OH \rightarrow Fe(OH)_2 + 2 \text{ electrons}$$

If two dissimilar metals are immersed in the same electrolyte, there is an interchange of ions between them or with the electrolytic solution, resulting in bimetallic corrosion. The iron in pipe may have impurities such as nickel, lead, tin, copper etc. which have a potential lower than that of pure iron. Theses impurities become cathode in presence of water, and iron becomes anode. The anode looses its ions which get deposited on the cathode, resulting in corrosion.

2. Hydrogenation. Since water is a mild electrolyte, it dissociates into H^+ and OH^- ions. The hydrogen ions will deposit on a metallic surface immersed in water. In order to maintain electrical neutrality in the solution, ions of the metal are discharged into the solution to neutralise OH^- ions released by the dissociation of water. This results in pipe corrosion. The hydrogen ions deposited on the metal surface retard the corrosion.

3. Electrolysis. When two different metals are immersed in an electrolytic solution (such as water), a cell is formed. When direct current is passed through this cell, the positively charged ions will travel from anode (pipe), to the cathode, resulting in the corrosion of cathode.

4. Chemical reaction. Corrosion may also take place due to chemical reaction when a metal is immersed in an acid. When water is acidic or when it contains carbon dioxide, corrosion of pipe takes place.

5. Direct oxidation. When a metal is exposed to an atmosphere containing oxygen, it combines with oxygen resulting in the formation of magnetite, Fe_3O_4. The rust deposited on the metal tends to inhibit further corrosion from direct oxidation. Also, oxygen helps in the electrolytic corrosion since it depolarises the cathode by removing hydrogen deposited there.

6. Biological action. The bacteria that cause corrosion of ferrous metals are (*i*) sulphate reducing bacteria, (*ii*) iron-consuming bacteria, and (*iii*) sulphur bacteria. Bacterial corrosion can be prevented by cathodic protection.

Prevention of pipe corrosion

Pipe corrosion can be prevented or retarded by the following methods : (*i*) cathodic protection (*ii*) proper selection of material (*iii*) providing protective linings and coatings (*iv*) treatment of water.

(*i*) **Cathodic Protection.** It has been observed that if the pipe line acts as cathode, the corrosion of pipe can be minimised. Cathodic protection is accomplished by connecting the metal in an electric circuit to the negative pole of a D.C. generator, the positive pole being connected to anodes buried in the ground. Fig. 15.7 shows the electric circuit. Due to such cathodic protection, electric potential of pipe metal will be lowered below that of the surroudings and currents will flow from the surroundings to the pipe. In this process, the anodes buried in the ground get corroded and have to be replaced. The voltage may vary between 1.5 to 30 volts. This method has been used by pipe-line companies to prevent corrosion of the exterior of pipes in unfavourable soils but in water works field it is used principally to prevent corrosion of steel tanks. Cathodic protection retards galvanic and electrolytic corrosion, hydrogenation, and to some extent bacterial corrosion of metals.

(*ii*) **Proper selection of pipe to material.** It has been found that homogenity of the pipe metal tend to minimise corrosion. Impurities of lower potential tend to create galvanic couples resulting

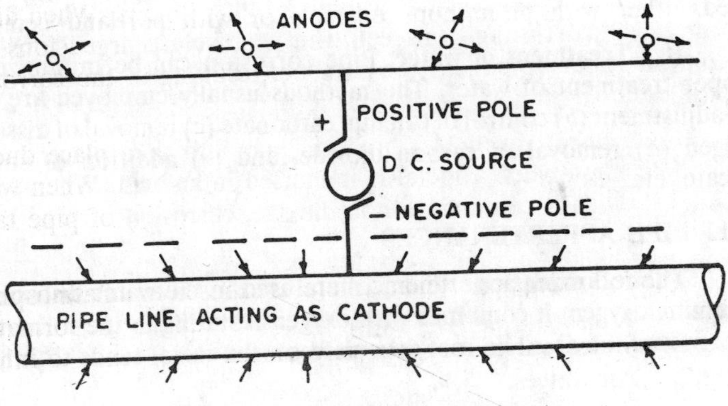

FIG. 15.7. CATHODIC PROTECTION

in corrosion. Alloys of iron or steel with chromium, copper or nick are more resistant to corrosion than iron or steel alone. Everd metal is a light weight corrosion-resistant alloy consisting of copp and silicon, with other metals.

(*iii*) **Protective linings and coatings.** If pipes are made non-corrosive materials, they either become very costly or they d not possess required strength. However, the corrosion in metal pip can be reduced by providing protective anti-corrosive linings. Thes linings will prevent corrosion and contamination of water, and wi also make the inside surface of pipe smooth. The common materia used for lining are : tar or asphaltic materials, enamels, resins, lacquer zinc (or galvanizing), metallizing, plastics and paints.

Coal tar enamel coatings are extensively used for steel pipe Cast iron pipes are coated with Angus Smith coating. Most coate steel pipe is also wrapped, frequently with asbestos felt saturate with bitumen. Asphaltic coatings are less permeable to oxidizin agents than to water. Resins may be prepared synthetically to produc thermosetting and thermoplastic coatings for the protection of metalli surfaces against corrosion. When lacquers and paints are used i connection with cathodic protection, they should be alkali resistan because of the alkali usually produced at the cathode. If paints ar used to protect the pipe from corrosion, they should be made o a metallic pigment, usually zinc or lead. Zinc is a particularly desirabl pigment on account of its electrolytic relation to iron. Vitreous coating for metal are prepared by running molten silicate into cold water The metal is coated with it by dipping or brushing.

Galvanizing is the process by which a zinc coating is applied to the metal. Nickel and copper plating are applied electrolytically to the cleaned metal. Tin or lead plating is applied in a manner similar to the hot-dip process of galvanizing. Metallizing consists in spraying a coat of molten metal on a metal surface. Pipes are lined either with bituminous enamels or with portland cement.

(*iv*) **Treatment of water.** Pipe corrosion can be minimized by proper treatment of water. The methods usually employed are : (*a*) *pH* adjustment (*b*) control of calcium carbonate (*c*) removal of dissolved oxygen (*d*) removal of carbon dioxide, and (*e*) addition of sodium silicate etc.

15.13. PIPE APPERTENANCES

The following appertenances are used in the water distribution system :

1. Sluice valves or gate valves.
2. Air valves.
3. Reflux valves.

4. Relief valves.
5. Altitude valves.
6. Scour valves.

Air valves, pressure relief valves, scour valves etc, are fixed at proper locations along the alignment of a water distribution line. Fig. 15.8 shows a typical alignment of a water supply main, showing the positions of these valves. These valves are necessary to isolate and drain pipeline sections for test, inspection, cleaning and repairs.

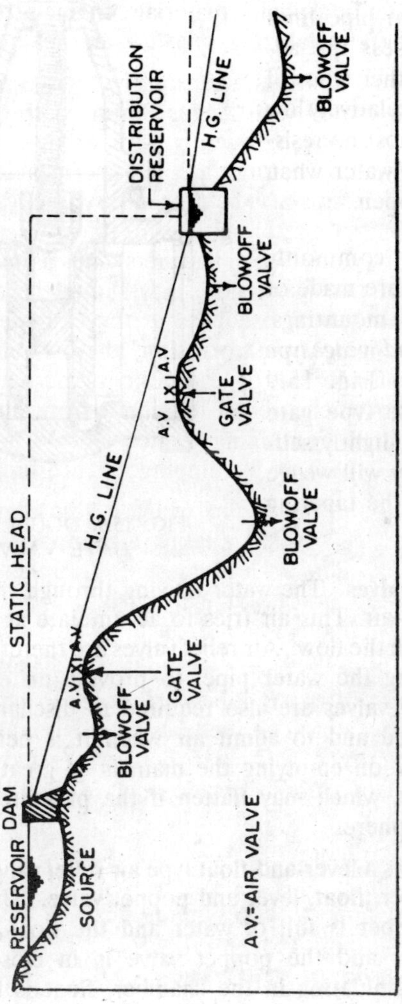

FIG. 15.8. PROFILE OF PIPE LINE SHOWING POSITIONS OF GATE VALVES, BLOW OFF VALVES AND AIR VALVES.

1. Sluice valve or gate valve. These are also known as shut off valves or stop valves. They are extensively used in the distribution system to shut off the supplies whenever desired. They are also helpful in dividing the water mains into suitable sections. The spacing of such valves may be between 150 to 300 metres. They are also placed at street corners or where two pipe lines intersect. They possess the advantage over most other types of valves, of combining relatively low cost and offering almost no resistance to the flow of water when the valve is wide open.

Gate valves commonly used in water works are made of cast iron with brass mountings. They are either of *solid-wedge type* or *double-disk type*. Fig. 15.9 shows a double disc type gate valve. The disks are slightly self-adjusting so that they will wedge themselves between the tapering seats.

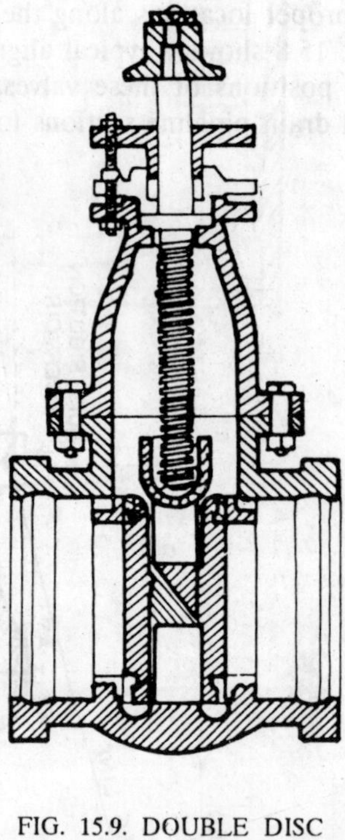

FIG. 15.9. DOUBLE DISC GATE VALVE

2. Air relief valves. The water flowing through the pipe lines always contain some air. This air tries to accumulate at high points, and may interfere with the flow. Air relief valves are therefore provided at the summits along the water pipe, to provide an exit for such accumulated air. Air valves are also required to discharge air when a main is being filled and to admit air when it is being emptied. The admission of air on emptying the main is of great importance on large steel mains, which may flatten if the pressure falls below that of the atmoshphere.

Fig. 15.10 shows a lever-and-float type air relief valve. It consists of a cast-iron chamber, float, lever and poppet valve. Under normal conditions, the chamber is full of water and the float touches the roof of the chamber and the poppet valve is in closed position. When air starts accumulating in the chamber, float is lowered and poppet valve is opened. Due to this, air is released.

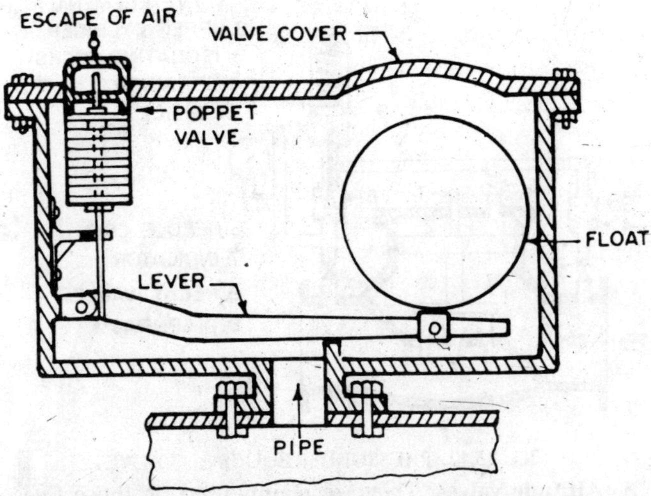

FIG. 15.10. AIR RELIEF VALVE

3. Reflux valves. Reflux valves are also known as *check valves* or *non-return valves.* It is an automatic device which allows water to flow in one direction only. They are placed in water pipes which obtain water directly from the pump. When the pump is stopped, the water in the pipe line does not rush back and damage the pump.

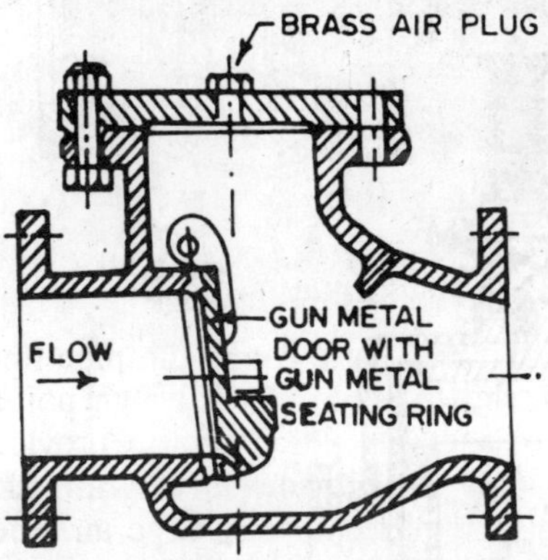

FIG. 15.11. REFLUX VALVE

Fig. 15.11 shows a non-return valve. It essentially consists of a flat disc within the pipe line, pivoted so that it is forced open when flow of water is in one direction and forced shut against a seating when flow tries to reverse.

4. Pressure relief valves. These are also known as *automatic cutoff valves* or *safety valves*. They are located at those points where pressure is likely to be maximum. When the line pressure increases above the pre-set value, the valve operates autometically, and the pressure is reduced. Fig. 15.12 shows a pressure reducing valves.

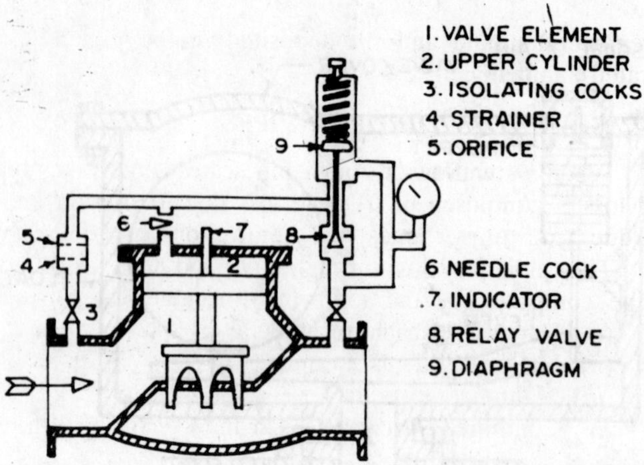

FIG. 15.12. PRESSURE REDUCING VALVE

5. Altitude valves. They are mainly used on those lines which supply water to elevated tanks or stand pipes. They close automatically when the tank is full and open when the pressure on the pump side is less than that on the tank side of the valve. Fig. 15.13 shows an altitude valve.

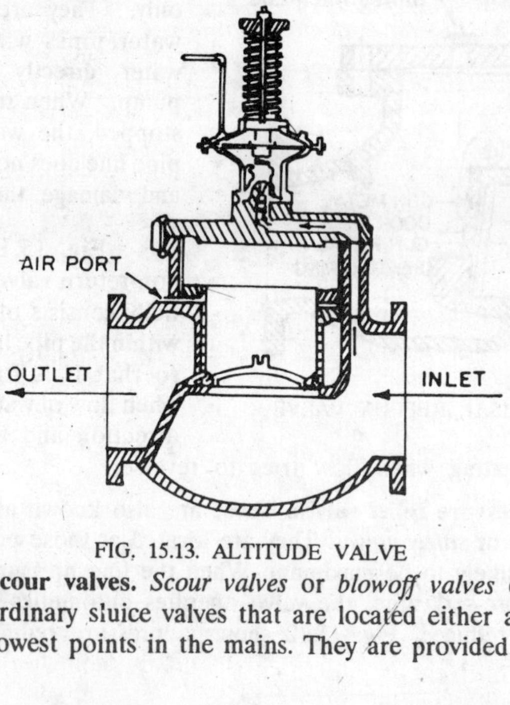

FIG. 15.13. ALTITUDE VALVE

6. Scour valves. *Scour valves* or *blowoff valves* or *washout valves* are ordinary sluice valves that are located either at the dead ends or at lowest points in the mains. They are provided to blowoff or remove the sand and silt deposited in the pipe line. They are operated manually.

15.14. HEAD LOSS THROUGH PIPES

When water flows through pipes, head loss takes place. Total head loss is composed of (*i*) head loss due to friction and (*ii*) head loss due to fittings, bends and contrictions.

The velocity of flow through pipes and head loss due to friction can be computed by the following formulae :

(*i*) Darcy Weisbach formula.
(*ii*) Hazen Williams formula.
(*iii*) Mannings formula.
(*iv*) Combined Darcy-Weisbach and Colebrook-White equation.

Detailed description and discussion about these formulae have been given in the next chapter, along with head loss due to fittings, bends and constrictions.

Example 15.1. *Water is delivered through a 90 cm dia. steel pipe, with a flow velocity of 1.8 m/sec. The pipe line contains a valve which is situated at 1600 m from the reservoir. Compute the water hammer pressure developed by the closure of the above valve, if*

(*a*) *the time of closure is 2 sec.*
(*b*) *the time of closure is 4 sec.*

The thickness of pipe shell is 22 mm.

Solution :

Given : $V_0 = 18$ m/sec

$$p_{h,max} = \frac{w}{g} \cdot U_p \cdot V_0 \qquad ...(15.15)$$

where

$$U_p = \frac{V_s}{\sqrt{1 + \dfrac{K_w}{E_p} \cdot \dfrac{d}{t}}} \qquad ...(15.16)$$

Here, $K_w = 2.07 \times 10^8$ kg/m^2

$E_p = 2.10 \times 10^{10}$ kg/m^2 (Table 15.6)

d = dia of pipe = 0.9 m

t = thickness of pipe = 0.022 m

V_s = velocity of sound wave in water = 1433 m/s

$$\therefore \quad U_p = \frac{1433}{\sqrt{1 + \dfrac{2.07 \times 10^8}{2.10 \times 10^{10}} \times \dfrac{0.9}{0.022}}} = 1209.7 \text{ m/s.}$$

$$\therefore \quad p_{h,max} = \frac{1000}{9.81} \times 1209.7 \times 1.8 \approx 22.2 \times 10^4 \text{ kg/m}^2 = 22.2 \text{ kg/cm}^2$$

This is the max. value of water hammer pressure. However, actual value of water hammer pressure is given by Eq. 15.21 :

$$p_h = p_{h,max} \cdot \frac{T_c}{T}$$

where $T_c = \dfrac{2L}{U_p} = \dfrac{2 \times 1600}{1209.7} = 2.65$ sec

(*a*) When actual time of closure, $T = 2$ sec :

$$\therefore \quad p_h = p_{h,max} = \mathbf{22.2 \ kg/cm^2}$$

(*b*) When actual time of closure is 4 sec. :

$$p_h = 22.2 \times \frac{2.65}{4} \approx \mathbf{14.7 \ kg/cm^2}$$

Example 15.2. *A steel penstock, 70 cm in diameter has a shell thickness of 12.5 mm. The modulus of elasticity of pipe shell material is 2.1×10^{10} kg/m^2 with Poisson's ratio of 1/4 and the volume modulus of elasticity of water is 2.1×10^8 kg/m^2. The pipe is designed to discharge water at mean velocity of 2 m/sec. Determine the water hammer pressure rise caused by sudden closure of valve at the downstrem end : (a) neglecting the elasticity of pipe material, and (b) considering also the elasticity of pipe material.*

Solution :

(a) Neglecting the elasticity of pipe

$$p_{h,max} = \frac{w}{g} . U_p . V_0$$

where $U_p = \sqrt{\dfrac{K_w}{\rho}} = V_s = 1433$ m/sec

$$\therefore \quad p_{h,max} = \frac{1000}{9.81} \times 1433 \times 2 = 29.22 \times 10^4 \text{ kg/m}^2$$

$$= \mathbf{29.22 \ kg/cm^2}$$

(b) Considering elasticty of pipe material

$$p_{h,max} = \frac{w}{g} . U_p . V_0$$

where $U_p = \dfrac{V_s}{\sqrt{1 + \dfrac{K_w}{K_p} . \dfrac{d}{t}}} = \dfrac{1433}{\sqrt{1 + \dfrac{2.1 \times 10^8}{2.1 \times 10^{10}} \times \dfrac{0.7}{0.0125}}}$

$$= \frac{1433}{1.249} = 1147.3 \text{ m/sec.}$$

$$\therefore \quad p_{h,max} = \frac{1000}{9.81} \times 1147.3 \times 2 \approx 23.4 \times 10^4 \text{ kg/m}^2$$
$$= \mathbf{23.4 \text{ kg/cm}^2}$$

Example 15.3. Design of a thrust block

Design a thrust block for a 1000 mm diameter main conveying water at a pressure of 10 kg/cm², at a location where the deviation angle is 45° in a horizontal plane. The subsoil has a density of 1800 kg/m³, angle of internal friction $\phi = 20°$ and zero cohesion. Assume velocity of flow as 2 m/sec. Take unit weight of concrete = 2400 kg/m³.

Solution

$$\text{Area } (A) \text{ of pipe} = \frac{\pi}{4}(100)^2 \approx 7854 \text{ cm}^2$$

$$P = 2A\left[\frac{wV^2}{g} + p\right]\sin\frac{\theta}{2}$$

$$\frac{wV^2}{g} = \frac{1000}{9.81}(2)^2 \approx 408 \text{ kg/m}^2 = 0.0408 \text{ kg/cm}^2$$

$$\therefore \quad P = 2 \times 7854\left[0.0408 + 10\right]\sin\frac{45°}{2}$$
$$= 60357 \text{ kg}$$

Let us try a thrust block of size 3.5 m × 3.5 m × 3.5 m. Let us assume a soil cover of 60 cm.

(i) Frictional resistance of block

$$W_b = 3.5 \times 3.5 \times 3.5 \times 2400 = 102900 \text{ kg}$$
$$W_w = \frac{\pi}{4}(1)^2 \times 3.5 \times 1000 = 2749 \text{ kg}$$
$$W_s = 3.5 \times 3.5 \times 0.6 \times 1800 = 13230 \text{ kg}$$
$$\therefore \text{ Total} \quad W = 102900 + 2749 + 13230 = 118879 \text{ kg}$$
$$\therefore \quad P_f = \mu W = 0.3 \times 118879 \approx 35664 \text{ kg} \quad ...(i)$$

(ii) Passive resistance of soil

$$k_p = \frac{1 + \sin\phi}{1 - \sin\phi} = \frac{1 + 0.5}{1 - 0.5} = 3$$
$$H = \text{height of block} = 3.5 \text{ m}$$
$$L = \text{length of block} = 3.5 \text{ m}$$
$$c = 0$$
$$\therefore \quad P_p = k_p \cdot \frac{\gamma H^2}{2} \cdot L = 3 \times \frac{1800}{2}(3.5)^2 \times 3.5$$
$$= 115762 \text{ kg} \quad ...(ii)$$

(iii) Active earth pressure

$$k_a = \frac{1 - \sin\phi}{1 + \sin\phi} = \frac{1}{3}$$
$$h = \text{dia. of pipe} = 1 \text{ m}$$

$$\therefore \qquad P_a = \frac{1}{3} \times 1800 \times 1 = 600 \text{ kg} \qquad ...(ii)$$

(This is negligibly small)

$$\therefore \qquad P_r = 35664 + 115762 + 600 = 152026 \text{ kg}$$

$$\therefore \quad \text{Factor of safety} = \frac{P_r}{P} = \frac{152026}{60357} = 2.52$$

O.K.

Minimum surface reinforcement should be at the rate of 5 kg/m^2, with spacing of bars not exceeding 500 mm. Hence provide 10 mm dia. bars @ 200 mm c/c as shown in Fig. 15.14. This reinforcement will be more than 5 kg/m^2.

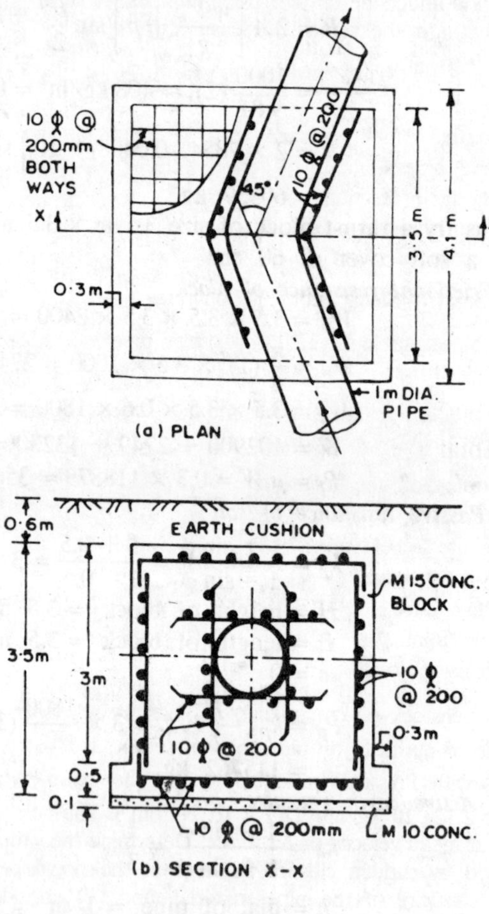

FIG. 15.14.

PROBLEMS

1. Discuss the advantages and disadvantages of various types of pipes used in water supply.

2. Describe C.I. pipes and steel pipes and give their comparison.

3. Describe with the help of sketches various types of joints used in C.I. pipes.

4. Discuss various typs of stresses to which a water supply main is subjected.

5. Explain how do you determine stresses in a water supply pipe, due to backfill and live load.

6. What do you understand by corrosion in pipes. What are the reasons of corrosion ? What are the remedial measures ?

7. Write notes on the following :

(*a*) Concrete pipes.
(*b*) Asbestos cement pipes.
(*c*) Plastic pipes.
(*d*) Cathodic protection.

8. State the comparative merits and demerits of the following materials used in the conveyance of water :

(*a*) Cast iron, (*b*) Steel, and (*c*) Concrete.

Describe a joint suitable for 1.20 m cast iron pipe used for conveyance of water

9. Describe measures that are commonly adopted to minimise pipe corrosion.

10. Write notes on the following :

(*a*) Gate valve.
(*b*) Pressure relief valve.
(*c*) Air relief valve.
(*d*) Reflux valve.
(*e*) Tyton joint.
(*f*) Expansion joint.

11. A steel penstock 60 cm in diameter has a shell thickness of 1.2 cm. The modulus of elasticity of pipe shell material is 2.1×10^6 kg/cm^2 (2.1×10^5 N/mm^2) with Poisson's ratio of 1/4 and the volume modulus of elasticity of water is 2.1×10^4 kg/cm^2 (2.1×10^3 N/mm^2). The pipe is designed to discharge water at mean velocity of 2.1 m/sec. Determine the water hammer pressure rise caused by sudden closure of valve at the downstream end : (*i*) neglecting the elasticity of the pipe material, and (*ii*) considering also the elasticity of the pipe material. (UPSC Engg. Services, 1990)

16

Distribution of Water

16.1. INTRODUCTION

Water distribution systems are ordinarily designed to adequately satisfy the water requirements for a combination of domestic, commerical, industrial and fire fighting purpose. The performance of a distribution system can be judged on the basis of the pressure available in the system for a specific rate of flow. The distribution system consists of a network of pipes with appertenances, for transporting water from the purification plant to the consumer's tap. It also includes the design and operation of storage, service or balancing reservoirs. A good distribution system should satisfy the following requirements :

1. The system should be capable of supplying water at consumers' tap at reasonable pressure head. Also, the head should not be excessive.
2. It should be capable of meeting the fire demand simultaneously.
3. It should maintain the degree of purity. The distribution system should be completely watertight.
4. It should be easy to operate and maintain.
5. Water should be available even during breakdown period.
6. It should be so laid that during repairs, it does not cause obstruction to traffic.
7. The intial cost of the distribution should be as low as possible.

16.2. METHODS OF DISTRIBUTION

The method of distribution depends upon the topography of the area. The following methods or systems may be adopted for distribution :

1. Gravity system.
2. Combined gravity and pumping system.
3. Pumping system.

1. Gravity system. In the gravity system, the source of water supply is so located with respect to the area of distribution that water is available with sufficient pressure at various points of the area. No pumping is normally required. However if the purification

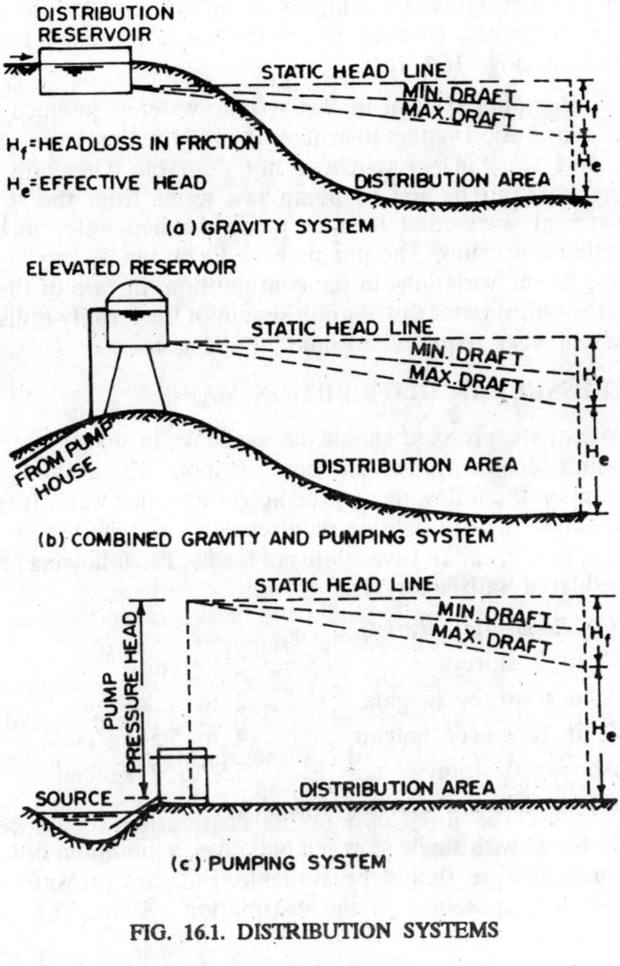

FIG. 16.1. DISTRIBUTION SYSTEMS

units are located on a hill, and the source of water supply (*i.e.* a lake or reservoir) is also located behind the hill, pumping may be required to convey water from the raw water source to the purification works. The purified water then flows entirely under gravity. This is the most reliable and economical distribution system.

Fig. 16.1 (*a*) shows the system in which H_e is the desired effective head.

2. Combined gravity anu pumping system. This is the most common system adopted in most of the cases. Generally, the water purification works are located almost at the same level as the area of distribution – specially when source of raw water supply is a river or a reservoir formed behind a dam. In order to obtain sufficient distribution pressure, filtered water is pumped into *clear water reservoir* (C.W.R.) located either on a higher ground or elevated on a tower. The water from the elevated reservoir then flows under gravity, as illustrated in Fig. 16.1 (*b*).

3. Pumping system. In this system, water is pumped directly into the distribution system to achieve the required pressure, as shown in Fig. 16.1 (*c*). Such a system is not desirable. Generally double pumping is required, first to pump raw water from the source to the treatment works and then to pump purified water direct into the distribution mains. The pumps have to be run at varying speeds according to the variations in the consumption. In case of the power failure, the entire water distribution system of the locality is disturbed. The system also requires constant attendance.

16.3. PRESSURE IN DISTRIBUTION MAINS

Adequate pressure should be available in distribution mains at all points located even at the remotest spots. The desired pressure depends upon the following : (*i*) the height to which water is required to be supplied (*ii*) fire fighting requirements (*iii*) whether the supply is metered or not, and (*iv*) availability of funds. The following pressures are considered satisfactory :

(*a*) ***Residential districts*** :

Upto 3 storeys : 2 kg/cm^2

3 to 6 storey heights : 2 to 4 kg/cm^2

6 to 10 storey heights : 4 to 5.5 kg/cm^2

Above 10 storeys : 5.5 to 7 kg/cm^2

(*b*) ***Commerical districts:*** 5 kg/cm^2.

In towns with single storeyed buildings, a minimum of 6 metres of residual pressure should be available. Full fire pressures should constantly be maintained in the distribution system. The National

Board of Fire Underwriters recommends a pressure of 5 kg/cm^2 at the fire nozzle, where more than ten buildings exceed three storeys in height; 4 kg/cm^2 in localities with less risk and 3.5 kg/cm^2 in thinly built-up districts. A pressure of 1.5 kg/cm^2 is considered satisfactory under most conditions where fire engines are used.

A minimum velocity of 0.6 m/sec should be maintained. The velocities in pipes of different diameters should be as follows :

Pipe dia. (*cm*)	*Velocity* (*m/sec*)
10	0.9
15	1.2
25	1.5
40	1.8

Manual Recommendations

The Manual on Water Supply and Treatment prepared by the Ministry of Urban Development, gives the following recommendations for minimum *residual pressures* at ferrule points :

Single storey building	:	7 m
Two storey building	:	12 m
Three storey building	:	17 m

Distribution system should not ordinarily be designed for residual pressures exceeding 22 metres. Multistoreyed buildings needing higher pressure should be provided with boosters.

Minimum pipe sizes

The manual recommends the following minimum pipe sizes:

Towns with population upto 50000	:	100 mm dia
Towns with population above 50000	:	150 mm dia.

For dead ends, less than 100 mm can be considered. If it is a grid, less than 100 mm can be used in situations where no further expansion is contemplated.

16.4. SYSTEMS OF WATER SUPPLY

Water may be supplied to the consumers by the following two systems :

(1) Continuous system
(2) Intermittent system

1. Continuous system. In the continuous system, water is available to the consumers for all the 24 hours of a day. No doubt, this is the best system since water is available as and when it is needed, but this leads to the wasteful use of water. If there are some minor leakages etc. in the system, great volume of water is wasted because of long duration of flow. In this system, water is not stagnant in the pipe at any instant, and hence fresh water is always available.

2. Intermittent system. In this system, water is supplied to the consumers only during some fixed hours of the day – say two to four hours in the morning and two to four hours in the evening. This is the most common system adopted in India. This method is adopted when either sufficient pressure is not available or when sufficient quantity is not available. Under these conditions, various distribution zones of the city are supplied water by turn. The normal supply timings may be between 6 A.M. to 10 A.M. and 4 P.M. to 8 P.M., though these timings may be changed to suit climatic or seasonal conditions.

The intermittent system has the following *drawbacks* :

(i) Fire demand. If fire breaks in a supply zone during non-supply period, the rescue operations cannot be effectively done. Water cannot be brought from other zones quickly, and fire damage will be more.

(ii) Domestic storage. The intermittent system requires the provision of small storage tanks in individual houses so that sanitary fittings in the house can work effectively during periods of no supply.

(iii) Pollution in supply. During the non-supply period, the pressure in the supply line may fall below atmospheric pressure. This may induce suction through leaking joints. When the pipe line is laid near the sewers etc., this may lead to severe pollution and contamination problems.

(iv) Size of pipes. Greater sizes of pipes will be required since the supply of whole day has to be made in a shorter period.

(v) Wastage from water taps. During the non-supply period, the water taps may be left open unknowingly or due to negligence. This will lead to large wastage of water during the supply period.

(vi) Staff requirements. Since a number of valves of different types are fitted on supply lines, many of which may be non automatic, extra staff will be required to operate and maintain these valves.

16.5. STORAGE AND DISTRIBUTION RESERVOIRS

Storage and distribution reservoirs are important units in a modern distribution system. *Clear water storage reservoirs* are required for storage of filtered water until it is pumped into the *service reservoirs* or *distribution reservoirs*. The pumps may generally work for 8-10 hours a day. Hence the clear water reservoirs should have a capacity to store filtered water corresponding to 14 to 16 hours average daily flow for storage when pumps are idle.

Distribution reservoirs provide service storage to meet the widely fluctuating demands often imposed on a distribution system, to provide storage for fire fighting and emergencies and to equalise operating pressures. They may be classified as 'surface reservoirs' or 'elevated reservoirs' according to their position, or classified according to the material of which they are built, such as steel, R.C.C. or masonry tanks. Distribution reservoirs are mostly of elevated type. They serve the following purposes :

1. They absorb the hourly variations in demand.

2. If pumps are used, the provision of reservoirs makes it possible to run the pumps at uniform rate.

3. Their provision results in an overall reduction in the sizes of pumps, pipes and treatment units. Thus the distribution system becomes economical.

4. They serve as storage for emergencies such as outbreak of fire, failure of pumps or bursting of mains.

5. They maintain the desired pressure in the main constantly. In absence of these, the pressure will fall as the demand of water increases. They provide and maintain desired pressure even in remote areas.

6. Operation of the distribution system becomes very easy.

16.6. TYPES OF STORAGE AND DISTRIBUTION RESERVOIRS

Storage and distribution reservoirs may be of three types :

(*i*) Surface reservoirs.
(*ii*) Elevated reservoirs.
(*iii*) Stand pipes.

(*i*) **Surface Reservoirs.** Surface reservoirs are made mostly of masonry or concrete. Common practice is to line surface reservoirs with concrete, gunite, asphalt or asphaltic membrane to check the leakage of water. Sometimes, these reservoirs may be built underground, specially when they are of large size, and a park may be constructed on its top. Such gardens are known as hanging gardens, as provided at Bombay. It is usual practice to construct surface reservoirs in two or more compartments so that one unit can be cleaned or repaired while other units are in operation. Surface reservoirs should be located at high points in the distribution system, so that gravity supply can be done directly. In some cases however, pumps are used to pump water, from the clear water storage surface reservoir to the elevated distribution reservoir.

(*ii*) **Elevated Reservoirs.** Elevated reservoirs are commonly known as overhead tanks. They may be constructed of stone masonry, reinforced concrete or steel. Earlier trend was to construct these of steel. However, the life of steel overhead tanks is very short and they corrode soon requiring constant maintenance and frequent replacement. Masonry overhead tanks require very heavy supporting tower. Hence modern trend is to provide the elevated reservoirs of reinforced concrete. These may be of rectangular, circular, conical or elliptical shape. Intze tank is the most common type used for water supply purposes. All elevated reservoirs are invariably provided with top cover, ladder, and man-holes for inspection and cleaning purposes.

(*iiii*) **Standpipes.** Standpipes are normally employed where the construction of a surface reservoir would not provide sufficient head. A standpipe is essentially a tall cylindrical tank whose storage volume includes an upper portion (the useful storage), which is above the entrance to the discharge pipe and a lower portion (supporting storage) which acts only to support the useful storage and provide the required head. It is often possible to take advantage of hills or high ground for a standpipe location, with a view toward making the entire capacity of the tank usable.

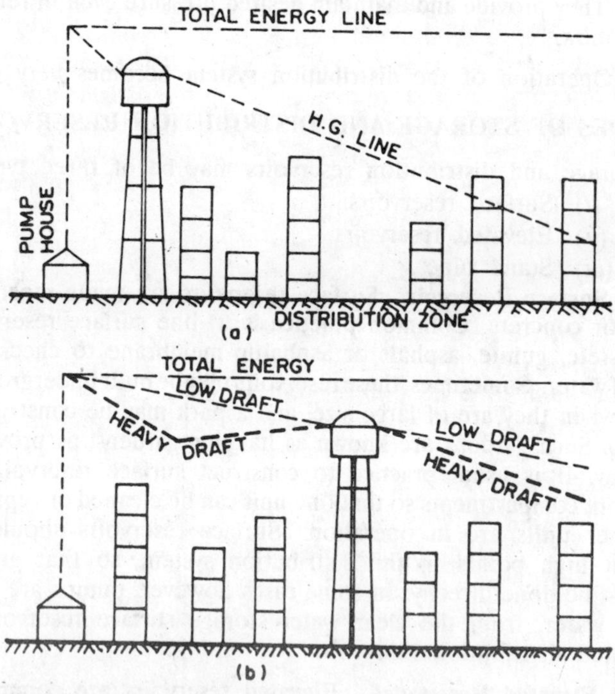

FIG. 16.2. EFFECT OF LOCATION OF DISTRIBUTION RESERVOIR

Location of distribution reservoirs. Distribution reservoirs should be located centrally or at least as near as possible to the zone that they serve. In large metropolitan areas, a number of distribution reservoirs may be located at key points. They should be located on high grounds, or at sufficient elevation to maintain adequate pressure. A central location of the reservoir will reduce friction losses in the distribution pipes by reducing the length of pipes. Positioning the reservoir so that pressures may be approximately equalized is an additional consideration of importance. This is illustrated in Fig. 16.2. If the tank is not located centrally (Fig. 16.2 *a*), there will be large head loss, and by the time water reaches the tail end of the supply, the pressue will be too low to serve the community usefully. However, if the reservoir is centrally located, as shown in Fig. 16.2 (*b*), the pressure over the whole distribution area is much more uniform during periods of both high and low demand. During the periods of peak demand, the tank supplies in both directions (being emptied) while during the periods of low demand, pump supplies water to both the tanks as well as the community.

16.7. CAPACITY OF DISTRIBUTION RESERVOIR

The storage capacity of the distribution reservoir is based on the following three requirements : (*i*) balancing or equalising reserve, (*ii*) breakdown reserve and, (*iii*) fire reserve. The magnitude of breakdown or emergency reserve depends upon the danger of interruption of reservoir inflow by failure of supply works and upon the time needed to make repairs. If repairs are expected to last only for a very short duration, the emergency reserve is sometimes made no more than 25% of the total storage. To compute the amount of storage required for balancing or equalising purposes, a mass diagram or hydrograph indicating the hourly rate of consumption is required. The National Board of Fire Underwriters recommends that distributing reservoirs be made large enough to supply water for fighting a serious conflagration for 10 hours in communities of more than 6000 people, and for 8, 6 and 4 hours in places with 4000, 2000 and 1000 people respectively. When reserve storage is elevated, the amount of fire reserve may be determined from the following expression :

$$R = [F - P]\,T \qquad ...(16.1)$$

where R = reserve storage (litres)
F = fire demand, litres/min
P = reserve fire pumping capacity in litres/min.
T = duration of fire, in minutes.

Sometimes, it is more economical to provide elevated storage for equalizing purposes only, fire reserve being stored in a surface reservoir. Such reserve storage must always be kept available for immediate use, so that it can be pumped into the distribution system during emergency, such as fire.

McDonald has suggested the following expression for determining storage capacity of reservoirs :

$$R = aD + bD + \frac{10}{24}(D + F - P) \quad ...(16.2)$$

where R = total storage capacity (million litres)
D = average domestic demand (m.l.d.) for maximum month
F = fire demand (m.l.d.)
P = capacity of pump (m.l.d)
a, b = coefficients, the values of which may be taken as 0.2 and 0.1 respectively.

When a storage or distribution reservoir is to be designed for the purpose of balancing or equalising the flow, its storage capacity can be determined by two methods :

(*a*) Hydrograph method.
(*b*) Mass curve method.

(*a*) **Hydrograph Method.** It is well known that the hourly water demand rate is not constant throughout the day. The demand

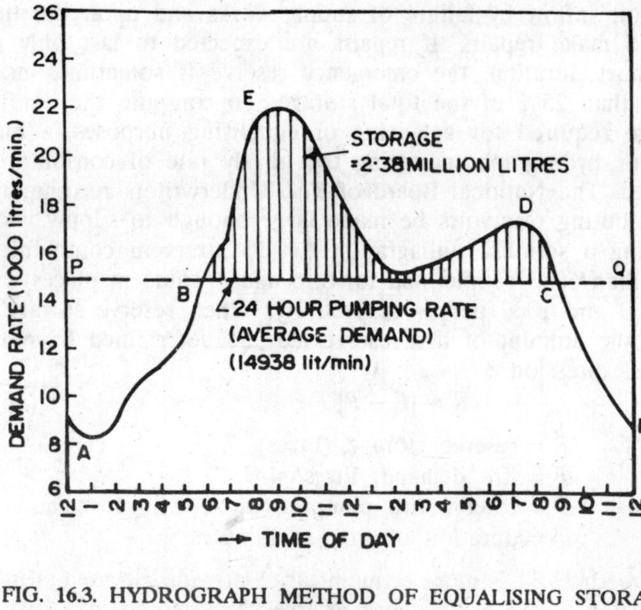

FIG. 16.3. HYDROGRAPH METHOD OF EQUALISING STORAGE (EXAMPLE 16.1)

is more during morning and evening and less during the other parts of the day. Fig. 16.3 shows a hydrograph of hourly demand for the maximum day. For a uniform 24 hour pumping, the pumping rate will be equal to the mean hourly demand, shown by the line *PQ*. The required storage is then obtained by planimetering or determining in some other manner the area (shown shaded) between curve *BEC* and the line *PQ*. This area can be easily converted into units of volume to yield the required storage in million litres.

From the above graph, it is clear that in absence of an equalizing reservoir, the maximum rate of pumping required would be 22000 lit/min. But by provision of an equalizing reservoir of 2.38 million litres capacity, the maximum rate of pumping required will be only 14938 lit/min. This is about 67.9% of that required with no storage.

(*b*) **Mass Curve Method.** A mass curve of demand is the cumulative demand curve, and is obtained by continuously adding

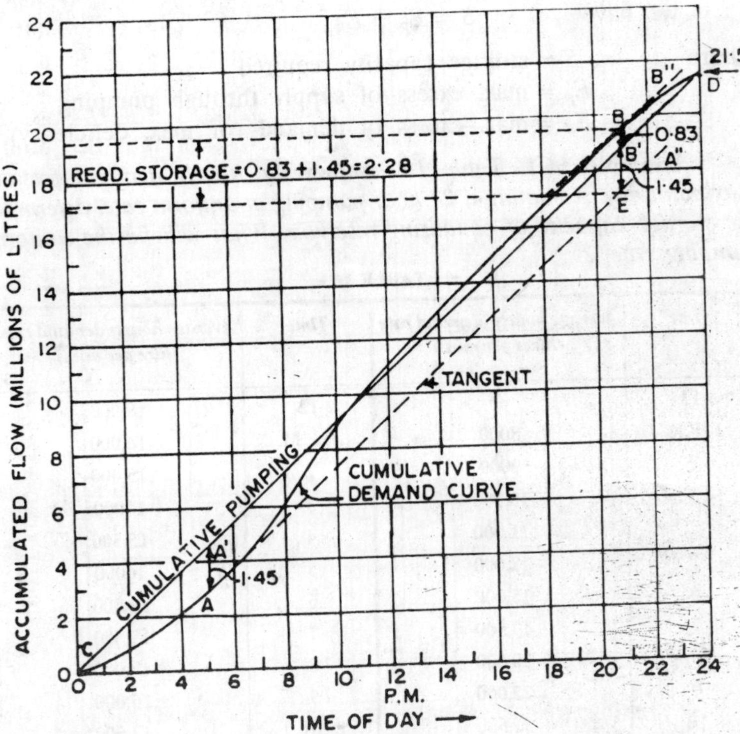

FIG. 16.4. MASS CURVE METHOD OF EQUALIZING STORAGE (SEE EXAMPLE 16.1)

the hourly demands and plotting these against time (hours) of the maximum day. Fig. 16.4 shows the mass demand curve *CAB* in thick line. A mass demand curve continuously rises. The steepness of the mass curve indicates a higher rate of demand, while flat portion shows the lower rate of demand. Line *CD* is the line indicating cumulative pumping at uniform rate. If pumping is to be done for all the 24 hours, the mass curve *CD* for pumping will be obtained by joining the two ends *C* and *D* of the mass curve of demand. In order to determine the required storage capacity, draw tangents, through lowest point *A* and highest point *B*, parallel to pumping rate line *CD*. The highest vertical distance *BE* between the two tangents will then give the required capacity of the equalizing reservoir. This is so because at *A* (6 A.M.), there is excess supply equal to *AA'* which should be stored, while at *B* (about 8.30 P.M.) there is deficit *B B'* which must be drawn from the storage that is replenished by mid-night. The required storage capacity will thus be the sum of morning excess *A A'* and evening deficiency *B B'*. In other words we have

$$S = E_p + E_d \quad ...(16.3)$$

where S = storage capacity required

E_p = max. excess of supply through pumping

E_d = max. excess of demand (or max. deficiency).

Example 16.1. *Table 16.1 shows the hourly demand for the maximum day. Assuming a 24 hour pumping at uniform rate, determine the storage capacity of equalizing reservoir. What will be the uniform pumping rate ?*

TABLE 16.1

Time	*Average hourly demand rate (litres per min.)*	*Time*	*Average hourly demand rate (litre per min.)*
12	0	12	18,000
1 P.M.	8000	1 P.M.	16,000
2	9000	2	15,000
3	10,500	3	15000
4	11,000	4	15,500
5	12,000	5	16000
6	15,000	6	16,500
7	17,500	7	17,000
8	20,500	8	17,000
9	22,000	9	15,000
10	21,500	10	12,000
11	20,000	11	10,000
		12	9,000

TABLE 16.2.

Time	*Average hourly demand rate l.p.m.*	*Hourly demand (litres)*	*Cumulative demand (litres)*	*Average hourly demand minus hourly demand i.e. 896250 – col. (3)*	
				–	+
(1)	*(2)*	*(3)*	*(4)*	*(5)*	
12	0	0	0	—	
1 A. M.	8000	480000	480000		416250
2	9000	540000	1020000		356250
3	10500	630000	1650000		266250
4	11000	660000	2310000		236250
5	12000	720000	3030000		176250
6	15000	900000	3930000	3750	
7	17000	1020000	4950000	123750	
8	20500	1230000	61800000	333750	
9	22000	1320000	7500000	423750	
10	21500	1290000	8790000	393750	
11	20000	1200000	9990000	303750	
12	18000	1080000	11070000	183750	
1 P. M.	16000	960000	12030000	63750	
2	15000	900000	12930000	3750	
3	15000	900000	13830000	3750	
4	15500	930000	14760000	33750	
5	16000	960000	15720000	63750	
6	16500	990000	16710000	93750	
7	17000	1020000	17730000	123750	
8	17000	1020000	18750000	123750	
9	15000	900000	19650000	3750	
10	12000	720000	20370000		176250
11	10000	600000	20970000		296250
12	9000	540000	21510000		356250
	Total	21510000		2280000	2280000
	Average	$\frac{21510000}{24}$ = 896250 litres/hour			

Solution :

Knowing the average hourly demand rate (l.p.m.) the hourly demand, in litres, can be calculated by multiplying it by 60. Column 3 of Table 16.2 shows the hourly demand, while column 4 shows the cumulative hourly demand. The average demand is obtained by taking the sum of column 3 and dividing it by 24. It comes out to be 896250 litres/hour. Thus, the pumping rate will be 896250 litres/hour or 14938 litres/min. Fig. 14.3 shows the hydrograph of demand in which the line *PQ* represents the pumping rate of 14938 litres/min. Column 5 shows the average hourly demand minus hourly demand.

The mass curve is shown in Fig. 16.4. *CD* is the line for cumulative pumping for 24 hours at a constant rate of 0.896 million litres/hour. Tangents are drawn at the lowest point *A* and apex *B*, parallel to *CD*. The maximum vertical ordinate *BE* between the two tangents gives the required storage, which comes out to be 2.28 million litres. Alternatively, the storage capacity is equal to $(AA'+BB') = 1.45+0.83 = 2.28$. This evidently is equal to the sum of negative (or positive) entries of column 5 of Table 16.2. This is about 10.6% of day's consumption.

Alternative analytical solution :

From Table 16.2 (column 3), total demand in 24 hours = 21510000 litres.

$$\therefore \quad \text{Pumping rate} = \frac{21510000}{24} = 896250 \text{ litres/hour.}$$

Table 16.3 shows the alternative computations :

From Table 16.3. we find that :

Max. Excess demand, $E_d = 828750\ l$

Max. excess pumping, $E_p = 1451250\ l$

$$\therefore \quad S = E_d + E_p = 828750 + 1451250 = \mathbf{2280000} \text{ litres.}$$

Example 16.2. *Find the reservoir capacity for the data of Table 16.1 if the pumping occurs between 6 A.M. and 6 P.M. (i.e. for 12 hours) only.*

Solution :

The total volume of water required to be pumped is 21.51 million litres, and this is to be accomplished in 12 hours, between 6 A.M. to 6 P.M. Hence in the mass curve shown in Fig. 16.5, draw the cumulative pumping line *AB* so that total pumping of 21.51 million litres is accomplished between 6 A.M. to 6 P.M. Project point *A* vertically upward to intersect with the cumulative demand

TABLE 16.3. (Example 14.1)

Time	*Hourly demand (l)*	*Cumulative demand (l)*	*Hourly pumping (l)*	*Cumulative pumping (l)*	*Excess demand*	*Excess pumping (l)*
(1)	*(2)*	*(3)*	*(4)*	*(5)*	*(6)=(3)−(5)*	*(7)=(5)−(3)*
12	0	0	–	–	–	–
1 A.M.	480000	480000	896250	896250	–	416250
2	540000	1020000	896250	1792500	–	772500
3	630000	1650000	896250	2688750	–	1038750
4	660000	2310000	896250	3585000	–	1275000
5	720000	3030000	896250	4481250	–	1451250
6	900000	3930000	896250	5377500	–	1447500
7	1020000	4950000	896250	6273750	–	1323750
8	1230000	6180000	896250	7170000	–	990000
9	1320000	7500000	896250	8066250	–	566250
10	1290000	8790000	896250	8962500	–	172500
11	1200000	9990000	896250	9858750	131250	–
12	1080000	11070000	896250	10755000	315000	–
1 P.M.	960000	12030000	896250	11651250	378750	–
2	900000	12930000	896250	12547500	382500	—
3	900000	13830000	896250	13443750	386250	–
4	930000	14760000	896250	14340000	420000	–
5	960000	15720000	896250	15236250	483750	–
6	990000	16710000	896250	16132500	577500	–
7	1020000	17730000	896250	17028750	701250	–
8	1020000	18750000	896250	17925000	825000	–
9	900000	19650000	896250	18821250	828750	–
10	720000	20370000	896250	19717500	652500	–
11	800000	20970000	896250	20613750	356250	–
12	540000	21510000	896250	21510000	–	–

curve at A_1. Construct line $A_1 B_1$ parallel to AB, where point B_1 is the intersection of line $A_1 B_1$ with the vertical extended upward from 6 P.M. on the abscissa. The required storage will then be ordinate $B_1 E$. This is evidently equal to ordinate $A A_1$ + ordinate $BE = 3.93 + 4.80 = 8.73$ million litres. This is about 40.6% of the day's consumption.

The analytical solution is given in Table 16.4. Since pumping is done only for 12 hours,

$$\text{Pumping rate} = \frac{21510000}{12} = 1792500 \text{ l/hour.}$$

TABLE 16.4 (Example 16.2)

Time	*Hourly demand (l)*	*Cumulative demand (l)*	*Hourly pumping (l)*	*Cumulative pumping (l)*	*Excess demand*	*Excess pumping (l)*
(1)	*(2)*	*(3)*	*(4)*	*(5)*	*(6) = (3) − (5)*	*(7) = (5) − (3)*
12	0	0	–	–	–	–
1 A.M.	480000	480000	–	–	480000	–
2	540000	1020000	–	–	1020000	–
3	630000	1650000	–	–	1650000	–
4	660000	2310000	–	–	2310000	–
5	720000	3030000	–	–	3030000	–
6	900000	3930000	–	–	3930000	–
7	1020000	4950000	1792500	1792500	3157500	–
8	1230000	6180000	1792500	3585000	2595000	–
9	1320000	7500000	1792500	5377500	2122500	–
10	1290000	8790000	1792500	7170000	1620000	–
11	1200000	9990000	1792500	8962500	1027500	–
12	1080000	11070000	1792500	10755000	315000	–
1 P.M.	960000	12030000	1792500	12547500	–	517500
2	900000	12930000	1792500	14340000	–	1410000
3	900000	13830000	1792500	16132500	–	2302500
4	930000	14760000	1792500	17925000	–	3165000
5	960000	15720000	1792500	19717500	–	3997500
6	990000	16710000	1792500	21510000	–	4800000
7	1020000	17730000		21510000	–	3780000
8	1020000	18750000		21510000	–	2760000
9	900000	19650000		21510000	–	1860000
10	720000	20370000		21510000	–	1140000
11	600000	20970000		21510000	–	540000
12	540000	21510000		21510000	–	0

From Table 16.4, we find that
Max. excess demand $E_d = 3930000$ litres
Max. excess pumping, $E_p = 4800000$ litres

$$\therefore \quad S = E_d + E_p = 3930000 + 4800000 = 8730000 \text{ litres}$$
$$= 8.73 \text{ million litres}$$

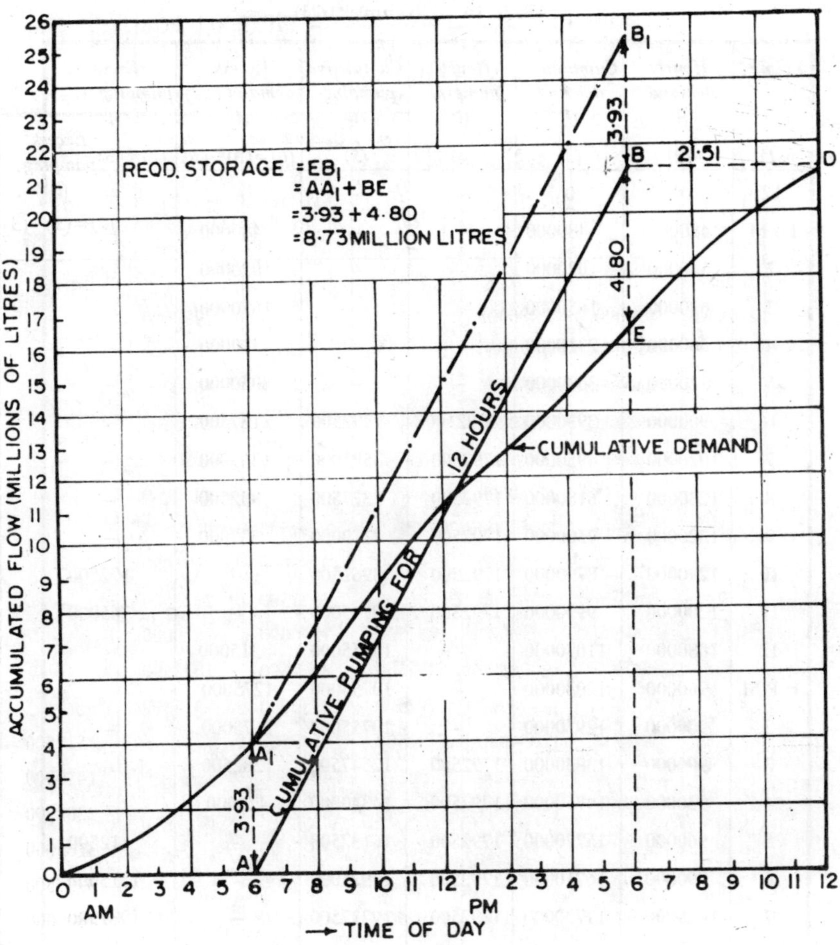

FIG. 16.5. (EXAMPLE 16.2)

Example 16.3. *Find the reservoir capacity for the data of table 16.1 if the pumping occurs between 5 AM to 11 AM (i.e. 6 hours) and again from 2 P.M. to 8 P.M. (6 hours), making a total of 12 hours pumping.*

Solution :

Since pumping takes place for a total duration of 6 + 6 = 12 hours, hourly pumping rate $= \frac{21510000}{12} = 1792500$ l/hour

The computations are shown in Table 16.5.

TABLE 16.5 (Example 16.3)

Time	*Hourly demand (l)*	*Cumulative demand (l)*	*Hourly pumping (l)*	*Cumulative pumping (l)*	*Excess demand* (E_d) *(l)*	*Excess pumping* (E_p) *(l)*
(1)	*(2)*	*(3)*	*(4)*	*(5)*	*(6)*=(3)−(5)	*(7)*=(5)−(3)
12	0	0	–	–	–	–
1A.M.	480000	480000	–	–	480000	–
2	540000	1020000	–	–	1020000	–
3	630000	1650000	–	–	1650000	–
4	660000	2310000	–	–	2310000	–
5	720000	3030000	–	–	3030000	–
6	900000	3930000	1792500	1792500	2137500	–
7	1020000	4950000	1792500	3585000	1365000	–
8	1230000	6180000	1792500	5377500	802500	–
9	1320000	7500000	1792500	7170000	330000	–
10	1290000	8790000	1792500	8962500	–	172500
11	1200000	9990000	1792500	10755000	-	765000
12	1080000	11070000	–	10755000	315000	–
1 P.M.	960000	12030000	–	10755000	1275000	–
2	900000	12930000	–	10755000	2175000	–
3	900000	13830000	1792500	12547500	1282500	–
4	930000	14760000	1792500	14340000	420000	–
5	960000	15720000	1792500	16132500	–	412500
6	990000	16710000	1792500	17925000	–	1215000
7	1020000	17730000	1792500	19717500	–	1987500
8	1020000	18750000	1792500	21510000	–	2760000
9	900000	19650000		21510000	–	1860000
10	720000	20370000		21510000	–	1140000
11	600000	20970000		21510000	–	540000
12	540000	21510000		21510000	–	0

From Table 16.5, we get

Max. excess demand, E_d = 3030000 litres

Max. excess pumping, E_p = 2760000 litres

$\therefore$ Required storage $= E_d + E_p = 3030000 + 2760000 = 5790000$

= 5.79 million litres

This is about 26.92% of day's consumption.

The graphical solution is shown in Fig. 16.6.

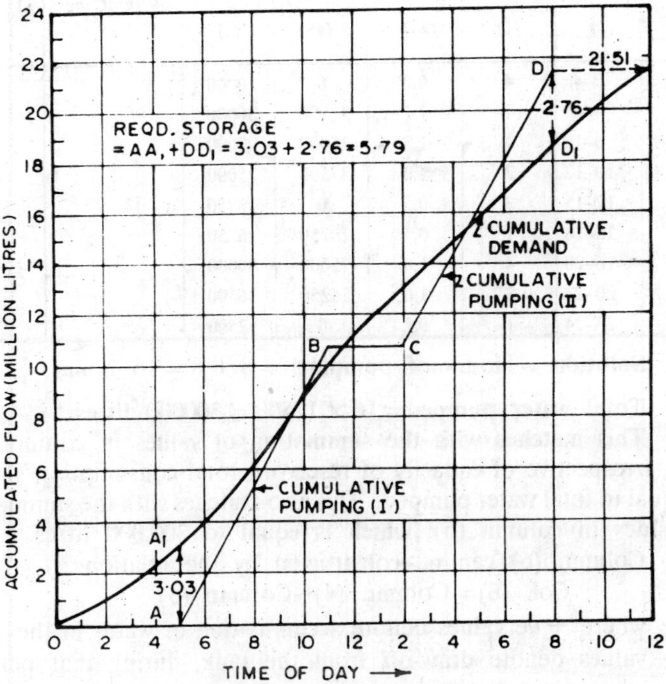

FIG. 16.6. (EXAMPLE 16.3)

From the graphical solution, required storage $= AA_1 + DD_1$
$= 3.03 + 2.76 = 5.79$ million litres.

Example 16.4. *An overhead tank is to be provided for a city water supply. Given the following data, calculate the minimum capacity of the tank without any fire demand storage. Assume the tanks to be empty between 12 and 15 hours. The draw off factors, water pumped and water consumed etc. are shown in the table below (Table 16.6)*

Compute columns 6 and 7 of the table to calculate the storage capacity of the tank.

Given (i) *Average water supply* = *12500 litres/h.*

(ii) *Rate of pumping* = *18750 litres/hr.*

(iii) *Hours of pumping* : *4 to 12 hours and 15 to 23 hours.*

TABLE 16.6

No.	*Time (hr.)* (1)	*Hours* (2)	*Drawoff factor* (3)	*Water pumped* (4)	*Water consumed* (5)	*Accumulation (+) drawoff (–)* (6)	*Water in Service Reservoir* (7)
A	0–4	4	0.2	0	10000		
B	4–6	2	1.2	37500	30000		
C	6–10	4	1.8	75000	90000		
D	10–12	2	1.0	37500	25000		
E	12–15	3	0.5	0	18750		
F	15–16	1	0.5	18750	6250		
G	16–20	4	1.6	75000	80000		
H	20–23	3	1.0	56250	36500		
I	23–24	1	0.2	0	2500		

Solution : Hours of pumping $= 8 + 8 = 16$ hours

Total water pumped $= 16 \times 12500 = 300000$ litres.

This matches with the summation of values in column (4). Also, irrespective of capacity of reservoir, total consumption should be equal to total water pumped. This also matches with the summation of values in column (5), which is equal to 300,000 litres.

Column (6) can be constructed by the relation :

$$\text{Col. (6)} = \text{Column (4)} - \text{Column (5)}$$

where + ve values denote accumulation of water in the tank, and – values denote draw-off from the tank, during that period.

TABLE 16.7. (Example 16.4)

No.	*Time (hr.)* (1)	*Hours* (2)	*Drawoff factor* (3)	*Water pumped* (4)	*Water consumed* (5)	*Column 6 =col. (4)--col. (5) Accumulation (+) Drawoff (–)* (6)	*Water in Service Reservoir* (7)
A	0–4	4	0.2	0	10000	– 10000	13750
B	4–6	2	1.2	37500	30000	+ 7500	21250
C	6–10	4	1.8	75000	90000	– 15000	6250
D	10–12	2	1.0	37500	25000	+ 12500	18750
E	12–15	3	0.5	0	18750	– 18750	0 ↓
F	15–16	1	0.5	18750	6250	+ 12500	12500
G	16–20	4	1.6	75000	80000	– 5000	7500
H	20–23	3	1.0	56250	37500	+ 18750	26250
I	23–24	1	0.2	0	2500	– 2500	23750

Now, to construct column 7, we start with the data given that the reservoir is empty between 12 – 15 hours. Hence

(*i*) Water in S.R. at E (12 – 15 hours) = 0
(*ii*) Water in S.R. at F = 0 + value in col. (6)
= 0 + 12500 = 12500 litres
(*iii*) Water in S.R. at G = 12500 – 5000 = 7500 litres
(*iv*) Water in S.R. at H = 7500 + 18750 = 26250 litres
(*v*) Water in S.R. at I = 26250 – 2500 = 23750 litres
(*vi*) Water in S.R. at A = 23750 – 10000 = 13750 litres
(*vii*) Water in S.R. at B = 13750 + 7500 = 21250 litres
(*viii*) Water in S.R. at C = 21250 – 15000 = 6250 litres
(*ix*) Water in S.R. at D = 6250 + 12500 = 18750 litres
Hence : (*x*) Water in S.R. at E = 18750 – 18750 = zero

From col. (8), we find that min. capacity of reservoir = **26250 litres.**

16.8. PIPE HYDRAULICS

When water flows through pipes, head loss takes place. Total head loss is composed of the following items : (1) head loss due to friction, (2) head loss due to bends and constrictions. (*i.e.* local losses)

1. HEAD LOSS DUE TO FRICTION

Head loss due to friction may be computed from the following formulae.

(*i*) Darcy-Weisbach formula
(*ii*) Hazen Williams formula
(*iii*) Manning's formula
and (*iv*) Combined Darcy-Weisbach and Colebrook-White equation.

(*a*) **Darcy-Weisbach formula**

Darcy and Weisbach suggested the first dimensionless equation for pipe flow problems as :

$$h_f = \frac{4f'LV^2}{2gD} = \frac{fLV^2}{2gD} \qquad \text{...(16.4)}$$

or

$$S = \frac{h_f}{L} = \frac{4f'V^2}{2gD} = \frac{fV^2}{2gD} \qquad \text{...(16.4 a)}$$

Expressed in terms of discharge $Q\left(=\frac{\pi}{4}D^2V\right)$,

we have $$h_f = \frac{fL}{2gD}\left(\frac{16Q^2}{\pi^2 D^4}\right)$$

or
$$h_f = \frac{f L Q^2}{12.1 D^5} \qquad ...(16.4\ b)$$

where h_f = head loss due to friction (m)

f' = Darcy's coefficient of friction.

f = coefficient of friction $= 4f'$

L = length of the pipe (m)

V = average velocity of flow (m/s)

D = internal diameter of pipe (m)

g = acceleration due to gravity $= 9.81\ \text{m/s}^2$

The friction factor f generally varies from 0.02 for new smooth pipes to 0.075 for old rough pipes. However, f depends upon (*i*) Reynolds number $R_e \left(= \frac{VD}{\nu} \right)$ and (*ii*) the *relative roughness* k/D, where k is the *effective roughness size* of the pipe wall.

Approximate value of f

The approximate value of f can be found from the following empirical relations :

$$f = 0.04 \left[1 + \frac{1}{35\,D} \right] \text{ for old pipes} \qquad ...16.5\ (a)$$

and
$$f = 0.02 \left[1 + \frac{1}{35\,D} \right] \text{ for new pipes} \qquad ...16.5\ (b)$$

Accurate value of f

(*i*) *For laminar flow* ($R_e \le 2000$) :

$$f = \frac{64}{R_e} \qquad ...(16.6)$$

(*ii*) *For smooth turbulent flow*: (f is a function of R_e only)

$$\frac{1}{\sqrt{f}} = 2 \log_{10} \frac{R_e \sqrt{f}}{2.51} \qquad ...(16.7\ a)$$

(*iii*) *For rough turbulent flow* : (f is function of k/D only)

$$\frac{1}{\sqrt{f}} = 2 \log_{10} \frac{3.75\,D}{k} \qquad ...(16.7\ b)$$

(*iv*) For smooth pipe and R_e varying from 20,000 to 20,00,000 Slitcher gave the following formula

$$f = 0.005 + \frac{0.396}{R_e^{0.3}} \qquad ...(16.8\ a)$$

(*v*) For smooth pipe and R_e from 20,000 to 32,40,000, Nickuradse gave the following expression :

$$f = 0.0032 + \frac{0.221}{R_e^{0.237}} \qquad ...(16.8\ b)$$

(*vi*) *For transition zone* (R_e between zero and 4000), f is a function of both R_e and k/D and is given by Colebrook-White equation :

$$\frac{1}{\sqrt{f}} = -2 \log_{10}\left[\frac{k}{3.7\,D} + \frac{2.51}{R_e\sqrt{f}}\right] \quad ...(16.9)$$

From study of Eqs. 16.5 to 16.9, we observe that the determination of f is very difficult. Since D may not be known to start with, it is a trial and error procedure.

(*b*) **Hazen William's Formula**

This is the most widely used formula, relating velocity of flow, hydraulic mean radius and hydraulic gradient.

$$V = 0.849\,CR^{0.63}\,S^{0.54} \quad ...(16.10)$$

where V = mean velocity of flow in pipe (m/sec)
R = hydraulic radius (mean depth) in metres
S = hydraulic gradient
C = coefficient of roughness of pipe.

In terms of diameter D of the pipe, the above formula reduces to

$$V = 0.354\,CD^{0.63}\,S^{0.54} \quad ...(16.11)$$

where D is the diameter of the pipe in metres. Expressed in terms of head loss h_f and the length L of the pipe, the Hazen Williams formula takes the following form :

$$h_f = \frac{6.843\,L}{D^{1.167}}\left(\frac{V}{C}\right)^{1.852} = 10.70\left(\frac{Q}{C}\right)^{1.852}\frac{L}{D^{4.87}} \quad ...(16.12)$$

The discharge Q (m/s) is given by

$$Q = 0.278\,CD^{2.63}\,S^{0.54} \quad ...(16.13)$$

If, however, D is expressed in mm, we have

$$V = 4.56 \times 10^{-3}\,CD^{0.63}\,S^{0.54} \text{ m/s} \quad ...(16.11\ a)$$

and $$Q = 3.581 \times 10^{-9}\,CD^{2.63}\,S^{0.54} \text{ m}^3\text{/s} \quad ...(16.13\ a)$$

or $$Q = 2.419 \times 10^{-7}\,CD^{2.63}\,S^{0.54} \text{ m}^3\text{/min} \quad ...(6.13\ b)$$

or $$Q = 1.289 \times 10^{-5}\,CD^{2.63}\,S^{0.54} \text{ m}^3\text{/hour} \quad ...(16.13\ c)$$

or $$Q = 3.09 \times 10^{-4}\,CD^{2.63}\,S^{0.54} \text{ kilolitres/day} \quad ...(16.13\ d)$$

The coefficient C depends upon several factors, such as (*i*) pipe material, (*ii*) age of the pipe, and (*iii*) diameter of the pipe (for cast iron pipes). Its value is high for smooth and new pipes, and goes on decreasing with age or roughness. Table 16.8 gives the value of C for different pipes.

Hazen William's formula is very complex, and is usually solved with the help of nomograms. Fig. 16.7 shows the nomogram for the solution of Hazen William's formula. There are four variables:

(*i*) discharge Q in m^3/sec or litres/sec; (*ii*) diameter of pipe in mm; (*iii*) loss of head in metres per 1000 m length of pipe and (*iv*) velocity of flow in m/sec. If out of the four quantities, any two are known, the other two can be found from the nomogram. These nomograms are valid for a value of $C = 100$. For other value of C, find the head loss from the chart and multiply it by the factor $\frac{C}{100}$.

TABLE 16.8 VALUES OF C IN HAZEN-WILLIAMS' FORMULA

Type of pipe	*Condition*	*C*		
		10 cm dia.	*30 cm dia.*	*60 cm dia. and over*
Cast iron				
	Smooth and new	125	130	135
	Ordinary	95	105	115
	Old	70	80	90
	Very rough	55	65	75
Spun Iron	Add 5 to value of C for cast iron pipe of same age or condition			
Welded Steel	Subtract 5 from value of C for cast iron pipe of same age or condition			
Riveted Steel	Subtract 10 from value of C for cast iron pipe of same age or condition			
Spun concrete or bitumen lined	Smoth and new	140–150		
Concrete or concrete lined	Average	120–140		
Asbestos cement	Average	140		
Wood stave	Smooth & clean	120		
Masonry	Smooth & clean	120		
Vitrified clay	Good condition	110		
Ordinary brick	Good condition	100		

For example, let the flow rate be 10 m^3/min. and the diameter of the pipe be 400 mm. It is required to find the head loss for a pipe of 800 metres length. Mark a point A corresponding to $Q = 10$ on the discharge line, and point B corresponding to diameter = 400 mm on the diameter line. Join points A and B by means of straight edge, and prolong the line AB to cut the head loss line at point C and the velocity line at point D. Thus, we get head loss per

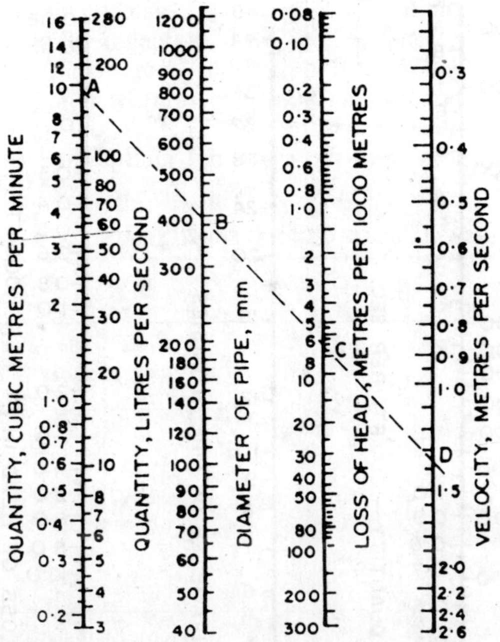

FIG. 16.7. NOMOGRAM FOR HAZEN WILLIAMS FORMULA (C= 100)

1000 m length ≈ 6.7 m. Hence the head loss for 800 m length $= 6.7 \times \frac{800}{1000} = 5.36$ m. The velocity of flow comes out to be 1.33 m/sec.

Fig. 16.8 shows the nomogram for F.P.S. units, for $C = 100$. To determine the loss of head for any value of C, multiply the loss of head obtained by Fig. 16.8 by a value of K as given in Table 16.9.

TABLE 16.9

C	40	60	80	90	100	110	120	130	140
K	5.46	2.58	1.51	1.22	1.00	0.838	0.713	0.615	0.536

(*c*) **Mannings Formula**

$$V = \frac{1}{n} R^{2/3} S^{1/2} \qquad ...(16.14)$$

For circular pipes, $R = \frac{D}{4}$, where D is the diameter.

$$\therefore \qquad V = \frac{0.397}{n} D^{2/3} S^{1/2} \qquad ...(16.15)$$

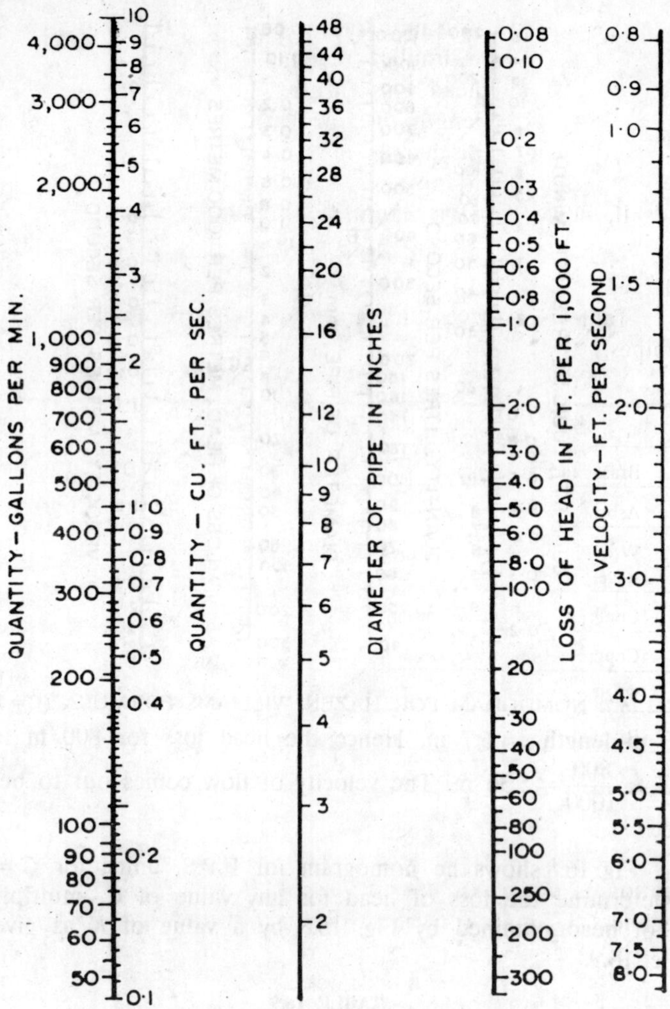

FIG. 16.8. NOMOGRAM FOR HAZEN-WILLIAMS FORMULA (C = 100)

Expressed in terms of head loss, the formula reduces to

$$h_f = \frac{n^2}{0.157} \cdot \frac{L V^2}{D^{4/3}} \qquad \text{...(16.16)}$$

The discharge Q (m/s) is given by

$$Q = \frac{0.312}{n} D^{8/3} S^{1/2} \qquad \text{...(16.17)}$$

where V = mean velocity of flow (m/s)

D = dia. of pipe (m)

R = hydraulic radius (m)
S = hydraulic gradient
h_f = head loss (m)
L = length of pipe (m)
Q = discharge (m/s)
n = Mannings coefficient of roughness

If, however, D is expressed in mm, we have

$$V = 3.97 \times 10^{-3} D^{2/3} S^{1/2} \quad ...(16.15\ a)$$

and

$$Q = 3.12 \times 10^{-9} D^{8/3} S^{1/2} \ (m^3/s) \quad ...(16.17\ a)$$

Table 16.10 gives the values of coefficent n for various types of pipes.

TABLE 16.10. VALUES OF MANNINGS' n

	Type of Pipe	n
1.	Brass and glass pipe	0.009 to 0.013
2.	Asbestos cement pipe	0.010 to 0.012
3.	Wrought iron, welded steel, wooden stave	0.010 to 0.014
4.	Concrete pipe, very smooth	0.011 to 0.012
5.	Concrete pipe with rough joints	0.016 to 0.017
6.	Vitrified sewer pipe	0.013 to 0.015
7.	Riveted steel pipe	0.013 to 0.017
8.	Corrugated iron pipe	0.020 to 0.022

(*d*) Combined Darcy-Weisbach and Colbrook-White Equation

We have seen that the friction factor f in the Darcy-Weisbach equation is dependant on the Reynolds No. (R_e) and the relative roughness (k/D), which can be expressed by Colebrook-White equation:

$$\frac{1}{\sqrt{f}} = -2 \log_{10}\left[\frac{k}{3.7D} + \frac{2.51}{R_e\sqrt{f}}\right] \quad ...(16.9)$$

The Colebrook-White equation (*i.e.* Eq. 16.9) was first plotted in the form of $f - R_e$ diagram by Moody (1944) and, hence such a plot, shown in Fig. 16.9 is known as 'Moody diagram' .

Combining Darcy-Weisbach equation (Eq. 16.4 *a*) and Colebrook-White equation (Eq. 16.9), the explicit expression for V is obtained as under

$$V = -2\sqrt{2gDS}\,\log_{10}\left[\frac{k}{3.7D} + \frac{2.51\nu}{D\sqrt{2gDS}}\right] \quad ...(16.18)$$

where S = friction slope.

If head loss due to other factors are ignored, we have

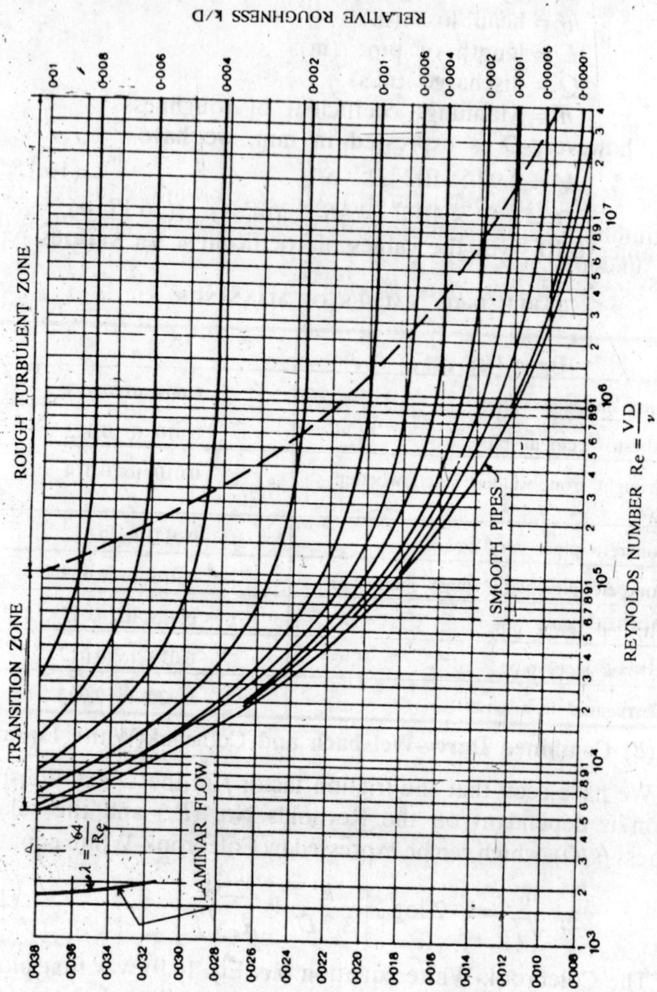

FIG. 16.9. MOODY DIAGRAM

$$S = \frac{h_f}{L}$$

In that case, Eq. 16.18 reduces to

$$V = 2\sqrt{2\,g\,D\frac{h_f}{L}}\ \log_{10}\left[\frac{k}{3.7\,D} + \frac{2.51\,\nu}{D\sqrt{2\,g\,D\frac{h_f}{L}}}\right] \quad \text{...(16.18 } a)$$

Due to implicit form of colebrook-white equation (Eq. 16.9) a number of approximations in *explicit form* in f have been proposed as under :

(i) Moody's formulation

$$f = 0.0055\left[1 + \left(20000\frac{k}{D} + \frac{10^6}{R_e}\right)^{1/3}\right] \quad \text{...(16.19)}$$

(ii) Barr's formulation

$$\frac{1}{\sqrt{f}} = -\,2\log_{10}\left[\frac{k}{3.7\,D} + \frac{5.1286}{R_e^{0.89}}\right] \quad \text{...(16.20)}$$

Thus, in order to solve the pipe flow problem by combined Darcy-Weisbach and Colebrook-White equation, there are three methods :

Method 1 :

Steps (*i*) Determination of f (Eq. 16.9) using Moody's diagram, by assuming any initial value of V.

(*ii*) Determination of V from the relation

$$H_L = K\frac{V^2}{2g} + \frac{f\,L\,V^2}{2\,g\,D} \quad \text{...(16.21)}$$

(*iii*) Knowing this value of V, again find f from Moody's diagram.

(*iv*) Find the final value of V from Eq. 16.21.

Method 2 :

Steps (*i*) Determination of approximate value of velocity of flow from Eq. 16.18 (*a*), by taking $S = h_f/L$ (*i.e.* by neglecting other losses).

(*ii*) Knowing V, compute other losses $\left(= K\frac{V^2}{2g}\right)$, and hence find total loss $h_L = h_f + K\frac{V^2}{2g}$

(*iii*) Compute $S = \frac{h_L}{L}$

(*iv*) Compute velocity V, from Eq. 16.18, using the above value of S.

(*v*) Repeat the cycle, if necessary.

Method 3

Steps : (*i*) Compute explicit value of f either from Eq. 16.19 (Moody's formulation) or from Eq. 16.20 (Barr's formulation)

(*ii*) Determine velocity V from Eq. 16.21.

2. LOCAL LOSSES

In addition to the friction head loss (h_f), which is quite prominent in a long pipe line, other losses take place, such as (*i*) entry loss $\left(= 0.5\, V^2/2g\right)$, (*ii*) velocity head loss $\left(V^2/2g\right)$, and (*iii*) head loss due to constrictions such as valves, pipe line transitions etc. These losses are also known as *minor losses* (h_m) and can be expressed as

$$h_m = K\frac{V^2}{2g} \qquad ...(16.21)$$

where V is in m/s and g is in m/sec^2.

The values of K are given in Table 16.11.

TABLE 16.11. VALUES OF K FOR DIFFERENT FITTINGS.

Type of fitting	*Value of K*
1. Sudden contractions	0.3* – 0.5
2. Entrance (shape well rounded)	0.5
3. Elbow (*i*) 90°	0.5–1.0
(*ii*) 145°	0.4–0.75
(*iii*) 22°	0.25–0.50
4. Tee (*i*) 90°	1.5
(*ii*) straight run	0.3
(*iii*) coupling	0.3
5. Gate valve (open)	0.3** – 0.4
6. With reducer and increaser	0.5
7. Globe	10.0
8. Angle	5.0
9. swing check	2.5
10. Meter venturi	0.3
11. Orifice	1.0

* Varying with area ratios   ** Varying with radius ratios.

Alternatively, the effect of these constrictions can be taken into account by considering *equivalent length of pipe*, the values of which are given in Table 16.12 for different sizes of fittings with $K = 1$.

TABLE 16.12. EQUIVALENT LENGTH OF PIPE FOR DIFFERENT SIZES OF FITTINGS WITH $K = 1$

Size (mm)	*Equivalent length (m)*	*Size (mm)*	*Equivalent length (m)*
10	0.3	65	2.4
15	0.6	80	3.0
20	0.75	90	3.6
25	0.90	100	4.2
32	1.20	125	5.1
40	1.50	150	6.0
50	2.1		

The local losses are also some times known as minor losses (h_m), since they are small in comparison to head loss (h_f) due to friction. Thus total $H_L = h_f + h_m$, where $h_m = h_L +$ entry loss + velocity head loss.

Example 16.5. *A large service reservoir supplies water to two colonies as under :*

Colony A : Population 12,000
Colony B : Population 60,000

Determine (a) the sizes of the supply conduits, and (b) the hydraulic gradient at which the pipe lines should be laid. Assume average daily water consumption as 200 litres/capita/day.

Solution : Let us assume that daily max. demand as 1.6 times the average demand.

(*i*) *For colony A,*

Quantity of water $= 200 \times 12000 = 24 \times 10^5$ l/day

$$\therefore \quad \text{Av. discharge} = \frac{(24 \times 10^5) \times 10^{-3}}{24 \times 3600} = 0.0278 \text{ cumecs}$$

$$\therefore \quad \text{Max.} \quad Q = 1.6 \times 0.0278 = 0.0445 \text{ cumes.}$$

Let us assume a flow velocity of 1.2 m/s.

$$A = \frac{0.0445}{1.2} = 0.0371 \text{ m}^2 = 371 \text{ cm}^2$$

$$\therefore \quad \text{dia.} \quad D = \sqrt{\frac{4 \times 371}{\pi}} = 21.7 \text{ cm} \approx \mathbf{22 \text{ cm}}$$

Actual velocity $V = \dfrac{0.0445 \times 4}{\pi\,(0.22)^2} = 1.171$ m/s.

For determining the hydraulic gradient (S), let us use Hazen William's formula, taking $C = 100$

$$V = 0.354\, C D^{0.63}\, S^{0.54} \qquad ...(16.11)$$

$$1.171 = 0.354 \times 100\,(0.22)^{0.63}\, S^{0.54}$$

From which $S^{0.54} = \dfrac{1}{11.649}$

$$\therefore \quad S = \left(\frac{1}{11.649}\right)^{1/0.54} = \left(\frac{1}{11.649}\right)^{1.852} \approx \frac{1}{94}$$

Alternatively, from Eq. 16.13,

$$Q = 0.278\, C\, D^{2.63}\, S^{0.54}$$

$$\therefore \quad 0.0445 = 0.278 \times 100\,(0.22)^{2.63}\, S^{0.54}$$

From which $S \approx \dfrac{\mathbf{1}}{\mathbf{94}}$

(ii) For colony B

Quantity of water $= 200 \times 60000 = 12 \times 10^6$ l/day

$\therefore$ Av. discharge $= \dfrac{(12 \times 10^6 \times 10^{-3})}{24 \times 3600} = 0.1389$ cumecs.

$\therefore$ Max. $Q = 1.6 \times 0.1389 = 0.2222$ cumecs.

Assuming a flow velocity of 1.2 m/s,

$$A = \frac{0.2222}{1.2} = 0.1852 \text{ m}^2 = 1852 \text{ cm}^2$$

$$\therefore \quad \text{Dia. } D = \sqrt{\frac{4 \times 1852}{\pi}} = 48.6 \text{ cm}$$

Provide a 50 cm dia. pipe.

Now, from Eq. 16.13,

$$Q = 0.278\, C\, D^{2.63}\, S^{0.54}$$

$$\therefore \quad 0.2222 = 0.278 \times 100\,(0.5)^{2.63}\, S^{0.54}$$

or $S^{0.54} = \dfrac{1}{20.211}$

From which $S \approx \dfrac{\mathbf{1}}{\mathbf{262}}$

From the above, we conclude that for equal velocities, head loss is more for smaller dia. pipes. Hence if we use smaller dia. pipes, the cost of pumping will increase though the cost of pipe will be less.

Example 16.6. *A 1 m dia. smooth concrete pipe carries a discharge of 0.9 cumecs at an averge temperature of 20°C. Compute* *the hydraulic gradient using (a) Darcy Weisbach formula (b) Mannings formula (c) Hazen-William's formula (d) Slitchers formula for f. Take $\nu = 1.011 \times 10^{-6}\ m^2/s$ at 20°C.*

Solution:

Area of pipe, $A = \frac{\pi}{4} \times 1 = 0.785\ \text{m}^2$

$$\therefore \quad V = \frac{Q}{A} = 0.9/0.785 = 1.146\ \text{m/sec}$$

(a) Darcy-Weisbach formula

$$S = \frac{h_f}{L} = \frac{fV^2}{2gD} \qquad \text{...(16.4 } a\text{)}$$

Taking $f = 0.02$ for new smooth pipe,

$$S = \frac{0.02\,(1.146)^2}{2 \times 9.81\,(1)} = \frac{1}{747}$$

Alternatively, using Eq. 16.5 (*b*) for new pipe,

$$f = 0.02\left[1 + \frac{1}{35\,D}\right] = 0.02\left[1 + \frac{1}{35 \times 1}\right] = 0.0206$$

$$\therefore \quad S = \frac{0.0206\,(1.146)^2}{2 \times 9.81\,(1)} = \frac{1}{726}$$

(b) Manning's formula

$$S = \frac{h_f}{L} = \frac{n^2 V^2}{0.157\,D^{4/3}} \qquad \text{...(16.16)}$$

Taking $n = 0.012$ (Table 16.10),

$$S = \frac{(0.012 \times 1.146)^2}{0.157\,(1)^{4/3}} = \frac{1}{830}$$

(c) Hazen-William's formula

$$V = 0.354\,CD^{0.63}\,S^{0.54} \qquad \text{...(16.11)}$$

Taking $C = 130$ (Table 16.8)

$$1.146 = 0.354 \times 130\,(1)^{0.63}\,S^{0.54}$$

$$\therefore \quad S^{0.54} = \frac{1}{40.157}$$

From which $S = \left(\frac{1}{40.157}\right)^{1.852} = \frac{1}{934}$

(d) Slitchers formula for f

$$R_e = \frac{VD}{\nu} = \frac{1.146 \times 1}{1.011 \times 10^{-6}} = 1.134 \times 10^6$$

Hence from Eq. 16.8 (*a*)

$$f = 0.005 + \frac{0.396}{R_e^{0.3}}$$

$$= 0.005 + \frac{0.396}{(1.134 \times 10^6)^{0.3}} = 0.005 + 0.006$$

$$= 0.011$$

$$\therefore \quad s = \frac{f V^2}{2 g D} = \frac{0.011\,(1.146)^2}{2 \times 9.8\,(1)} = \frac{\mathbf{1}}{\mathbf{1358}}$$

Example 16.7. *A pipe line 4000 m long, 200 mm in diameter and roughness size 0.04 mm conveys water at 20°C between two reservoirs, the difference in level between which is maintained constant at 40 m. In addition to the entry loss of 0.5* $\frac{V^2}{2g}$, *a valve produces a head loss of* $\frac{8V^2}{2g}$. *Determine the steady discharge between the reservoirs using*

(*a*) *Colebook-White equation*
(*b*) *The Moody diagram*
(*c*) *An explicit function for f*

Take $\nu = 1.011 \times 10^{-6}\ m^2/s$ *at* $20°\ C$.

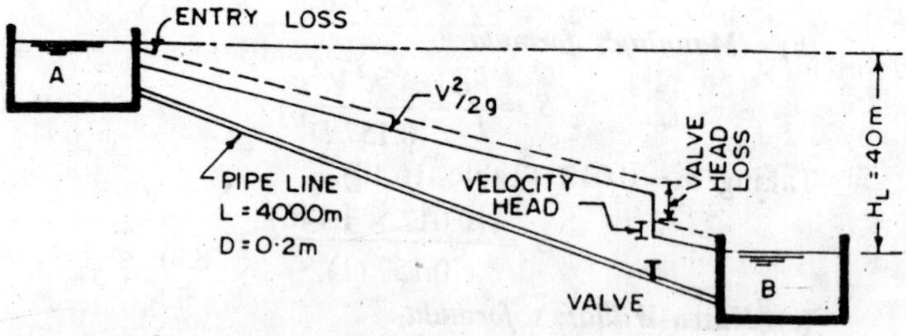

FIG. 16.10.

Solution :

Fig 16.10 shows the H.G. line along with various losses. Applying Bernoulli's equation to *A* and *B*

$$H_L = \frac{0.5\,V^2}{2g} + \frac{V^2}{2g} + \frac{8\,V^2}{2g} + \frac{f L V^2}{2 g D} \quad ...(i)$$

Gross Head	Entry Loss	Velocity head	Valve Head Loss	Friction Head Loss

or $$H_L = h_m + h_f \quad ...(i)$$

where $$h_m = \text{minor head loss} = 10.5 \frac{V^2}{2g}$$

(a) By Colebrook-White Equation

$$\frac{1}{\sqrt{f}} = -2\log_{10}\left[\frac{k}{3.7D} + \frac{2.51}{R_e\sqrt{f}}\right] \qquad ...(ii)$$

Combining Darcy-Weisbach Eq. and Colebrook-White equation,

$$V = -\sqrt{2gDS}\,\log_{10}\left[\frac{k}{3.7D} + \frac{2.51\,\nu}{D\sqrt{2gDS}}\right] \qquad ...(16.18)$$

To start with, ignoring all other terms of Eq. (*i*) except the friction term (h_f) we have

$$S = H_L/L$$

where $$h_f = \frac{fLV^2}{2gD}$$

$$\therefore \quad V = -2\sqrt{2gD.\frac{h_f}{L}}\,\log_{10}\left[\frac{k}{3.7D} + \frac{2.51\,\nu}{D\sqrt{2gD.\frac{h_f}{L}}}\right] \qquad ...(iii)$$

Here, $h_f \approx H_L = 40$ m; $\therefore \frac{h_f}{L} = \frac{40}{4000} = 0.01$

$$\therefore \quad V = -2\sqrt{19.62 \times 0.2 \times 0.01}\,\log_{10}\left[\frac{0.04 \times 10^{-3}}{3.7 \times 0.2} + \frac{2.51 \times 1.011 \times 10^{-6}}{0.2\sqrt{19.62 \times 0.2 \times 0.01}}\right]$$

$$= 1.556 \text{ m/s}$$

Knowing V, we have

Minor head loss $h_m = 10.5\frac{V^2}{2g} = 10.5\frac{(1.556)^2}{2 \times 9.81} \approx 1.30$ m

Hence the better estimate of h_f is

$$h_f = H_L - h_m = 40 - 1.30 = 38.7 \text{ m}$$

$$\therefore \quad \frac{h_f}{L} = \frac{38.7}{4000} = 0.009675$$

Hence from Eq. (*iii*), we get

$$V = 1.529 \text{ m/s}$$

$$\therefore \quad h_m = 10.5\frac{(1.529)^2}{2 \times 9.81} = 1.251 \text{ m}$$

$$\therefore \quad h_f = 40 - 1.251 \text{ m} = 38.749 \text{ m}$$

and $$\frac{h_f}{L} = \frac{38.749}{4000} = 0.009687$$

Finally, substituting this value of h_f/L in Eq. (*iii*), we get

$$V = 1.53 \text{ m/s} ; \; h_m = 1.253 ; \; h_f = 38.747$$

Hence $Q = 1.53 \times \frac{\pi}{4}(0.2)^2 = \mathbf{0.0481\ m^3/s = 48.1\ litres/s}$

(b) By Moody chart

In order to use Moody chart, let us *assume* $V = 1.2$ m/s

$$\therefore \quad R_e = \frac{VD}{\nu} = \frac{1.2 \times 0.2}{1.011 \times 10^{-6}} = 2.37 \times 10^5$$

$$\frac{k}{D} = \frac{0.04}{200} = 0.0002$$

Hence from Moody chart (Fig 16.7), we get

$$f = 0.017$$

Rearranging Eq. (*i*), we get

$$V = \sqrt{\frac{2gH_L}{10.5 + \frac{fL}{D}}} \qquad \text{...}(iv)$$

$$\therefore \quad V = \sqrt{\frac{2 \times 9.81 \times 40}{10.5 + \frac{0.017 \times 4000}{0.2}}} = 1.496 \text{ m/s.}$$

Using this *better value of V*, we have

$$R_e = \frac{1.496 \times 0.2}{1.011 \times 10^{-6}} = 2.96 \times 10^5$$

$$k/D = 0.0002, \text{ as before.}$$

Hence from Moody diagram, we get $f = 0.0165$

From from Eq. (*iv*), we get $V = 1.518$ m/s.

Further trial is not essential, and one can take $V \approx 1.52$ m/s

$$\therefore \quad Q = 1.52 \times \frac{\pi}{4}(0.2)^2 = \mathbf{0.0478\ m^3/s = 47.8\ l/s}$$

(c) By explicit function of f

Using Barr's formulation (Eq. 16.20) :

$$\frac{1}{\sqrt{f}} = -2\log_{10}\left(\frac{k}{3.7D} + \frac{5.1286}{R_e^{0.89}}\right) \qquad \text{...}(v)$$

Assuming $V = 1.2$ m/s, to start with,

$$R_e = \frac{1.2 \times 0.2}{1.011 \times 10^{-6}} = 2.37 \times 10^5; \text{ Also } \frac{k}{D} = \frac{0.04}{200} = 0.0002$$

Hence from Eq. (*v*), we get

$$\frac{1}{\sqrt{f}} = -2\log_{10}\left[\frac{0.0002}{3.7} + \frac{5.1286}{(2.37\times 10^5)^{0.89}}\right]$$

$$= 7.717$$

From which $f = \left(\frac{1}{7.717}\right)^2 = 0.0168$

Hence from Eq. (*iv*)

$$V = \sqrt{\frac{2gH_L}{10.5 + \frac{fL}{D}}} = \sqrt{\frac{2\times 9.81\times 40}{10.5 + \frac{0.0168\times 4000}{0.2}}} = 1.505 \text{ m/s}$$

$\therefore$ Revised $R_e = \frac{1.505\times 0.2}{1.011\times 10^{-6}} = 2.978\times 10^5$

Revised $f = 0.0164$ From Eq. (*v*)

Revised $V = 1.525$ m/s.

Taking this value of V,

$$Q = 1.525\times\frac{\pi}{4}(0.2)^2 = \mathbf{0.0479\ cumecs = 47.9\ l/s}$$

Thus we get almost identical results by all the three methods.

Example 16.8. *A storage reservoir is situated at a distance of 6 km from a city of 3 lakh population. The total loss of head from the source to the city is not to exceed 20 m. Taking the daily demand of 200 litres/capita/day, pumping is to be done for 12 hours only, determine the size of supply main by(a) Darcy-Weisbach formula taking coefficient of friction as 0.015 (b) Hazen-Willams formula, taking C = 130. Assume minor losses equal to $10V^2/2g$.*

Solution:

Max. demand $= 300000\times 200 = 60\times 10^6$ litres/day

Since puming is to be done only for 12 hours a day,

$$Q = \left(\frac{60\times 10^6}{12\times 3600}\right)\times 10^{-3} = 1.389 \text{ cumecs.}$$

Also, $H_L = 20$ m, $L = 6000$ m

(a) Darcy-Weisbach formula

$$h_f = f\frac{LV^2}{2gD}$$

But $V = Q/A = \frac{4Q}{\pi D^2}$

$$\therefore \quad h_f = \frac{fL}{2gD}\left(\frac{16Q^2}{\pi^2 D^4}\right) = \frac{fLQ^2}{12.1D^5} \qquad ...(16.4\ b)$$

Now, neglecting the minor losses, $h_f \approx H_L = 20$ m.

$$\therefore \quad 20 = \frac{0.015 \times 6000\,(1.389)^2}{12.1\,D^5}$$

From which $D = 0.936$ m.

Hence $$V = \frac{4 \times 1.389}{\pi\,(0.936)^2} = 2.02 \text{ m/s.}$$

$$\therefore \text{ Minor losses, } h_m = 10\,\frac{(2.02)^2}{2 \times 9.81} = 2.08 \text{ m.}$$

$\therefore$ Better value of $h_f = H_L - h_m = 20 - 2.08 = 17.92$ m.

Hence from Eq. 16.4 (*b*)

$$17.92 = \frac{0.015 \times 6000\,(1.389)^2}{12.1\,D^5}$$

From which $D \approx \mathbf{0.96}$ **m**

However, use nearest standard available pipe dia. of 1 m.

(b) Hazen-William's formula

$$V = 0.354\,CD^{0.63}\,S^{0.54} \qquad ...(16.11)$$

Expressed in terms of discharge

$$Q = 0.278\,CD^{2.63}\,S^{0.54} \qquad ...(16.12)$$

or $$Q = 0.278\,CD^{2.63}\,(h_f/L)^{0.54} \qquad ...(16.12\,a)$$

Neglecting minor losses (h_m), $h_f \approx H_L = 20$ m.

$$\therefore \quad 1.389 = 0.278\,(130)\,D^{2.63}\left(\frac{20}{6000}\right)^{0.54}$$

From which $D = (0.8363)^{1/2.63} = 0.934$ m

$$\therefore \quad V = \frac{4 \times 1.389}{\pi\,(0.934)^2} = 2.027 \text{ m/s}$$

$$\therefore \quad h_m = \frac{10\,(2.027)^2}{2 \times 9.81} = 2.09 \text{ m/s}$$

$$\therefore \quad h_f = H_L - h_m = 20 - 2.09 = 17.91 \text{ m}$$

Hence from Eq. 16.12 (*a*)

$$1.389 = 0.278\,(130)\,D^{2.63}\left(\frac{17.91}{6000}\right)^{0.54}$$

From which $D = \mathbf{0.96}$ **m.**

However, use nearest standard available pipe dia. of 1 m.

Example 16.9. *Water flows in a 100 mm pipe at Reynolds number 90000. The pipe is estimated to have an equivalent sand grain roughness of size 0.12 mm. Determine the head loss expected in 1200 m length of pipe. How much head loss would be expected if this pipe were smooth ? Take kinematic viscosity of water as $1 \times 10^{-6}\ m^2/s$. The following explicit equation for friction factor may be used.*

$$\frac{1}{\sqrt{f}} = 1.14 - 2 \log_{10}\left(\frac{k}{d} + \frac{21.25}{R^{0.9}}\right)$$

in which k is the equivalent sand grain roughness, d is the pipe diameter and R the pipe Reynolds number.

Solution:

Darcy-Weisbach friction factor is given by the explicit equation

$$\frac{1}{\sqrt{f}} = 1.14 - 2 \log_{10}\left(\frac{k}{d} + \frac{21.25}{R^{0.9}}\right) \quad ...(i)$$

It is to be noted that this equation is practically the same as Barr's equation (Eq. 16.20) in which factor 3.7 has been taken out of the log term.

(a) *Rough pipe*

When the pipe is rough, f is a function of both k and R.

Here $\frac{k}{d} = \frac{0.12}{100} = 1.2 \times 10^{-3}$

$$\therefore \quad \frac{1}{\sqrt{f}} = 1.14 - 2 \log_{10}\left[1.2 \times 10^{-3} + \frac{21.25}{(90000)^{0.9}}\right] = 6.5649$$

$$\therefore \quad f = 0.0232$$

Also, $R = \frac{V d}{\nu}$

$$\therefore \quad V = \frac{R \nu}{d} = \frac{90000 \times 1 \times 10^{-6}}{0.1} = 0.9 \text{ m/s.}$$

$$\therefore \quad h_f = \frac{f L V^2}{2 g d} = \frac{0.0232 \times 1200\,(0.9)^2}{2 \times 9.81\,(0.1)}$$

$$= \mathbf{11.49\ m}$$

(b) *Smooth pipe*

If the pipe were smooth, $k = 0$.

$$\therefore \quad \frac{1}{\sqrt{f}} = 1.14 - 2 \log_{10}\left[\frac{21.25}{(90000)^{0.9}}\right] = 7.403$$

From which $f = 0.0182$

$$V = 0.9 \text{ m/s, as before}$$

$$\therefore \quad h_f = \frac{0.0182 \times 1200\,(0.9)^2}{2 \times 9.81\,(0.1)} = \mathbf{9.04\ m}$$

16.9. PIPES IN SERIES AND PARALLEL

(*a*) Pipes in series

When pipes of different diameter are connected in series, the total head loss is equal to the summation of the individual head losses in different pipes. Such an arrangement is quite common since draw-off usually takes place along the length of the service main, necessitating the decrease in diameter as the discharge decreases. In addition to the frictional head loss, the *minor losses* in bends, valves, gates, contraction/expansion also come into play, and if the length of pipe line is short, these minor losses may be quite significant in comparison to friction losses.

(*b*) Pipes in parallel

Laying the pipes in parallel may be necessary due to the following reasons :

(*i*) To increase the capacity of the line.

(*ii*) To facilitate repairs without closing down the complete supply

(*iii*) To use smaller diameter pipes, by laying them in parallel, if one single large diameter pipe is not available.

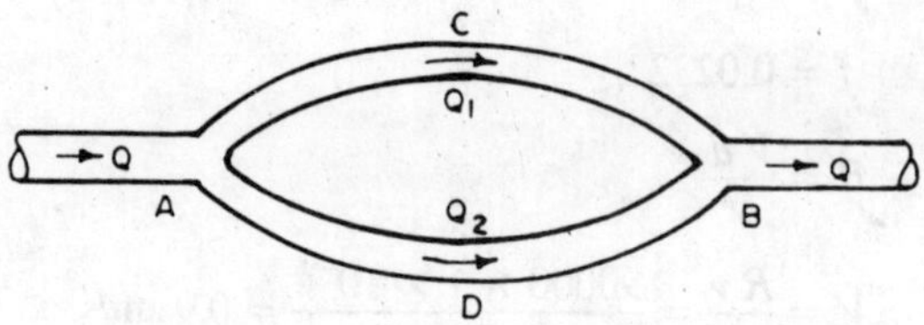

FIG. 16.11. PIPES IN PARALLEL

When pipes are laid in parallel, as shown in Fig. 16.11, the head loss through each pipe will be the same. Also, the total discharge Q will be equal to the sum of the discharge in each pipe.

Example 16.10. *A service reservoir receives water from an impounding reservoir, distant 40 km. The difference in levels between the water levels of these reservoirs is 120 m. The two reservoirs were originally connected by a single pipe line designed to carry a max. daily requirement of 18 million litres per day. It is now desired to increase the supply to 25 million litres per day, by laying another pipe* *line of the same diameter along side the first one, in a certain required length and cross-connecting the two. Using Hazen-William equation (C = 120), determine (a) the diameter of the pipes and (b) the length of second pipe required to be laid.*

Solution :

$$\text{Initial} \quad Q = \frac{(18 \times 10^6) \times 10^{-3}}{24 \times 60 \times 60} = 0.2083 \text{ cumecs}$$

$$\text{New} \quad Q = \frac{(25 \times 10^6) \times 10^{-3}}{24 \times 60 \times 60} = 0.2894 \text{ cumecs}$$

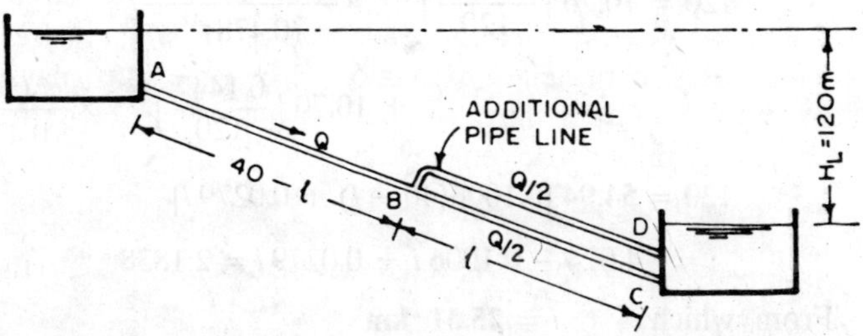

FIG. 16.12.

Also, $h_f \approx H_L = 120$ m, neglecting the minor losses, since in a very long supply conduit minor losses are negligibly small in comparison to friction loss.

Hazen-William's formula for the circular pipe is

$$V = 0.354\, C D^{0.63}\, S^{0.54} \qquad ...(16.11)$$

Let us first reduce the above formula expressing head loss (h_f) in terms of discharge Q.

$$\therefore \quad \frac{4Q}{\pi D^2} = 0.354\, C\, D^{0.63} \left(\frac{h_f}{L}\right)^{0.54}$$

or

$$Q = 0.278\, C\, D^{2.63} \left(\frac{h_f}{L}\right)^{0.54}$$

or

$$h_f = 10.70 \left(\frac{Q}{C}\right)^{1.852} \frac{L}{D^{4.87}} \qquad ...(16.12)$$

This same as Eq. 16.12 given earlier.

Now, initially, $Q = 0.2083$ and $h_f = 120$ m

$$\therefore \quad 120 = 10.70 \left(\frac{0.2083}{120}\right)^{1.852} \frac{40000}{D^{4.87}}$$

or

$$D^{4.87} = 0.02756$$

From which $D = 0.478$ m

Now let the additional pipe line of the same dia. D (= 0.478 m) be laid for a length l as shown in Fig. 16.12. The head loss through each pipe of length l will be equal. Thus we have the following head loss equation :

Total head loss (= 120 m)= Head loss in AB for Q (= 0.2894)

+ head loss in BC (or BD) for $Q/2$

$$\therefore \quad H_L = h_{fab} + h_{fbc}$$

Substituting the values, we get

$$120 = 10.70\left(\frac{0.2894}{120}\right)^{1.852} \times \frac{(40 - l)\,1000}{(0.478)^{4.87}}$$

$$+\ 10.70\left(\frac{0.1447}{120}\right)^{1.852} \times \frac{l\,(1000)}{(0.478)^{4.87}}$$

or $$120 = 54.94\,[0.1006(40 - l) + 0.0279\,l]$$

or $$4.024 - 0.1006\,l + 0.0279\,l = 2.1838$$

From which $\quad l$ = **25.31 km**

Check

$$h_{fbc} = 10.70\left(\frac{0.1447}{120}\right)^{1.852} \frac{25.31 \times 1000}{(0.478)^{4.87}} = 38.76$$

$$h_{fab} = 10.70\left(\frac{0.2894}{120}\right)^{1.852} \frac{(40 - 25.31)\,1000}{(0.478)^{4.87}} = 81.22$$

$\therefore$ Total $\quad h_f = 38.76 + 81.22 = 119.98 \approx 120$ m.

Example 16.11. *A pipe line 0.8 m diameter is 2 km long. To augment the discharge, another pipe line of the same diameter is introduced parallel to the first in the second half of its length. Find the increase in discharge taking Darcy-Weisbach friction factor as 0.04 and head at the inlet as 40 m.*

Solution

Darcy-Weisbach equation is

$$h_f = \frac{f\,L\,V^2}{2\,g\,D} = \frac{f\,L\,Q^2}{12.1\,D^5} \qquad \text{...(16.4 } b\text{)}$$

Here, $\quad h_f = 40$ m, $f = 0.04$, $D = 0.8$ and $L = 2000$

$$\therefore \text{ Initial } Q = Q_1 = \left[\frac{40 \times 12.1\,(0.8)^5}{0.04 \times 2000}\right]^{1/2} = 1.408 \text{ cumes.}$$

After the addition of a parallel pipe line of length = 2/2 = 1 km, let the discharge be Q_2. Hence the total head loss equation will be.

$$\begin{bmatrix}\text{Total head}\\ \text{loss}\end{bmatrix} = \begin{bmatrix}\text{Head loss for 1 km}\\ \text{length with discharge } Q_2\end{bmatrix} + \begin{bmatrix}\text{Head loss for 1 km}\\ \text{length with discharge } \frac{1}{2} Q_2\end{bmatrix}$$

$$\therefore \quad 40 = \frac{0.04 \times 1000\, Q_2^2}{12.1\,(0.8)^5} + \frac{0.04 \times 1000\,(Q_2/2)^2}{12.1\,(0.8)^5}$$

or $$40 = 10.0885\, Q_2^2 + 2.5221\, Q_2^2$$

From which $Q_2 = 1.781$ cumecs.

$\therefore$ Increase in discharge = 1.781 − 1.408 = **0.373 cumes**

$$\% \text{ Increase} = \frac{0.373}{1.408} \times 100 \approx 26.5\ \%$$

Example 16.12. *An upper reservoir delivers water to a lower reservoir through two uniform pipe lines AC and CB of diameters 250 mm and 150 mm respectively. Just upstream of the change in the diameter, which is assumed to be gradual, a controlled discharge of 20 l/s is taken off. Determine the discharge to the lower reservoir, given the following :*

Length of pipe line AC= 2000 m; effective roughness = 0.0075 mm

Length of pipe line CB= 2500 m; effective roughness = 0.006 mm

Gross head = 30 m.

Neglect minor losses. Use the following explicit function for f:

$$\frac{1}{\sqrt{f}} = -2 \log \left(\frac{k}{3.7\,D} + \frac{5.1286}{R_e^{0.89}} \right)$$

Take $\nu = 1 \times 10^{-6}\ m^2/sec$

Solution :

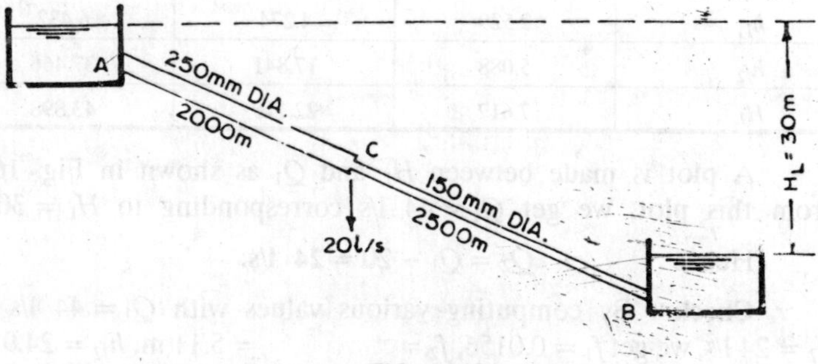

FIG. 16.13.

If the minor losses are neglected,

$$H_L = h_{f1} + h_{f2}$$

From Darcy-Weisbach equation expressed in terms of discharge, we have

$$h_f = \frac{f L Q^2}{12.1 D^5} \quad ...(16.4\ b)$$

$$\therefore \quad H_L = \frac{f_1 L_1 Q_1^2}{12.1 D_1} + \frac{f_2 L_2 Q_2^2}{12.1 D_2^5} \quad ...(i)$$

Here both f_1 and f_2 are different.

Discharge take off at $C = 20$ l/s $= 20 \times 10^{-3}$ m³/s

Also, $\quad Q_2 = Q_1 - 20 \times 10^{-3} \quad ...(ii)$

Since f_1 and f_2 are initially unknown, the simplest method of solution is to input a series of trial values of Q_1. The corresponding values of various parameters, including h_{f1}, h_{f2} and H_L are tabulated below (Table 16.13).

$$\frac{k_1}{D_1} = \frac{0.0075}{250} = 3 \times 10^{-5}; \quad \frac{k_2}{D_2} = \frac{0.006}{150} = 4 \times 10^{-5};$$

$$A_1 = 0.04909 \text{ m}^2 \quad ; \quad A_2 = 0.01767 \text{ m}^2$$

TABLE 16.13

Q_1 (l/s)	30	40	50
V_1 (m/s)	0.611	0.815	1.019
V_2 m/s	0.566	1.132	1.698
Re_1 (× 10⁵)	1.528	2.038	2.548
Re_2 (× 10⁵)	0.849	1.698	2.547
f_1	0.0166	0.0158	0.0152
f_2	0.0187	0.0164	0.0153
h_{f1}	2.529	4.274	6.432
h_{f2}	5.088	17.841	37.466
H_L	7.617	22.115	43.898

A plot is made between H_L and Q_1 as shown in Fig. 16.14. From this plot, we get $Q_1 \approx 44$ l/s corresponding to $H_L = 30$ m.

Hence $\quad Q_2 = Q_1 - 20 = \mathbf{24}$ **l/s.**

Check : By computing various values with $Q_1 = 44$ l/s and $Q_2 = 24$ l/s, we get $f_1 = 0.0156$, $f_2 =$ $h_{f1} = 5.11$ m, $h_{f2} = 24.91$ m.

$$\therefore \quad H_L = 5.11 + 24.91 = 30.02 \text{ m} \approx 30 \text{ m}.$$

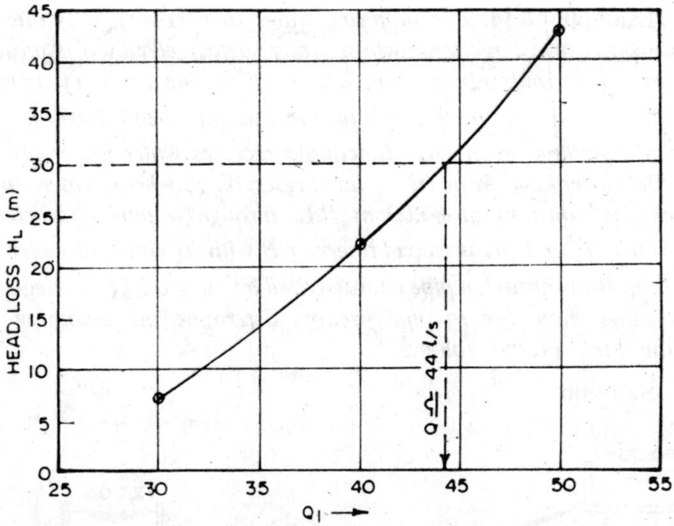

FIG. 16.14.

Example 16.13. *Determine the discharge to the lower reservoir for the data of example 16.12, taking a uniform value of f = 0.016 for both the pipes at all the discharges.*

Solution :

The govening equation is

$$H_L = \frac{f L_1 Q_1^2}{12.1\, D_1^5} + \frac{f L_2 Q_2^2}{12.1\, D_2^5}, \text{ where } Q \text{ is in cumecs.}$$

Expressing Q_1 and Q_2 in litres per second and noting that $Q_2 = Q_1 - 20$, we get

$$H_L = \frac{f L_1}{12.1\, D_1^5}\left(\frac{Q_1}{1000}\right)^2 + \frac{f L_2}{12.1\, D_2^5}\left(\frac{Q_1 - 20}{1000}\right)^2$$

or $$30 = \frac{0.016}{12.1 \times 10^6}\left[\frac{2000\, Q_1^2}{(0.25)^5} + \frac{2500\,(Q_1 - 20)^2}{(0.15)^5}\right]$$

or $$2.048\, Q_1^2 + 32.922\,(Q_1 - 20)^2 = 22687.5$$

or $$34.97\, Q_1^2 - 1316.9\, Q_1 + 13168.8 = 22687.5$$

or $$Q_1^2 - 37.66\, Q_1 - 272.2 = 0$$

From which $Q_1 = 43.86$ l/s.

Hence $Q_2 = 43.86 - 20 =$ **23.86 l / s**

Example 16.14. *For supplying water to a colony, it is intended to store water in a large tank, whose water surface elevation is presently 6800 m. This tank is being supplied by a pipe line from a junction J, the friction loss in meters of water in this pipe being $0.6Q^{1.95}$, where Q is the discharge in m^3/sec. Junction J receives water under pressure and without leakage from (i) a large reservoir A whose water surface elevation is constant at 80.00 m R.L, through a pipe characterised by $h_f = 0.4\,Q^2$, and (ii) a second reservoir B with its constant water level at 77.0 m R.L. through a pipe characterised by $h_f = 0.5\,Q^{1.90}$. Neglecting losses other than due to pipe friction, determine the incoming flow into the tank in the colony.*

Solution :

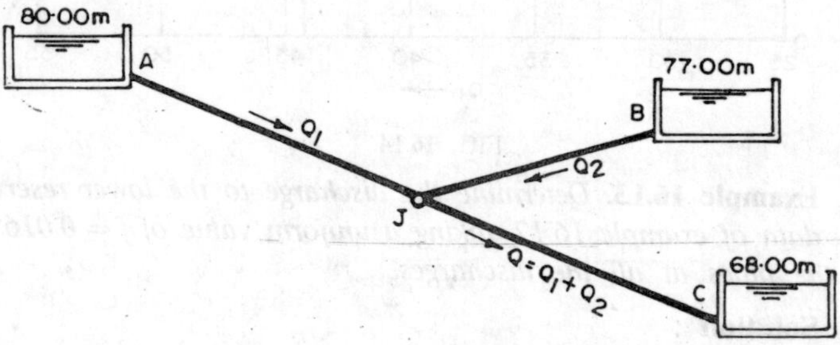

FIG. 16.15.

Fig 16.15 shows the arrangement. Let line *AJ* carry a discharge Q_1 to *J* and line *BJ* carry a discharge Q_2 to *J*. Hence the discharge flowing through *JC* (and hence to the tank) will be $Q = Q_1 + Q_2$.

We have the following head loss equations between *AJC* and *BJC* :

$$H_{L1} = 80.00 - 68.00 = 0.4\,Q_1^2 + 0.6\,(Q_1 + Q_2)^{1.95} \quad ...(i)$$

and

$$H_{L2} = 77.00 - 68.00 = 0.5\,Q_2^{1.90} + 0.6\,(Q_1 + Q_2)^{1.95} \quad ...(ii)$$

Subtracting (*i*) from (*ii*) we get

$$0.5\,Q_2^{1.9} - 0.4\,Q_1^2 + 3 = 0$$

$$\therefore \quad Q_1^2 = 1.25\,Q_2^{1.9} + 7.5 \quad ...(iii)$$

Substituting in (*i*), we get

$$0.4\,(1.25\,Q_2^{1.9} + 7.5) + 0.6\left[(1.25\,Q_2^{1.9} + 7.5)^{1/2} + Q_2\right]^{1.95} = 12$$

or

$$0.5\,Q_2^{1.9} + 0.6\left[(1.25\,Q_2^{1.9} + 7.5)^{1/2} + Q_2\right]^{1.95} = 9 \quad ...(iv)$$

The above equation can be solved by trial and error, using graphical procedure. For that, it can be rewritten as

$$F = 0.5\,Q_2^{1.9} + 0.6\left[(1.25\,Q_2^{1.9} + 7.5)^{1/2} + Q_2\right]^{1.95}$$

The value of F is computed for three or four values of Q_2, as shown in Table below (Table 16.14).

TABLE 16.14.

Q_2 (cumecs)	F
0.4	5.81
0.6	6.80
0.8	7.96
1	9.27

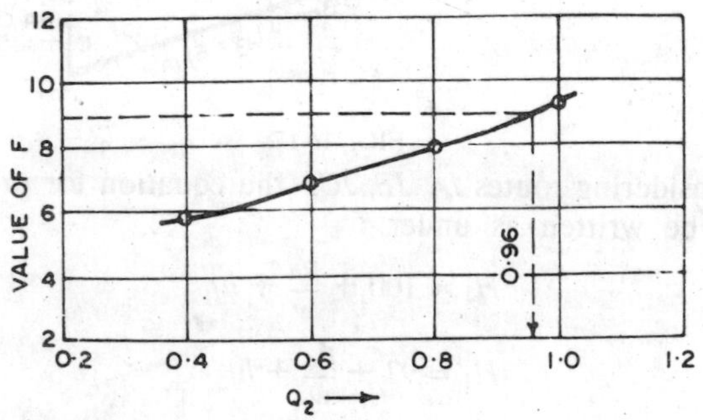

FIG. 16.16.

A curve is plotted between Q_2 and F (Fig. 16.16). From this curve, we get $Q_2 = 0.96$ cumecs corresponding to the value of $F = 9$.

Hence from Eq. (*iii*), $Q_1 = \left[1.25\,(0.96)^{1.9} + 7.5\right]^{1/2} = 2.94$ cumecs.

Hence discharge in the colony tank = 2.94 + 0.96 = **3.90 cumecs**

Check : Substituting in (*i*), we get $H_{L1} = 0.4\,(2.94)^2 + 0.6\,(3.90)^{1.95}$ $= 11.98$ m ≈ 12 m.

Example 16.15. *A water supply main trifurcates at a junction point J into three branches each feeding a separate reservoir. The details of the pipes and reservoirs are as follows.*

Pipe	*Dia. (mm)*	*length (km)*	*Darcy Weisbach friction factor (f)*	*Terminal reservoir*	*Reservoir elevation (m)*
JA	300	4 km	0.018	*A*	100.00
BJ	300	5 km	0.018	*B*	92.00
JC	300	6 km	0.018	*C*	84.00

If the inflow from the main at the junction is 0.4 m^3/s, determine the delivery into each reservoir.

Solution:

Fig. 16.17 shows the arrangement as per data given.

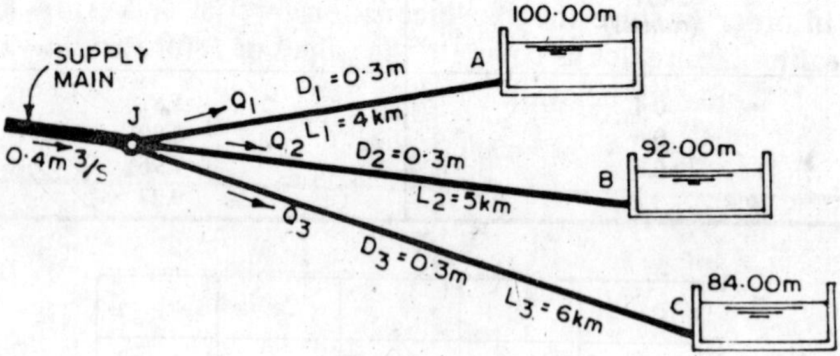

FIG. 16.17.

Considering routes *JA, JB, JC,* the equation for pressure (H_J) at *J* can be written as under :

$$H_J = 100 + \frac{p_a}{w} + h_{f1} \quad ...(a)$$

$$H_J = 92 + \frac{p_a}{w} + h_{f2} \quad ...(b)$$

and

$$H_J = 84 + \frac{p_a}{w} + h_{f3} \quad ...(c)$$

Hence we have

$$h_{f1} + 8 = h_{f2} \quad ...(i)$$

and

$$h_{f2} + 8 = h_{f3} \quad ...(ii)$$

Now

$$h_{f1} = \frac{f_1 L_1 Q_1^2}{12.1\, D_1^5} = \frac{0.018 \times 4000\, Q_1^2}{12.1\,(0.3)^5} = 2448.7\, Q_1^2$$

$$h_{f2} = \frac{0.018 \times 5000\, Q_2^2}{12.1\,(0.3)^5} = 3060.9\, Q_2^2$$

Hence

$$2448.7\, Q_1^2 + 8 = 3060.9\, Q_2^2, \quad \text{From } (i)$$

From which

$$Q_1^2 = 1.25\, Q_2^2 - 3.267 \times 10^{-3} \quad ...(iii)$$

Also,

$$h_{f3} = \frac{0.018 \times 6000\, Q_3^2}{12.1\,(0.3)^5} = 3673.1\, Q_3^2$$

Hence

$$3060.9\, Q_2^2 + 8 = 3673.1\, Q_3^2, \quad \text{From } (ii)$$

or

$$Q_2^2 = 1.2\, Q_3^2 - 2.614 \times 10^{-3} \quad ...(iv)$$

But $\qquad Q_3 = 0.4 - (Q_1 + Q_2)$

$$\therefore \qquad Q_2^2 = 1.2\left[0.4 - (Q_1 + Q_2)\right]^2 - 2.614 \times 10^{-3}$$

or $\quad Q_2^2 = 1.2\left[0.4 - \left\{(1.25\,Q_2^2 - 3.26 \times 10^{-3})^{1/2} + Q_2\right\}\right]^2 - 2.614 \times 10^{-3}$

or $1.2\left[0.4 - \left\{(1.25\,Q_2^2 - 3.267 \times 10^{-3}) + Q_2\right\}\right]^2 - Q_2^2 = 2.614 \times 10^{-3} = F$ (say)

In order to solve the above equation by trial and error, using graphical procedure, let us compute the values of F for various values of Q_2, as tabulated below.

TABLE 16.15.

Q_2 *(cumecs)*	F
0.10	39.9×10^{-3}
0.11	27.3×10^{-3}
0.12	15.8×10^{-3}
0.13	5.4×10^{-3}
0.14	-3.9×10^{-3}

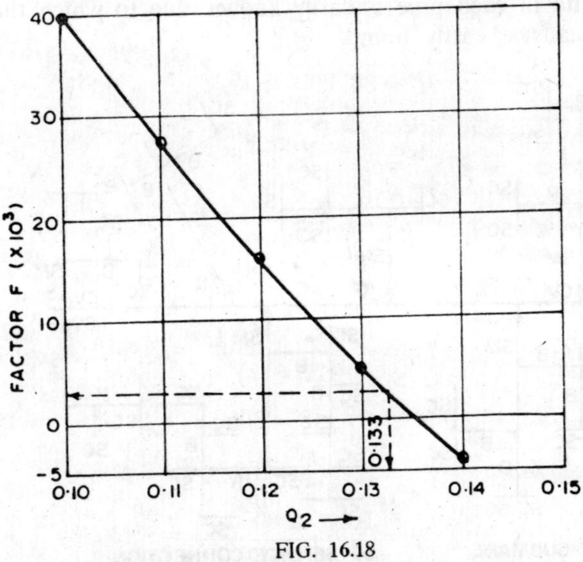

FIG. 16.18

Fig. 16.18 shows the plot between Q_2 and F, from which we get

$$Q_2 = \mathbf{0.133\ cumecs} \text{ corresponding to } F = 2.614 \times 10^{-3}.$$

Hence from (*iii*),

$$Q_1 = \left[1.25\,(0.133)^2 - 3.267 \times 10^{-3}\right]^{1/2} = \mathbf{0.1373\ cumecs.}$$

$$\therefore \quad Q_3 = 0.4 - (0.133 + 0.1373) = \mathbf{0.1279\ cumecs}$$

16.10. LAYOUT OF DISTRIBUTION SYSTEM

There are four principal methods of laying out distribution system :

(*i*) Dead-end system or Tree system
(*ii*) Grid-iron system or Reticulation system
(*iii*) Circular system or Ring system
(*iv*) Radial system.

(*i*) **Dead end or Tree system.** In this system, one main pipe line runs through the centre of the populated area and submains takeoff from this to both the sides. The submains divide into several branch lines from which service connections are given. Thus the entire distribution area is covered by pipe lines running like a tree, as shown in Fig. 16.19. There are no cross-connections between the branches and submains. Due to several dead ends, there is accumulation of sediment there, and stagnation of water. However, the flow rate in each pipe is easily known, due to which the pipe diameters can be easily found.

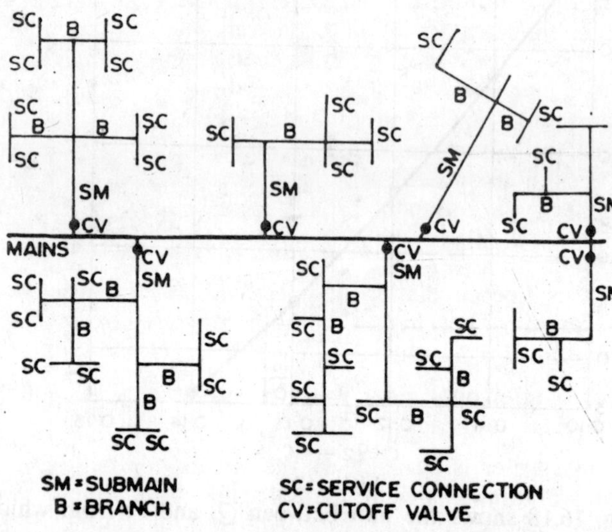

FIG. 16.19.

The dead end system has following *advantages* :

1. The design calculations are simple and easy. It is possible to determine the discharge and pressure in each pipe very accurately.
2. The pipe diameters are to be designed for the pupulation likely to be served by them. This leads to cheap and economical design.
3. The laying of pipe is simple.
4. The number of cutoff valves employed in the system are less.

The system has however the following *disadvantages* :

1. Due to a number of dead ends in the system, there is stagnation of water and accumulation of sediment at dead end.
2. A large number of scour valves are required at the dead ends. Also, staff is required to open these valves periodically for the removal of stale water and sediment.
3. In case of repairs to any of the pipe sections in the system, the whole of the portion beyond that point to the end will be required to be cutoff completely.
4. The system is less successful in maintaining satisfactory pressure in the remote parts.
5. The discharge available for fire fighting in the streets will be limited.

(*ii*) **Grid-iron or Reticulation system.** If the dead ends of the previous system are inter-connected, water can be made to circulate continuously through the whole of the distribution system. This system is therefore also known as the interlaced system. As shown in Fig. 16.20, the main supply pipe runs through the centre of the rectangular area and submains takeoff from these in perpendicular directions. The branch lines interconnect all the submains. The system is ideal for cities laid out on rectangular plan resembling a grid-iron.

The system has the following *advantages* :

1. There is free circulation of water, without any stagnation or sediment deposit. The chances of pollution due to stagnation are not there.

2. Due to interconnection, water is available at each point with minimum loss of head.

3. Enough water is available in the streets, for fire fighting, since hydrant will draw water from more than one side.

4. In case of repairs, only very small area of the distribution system is affected.

The system has the following *disadvantages* :

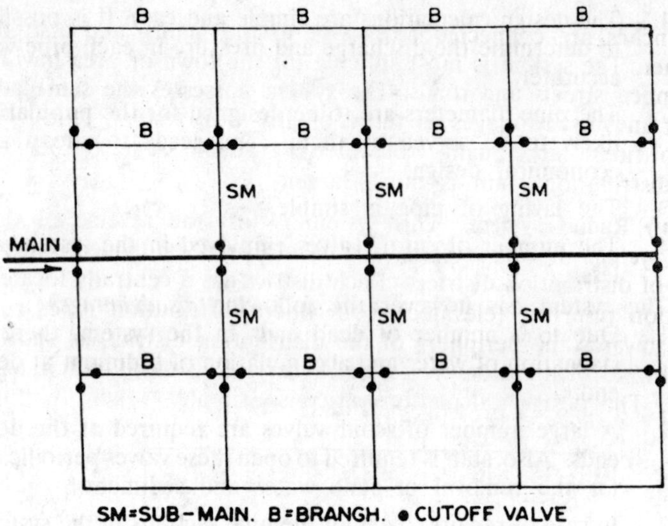

FIG. 16.20. GRID-IRON SYSTEM

1. A large number of cutoff valves are required.
2. The system requires longer pipe lengths and bigger diameters.
3. The analysis of discharge, pressure and velocities in the pipes is difficult and cumbersome.
4. The cost of laying water pipes is more.

(*iii*) **Circular or Ring system.** In this system, the supply main forms a ring around the distribution district, as shown in Fig. 16.21.

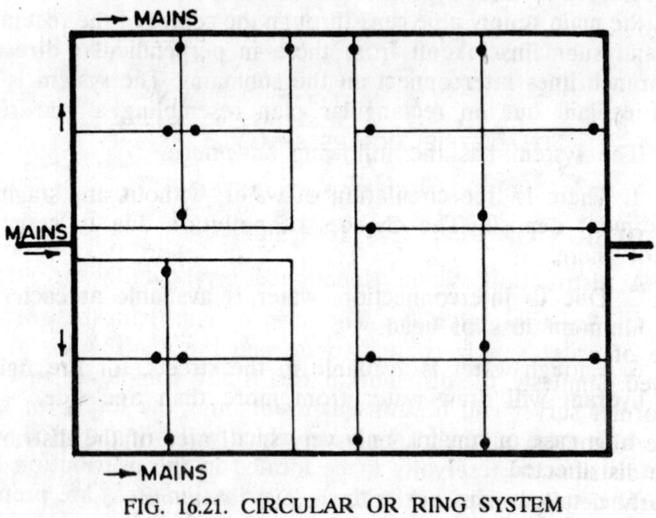

FIG. 16.21. CIRCULAR OR RING SYSTEM

The branches are connected cross-wise to the mains, and also to each other. The system is most suitable for the town or area having well planned streets and roads. The system possesses the same advantages and disadvantages as those of grid iron system. However, in case of fire, a large quantity of water is available. Also, the length of the distribution main is much larger.

(*iv*) **Radial system.** This system is just the reverse of the circular system. In this system, the whole area is divided into a number of distribution districts. Each district has a centrally located distribution reservoir (elevated) from where distribution pipes run radially towards the periphery of the distribution district, as shown in Fig. 16.22. This system gives quick service, without much loss of head. The design calculations are also simple.

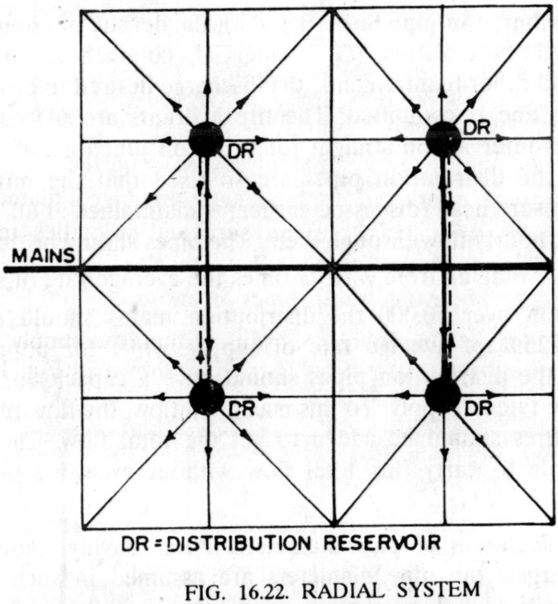

FIG. 16.22. RADIAL SYSTEM

16.11. DESIGN OF DISTRIBUTION SYSTEM

A distribution system is designed in the following steps :

1. Surveys and maps. The strip of land lying between the source of water supply (water works) and the distribution area is surveyed. Similarly, the distribution area is also completely surveyed. The former survey will be utilised in obtaining the levels for fixing up the alignment of the rising main which will bring treated water to the distribution reservoirs to be located in the distribution area. Similarly, detailed maps of the town (distribution area) are prepared to show the position of roads, streets, lanes, commercial locality, industrial area, gardens etc. A topographical map of the area is also prepared to locate the high and low areas. The cross-section of the streets, roads, lanes etc. are also prepared, showing the positions of existing underground service lines such as gas lines, electric and telephone lines, sever lines, existing water supply lines, if any.

2. Tentative layout. A tentative layout of the distribution lines is then marked, showing the position of treatment plant, distribution mains, distribution and balancing reservoirs, valves, hydrants etc. The area is divided into various distribution districts. The density of population (*i.e.* average number of persons per hectare area) is also marked. The various systems of distribution net work have already been discussed in the previous article. The lengths of pipe lines should be kept as short as possible.

3. Discharge in pipe lines. Based on the density of population, type of distribution district (*i.e.* residential, commerical etc.) and on the fire and other requirements, the discharge desired to be carried by each pipe line is computed. The fire hydrants are placed at 50 to 100 metres intervals on straight runs, and on junctions of streets. The size of the distribution pipes are so fixed that the minimum necessary pressure head (discussed earlier) is maintained at all points, carrying peak hourly flow through them. The pipes should be designed for a discharge ranging from $2\frac{1}{4}$ to 3 times the average rate of supply. For population over 50,000 the distribution mains should have a capacity of 225% of average rate of supply, while for population below 5000, the distribution pipes should have a capacity of 300% of the average rate of supply. To this maximum flow, the flow required for fighting fires should be added to get the total flow. The pipes should be able to carry this total flow without excessive pressure drops.

4. Calculation of pipe diameters. After having known the design discharges, the pipe diameters are assumed in such a way that the velocities of flow in pipes remain between 0.6 to 3 m/sec. Smaller velocity is assumed for pipes of smaller diameter and larger velocity for pipes of larger diameter. The loss of head in the pipes is then calculated using Hazen-Williams formula (or nomogram) or any other formula. Alternatively, Table 16.16 can be used for selecting pipe diameter and estimating loss of head due to friction per 100 m length of pipe. This table is based on Crimps and Burges Tables.

5. Computation of pressures. The procedure for the computation of pressure etc. in the pipe net work is explained in the next article.

TABLE 16.16 DESIGN OF C.I. PIPES FOR WATER SUPPLY

(Based on Crimps and Burges Tables) Maximum Flow = $2\frac{1}{2}$ Average Flow

Pipe Dia. (mm)	*Capacity m^3/min*	*Velocity m/sec*	*Loss of head m/100 m*	*Population served @*		
				225 l/phd	*180 l/phd*	*135 l/phd*
75	0.251 0.291 0.157	0.90 0.75 0.60	23.80 16.40 10.52	640 531 425	800 665 531	1068 885 790
100	0.444 0.370 0.297	0.90 0.75 0.60	16.11 11.22 7.14	1130 994 756	1412 1180 945	1880 1570 1262
125	0.691 0.580 0.466	0.90 0.75 0.60	11.90 8.33 5.40	1765 1478 1180	2205 1850 1475	2940 2460 1968
150	1.17 1.05 0.9828	1.05 0.90 0.75	12.80 9.53 6.45	2980 2562 2110	3720 2305 2640	4960 4275 3520
175	1.59 1.367 1.131	1.05 0.90 0.75	10.41 7.69 5.26	4050 3430 2880	5060 4350 3600	6750 5800 4800
200	2.075 1.79 1.49	1.05 0.90 0.75	8.70 6.45 4.45	5280 4550 3790	6600 5690 4740	8800 7580 6310
225	3.06 2.62 2.24	1.20 1.05 0.90	10.0 7.40 5.40	7780 6690 5710	9740 8360 7150	12900 11150 9470
250	3.68 3.255 2.78	1.20 1.05 0.90	8.33 6.45 4.76	9360 8280 7100	11700 10350 8000	15600 13800 11800
300	5.35 4.70 3.95	1.20 1.05 0.90	6.67 5.13 3.70	13600 11970 10160	17000 14960 12700	22650 19950 16920
350	7.25 6.39 5.45	1.20 1.05 1.90	5.40 4.17 3.33	18500 16280 13900	23150 20350 17400	30800 27150 23150
375	9.40 8.31 7.30	1.35 1.20 1.05	6.25 4.89 3.77	23900 21200 18600	27400 26450 23200	39150 35350 31000
400	10.70 9.53 3.29	1.35 1.20 1.05	5.72 4.54 3.45	27200 24250 21100	34000 30300 26400	45400 40500 35200
450	15.08 13.51 12.00	1.50 1.35 1.20	6.06 4.88 3.85	38300 34400 30500	47900 43000 38150	63800 57300 50900

16.12. ANALYSIS OF PRESSURE IN DISTRIBUTION SYSTEM

The following methods are used for the analysis of pressure in the distribution system :

1. Equivalent pipe method
2. Hardy cross method
3. Method of sections and circle method
4. Graphical method
5. Electric network analyser method
6. Pitometer distribution studies method.

Out of these, the first two methods are very common, and will be discussed here.

EQUIVALENT PIPE METHOD

In this method, a complex system of pipes is replaced by a single hydraulically equivalent pipe. *The equivalent pipe is one which will replace a given system of pipes with equal head loss for a given flow.* The method of equivalent pipes make use of two hydraulic axioms : (1) the head losses through pipes in series, such as *AB* and *BD* in Fig. 16.23 are *additive*, and (2) that flows through pipes in parallel, such as *ABD* and *ACD* in Fig. 16.23 must be so distributed that the head losses are identical. Where the equivalent pipe is required to replace a series system of pipes, the equivalent pipe may be designed

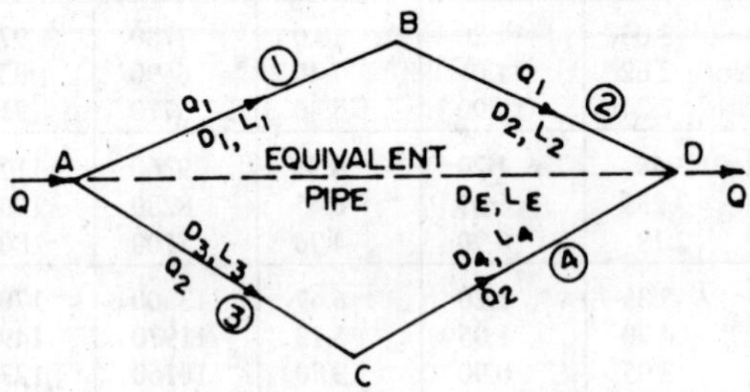

FIG. 16.23.

to have the same length as the system, in which case it is probable that the diameter will be non-standard, or the equivalent pipe may be designed for a specified standard diameter but of length differing from the system. Similarly in the case of a parallel system, the diameter of an equivalent pipe of given or specified length can be formed, or the length of an equivalent pipe of given diameter can be found.

Any pipe flow formula may be used for the computations. If Hazen-Williams formula is used, the length L_E of the equivalent pipe of diameter D_E and discharge coefficient $C_E = 100$ is given by

$$L_E = \left(\frac{100}{C}\right)^{1.85} \left(\frac{D_E}{D}\right)^{4.87} L \qquad ...(16.22)$$

where C, D and L are the corresponding properties of the existing pipe.

If the head loss equation is of the form

$$h_F = \frac{K L Q^n}{D^x} \quad \text{or} \quad Q = K D^a S^b$$

The length or diameter of equivalent pipe will be as given in Table 16.17.

TABLE 16.17.

	Dia. of equivalent pipe of equal length (cm)	*Length of equivalent pipe of diameter D_E (m)*
System in series	$\sqrt[x]{L_t \sum \frac{D^x}{L}}$ where L_t = total length of system	$D_E^x \sum \frac{L}{D^x}$
System in parallel	Dia. of equivalent pipe of length L_E (cm)	Length of equivalent pipe of dia. D_E (m)
	$\sqrt[a]{L_E^b \sum \frac{D^a}{L^b}}$	$\sqrt[b]{D_E^a \sum \frac{L^b}{D^a}}$

Note. For Hazen-Wiliams formula,

$x = 4.87$, $a = 2.63$ and $b = 0.54$.

For the flow net-work shown in Fig. 16.22 let us use the simple head loss formula

$$h_f = \frac{f L V^2}{2 g D}$$

or

$$h_f = \frac{16 f L Q^2}{2 g \pi^2 D^5} = \frac{K L Q^2}{D^5} = \frac{K L Q^n}{D^x}$$

Hence $n = 2$ and $x = 5$

The length of equivalent pipe, for pipes AB and BD in series is given by the relation :

$$\frac{L_E}{D_E^5} = \frac{L_1}{D_1^5} + \frac{L_2}{D_2^5}$$

or

$$L_E = D_E^5 \left[\frac{L_1}{D_1^5} + \frac{L_2}{D_2^5}\right]$$

or in general,

$$L_E = D_E^x \Sigma \frac{L}{D^x} \qquad ...(16.23)$$

This is same as tabulated in Table 16.17.

Similarly, for the pipes in parallel (*i.e.* pipes *ABD* and *ACD*),

$$(h_f)_{ABD} = (h_f)_{ACD}$$

$$\therefore \quad \frac{L_1 Q_1^2}{D_1^5} + \frac{L_2 Q_1^2}{D_2^5} = \frac{L_3 Q_2^2}{D_3^5} + \frac{L_4 Q_2^2}{D_4^5}$$

or
$$\frac{Q_1}{Q_2} = \frac{\sqrt{\dfrac{L_3}{D_3^5} + \dfrac{L_4}{D_4^5}}}{\sqrt{\dfrac{L_1}{D_1^5} + \dfrac{L_2}{D_2^5}}}$$

or
$$Q_1 = A\, Q_2 \qquad \text{...}(a)$$

Also, $Q = Q_1 + Q_2 = Q_2 (A + 1)$

Now if L_E and D_E are the properties of equivalent pipe *AD*, we get

$$\frac{L_E Q^2}{D_E^5} = \frac{L_3 Q_2^2}{D_3^5} + \frac{L_4 Q_2^2}{D_4^5}$$

or
$$\frac{L_E}{D_E^5}(A+1)^2 = \frac{L_3}{D_3^5} + \frac{L_4}{D_4^5}$$

$$\therefore \quad L_E = \frac{D_E^5}{(A+1)^2}\left(\frac{L_3}{D_3^5} + \frac{L_4}{D_4^5}\right) \qquad \text{...(16.24)}$$

Example 16.16. *Find the equivalent length of 30 cm diameter pipe for the net work shown in Fig. 16.24, using (a) Darcy's formula (b) Hazen Williams formula.*

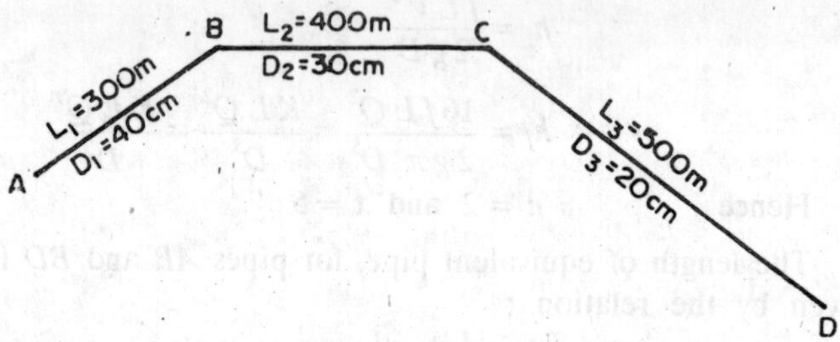

FIG. 16.24.

Solution :

(a) Darcy's formula

$$h_f = \frac{16 f L Q^2}{2 g \pi^2 D^5} = \frac{K L Q^2}{D^5}$$

$$L_E = D_E^5 \left[\frac{L_1}{D_1^5} + \frac{L_2}{D_2^5} + \frac{L_3}{D_3^5} \right]$$

$$= (0.30)^5 \left[\frac{300}{(0.40)^5} + \frac{400}{(0.30)^5} + \frac{500}{(0.20)^5} \right]$$

$$= \mathbf{4268 \ m.}$$

(*b*) ***Hazen Williams formula***

From Eq. 16.11

$$V = 0.355\, CD^{0.63}\, S^{0.54}$$

or
$$Q = 0.355\, CD^{0.63} \left(\frac{h_f}{L} \right)^{0.54} \left(\frac{\pi}{4} D^2 \right)$$

or
$$Q = K_1 \frac{D^{2.63}\, h_f^{0.54}}{L^{0.54}}$$

or
$$h_f = \frac{KL\, Q^{1.85}}{D^{4.87}} \qquad ...(16.25)$$

$$\therefore \quad \frac{L_E}{D_E^{4.87}} = \frac{L_2}{D_1^{4.87}} + \frac{L_2}{D_2^{4.87}} + \frac{L_3}{D_3^{4.87}} = \Sigma \frac{L}{D^{4.87}}$$

$$\therefore \quad L_E = (0.30)^{4.87} \left[\frac{300}{(0.4)^{4.87}} + \frac{400}{(0.3)^{4.87}} + \frac{500}{(0.2)^{4.87}} \right]$$

$$= \mathbf{4080.5 \ m}$$

Example 16.17. *Find the equivalent length of 30 cm dia. pipe for the network shown in Fig. 16.25. Use Hazen Williams formula.*

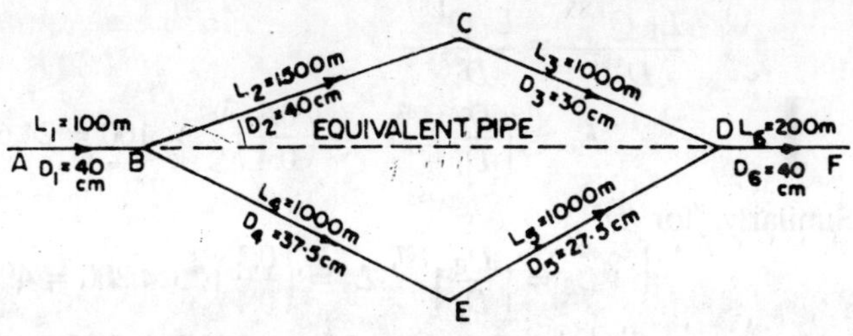

FIG. 16.25.

Solution :

Let the loops *BCD* (Discharge $= Q'$) and *BED* (discharge Q'') be replaced by an equivalent pipe *BD* of length L_E and diameter D_E (= 0.30 m).

$$(h_f)_{BCD} = (h_f)_{BED}$$

But
$$h_f = \frac{KL\, Q^{1.85}}{D^{4.87}} \qquad ...(16.25)$$

$$\frac{L_2 (Q')^{1.85}}{D_2^{4.87}} + \frac{L_3 (Q')^{1.85}}{D_3^{4.87}} = \frac{L_3 (Q'')^{1.85}}{D_4^{4.87}} + \frac{L_5 (Q'')^{1.85}}{D_5^{4.87}}$$

$$\therefore \quad \left(\frac{Q'}{Q''}\right)^{1.85} = \left(\frac{L_4/D_4^{4.87} + L_5/D_5^{4.87}}{L_2/D_2^{4.87} + L_3/D_3^{4.87}}\right)$$

or
$$\left(\frac{Q'}{Q''}\right)^{1.85} = \frac{\dfrac{1000}{(0.375)^{4.87}} + \dfrac{1000}{(0.30)^{4.87}}}{\dfrac{1500}{(0.40)^{4.87}} + \dfrac{1000}{(0.30)^{4.87}}}$$

or
$$\frac{Q'}{Q''} = \left(\frac{470573}{481903}\right)^{1/1.85} = 0.987 \text{ (on simplification)}$$

$$\therefore \quad Q' = 0.987\, Q''$$

Now $\quad Q = Q' + Q'' = 0.987\, Q' + Q'' = 1.987\, Q''$

For the equivalent pipe BD,

$$\frac{L_E Q^{1.85}}{D_E^{4.87}} = \frac{L_4 (Q'')^{1.85}}{D_4^{4.87}} + \frac{L_5 (Q'')^{1.85}}{D_5^{4.87}}$$

or
$$\frac{L_E}{D_E^{4.87}} (1.987)^{1.85} = \frac{L_4}{D_4^{4.87}} + \frac{L_5}{D_4^{4.87}}$$

$$\therefore \quad L_E = \frac{(0.30)^{4.87}}{(1.987)^{1.85}} \left[\frac{1000}{(0.375)^{4.87}} + \frac{1000}{(0.30)^{4.87}}\right]$$

$$= 375.5$$

Equivalent length for AB

$$\frac{L_E Q^{1.85}}{D_E^{4.87}} = \frac{L_1 Q^{1.87}}{D_1^{4.87}}$$

$$\therefore \quad L_E = \left(\frac{D_E}{D_1}\right)^{4.87} L_1 = \left(\frac{0.3}{0.4}\right)^{4.87} \times 100 = 24.6 \text{ m}$$

Similarly, for *DF*,

$$L_E = \left(\frac{D_E}{D_6}\right)^{4.87} L_6 = \left(\frac{0.3}{0.4}\right)^{4.87} \times 200 = 49.2 \text{ m}$$

$\therefore$ Total equivalent length $= 24.6 + 375.5 + 49.2 =$ **449.3**.

16.13. HARDY CROSS METHOD

Hardy Cross method is essentially a relaxation technique involving controlled trial and error. In any system of pipes, connected together and conveying water under pressure, the following three laws are applicable :

(*i*) In each separate pipe or element comprising the system, there will be a relation between the head loss in the element and the quantity of water flowing through it.

(*ii*) At each junction, the algebraic sum of the quantities of water entering and leaving the junction is zero, *i.e.* $\Sigma Q = 0$.

(*iii*) In any closed path or circuit, the algebraic sum of the head loss in the individual elements is zero, *i.e.* $\Sigma h = 0$

R.J. Cornish has given two methods based on Hardy Cross method : (*a*) method of balancing heads by correcting assumed flows, and (*b*) method of balancing flows by correcting assumed heads. The modifications suggested by Cornish ensure that the third law is complied with at each step in the calculations.

Method (*a*) : Balancing Heads by Correcting Assumed Flows

In this method, *clockwise* flows and associated head losses are assigned *positive sign* while the *counter clockwise* flows and associated head losses are assigned *negative sign*. Reasonable values of flows in all elements of the system are assumed, subject, of course, to the second law mentioned above. The corresponding head losses in one closed circuit are next calculated by using any flow formula. The total gain or loss of head Σh_f round the circuit is calculated with due regard to sign. By the third law above, correct balance has been achieved when $\Sigma h_f = 0$. The value of the quantity $n\, h_f/Q$ for each element is computed and results added together *without* regard to sign. The initial assumed flow is then corrected by the following formula :

$$\Delta Q = -\frac{\Sigma h_f}{n\, \Sigma h_f/Q}$$

The proof of the above formula is given below :

Any head loss formula can be expressed in the form :

$$h_f = K Q^n \qquad \text{...(16.27)}$$

For Hazen-Williams formula

$$n = 1.85$$

Fig. 16.26 shows a small circuit in which inflow at junction B is Q_I and outflow at D is $Q_0 = Q_I$. At B, let the *assumed* flows in BCD and BED be Q_1 and Q_2 respectively. The value of Q_1 and Q_2 should finally be such that $h_{f\text{BCD}}$ and $h_{f\text{BED}}$ are equal. If, however, $h_{f\text{BCD}} > h_{f\text{BED}}$, then a new value for Q_1 will be $Q_1 - \Delta Q = Q_1'$ and

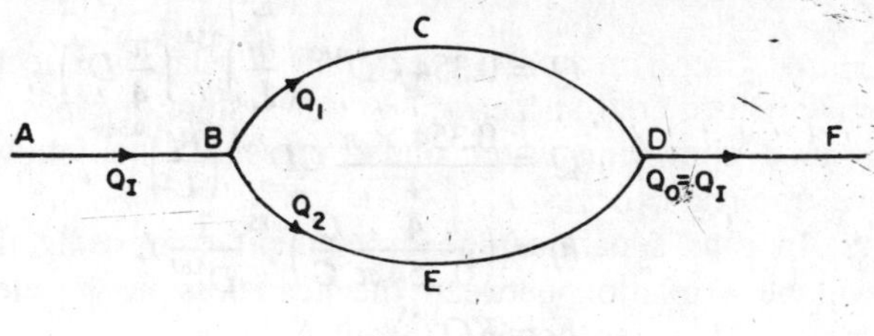

FIG. 16.26.

the new value of Q_2 must be $Q_2 + \Delta Q = Q_2'$. The corresponding values of head loss will be $h'_{f\text{BCD}}$ and $h'_{f\text{BED}}$. If ΔQ is *true correction*, then

$$h'_{f\text{BCD}} - h'_{f\text{BED}} = 0 = K_1 (Q_1 - \Delta Q)^n - K_2 (Q_2 + \Delta Q)^n$$

Expanding the Binomials,

$$K_1 (Q_1^n - n\,\Delta Q . Q_1^{n-1} + ...) - K_2 (Q_2^n + n\,\Delta Q\,.\,Q_2^{n-1} + ...) = 0$$

Since ΔQ is small, its higher powers can be neglected.

$$\therefore\quad K_1 Q_1^n - n K_1 \Delta Q\,.\,Q_1^{n-1} - K_2 Q_2^n - n K_2 \Delta Q\,.\,Q_2^{n-1} = 0$$

Substituting $h_{f\text{BCD}}$ for $K_1 Q_1^n$ and $h_{f\text{BED}}$ for $K_2 Q_2^n$ and rewriting the terms KQ^{n-1} as $K(Q^n/Q)$, we get

$$h_{f\text{BCD}} - \Delta Q\, n K_1 \frac{Q_1^n}{Q_1} - h_{f\text{BED}} - \Delta Q\,.\,n K_2 \frac{Q_2^n}{Q_2} = 0$$

or $$h_{f\text{BCD}} - h_{f\text{BED}} = \Delta Q\,.\,n \left(\frac{H_{f\text{BCD}}}{Q_1} + \frac{H_{f\text{BED}}}{Q_2} \right)$$

or $$\Delta Q = \frac{h_{f\text{BCD}} - h_{f\text{BED}}}{n \left(\frac{H_{f\text{BCD}}}{Q_1} + \frac{H_{f\text{BED}}}{Q_2} \right)}$$

Extending this to a more general case, the equation for the *flow correction* ΔQ becomes :

$$\Delta Q = -\frac{\Sigma h_f}{n\,\Sigma h_f/Q} \qquad ...(16.26)$$

It should be noted that in computing the numerators, proper sign of individual h_f in each pipe should be taken into consideration, while the denominator is the absolute sum without regard to sign convention.

Since $h_f = KQ^n$, Eq. 16.26 can also be expressed in the following form :

$$\Delta Q = \frac{\Sigma h_f}{n\,\Sigma h_f/Q} = -\frac{\Sigma K Q^n}{n\,\Sigma K Q^{n-1}} \qquad ...(16.28)$$

In Eq. 16.27 or 16.28, K represents the *relative resistance* of the pipe. From Eq. 16.11, we have

$$V = 0.354\, C D^{0.63} \left(\frac{h_f}{L}\right)^{0.54}$$

$$\therefore\quad Q = 0.354\, C D^{0.63} \left(\frac{h_f}{L}\right)^{0.54} \left(\frac{\pi}{4} D^2\right)$$

or $$Q = \frac{0.354 \times \pi}{4} C D^{2.63} \left(\frac{h_f}{L}\right)^{0.54}$$

$$\therefore\quad h_f = \left(\frac{4}{0.354\,\pi}\frac{Q}{C}\right)^{1.85} \frac{1}{D^{4.87}} L$$

or $$h_f = KQ^{1.85} \qquad ...(16.29\ a)$$

where $$K = \frac{10.62\,L}{C^{1.85}\,D^{4.87}} \quad ...(16.29)$$

In the above method, it is very essential that the values of Q_1 and Q_2 should be reasonable so that the trials are minimised. If two pipes are meeting at a junction, the following relationship may be used :

$$\frac{Q_1}{Q_1 + Q_2} = \frac{K_2^{1/2}}{K_1^{1/2} + K_2^{1/2}} \quad ...(16.30)$$

where K_1 and K_2 are the relative resistances of the two pipes, defined by Eq. 16.29. If there are three pipes meeting at a junction, the following relation may be used :

$$\frac{Q_1}{Q_1 + Q_2 + Q_3} = \frac{K_2^{1/2}\,K_3^{1/2}}{K_1^{1/2}\,K_2^{1/2} + K_1^{1/2}\,.\,K_3^{1/2} + K_2^{1/2}\,K_3^{1/2}} \quad ...(16.31)$$

Method (*b*). Balancing Flows by Correcting Assumed Heads

In this method, heads at all junctions where these are not known are assumed. *Positive signs* are arbitrarily assigned to *flow towards junctions* other than inlet and outlet junctions (for which heads are known) and *negative signs* to *flows away from* these *intermediate junctions.* The assumed values of heads should be near the correct value since the method of correction assumes that the difference between the assumed and correct values is small. Using any pipe formula, the flow in all pipes connected to one junction is calculated from the head differences which have been assumed. The excess or deficiency of inflow over outflow is calculated, *i.e.* $\Sigma\, Q$ is calculated. The quantity $Q/n\, h_f$ for each element connected to the junction is then calculated and the results added *without* regard to sign. The correction h_f, to be applied to the assumed head losses is computed from the following equation :

$$h_f = \frac{n\,\Sigma\, Q}{\Sigma\, Q/h} \quad ...(16.32)$$

Each junction is corrected in turn in a similar manner, thereby completing the first correction of the whole system.

Example 16.18. *Analyse the pipe network shown in Fig. 16.27 Take C = 100 in the Hazen Williams' formula.*

Solution :

The head loss is given by Eq. 16.29 (*a*)

$$h_f = KQ^{1.85} \text{ where } Q \text{ is in m}^3/\text{sec}$$

where $$K = \frac{10.62\,L}{C^{1.85}\,D^{4.87}} \quad ...(16.29)$$

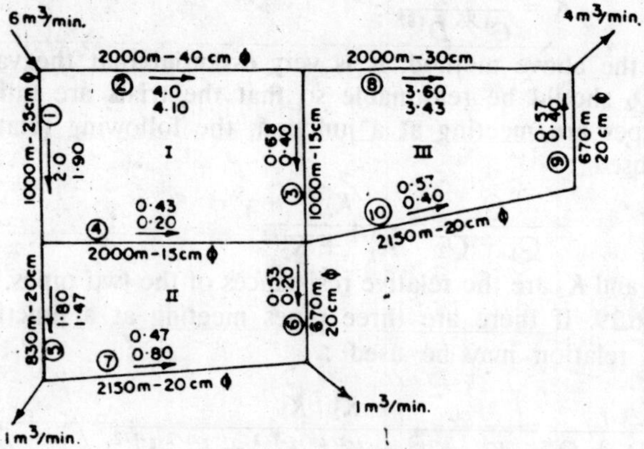

FIG. 16.27

In our problem, discharge Q is expressed in m^3/min. This can be reduced to the units of m^3/sec by dividing it by 60.

Hence $$h_f = K\left(\frac{Q}{60}\right)^{1.85} = \frac{K}{1948}Q^{1.85} = K_1 Q^{1.85}$$

where $$K_1 = \frac{K}{1948}$$

It is essential to calculate K_1 for various pipes. The values of K_1 for $C = 100$ are tabulated in Table 16.18.

TABLE 16.18

Pipe No.	*Length (m)*	*Dia. (cm)*	*K*	K_1
1	1000	25	1812	0.93
2	2000	40	367	0.188
3	1000	15	21841	11.21
4	2000	15	43682	22.42
5	830	20	4463	2.29
6, 9	670	20	3620	1.86
7, 10	2150	20	11560	5.93
8	2000	20	1491	0.765

The initially *assumed* values of the flow are marked near the arrows in Fig. 16.27, along with the directions of flow. Pipes 3 and 4 appear in more than one loop. Hence they are subject to combined corrections for loops I-III and I-II respectively. The computations are arranged in Table 16.19, spread over two pages. The final values of flow are marked in Fig. 16.27, near the initially assumed value of each flow.

TABLE 16.19
(EXAMPLE 16.5) HARDY CROSS METHOD

Loop	Trial–1								Trial–2						
	Pipe No.	K_1	Q	h_f	h_f/Q	$n\Sigma h_f/Q$	Σh_f	ΔQ	Q	h_f	h_f/Q	$n\Sigma h_f/Q$	Σh_f	ΔQ	
(1)	(2)	(3)	(4)	(5)	(6)	(7)	(8)	(9)	(10)	(11)	(12)	(13)	(14)	(15)	(16)
I	1	0.93	− 2.0	− 3.35	1.68			0.00	− 2.00	− 3.35	1.68			+ 0.07	
	2	0.188	− 4.0	+ 2.44	0.61			0.00	+ 4.00	+ 2.44	0.61			+ 0.07	
	3	11.21	+ 0.4	+ 2.06	5.15			+ 0.24	+ 0.64	+ 4.91	7.67			+ 0.03	
	4	22.42	− 0.20	− 1.14	5.70	24.31	+0.01	− 0.34	− 0.54	− 7.17	13.28	42.98	− 3.17	+ 0.12	
II	4	22.42	+ 0.20	+ 1.14	5.70			+ 0.34	+ 0.54	+ 7.17	13.28			− 0.12	
	6	1.86	+ 0.20	+ 0.10	0.50			+ 0.34	+ 0.54	+ 0.59	1.09			− 0.05	
	7	5.93	− 0.80	− 3.92	4.90			+ 0.34	− 0.46	− 1.41	3.06			− 0.05	
	5	2.29	− 1.80	− 6.79	3.77	27.51	− 9.47	+ 0.34	− 1.46	− 4.61	3.16	38.10	+ 1.74	− 0.05	
III	3	11.21	− 0.40	− 2.06	5.15			− 0.24	− 0.64	− 4.91	7.67			− 0.03	
	8	0.765	+ 3.60	+ 8.18	2.27			− 0.24	+ 3.36	+ 7.20	2.14			+ 0.04	
	9	1.86	− 0.40	− 0.34	0.85			− 0.24	− 0.64	− 0.81	1.26			+ 0.04	
	10	5.93	− 0.40	− 1.09	2.72	19.78	+ 4.69	− 0.24	− 0.64	− 2.60	4.06	28.0	− 1.12	+ 0.04	

TABLE 16.19. (CONTD.)

Loop	Trial–3								Trial–4						Final Q
(1)	(2)	(3)	(4)	(5)	(6)	(7)	(8)	(9)	(10)	(11)	(12)	(13)	(14)	(15)	(16)
I	1		− 1.93	− 3.14	1.63			− 0.01	− 1.94	− 3.17	1.63			+ 0.04	− 1.90
	2		+ 4.07	+ 2.52	0.62			− 0.01	+ 4.06	+ 2.51	0.62			+ 0.04	+ 4.10
	3		+ 0.67	+ 5.34	7.97			− 0.04	+ 0.63	+ 4.73	7.66			− 0.05	+ 0.48
	4		− 0.42	− 4.50	10.70	38.72	+ 0.22	− 0.06	− 0.48	− 5.77	12.02	40.6	− 1.66	+ 0.05	− 0.43
II	4		+ 0.42	+ 4.50	10.70			+ 0.06	+ 0.48	+ 5.77	12.02			− 0.05	+ 0.43
	6		+ 0.49	+ 0.50	1.02			+ 0.05	+ 0.54	+ 0.59	1.09			− 0.01	+ 0.53
	7		− 0.51	− 1.71	3.35			+ 0.05	− 0.46	− 1.41	3.06			− 0.01	− 0.47
	5		− 1.51	− 4.91	3.25	33.9	− 1.62	+ 0.05	− 1.46	− 4.61	3.16	35.8	+ 0.34	− 0.01	− 1.47
III	3		− 0.67	− 5.34	7.97			+ 0.04	− 0.63	− 4.77	7.57			− 0.05	− 0.68
	8		+ 3.40	+ 7.36	2.16			+ 0.03	+ 3.43	+ 7.48	2.18			0.00	+ 3.43
	9		− 0.60	− 0.72	1.20			+ 0.03	− 0.57	− 0.66	1.16			0.00	− 0.57
	10		− 0.60	− 2.30	3.83	28.06	− 1.00	+ 0.03	− 0.57	− 2.01	3.52	26.7	+ 0.04	0.00	− 0.57

PROBLEMS

1. Discuss in brief various methods of water distribution. Which method do you prefer ?

2. Compare the continuous and intermittent systems of water supply.

3. Write a note on distribution reservoirs. Where are these located ?

4. How do you determine the capacity of reservoir ? Discuss the hydrograph method for determining the storage capacity of equalising reservoir.

5. Exaplain how do you determine the storage capacity of balancing reservoir by the mass curve method (*a*) for 24 hour pumping, (*b*) for 15 hour pumping from 6 A.M. to 9 P.M.

6. Explain various formulae commonly used for determining the head loss due to friction in pipes.

7. Write a note on coefficient *C* used in the Hazen Williams formula.

8. Write down the Hazen Williams formula for (*a*) velocity of flow, (*b*) discharge *Q*, (*c*) head los due to friction. Taking $C = 100$, find the value of various constants used in the above formulae.

9. Discuss with the help of diagrams, various methods of laying out the distribution system.

10. What do you understand by an equivalent pipe ? How do you determine its length when the pipes are (*a*) in series (*b*) in parallel.

11. Taking the flow equation $h_f = \dfrac{KLQ^n}{D^x}$, shows that the equivalent length (L_E) of diameter D_E, for pipes arranged in series is given by

$$L_E = D_E^x \Sigma \frac{L}{D^x}$$

What will be the value of *x* if Hazen-Willams formula is used ?

12. Taking the flow equation $h_f = \dfrac{KLQ^n}{D^x}$, show that the diameter D_E for equivalent pipe is given by

$$D_E = \sqrt[x]{L_t \Sigma \frac{D^x}{L}}$$

where L_t is the total length of system.

13. Explain Hardy Cross method of solving the net work, by balancing heads by correcting flow. Derive the expression you use for correction of the assumed flow.

17

Water Supply For Buildings

17.1 MATERIALS FOR SERVICE PIPES

The pipe leading from the distribution main of the municipal water supply to the plumbing system of the house is known as the *service pipe*. The following materials are commonly used for service pipes :

(*i*) Copper pipe or brass pipe
(*ii*) Galvanised iron, either lined or unlined
(*iii*) Lead pipe, either lined or unlined
(*iv*) Polythene pipe.

Copper Tubing. Copper tubing is non-corrosive with most waters. It is used extensively in better grade houses and where ground water is highly corrosive to steel pipe. It has considerable strength, reasonable ductility, and is obtainable in long lengths. There are two types of copper water service pipes : *heavy gauge* and *light gauge*. The former can be threaded and is used for high pressure work in industrial layouts. For general purpose work where pressure does not exceed 1.5 kg/cm^2 and for internal domestic work, light gauge tube is used. For underground work, as for the pipe from the main to the building, special copper tube is used having a heavier gauge. The copper tube can be attached to the main without the use of conventional gooseneck, the flared end of the tubing being connected directly to the corporation cock without threading.

G.I. pipes. These are used where water is suitable. In India, these are extensively used because of their low cost in comparison to copper tubes. Galvanised iron or steel tubes corrode more readily in soft and acid waters and are not so easily manipulated as copper although they are stronger and can be used in hard water area when they withstand the hammering needed to remove the scale.

Lead Pipe. Lead pipe has the following advantages : (*i*) highly resistant to corrosion, (*ii*) highly flexible, and (*iii*) high hydraulic coefficient of flow. However, lead has cumulative poisoning effect, specially when it goes into solution. Due to this, it is not preferred.

Polythene tube. These are being increasingly used internally and externally for cold water service pipes only. Plastic pipe has the following advantages over metal piping : (*i*) it is lower in cost, (*ii*) it is non-corrosive, (*iii*) it is light in weight, (*iv*) it can be installed with ordinary tools, (*v*) it does not require threading, (*vi*) it is more resistant to live and bacterial scale, and (*vii*) it has some insulating value. However, these are not useful where temperature is high.

17.2. SERVICE CONNECTION

A service connection is primarily a connection from the distribution system to the consumer. A consumer may be a single house, an apartment house, a planned block development or a water district buying water 'wholesale'. A connection for a single house will normally involve tapping the main while it is under pressure and installing a corporation cock.

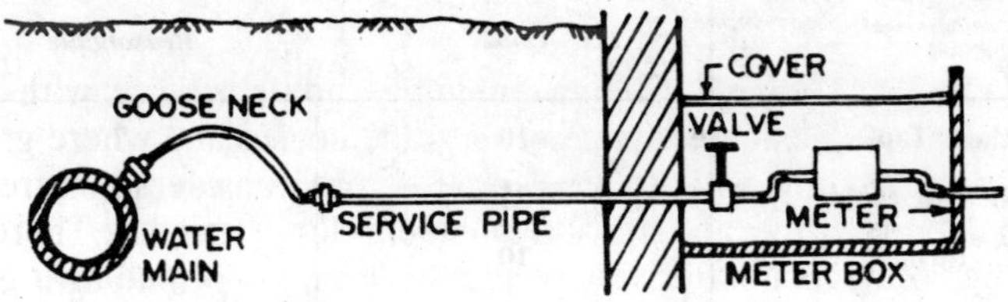

FIG. 17.1. SERVICE CONNECTION

A domestic service connection includes the following components :

1. Brass or bronze ferrule. For a typical service connection to a house a ferrule is inserted in the main, most usually an 'under pressure, connection which can be inserted without shutting down the main.

2. Goose neck. This consists of a 40 to 50 cm long curved piece of flexible pipe made of brass, copper or lead. The goose neck prevents the breaking of the main service pipe due to movement that takes place between the water main and the service pipe, thus providing flexibility at the junction.

3. Stop cock or curb valve. It is installed in a suitable chamber with cover, to close down the supply, for repairs of the plumbing system.

4. Main service pipe. It may be of various materials discussed in § 17.1. Its diameter may vary from 12 mm to 40 mm.

5. Water meter. It is also installed in a suitable chamber with cover. It measures the quantity of water used by the consumer.

17.3. SIZE OF SERVICE PIPES

The sizes of service pipes are determined on the basis of the following : (*i*) the minimum pressure in the distribution main at the proposed point of connection, (*ii*) length of service pipe required, (*iii*) elevation of the highest point of delivery above the distribution main (*iv*) number and types of plumbing fixtures in the building and the maximum rate of flow required. The maximum rate of flow may be estimated by considering what the average householder expects at his plumbing fixtures. Guidance may be taken from the following table :

TABLE 17.1

Plumbing fixture	*Flow Reqd. (litres/min.)*	
	Good	*Reasonable*
Kitchen Tap	10	7
Bath Tap (cold)	25	15
W.C. flushing cistern	10	7

If it is assumed that one of each of the these is in operation simultaneously, a maximum demand rate of 29 to 45 litres/min. is obtained. Knowing the rate of flow and the desired velocity, pipe diameter can be found. Alternatively, the diameter of service pipe can also be fixed on the basis of occupants in the house, as given in Table 17.2.

TABLE 17.2

No. of occupants	4	8	24	60
Dia. of service pipe (mm)	12.5	20	25	30

17.4. WATER METERS

Water meters are normally used for measuring flows to domestic buildings. A water meter should possess the following characteristics:

1. It should accurately measure and register both small and large flows.

2. It should be easy to maintain and repair. Spare parts should be readily available.

3. It should have good capacity with reasonable head loss.

4. It should be capable of working at all pressures efficiently.

5. It should be durable. Its parts should not be affected by chemicals used for purification and the impurities in water.

6. It should be rugged.

7. It should prevent back flow passing through it and should not be liable to clogging.

8. It should have low cost.

Meters used on water distribution systems may be of two types: (*i*) inferential or velocity meters, and (*b*) positive or displacement meters.

Inferential meter measures the velocity of flow across a cross-section whose area is known. They are used only for high flows. Common examples of this type of meters are the rotary and the turbine meters.

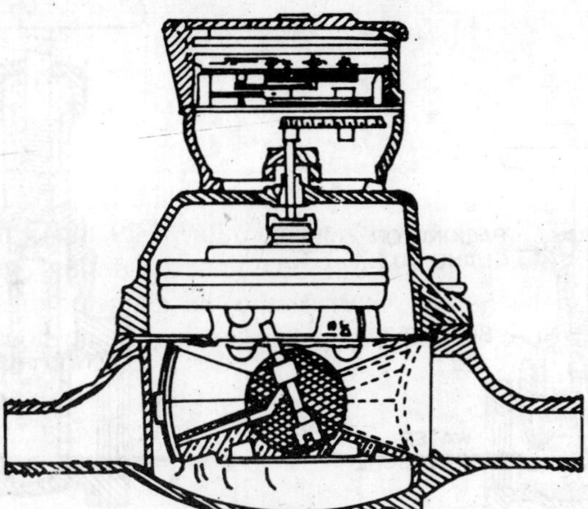

FIG. 17.2. NUTATING-DISC METER

Displacement meters are primarily used for relatively low flows, as for the residential buildings. In this meter, the quantity of water actually passing through it is measured by filling and emptying the chamber of known capacity. Types of displacement meters in use include reciprocating, rotary, oscillating and nutating disc meters, depending upon the motion of the moving part in the measuring device.

Disc-meter, shown in Fig. 17.2 consists of a disc of hard rubber placed inside a chamber which is provided with inlets and outlets. The water entering the chamber oscillates the disc about its centre with a spiral motion. The oscillations imparted by one complete filling and emptying are recorded by the disc meter in terms of volume of water.

17.5. VALVES

For domestic water supply, two types of valves are commonly used : (*a*) globe valve, (*b*) gate valve.

Globe valve. It is used in pipe lines for convenience in manually closing the pipes to control the flow of water. Fig 17.3 shows the section of a globe valve. It should be installed with water pressure under the valve seat as shown. The globe valve has advantage of quicker opening and closing, of longer life and of being more easily repaired.

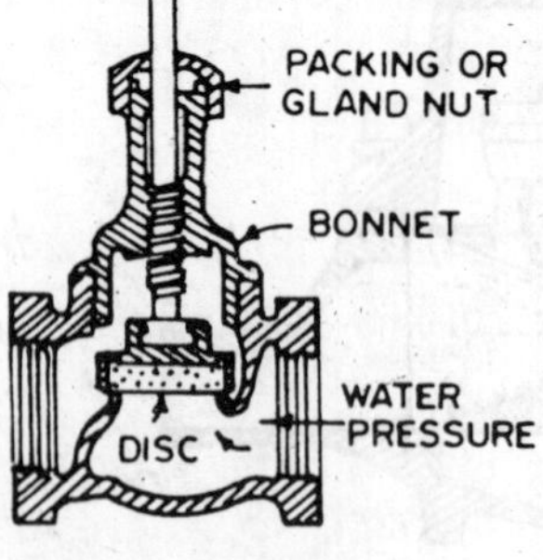

FIG. 17.3. GLOBE VALVE

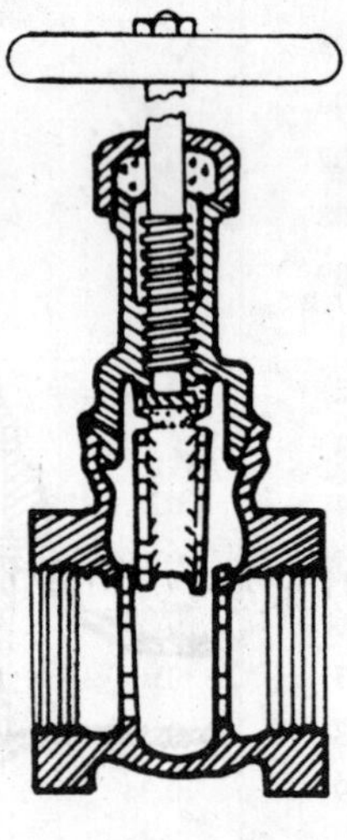

FIG. 17.4. GATE VALVE

Gate valve. Gate valve is also used in pipe lines for convenience in manually closing the pipes. The gate valve has an advantage over the globe valve in that it offers less resistance to flow. It is therefore used in preference to the globe valve where resistance to the flow of water is to be kept at a minimum. It also controls the flow equally well from either direction.

17.6. LOSS OF HEAD THROUGH PIPES AND PIPE FITTINGS

The loss of head of water due to friction in a pipe depends on its length, number of bends and other fittings, its diameter and smoothness and the speed of water. Table 17.3 gives the flow rate in litres/minute, in copper and plastic pipes of different diameters, for losses in head of *metres per metre run of pipe.* Table 17.4 gives the resistance offered by various pipe fittings, expressed as equivalent lengths of straight pipe.

TABLE 17.3 RATE OF FLOWS FOR LOSSES IN HEAD (Copper and Plastic pipes)

Loss of head (m per m run of pipe)	*Flow in litres per second, for pipe diameters (mm)*					
	12	*15*	*18*	*22*	*28*	*35*
0.01	–	–	0.065	0.12	0.2	0.42
0.02	–	0.064	0.103	0.19	0.325	0.67
0.03	–	0.082	0.126	0.25	0.41	0.86
0.04	0.05	0.097	0.156	0.28	0.49	1.0
0.05	0.06	0.11	0.177	0.33	0.56	1.16
0.06	0.067	0.125	0.20	0.37	0.62	1.30
0.07	0.073	0.136	0.22	0.395	0.68	1.41
0.08	0.078	0.147	0.24	0.43	0.73	1.52
0.09	0.084	0.16	0.256	0.46	0.79	1.65
0.1	0.09	0.17	0.267	0.49	0.84	1.75
0.2	0.137	0.26	0.4	0.73	1.27	2.7
0.3	0.176	0.32	0.52	0.93	1.6	3.4
0.4	0.21	0.38	0.62	1.1	1.9	4.0
0.5	0.25	0.43	0.70	1.26	2.2	4.6
0.6	0.275	0.48	0.78	1.4	2.5	5
0.7	0.3	0.53	0.87	1.54	2.7	5.7
0.8	0.32	0.58	0.93	1.66	2.9	6
0.9	0.34	0.62	1.0	1.76	3.1	6.5
1	0.37	0.67	1.07	1.9	3.4	7

TABLE 17.4
RESISTANCE OF FITTINGS IN EQUIVALENT LENGTHS

Fitting	*Resistance to flow by different fittings expressed as equivalent lengths in metres of straight pipes for various dia. (mm)*					
	12	***15***	***18***	***22***	***28***	***35***
Bend	0.21	0.27	0.32	0.42	0.6	0.75
Elbow	0.47	0.60	0.74	0.95	1.2	1.50
Tee	0.57	0.74	0.93	1.23	1.7	2.25
Reduction	0.11	0.13	0.2	0.23	0.33	0.44
Enlargement	0.18	0.27	0.34	0.51	0.94	0.95
Valve or Tap	2.7	4	5	7	9	12

PROBLEMS

1. Discuss in brief various types of service pipes commonly used.

2. Sketch a typical service connection from the water mains.

3. State various requirements of a good domestic water meter. Explain various types of water meters.

4. Write notes on : (1) Globe valve, (2) Gate valve.

Appendix

TABLE A–1 VISCOSITY AND DENSITY OF WATER

Temperature, C	*Density (grams/cm³),*	*Absolute Viscosity μ, centipoises*	*Kinematic Viscosity ν, centistokes*	*Temperature, F*
0	0.99987	1.7921	1.7923	32.0
2	0.99997	1.6740	1.6741	35.6
4	1.00000	1.5676	1.5676	39.2
6	0.99997	1.4726	1.4726	42.8
8	0.99988	1.3872	1.3874	46.4
10	0.99973	1.3097	1.3101	50.0
12	0.99952	1.2390	1.2396	53.6
14	0.99927	1.1748	1.1756	57.2
16	0.99897	1.1156	1.1168	60.8
18	0.99862	1.0603	1.0618	64.4
20	0.99823	1.0087	1.0105	68.0
22	0.99780	0.9608	0.9629	71.6
24	0.99733	0.9161	0.9186	75.2
26	0.99681	0.8746	0.8774	78.8
28	0.99626	0.8363	0.8394	82.4
30	0.99568	0.8004	0.8039	86.0

Notes : (*i*) 1 Centipoise $= 10^{-2}$ (gram mass)/(cm)(sec).

(*ii*) To convert to (lb force) (sec)/(sq.ft), multiply centipoise by 2.088×10^{-5}.

(*iii*) To convert to (kg f) (sec)/m², multiply centipoise by 1.019×10^{-4}.

(*iv*) 1 (N-s)/m² = 10 poise = 1000 centipoises.
1 centipoise $= 10^{-3}$ (N-s)/m².

(*v*) 1 centistoke $= 10^{-2}$ cm²/sec. To convert to (sq. ft)/(sec), multiply centistoke by 1.075×10^{-5}

TABLE A–2 VAPOUR PRESSURE OF WATER AND SURFACE TENSION OF WATER IN CONTACT WITH AIR

Temperature, C	0	5	10	15	20	25	30
Vapour pressure, mm Hg	4.58	6.54	9.21	12.8	17.5	23.8	31.8
Surface tension, dyne/cm	75.6	74.9	74.2	73.5	72.8	72.0	71.2

TABLE A–3 SELECTED ENGLISH–METRIC CONVERSION FACTORS

	English Unit	*Multiplier*	*Metric Unit*
Mass	lb	0.4536	kg
Length	ft	0.3048	m
Area	ft^2	0.09290	m^2
	acre	4047	m^2
	acre	0.4047	ha
Volume	gal	0.003785	m^3
	gal	3.785	litre
	ft^3	0.02832	m^3
	ft^3	28.32	litre
Velocity	fpm	0.00508	m/sec
Flow	Mgd**	3785	m^3/sec
	gpm	5.450	m^3/day
	gpm	0.06309	l/sec
	cfs	0.02832	m^3/day
BOD loading	lb/1000 ft^3/day	16.02	g/m^3/day
	lb/acre/day	1.121	kg/ha / day
Solids loading	lb/ft^2/day	4.883	kg/m^2/day
	lb/ft^3/day	16.02	kg/m^3/day
Hydraulic loading	gpd/ft^2	0.04075	m^3/m^2/day
	Mg*/acre/day	0.9353	m^3/m^2/day
	Mg*/acre/day	9353	m^3/ha/day
Concentration	lb/Mg*	0.1198	mg/l

Mg* = million gallons ; Mgd** = million gallons/day

INDEX

———*———